VOLUME ONE HUNDRED AND THIRTY NINE

INTERNATIONAL REVIEW OF NEUROBIOLOGY

Neurobiology of the Placebo Effect Part II

INTERNATIONAL REVIEW OF NEUROBIOLOGY

VOLUME 139

VOLUME ONE HUNDRED AND THIRTY NINE

INTERNATIONAL REVIEW OF NEUROBIOLOGY

Neurobiology of the Placebo Effect Part II

Edited by

LUANA COLLOCA

Departments of Anesthesiology and Psychiatry,
School of Medicine;
Department of Pain Translational Symptom Science,
School of Nursing;
Center to Advance Chronic Pain Research,
University of Maryland, Baltimore, MD, United States

ELSEVIER

ACADEMIC PRESS
An imprint of Elsevier

Academic Press is an imprint of Elsevier
50 Hampshire Street, 5th Floor, Cambridge, MA 02139, United States
525 B Street, Suite 1650, San Diego, CA 92101, United States
The Boulevard, Langford Lane, Kidlington, Oxford OX5 1GB, United Kingdom
125 London Wall, London, EC2Y 5AS, United Kingdom

First edition 2018

ISBN: 978-0-12-815416-8
ISSN: 0074-7742

For information on all Academic Press publications
visit our website at https://www.elsevier.com/books-and-journals

Publisher: Zoe Kruze
Acquisition Editor: Sam Mahfoudh
Editorial Project Manager: Teresa Pons-Ferrer
Production Project Manager: Abdulla Sait
Cover Designer: Mark Rogers
Cover Image Courtesy: Giancarlo Colloca

Typeset by SPi Global, India

CONTENTS

CONTRIBUTORS

Marco Annoni
Consiglio Nazionale delle Ricerche (CNR)—Istituto di Tecnologie Biomediche (ITB), Rome; Fondazione Umberto Veronesi, Milan, Italy

Claire E. Ashton-James
Pain Management Research Institute, The University of Sydney, Sydney, NSW, Australia

Carmen-Édith Bellei-Rodriguez
Research Center of the Centre hospitalier universitaire de Sherbrooke (CHUS), Department of Surgery, Faculty of Medicine and Health Sciences, Université de Sherbrooke, Sherbrooke, QC, Canada

Maxie Blasini
Acupuncture and Chinese Medicine Program, Pacific College of Oriental Medicine, San Diego, CA, United States

Ellen Broelz
Department of Internal Medicine VI: Psychosomatic Medicine and Psychotherapy, University Hospital Tübingen, Tübingen, Germany

Florian Chouchou
PERFORM Center & Department of Exercise Science, Concordia University; Centre de Recherche de l'Institut Universitaire de Gériatrie de Montréal, Université de Montréal, Montreal, QC, Canada

Ben Colagiuri
School of Psychology, University of Sydney, Sydney, NSW, Australia

Matthew J. Coleshill
School of Psychology, University of Sydney; St Vincent's Clinical School, St Vincent's Hospital, University of New South Wales; Department of Clinical Pharmacology and Toxicology, St Vincent's Hospital, Sydney, NSW, Australia

Luana Colloca
Departments of Anesthesiology and Psychiatry, School of Medicine; Department of Pain Translational Symptom Science, School of Nursing; Center to Advance Chronic Pain Research, University of Maryland, Baltimore, MD, United States

Janie Damien
Research Center of the Centre hospitalier universitaire de Sherbrooke (CHUS), Department of Surgery, Faculty of Medicine and Health Sciences, Université de Sherbrooke, Sherbrooke, QC, Canada

Thien Thanh Dang-Vu
PERFORM Center & Department of Exercise Science, Concordia University; Centre de Recherche de l'Institut Universitaire de Gériatrie de Montréal, Université de Montréal, Montreal, QC, Canada

Beth D. Darnall
School of Medicine, Department of Anesthesiology, Perioperative and Pain Medicine, Division of Pain Medicine, Psychiatry and Behavioral Sciences (by courtesy), Stanford University, Palo Alto, CA, United States

Paul Enck
Department of Internal Medicine VI: Psychosomatic Medicine and Psychotherapy, University Hospital Tübingen, Tübingen, Germany

Kate Faasse
School of Psychology, University of New South Wales, Sydney, NSW, Australia

Damien G. Finniss
Department of Anaesthesia; University of Sydney Pain Management Research Institute, Royal North Shore Hospital, Sydney, NSW; School of Allied Health Sciences, Griffith University, Gold Coast, QLD, Australia

Mirta Fiorio
Department of Neuroscience, Biomedicine and Movement Sciences, University of Verona, Verona, Italy

Ilona Fridman
Fuqua School of Business, Duke University, Durham, NC, United States

Paul Glare
Pain Management Research Institute, The University of Sydney, Sydney, NSW, Australia

Javeria Ali Hashmi
Department of Anesthesia, Pain Management and Perioperative Medicine, Dalhousie University, Nova Scotia Health Authority, Halifax, NS, Canada

Bjoern Horing
Institute of Systems Neuroscience, University Medical Center Hamburg-Eppendorf, Hamburg, Germany

John M. Kelley
Distinguished Professor of Psychology, Endicott College, Beverly; Deputy Director, Program in Placebo Studies, Harvard Medical School, Boston, MA, United States

Regine Klinger
Department of Anesthesiology, Section Pain Medicine and Pain Psychology, University Medical Center Hamburg-Eppendorf, Hamburg, Germany

Gilles Lavigne
Faculté de Médecine Dentaire, Université de Montréal; Centre d'Etudes Avancées en Médecine du Sommeil, Hôpital du Sacré-Coeur de Montréal, Montreal, QC, Canada

Klaus Linde
Institute of General Practice, Technical University of Munich, München, Germany

Serge Marchand
Research Center of the Centre hospitalier universitaire de Sherbrooke (CHUS), Department of Surgery, Faculty of Medicine and Health Sciences, Université de Sherbrooke, Sherbrooke; Fonds de Recherche du Québec—Santé (FRQS), Montréal, QC, Canada

Leslie R. Martin
Department of Psychology and Neuroscience, La Sierra University, Riverside, CA, United States

Karin Meissner
Division of Integrative Health Promotion, Coburg University of Applied Sciences, Coburg, Germany

Dimos-Dimitrios Mitsikostas
1st Neurology Department, Aeginition Hospital, National and Kapodistrian University of Athens, Athens, Greece

Nathalie Peiris
Department of Anesthesiology, School of Medicine, University of Maryland, Baltimore, MD, United States

Pierre Rainville
Centre de Recherche de l'Institut Universitaire de Gériatrie de Montréal; Faculté de Médecine Dentaire, Université de Montréal, Montreal, QC, Canada

Julia Schmitz
Department of Anesthesiology, Section Pain Medicine and Pain Psychology, University Medical Center Hamburg-Eppendorf, Hamburg, Germany

Marie Schwartz
Department of Anesthesiology, Section Pain Medicine and Pain Psychology, University Medical Center Hamburg-Eppendorf, Hamburg, Germany

Louise Sharpe
School of Psychology, University of Sydney, Sydney, NSW, Australia

Julia Stuhlreyer
Department of Anesthesiology, Section Pain Medicine and Pain Psychology, University Medical Center Hamburg-Eppendorf, Hamburg, Germany

Roland Sussex
School of Languages and Cultures & Institute for Teaching and Learning Innovation, The University of Queensland, St Lucia, QLD, Australia

Bruce E. Wampold
Modum Bad Psychiatric Center, Vikersund, Norway; University of Wisconsin—Madison, Madison, WI, United States

Katja Weimer
Clinic for Psychosomatic Medicine and Psychotherapy, University Hospital Ulm, Ulm, Germany

Thelma Wright
Department of Anesthesiology, School of Medicine, University of Maryland, Baltimore, MD, United States

Robert Zachariae
Aarhus University Hospital and Department of Psychology and Behavioural Science, Aarhus University, Aarhus, Denmark

Panagiotis Zis
Academic Department of Neurosciences, Sheffield Teaching Hospitals NHS Foundation Trust, Sheffield, United Kingdom

PREFACE

Part II: The Fascinating Mechanisms and Implications of the Placebo Effect

Neurobiology of the Placebo Effect, Part II is the second part of a two-volume set that examines mechanisms and translational aspects derived from the most recent developments in the area of placebo and nocebo research (Colloca, 2018a, 2018b).

We are seeing a gradual development of placebos and placebo effects from simple tools used to please patients or to validate new drugs to a complex neurobiological phenomenon.

The evaluation of key historical events helps us come to a full appreciation of placebo and its meanings (Finniss, 2018). Moreover, a taxonomy of placebo effects helps us navigate the broadest and narrowest theories by either lumping the most important common features of placebo conceptualizations, or conversely, by splitting their differential elements (Kelley, 2018). Additionally, a linguistic approach by looking at the terms "placebo" and "placebo effects" via Internet-based language corpora helps us discover the impact of placebos not only in medicine but also on our society (Sussex, 2018), while a bibliometric analysis helps us narrow down limitations and gaps in current research (Enck, Horing, Broelz, & Weimer, 2018).

Placebo research started primarily with early 19th century basic clinical and translational studies that are today seen as significant milestones to enhance human health and well-being. These studies include sham surgical interventions such as the skull's removal by Hildred Carlill to treat a refractory case of insomnia (Cardill, 1919) to Wolff's advocacy for prescribing placebos in accordance with patients' preferences and choices (Wolff, Dubois, et al., 1946). Pioneering placebo research delved into an exploration of clinical and demographic characteristics of placebo responders (Lasagna, Mosteller, Von Felsinger, & Beecher, 1954). Lasagna and colleagues evaluated patients suffering from postoperative pain who were treated with doses of the opioid morphine interspersed with placebos. Patients were categorized by the frequency at which they responded to the placebo—consistently, intermittently or barely. Translational investigation of placebo continued with the use of "active placebo action" to reduce postoperative pain in patients undergoing major abdominal surgery

(Egbert, Battit, Welch, & Bartlett, 1964). Such an active placebo action consisted of interactive preoperative and postoperative visits with increased attention to patients' expectations and pain coping. As compared to the usual care arm (control), the active placebo action resulted in a significant larger reduction of daily morphine intake (30–50% reduction) for postoperative pain (Egbert et al., 1964).

These pioneering studies have paved the way for studying potential factors linked to placebo responsiveness, opening up a set of questions that are timely and of current interest. These findings are among the first attempts to outline the clinical relevance of shaping patients' expectancies (Darnall & Colloca, 2018; Klinger, Stuhlreyer, Schwartz, Schmitz, & Colloca, 2018) and to use dose-extending placebos along with pharmacological conditioning as a potential pain management tool to taper opioids (Colloca, Enck, & DeGrazia, 2016; Colloca & Howick, 2018).

The United States currently faces an epidemic of opioids (Nahin, 2015; Sullivan & Howe, 2013). Opioid analgesics relieve many types of pain and improve function, but the benefits of opioids when prescribed for chronic pain are questionable (Dowell, Haegerich, & Chou, 2016). Indeed, opioid therapy may actually complicate chronic pain management (Chou, 2016; Chou, Clark, & Helfand, 2003) by causing the development of opioid use disorder (Bohnert et al., 2011) and by increasing the risk of opioid-overdose death (Ray, Chung, Murray, Hall, & Stein, 2016). Placebo research can promote the development and integration of placebos into therapeutic strategies to address both the burden of pain management and the opioid crisis (Darnall & Colloca, 2018).

There is increasing consensus that clinicians can actively set positive patients' expectations (Darnall & Colloca, 2018; Glare, Fridman, & Ashton-James, 2018; Klinger et al., 2018). Educating patients about pain treatments and their effects can strengthen their overall pain reduction, coping, and resilience skills. At the same time, careful attention to negative thoughts, expectations, and beliefs about pain, and pain therapies can help develop strategies for nonpharmacological approaches to be used in conjunction with pharmacological medications. Current understanding of pain management urges us to equip patients and clinicians with "toolkits" (for an example, see Darnall & Colloca, 2018) that can be used to train pain specialists and to strengthen patients' inhibitory endogenous pain modulation (EPM) mechanisms.

A fundamental part of this research is represented by the impact of words and delivered information about treatment responses and their adverse effects (Glare et al., 2018). In clinical research, symptoms and complaints in patient populations, medication nonadherence, and need for additional drug prescriptions, among other topics, are often grouped together into a category called *risk communication and framing effects*. For example, the mere mention of headaches as a common side effect during risk communication for studies on antidepressants and other medications can increase the likelihood that headaches are experienced during the study (Blasini, Corsi, Klinger, & Colloca, 2017; Colloca, 2017a, 2017b; Klinger, Blasini, Schmitz, & Colloca, 2017). Risk communication also leads to an increased rate of withdrawal from the studies, making it difficult to recruit and retain study participants in clinical trials. Communication strategies and framing effects may therefore contribute to the occurrence of side effects and shape decision-making processes, clinical outcomes, and patients' adherence to research protocols and medication (Colloca, 2017a; Colloca & Miller, 2011).

Preventing harmful communication while still protecting patients' rights and preferences is critical. A potential area of clinical relevance for its applicability is the use of message framing to remodel and retain expectations of patients in clinical contexts (Glare et al., 2018). In the era of automation of health care and telemedicine, message framing can be powerfully used to engage patients during the delivery of treatment and long-term follow-ups (Darnall & Colloca, 2018; Glare et al., 2018). The implementation of such toolkits, automation of health care, and telemedicine should occur in clinical settings in which the patient-clinician interaction and the overall encounter around the patient are optimized. For example, research has linked the relationship between health care providers and outcomes to placebo and nocebo effects (Wampold, 2018). Rooted in ancient medicine, the patient-clinician interaction and its anthropological elements can meaningfully contribute to the overall improvement of outcomes (Blasini, Peiris, Wright, & Colloca, 2018).

Outcome changes do not occur in a vacuum; rather, they can shape brain responses favoring pain relief and well-being. Brain responses and perceptions generate predictions that can in turn affect the process of interpreting pain-related sensory inputs received from the periphery (Hashmi, 2018), further raising the interest in investigating the EPM mechanisms in physiological conditions and pathological disorders. How does the central nervous system (CNS) inhibit pain signaling? Which are the CNS mechanisms that are in common between placebo effects and other forms of pain modulation such as conditioned pain modulation (CPM) and the diffuse noxious inhibitory control (DNIC)? These and other questions are addressed in a comprehensive overview of the CPM mechanisms and their clinical relevance (Damien, Colloca, Belleï-Rodriguez, & Marchand, 2018). It has been speculated that patients who present less efficient CPM might need treatments that restore central and peripheral pain inhibitory processes (Colloca et al., 2017). The prospect of using individual patterns of pain modulation may lead the way to a form of personalized pain management.

Indeed, EPM profiles might predict the development of neuropathic and chronic pain components, the responsiveness to specific classes of pain treatments, and the ability to harness self-healing pain processes.

Placebo effects are observed not only in pain processes but also in other systems. For example, placebo effects and related analyses of patterns of responsiveness have been described in the motor system (Fiorio, 2018; Pollo, Carlino, & Benedetti, 2011) and other neurological disorders (Panagiotis & Mitsikostas, 2018). Also, placebo effects are likely influenced by sleep patterns whereby sleep deprivation may improve expectations and placebo-induced pain reduction (Chouchou, Thien Thanh, Rainville, & Gilles, 2018).

Factors that are external to the patient can also be relevant in triggering placebo and nocebo effects. These factors can range from manufacturing characteristics (e.g., blue versus green pills) to marketing features (Faasse & Martin, 2018; Meissner & Linde, 2018). Generic, as opposed to brand labelling, is one of the most common marketing factors that influence placebo and nocebo effects. Generic medicines that are pharmaceutically equivalent to their branded medicines are often met with distrust and perceived as less effective than branded medicines. Negative perceptions of generic pharmaceuticals may contribute to reduced treatment efficacy via enhanced nocebo effects (Faasse & Martin, 2018). These effects are relevant when medications are released on the market.

Depending on the medication and its mechanisms of action, placebo and treatment effects can be additive, subadditive, or superadditive (Coleshill, Sharpe, Colloca, Zachariae, & Colagiuri, 2018). Additivity implies that the treatment effect and the placebo component do not interact; that is, the placebo component is identical for both the placebo and the treatment arms and the overall outcome is the sum of the treatment plus the placebo component. As such, additivity cannot explain the variety of results observed in clinical trials. Subadditivity, the combined placebo and treatment effects that are less than the summed size of the isolated active treatment and isolated placebo effect, can result in considering as ineffective a treatment that is effective. On the other hand, superadditivity, a synergistic interaction between the active treatment and placebo effects, can overestimate the efficacy of the treatment in a randomized clinical trial (Coleshill et al., 2018).

The attentive consideration of the variety of additive- versus nonadditive-related placebo mechanisms, the use of ethical approaches (Annoni, 2018), the development of toolkits (Darnall & Colloca, 2018), and automation of message framing to engage patients (Glare et al., 2018) can maximize placebo effects in medical practice and help implement novel strategies to improve patients' well-being and drug validation and contribute to clinically relevant health outcomes.

We acknowledge support from the MPowering the State funds and NIDCR 1R01DE025946.

Luana Colloca
Department of Pain Translational Symptom Science, School of Nursing, University of Maryland, Baltimore, MD, United States
Departments of Anesthesiology and Psychiatry, School of Medicine, University of Maryland, Baltimore, MD, United States
Center to Advance Chronic Pain Research, University of Maryland, Baltimore, MD, United States

REFERENCES

Annoni, M. (2018). The ethics of placebo effects in clinical practice and research. In L. Colloca (Ed.), *Neurobiology of the placebo effect, part II: Vol. 139* (pp. 463–484). New York: Academic Press.

Blasini, M., Corsi, N., Klinger, R., & Colloca, L. (2017). Nocebo and pain: An overview of the psychoneurobiological mechanisms. *Pain Reports*, *2*(2), e585. https://doi.org/10.1097/PR9.0000000000000585.

Blasini, M., Peiris, N., Wright, T., & Colloca, L. (2018). The role of patient-practitioner relationships in placebo and nocebo phenomena. In L. Colloca (Ed.), *Neurobiology of the placebo effect, part II: Vol. 139* (pp. 211–231). New York: Academic Press.

Bohnert, A. S., Valenstein, M., Bair, M. J., Ganoczy, D., McCarthy, J. F., Ilgen, M. A., et al. (2011). Association between opioid prescribing patterns and opioid overdose-related deaths. *JAMA*, *305*(13), 1315–1321. https://doi.org/10.1001/jama.2011.370.

Cardill, H. (1919). Hysterical sleeping attacks: Treatment by gross suggestion, reinforced by the operation of trephining, with notes on behavior in hypnotic sleep. *The Lancet*, *194*(5025), 1128–1131.

Chou, R. (2016). Long-acting opioids for chronic noncancer pain were linked to mortality. *Annals of Internal Medicine*, *165*(6), JC34. https://doi.org/10.7326/ACPJC-2016-165-6-034.

Chou, R., Clark, E., & Helfand, M. (2003). Comparative efficacy and safety of long-acting oral opioids for chronic non-cancer pain: A systematic review. *Journal of Pain and Symptom Management*, *26*(5), 1026–1048.

Chouchou, F., Dang-Vu, T. T., Rainville, P., & Lavigne, G. (2018). The role of sleep in learning placebo effects. In L. Colloca (Ed.), *Neurobiology of the placebo effect, part II: Vol. 139* (pp. 321–355). New York: Academic Press.

Coleshill, M. J., Sharpe, L., Colloca, L., Zachariae, R., & Colagiuri, B. (2018). Placebo and active treatment additivity in placebo analgesia: Research to date and future directions. In L. Colloca (Ed.), *Neurobiology of the placebo effect, part II: Vol. 139* (pp. 407–441). New York: Academic Press.

Colloca, L. (2017a). Nocebo effects can make you feel pain. *Science*, *358*(6359), 44. https://doi.org/10.1126/science.aap8488.

Colloca, L. (2017b). Tell me the truth and I will not be harmed: Informed consents and nocebo effects. *The American Journal of Bioethics*, *17*(6), 46–48. https://doi.org/10.1080/15265161.2017.1314057.

Colloca, L. (2018a). Neurobiology of the placebo effect, part I. L. Colloca Ed. (Vol. 138). Cambridge: Elsevier.

Colloca, L. (2018b). Neurobiology of the placebo effect, part II L. Colloca Ed. (Vol. 139). Cambridge: Elsevier.

Colloca, L., Enck, P., & DeGrazia, D. (2016). Relieving pain using dose-extending placebos: A scoping review. *Pain*, *157*(8), 1590–1598. https://doi.org/10.1097/j.pain.0000000000000566.

Colloca, L., & Howick, J. (2018). Placebos without deception: Outcomes, mechanisms, and ethics. In L. Colloca (Ed.), *Neurobiology of the placebo effect, part I: Vol. 138* (pp. 219–240). New York: Academic Press.

Colloca, L., Ludman, T., Bouhassira, D., Baron, R., Dickenson, A. H., Yarnitsky, D., et al. (2017). Neuropathic pain. *Nature Reviews. Disease Primers*, *3*, 17002. https://doi.org/10.1038/nrdp.2017.2.

Colloca, L., & Miller, F. G. (2011). The nocebo effect and its relevance for clinical practice. *Psychosomatic Medicine*, *73*(7), 598–603. https://doi.org/10.1097/PSY.0b013e3182294a50.

Damien, J., Colloca, L., Belleï-Rodriguez, C.-É., & Marchand, S. (2018). Pain modulation: From conditioned pain modulation to placebo and nocebo effects in experimental and clinical pain. In L. Colloca (Ed.), *Neurobiology of the placebo effect, part II: Vol. 139* (pp. 255–296). New York: Academic Press.

Darnall, B. D., & Colloca, L. (2018). Optimizing placebo and minimizing nocebo to reduce pain, catastrophizing, and opioid use: A review of the science and an evidence-informed clinical toolkit. In L. Colloca (Ed.), *Neurobiology of the placebo effect, part II: Vol. 139* (pp. 129–157). New York: Academic Press.

Dowell, D., Haegerich, T. M., & Chou, R. (2016). CDC guideline for prescribing opioids for chronic pain—United States, 2016. *MMWR. Recommendations and Reports: Morbidity and Mortality Weekly Report. Recommendations and Reports*, *65*(1), 1–49. https://doi.org/10.15585/mmwr.rr6501e1.

Egbert, L. D., Battit, G. E., Welch, C. E., & Bartlett, M. K. (1964). Reduction of postoperative pain by encouragement and instruction of patients. A study of doctor-patient rapport. *The New England Journal of Medicine*, *270*, 825–827.

Enck, P., Horing, B., Broelz, E., & Weimer, K. (2018). Knowledge gaps in placebo research: With special reference to neurobiology. In L. Colloca (Ed.), *Neurobiology of the placebo effect, part II: Vol. 139* (pp. 85–106). New York: Academic Press.

Faasse, K., & Martin, L. R. (2018). The power of labeling in nocebo effects. In L. Colloca (Ed.), *Neurobiology of the placebo effect, part II: Vol. 139* (pp. 379–406). New York: Academic Press.

Finniss, D. G. (2018). Placebo effects: Historical and modern evaluation. In L. Colloca (Ed.), *International review of neurobiology: Vol. 139* (pp. 1–27). New York: Academic Press.

Fiorio, M. (2018). Modulation of the motor system by placebo and nocebo effects. In L. Colloca (Ed.), *Neurobiology of the placebo effect, part II: Vol. 139* (pp. 297–319). New York: Academic Press.

Glare, P., Fridman, I., & Ashton-James, C. E. (2018). Choose your words wisely: The impact of message framing on patients' responses to treatment advice. In L. Colloca (Ed.), *Neurobiology of the placebo effect, part II: Vol. 139* (pp. 159–190). New York: Academic Press.

Hashmi, J. A. (2018). Placebo effect: Theory, mechanisms and teleological roots. In L. Colloca (Ed.), *Neurobiology of the placebo effect, part II: Vol. 139* (pp. 233–253). New York: Academic Press.

Kelley, J. M. (2018). Lumping and splitting: Toward a taxonomy of placebo and related effects. In L. Colloca (Ed.), *Neurobiology of the placebo effect part II: Vol. 139* (pp. 29–48). New York: Academic Press.

Klinger, R., Blasini, M., Schmitz, J., & Colloca, L. (2017). Nocebo effects in clinical studies: Hints for pain therapy. *Pain Reports*, *2*(2), e586. https://doi.org/10.1097/PR9.0000000000000586.

Klinger, R., Stuhlreyer, J., Schwartz, M., Schmitz, J., & Colloca, L. (2018). Clinical use of placebo effects in patients with pain disorders. In L. Colloca (Ed.), *Neurobiology of the placebo effect, part II: Vol. 139* (pp. 107–128). New York: Academic Press.

Lasagna, L., Mosteller, F., Von Felsinger, J. M., & Beecher, H. K. (1954). A study of the placebo response. *The American Journal of Medicine*, *16*(6), 770–779.

Meissner, K., & Linde, K. (2018). Are blue pills better than green? How treatment features modulate placebo effects. In L. Colloca (Ed.), *Neurobiology of the placebo effect, part II: Vol. 139* (pp. 357–378). New York: Academic Press.

Nahin, R. L. (2015). Estimates of pain prevalence and severity in adults: United States, 2012. *The Journal of Pain*, *16*(8), 769–780. https://doi.org/10.1016/j.jpain.2015.05.002.

Panagiotis, Z., & Mitsikostas, D. (2018). Nocebo responses in brain diseases: A systematic review of the current literature. In L. Colloca (Ed.), *Neurobiology of the placebo effect, part II: Vol. 139* (pp. 443–462). New York: Academic Press.

Pollo, A., Carlino, E., & Benedetti, F. (2011). Placebo mechanisms across different conditions: From the clinical setting to physical performance. *Philosophical Transactions of the Royal Society of London. Series B, Biological Sciences*, *366*(1572), 1790–1798. https://doi.org/10.1098/rstb.2010.0381.

Ray, W. A., Chung, C. P., Murray, K. T., Hall, K., & Stein, C. M. (2016). Prescription of long-acting opioids and mortality in patients with chronic noncancer pain. *JAMA*, *315*(22), 2415–2423. https://doi.org/10.1001/jama.2016.7789.

Sullivan, M. D., & Howe, C. Q. (2013). Opioid therapy for chronic pain in the United States: Promises and perils. *Pain*, *154*(Suppl. 1), S94–100. https://doi.org/10.1016/j.pain.2013.09.009.

Sussex, R. (2018). Describing placebo phenomena in medicine: A linguistic approach. In L. Colloca (Ed.), *Neurobiology of the placebo effect, part II: Vol. 139* (pp. 49–83). New York: Academic Press.

Wampold, B. E. (2018). The therapeutic value of the relationship for placebo effects and other healing practices. In L. Colloca (Ed.), *Neurobiology of the placebo effect, part II: Vol. 139* (pp. 191–210). New York: Academic Press.

Wolff, H. G., Dubois, E. F., et al. (1946). The use of placebos in therapy. *New York State Journal of Medicine*, *46*, 1718–1727.

CHAPTER ONE

Placebo Effects: Historical and Modern Evaluation

Damien G. Finniss[*,†,‡,1]
[*]Department of Anaesthesia, Royal North Shore Hospital, Sydney, NSW, Australia
[†]University of Sydney Pain Management Research Institute, Royal North Shore Hospital, Sydney, NSW, Australia
[‡]School of Allied Health Sciences, Griffith University, Gold Coast, QLD, Australia
[1]Corresponding author: e-mail address: damien.finniss@sydney.edu.au

Contents

Abstract

The history of the placebo phenomenon is both interesting and informative. The placebo effect is a core component of health care, and if one considers a contemporary view of placebo effects, this has likely been the case for as long as records exist. An exploration of placebos and placebo effects across several centuries demonstrates many interesting facts, ranging from very early writings and experiments to the modern investigation of placebo mechanisms. A longitudinal review reveals the many challenges that have been faced, whether this is understanding *what* a placebo is, *why* it may work, and the meaning of such responses for routine clinical practice. At the core of this process are the concepts surrounding placebo, as these are imperative in the application of placebo to both routine clinical care and in the context of clinical trial design. This narrative review will look at the history of placebo, particularly key aspects over time, with a view to presenting a modern re-conceptualization of this field so as to facilitate ongoing scientific inquiry and to improve health care.

International Review of Neurobiology, Volume 139
ISSN 0074-7742
https://doi.org/10.1016/bs.irn.2018.07.010

1. INTRODUCTION

The placebo phenomenon has been a topic of much interest in recent times, and the literature provides countless examples of the many wide-ranging views that have been taken on the topic over generations of medical practitioners. Placebo effects have been given both negative and positive connotations, and these have stemmed from a variety of issues surrounding definition and language, historical meaning, and utility in research or in clinical practice. The topic is rich in its depth and demonstrates the importance of the biopsychosocial appraisal of the health care encounter (Engel, 1977). To this extent, it represents an important pillar in medical and broader health care practice (Finniss, 2013).

The study of the evolution of the word "placebo," its use and meaning, provide an important insight into not only the topic itself but into health care practice. At the core of this topic is the concept of what a placebo is, and what placebo administration actually represents. A modern re-conceptualization of the field of placebo effects has been crafted, based on historical examination and on new psychological and neurobiological research. Despite the substantial rise in literature concerning placebos and placebo effects, for many its meaning is still defined by its lengthy and somewhat controversial history. On this basis, exploration of key historical events and concepts allows for a potentially new and practical appraisal of this important topic.

2. THE ORIGINS OF THE TERM PLACEBO

It is believed that the word "placebo" originated in the religious context, in an early Latin translation of the Hebrew Bible (around the thirteen hundreds) (Kerr, Milne, & Kaptchuk, 2008; Shapiro, 1964). In this translation, the ancient Hebrew word "ethalech," meaning "I shall walk," was translated to "Placebo Domino in regione vivorum" (Hart, 1999; Jacobs, 2000). This is the opening phrase of the Vespers for the Dead (Psalm 116, 9th verse) which when translated states "I shall walk before the Lord." At some stage there was a mistranslation, with the word "Placebo" being used to mean "to please," and therefore the translation states "I will please the Lord" (Lasagna, 1986). This original mistranslation played an important role in both conceptual and definitional aspects of placebo effects for many hundreds of years, and remains the case in current times.

There was further mention of the word placebo in the late thirteen hundreds. At this time, placebo had taken a rather disparaging meaning. One example was the hiring of mourners at burials, who would "sing placebos" of false praise to flatter the dead (Kerr et al., 2008). The notion of "pleasing" was extended in Geoffrey Chaucer's Canterbury Tales. In the Merchant's Tale, a character "Placebo" was depicted as a sycophant, whereas in the Parson's Tale, flatterers would sing placebo and were referred to as the Devil's Chaplains (Aronson, 1999; de Craen, Kaptchuk, Tijssen, & Kleijnen, 1999). These early uses of the word placebo were an interesting interpretation of the notion "to please," given that its likely biblical meaning was not negative. Nevertheless, the idea that pleasing or placating someone was somehow a negative act was an important one, and these initial uses of the word in a negative context undoubtedly played an important role in shaping the meaning of placebo effects for centuries to come.

3. PLACEBO AND MEDICINE

The word "placebo" is likely to have entered the medical lexicon in the mid to late seventeen hundreds. Interestingly, despite the initial negative meaning attributed to placebo, the first use in health seems to have been a positive one. In a series of lectures delivered in 1772, a British physician by the name of Willian Cullen used the word "placebo" when describing situations where he would administer (a placebo) in the effort to please and comfort patients who were suffering from incurable illnesses. This was an important and arguably positive conceptualization. Cullen wrote that "I did not trust much to it, but I gave it because it is necessary to give a medicine, and as what I call a placebo" (Kerr et al., 2008, p. 90). Cullen had been able to recognize a potential utility in using a placebo to aid the symptomatic management of patients. Interestingly, Cullen's intent was not to administer what one would now deem to be a traditional placebo (for example a sugar tablet or saline injection), rather he used compounds for which there was genuine belief that the compound was reasonably effective for the patient's symptomatology. Interestingly, Cullen advocated the use of lower doses of these compounds, with the goal of modulating symptomatology without necessarily affecting the disease process. This was a particularly insightful direction for a physician with a strong background in clinical chemistry, and the idea of using a so called "active placebo" (where a real drug is given at a sub-therapeutic dose) persisted until the present day (see Kerr et al., 2008 for comprehensive review).

It is likely that Cullen's work resulted in the word "placebo" gaining mention in a medical dictionary several years later (1785) (Aronson, 1999). Placebo was defined as "a commonplace method or medicine" in Motherby's New Medical Dictionary, a definition that probably did not reflect either the origin of the word or Cullen's viewpoint. There was some further clarification in 1811 in Hooper's Medical Dictionary, where placebo was defined as "any medicine adapted more to please than benefit the patient," a definition which was far better at integrating the initial meaning of the word and its then recent application in Cullen's work. In retrospect this was a very important conceptualization of a placebo, whereby it placed focus on therapeutic intent; that is, the adaptation of therapy to please (symptomatic management) versus cure. There was a principle that placebo (or pleasing) was something that was done *instead* of a routine therapy, sometimes when a conventional therapy had not worked, although there was little written specifically outlining such use with much of the discussion being in the form of debates between physicians (Brody, 1982; Kerr et al., 2008; Risse, 1986). This lack of acknowledgment of a beneficial role for placebo in symptom management, or the potential role for concurrent placebo and "real" drug administration, has been a conceptual issue for placebo and its clinical utility ever since this definition in 1811.

4. A DIFFERENT ROLE FOR PLACEBO: THE "CONTROL"

In the late seventeen hundreds, Franz Anton Mesmer, a German physician, believed that he had found a new curative technique for many diseases. Mesmer believed that humans had internal "invisible" fluid channels, and that these channels could be accessed to promote healing. This was a property of humans he called Animal Magnetism, and he had developed various techniques to promote change in these fluid channels. These techniques (so called "mesmerism") were conducted on both individuals and groups, and typically involved elaborate therapeutic rituals (Kaptchuk, Kerr, & Zanger, 2009). Mesmer had a belief that there was an interaction between animated and inanimate objects, and one component of his rituals would involve using objects which had been mesmerized, allowing then to pass over a patient's body, accessing and unblocking the fluid channels. There was much interest in Mesmer's work, particularly due to some of the dramatic responses to his work. In 1784, King Louis XVI appointed a Royal Commission to investigate Mesmer's claims. Two of the key commissioners were Benjamin Franklin and Antione Lavoisier (Kaptchuk et al., 2009). The exact details of their work are not clear, but nevertheless the commission conducted a series of single blinded experiments on patients. Of note was the use of sham (or placebo) objects instead of those with mesmerizing power. A large proportion of patients responded to both the "real" and placebo rituals, and this represents the first use of a placebo control to assess the veracity of a proposed treatment. The commission concluded that there was no scientific evidence to support Mesmer's theories, and that any therapeutic benefit was likely due to a patients "imagination" (Kaptchuk et al., 2009). It should be noted that "ritual" in the setting of this paper refers to a component of the health care encounter pertaining to the delivery or act of a treatment (such as taking a tablet, application of a cream or the giving of an injection), and this term can have different meanings outside this context. This remains a concept that requires further exploration.

The idea that a response to a placebo intervention was in some way not *real* (as implied by the use of the word "imagination") extended from this time. Another placebo-controlled trial had been conducted some 25 years later in 1799 by a British physician, John Haygarth (Kaptchuk, 1998a, 1998b). In this surgical trial, patients were implanted with metallic rods, or "Perkins Tractors." It was believed that the metallic properties of the tractors were able to alleviate a wide range of symptoms of disease (Booth, 2005). In this small trial, Haygarth implanted five patients with the tractors and assessed their response. He then repeated the procedure the following day, removing the tractor and replacing it with a placebo (a dummy wooden rod). Patients were blinded on both occasions, and Haygarth noted improvements in three of the five patients to *both* the Perkins tractor and the placebo. He quotes "an important lesion in the physic is here to be learnt, the wonderful and powerful influence of the passions of the mind upon the state and disorder of the body" (de Craen et al., 1999). Despite making the important link between the mind and the brain, there was essentially a negative connotation to both the responsiveness to placebo and to the intervention under scrutiny. Haygarth presented his work at the Literary and Philosophical Society of Bath in 1800, with the title "Of the Imagination, as a cause and a cure of disorders of the body; exemplified by fictitious tractors, and epidemical convulsions" (http://www.jameslindlibrary.org/haygarth-j-1800/).

These two early uses of placebo as a control have continued to shape contemporary thinking about placebo. Both trials resulted in similar conceptual conclusions. Firstly, a response to the placebo intervention meant that the index intervention was not scientifically valid (for the hypothesis would be that one would not respond to the placebo but would respond to the "real" treatment). Secondly, there was great appreciation of the magnitude of response to the placebo groups, but this was dismissed as a fictitious response due to the imagination rather than a genuine biological event. Despite the writings of Cullen at a similar time, the setting of a "trial" brought about a more linear appraisal of placebo administration and response to placebo.

The use of the "placebo control" was further developed in the 1800s. Austin Flint, an American physician, conducted a small trial in 1863 for the treatment of articular rheumatism. He used a dilute remedy as a placebo control. Flint believed that the placebo may provide symptomatic improvement as the disease progressed according to its natural history, whereas the trial drug should alter the course of the disease (de Craen et al., 1999; Kaptchuk, 1998b). This was a rather progressive understanding of placebo, particularly as there was no such focus placed on response to placebo being an "imagination," rather Flint introduced the idea that a placebo may be a useful way to reduce symptoms of disease rather than to modify the disease process itself.

Flints' work had extended the work of Haygarth in a medical trial, likely influencing further trials of medicine in the early 20th Century. The new "blinded method" was introduced, with the first small trial in 1913 (Hewlett, 1913) and a larger one in 1918 (Sollmann, 1917). In the larger trial of almost 1000 subjects, Adolf Bingel tested an "antitoxin" for the treatment of diphtheria (against a normal serum). While a relatively minor point, Bingel used assessors who were blinded to the allocation of the drug, potentially underscoring a gradual appreciation of the role of the psychosocial context around the patient on treatment outcomes. Further famous blinded trials looked at different ether preparations (Hediger & Gold, 1935) and Aminophylline for angina pectoris (Gold, Kwit, & Otto, 1937). In addition to the blinded trials (where no placebo was given), Gold and colleagues introduced placebo administration to create the placebo-controlled double-blinded trial which is still a benchmark in modern medicine (Shapiro & Shapiro, 1997). This method was made famous in 1950 with the study of a treatment, khellin, for angina pectoris. Investigators specifically wanted to control for the effect of placebo to determine a genuine drug effect (Greiner et al., 1950). This paradigm has extended to current clinical trial methodology.

5. PLACEBO AS A TREATMENT

The notion of using placebos as a treatment was discussed by Cullen in his lectures, and as mentioned, this appears to be first use of the word "placebo" in the medical setting. Historical assessment of placebo treatments is somewhat complex, as it is reliant on several factors. Firstly, it is unclear whether physicians would report placebo use in the literature, particularly given the negative connotations placed on it. Secondly, in the absence of a clear indication of placebo use (by means of a traditional placebo or by the intent of the administrator), assessment of placebo treatments then becomes more complex as it is reliant on appraisal of ancient treatments within our current scientific framework. Nevertheless, several historians and anthropologists have attempted to describe placebo use in ancient medicine (Moerman, 1997, 2002; Shapiro & Shapiro, 1997). When faced with this task, anthropologists have looked at both conventional pharmacopoeias and also the remedies of different cultures. In the former, analysis demonstrated an estimated 5000 ancient remedies with over 16,000 different prescribing practices (Shapiro & Shapiro, 1997). Using a modern scientific approach, one may conclude that many if not all of these remedies were in fact placebo treatments. For example, the London Pharmacopoeia included such remedies such as Gascoyne's Powder (amber, bezoar, crabs eyes & claws, coral and pearls) and "Usnea" (the moss from the skull of a victim of violent death) (Garrison, 1921). Other remedies included various animal derivatives such as bezoar stones (animal gallstones) and crushed elephant horns.

The assessment of treatments across some cultures has also given further perspective on ancient medicine. The study of Native American healing rituals identified a wide range of medicinal plants and prescriptions (Moerman, 1997). In fact, Moerman concluded that over 219 different cultures of Native Americans used more than 2800 species of medicinal plants in over 25,000 ways. Undoubtedly many if not all of these remedies would constitute what is called a placebo. Such an extensive and exotic array of therapies were undoubtedly administered as part of a therapeutic ritual, and hence it is not unreasonable on both conceptual and scientific grounds to suggest that these treatments were in fact placebo treatments, even though it is unlikely the clinician's intent was "to please" the patient.

When one broadens the analysis beyond administration of a medicine, the complexity becomes greater. For example, many of the above remedies would have been administered in very specific rituals. In fact, healing rituals were used in many cultures, and were not always associated with medication administration. As an example, the "Royal Touch" was a procedure of placing hands on a patient in an attempt to treat illness, and this ritual extended for centuries (Shapiro, 1959). One could easily conclude that these rituals also constituted placebo administration.

Further assessment is made difficult, particularly when the intent of the clinician is not evident. Cullen was very clear that giving a placebo (a drug with pharmacological properties but at an ineffective dose) was specifically done to appease a patient when he felt that there was no further chance of disease cure with a drug. There is a relative lack of further publications about clinical placebo use in history, and again this is possibly due to the negative meaning given to the topic. One example was a letter written by Thomas Jefferson referring to "pious fraud," where he recounts a successful physician using more bread pills, drops of colored water and powders of hickory ashes than all other medicines combined (de Craen et al., 1999). Such limited description of placebo use as a treatment persisted from Cullen's time, and has only been studied and discussed very recently.

5.1 The 20th Century and "The Powerful Placebo"

Discussion of placebo increased in the literature in the early 20th century. This was in parallel with the developments in the blinded and ultimately placebo-controlled double-blinded trials, as previously mentioned. There seemed to be a continual tension between acknowledgement of the possible therapeutic utility of placebo prescription and the ethical issues surrounding its use. One notable example was the work by a physician Dr. Hildred Carlill, who conducted a relatively invasive placebo (sham) operation, involving the removal of part of the skull, for a refractory case of "sleepiness" in a young man (Carlill, 1918). Carlill reported a complete cure of symptoms and noted that there was no clear physiological basis for its effect; rather it created a psychological effect on the patient. There was an ensuing debate about the power of the sham intervention and defense of a clinician whose intention was to solely benefit the patient (Cockayne & Ward, 1920). Others criticized the work, demonstrating the ethical issues, particularly that of deception. One such letter acknowledges the ability to "emulate the atmosphere of old Esculapian temples for curative effect," however, states that this should not "descend into the crude charlatanism of their ceremonies" (Ready, 1920, p. 289). These discussions acknowledged the power of the placebo and linked it to therapeutic ritual's, but also demonstrated concern for the deceptive practice. Furthermore, there was a suggestion that placebo ritual's may be beneficial in "functional" conditions, but the downside may be propagation of these conditions (Ready, 1920). This was an example of an important step in biomedicine's view on placebo. It had progressed from an "effect" which was due to "imagination," to a psychological response, and then to a therapeutic ritual or health care encounter, however there was belief that it was only useful for conditions which were psychologically based or "functional," that is, there was no acknowledgement of a genuine physiological change.

In 1938, Houston published a paper titled "The doctor himself as a therapeutic agent" (Houston, 1938). Houston's work was particularly interesting. He acknowledges the concept of placebo, yet primarily did not advocate placebo use per se, rather the principles of what placebo was doing. Houston believed that above expectation of effect, there was a critical importance of the doctor–patient relationship, and that this relationship had "dynamic power." One liberal interpretation of his work was that he acknowledged that placebo administration clearly had benefit, and he seemed to dissect the components of why this was the case (that is, both the setting up of an expectation of benefit plus the therapeutic context in which a placebo is given). At the center of this treatment ritual was a powerful interaction between the doctor and patient. Houston actually cautioned the use of placebo as a simple tool, yet saw the therapeutic relationship as an additional "drug" therapy of great benefit to the patient.

The discussion and advocacy around aspects of placebo use extended into the mid-20th century. One such example was that of Wolf and colleagues in 1947, discussing conscious prescription of placebo. In this paper, there was clear advocacy for placebo prescription, with validation given by the fact that if a patient was pleased by a placebo treatment, then this was actually a positive therapeutic outcome (Wolfe, 1946). This paper was seemingly the first to use placebo treatments in the title with the specific discussion about using placebo as part of clinical practice.

Following the first half of the 20th century, there was more literature discussing either the use of placebo in practice or enhancement of the elements of what placebo was representing, such as the therapeutic ritual and doctor–patient relationship (note this conceptualization of placebo did not seem to be generally appreciated at the time, based on a review of the published literature). Lasagna and colleagues conducted an interesting study in placebo analgesia in 1954 (Lasagna, Mosteller, von Felsinger, & Beecher,

1954). The goal of the study was to characterize the magnitude of response to placebo and the pattern of responsiveness. In this study, 162 patients with post-operative pain were given alternating doses of an opioid (morphine) and placebo. The clinically meaningful effect was determined as 50% reduction in pain. The multiple doses permitted assessment of response over time. It was found that only 14% of patients responded consistently to placebo, with 30% not responding at all. The majority of patients (55%) were inconsistent responders. Response was greater to placebo in those who also responded to morphine. The authors made some qualitative analyses, suggesting that higher somatic symptoms, higher anxiety and more positive opinions of the hospital were associated with response to placebo. This was a pioneering study which attempted to quantify the "power" of placebo, not to mention some of the response characteristics. It also paved the way for exploration into some of the potential psycho-behavioral factors surrounding responsiveness.

Momentum had continued with regard to the exploration of placebo in the clinical setting, even if no placebo was given. In a paper by Egbert and colleagues in 1964, the investigators assessed whether "active placebo action" could be used to improve the post-operative course in patients undergoing major abdominal surgery (Egbert, Battit, Welch, & Bartlett, 1964). This paper built on some of the previously discussed work, particularly the acknowledgement of the doctor–patient interaction in the outcome of therapy. Although no placebo was given, there was a clear link between the idea that placebo administration causes positive effects in some patients, and that part of this reason was the interaction between doctor and patient. In this trial, 97 patients were studied. A control group had standard care. The experimental group or "special care" group received additional interaction with the doctor (an anesthesiologist). This interaction was in the form of both pre-operative and post-operative visits, with increased attention, time and shaping of post-operative expectations relating to pain. The study showed that patients in the experimental group required a significantly reduced amount of daily morphine (between 30% and 50% less) for post-operative pain than the control group, which led to the conclusion that the doctor–patient interaction had contributed to this outcome.

As previously discussed, the new "gold standard" placebo-controlled double-blinded study had entered research practice in the mid-20th century. Arguably, it was these trials that led to a significant milestone in the history of the field of placebo. In 1955, Beecher published a paper titled "The powerful placebo." In his seminal paper, Beecher looked at the placebo groups in these new trials (15 in total), pooling the responses together to estimate the power of the placebo effect. He estimated "the effect" at 35.2% (range 21–58%) (Beecher, 1955, 1959). Although there had already been an attempt by Lasagne and colleagues to quantify the placebo effect, this paper was significant in that it used an empirical approach and had data from multiple studies, demonstrating a meaningful effect and flagging "the placebo effect" as a phenomenon worthy of further inquiry.

Placebo-controlled trials would continue to serve as important reminders of the power of the placebo effect. This was particularly noted in the more interventional trials, such as those conducted soon after Beecher's paper. Internal mammary artery ligation was an accepted surgical treatment for angina pectoris, a painful condition attributed to a reduction in blood supply to the myocardium. Ligation of the mammary artery was performed, as it was believed to increase myocardial blood flow, and thus oxygen supply (Moerman, 2002). On the basis of positive results, a series of placebo "sham" controlled trials were conducted (Cobb, Thomas, Dillard, Marendino, & Bruce, 1959; Dimond, Kittle, & Crockett, 1960). The biomedical model remained unchanged from the very early placebo-controlled surgical trial by Haygarth (de Craen et al., 1999). That is, response to the sham surgery likely meant that the surgery itself was due to a "placebo effect" rather than modification of a pathological process. In these small surgical trials, patients did equally well as with the sham surgery (skin incision only), effectively eradicating the surgery from practice. This is a trend which continues, particularly in surgical or procedural trials where large placebo effects exist. More recently, impressive and long-lasting responses (up to 1 year later) were seen with sham arthroscopy versus real arthroscopy (where the knee was "cleaned out") (Moseley et al., 2002). Similarly, a technique called "vertebroplasty" has been purported to be useful for the treatment of crush fractures of the vertebral bodies. In this technique, patients remain conscious and have a cement injected into the vertebral bone. In a well-designed placebo-controlled trial, where the placebo group did not receive cement but an injection (with the smell of cement in the room), there was no difference between index treatment and the placebo (Buchbinder et al., 2009; Buchbinder, Osborne, & Kallmes, 2009). Both groups did well. Most recently, a well-designed placebo-controlled trial of percutaneous stenting for stable angina demonstrated no difference in exercise tolerance between actual stenting and a placebo procedure (Al-Lamee et al., 2018). These recent studies have caused great discussion in the literature, particularly with regard to their interpretation for clinical care. For example, vertebroplasty is no better than placebo vertebroplasty, yet both are better than routine care. This raises the question of whether one can perform a procedure when the procedure is more beneficial than other available therapies, yet its primary mechanism of action is placebo (Finniss, Kaptchuk, Miller, & Benedetti, 2010).

A final component of the evolution of "the powerful placebo" in the 20th century was the attempt to extend Beecher's work. Several analyses have been conducted in recent times, attempting to quantify "the placebo effect" across many conditions (Turner, Deyo, Loeser, Von Korff, & Fordyce, 1994). These papers brought with them great scrutiny and criticism, with the original meta-analysis suggesting that placebo effects were in fact not so powerful (Hróbjartsson & Gøtzsche, 2001). Further work subsequently showed that placebo effects were larger in experimental trials (which are aimed at maximizing placebo effects) than control groups in clinical trials, and there are significant methodological issues when assessing placebo effect(s) as a single entity across many conditions (Hróbjartsson & Gøtzsche, 2004; Vase, Riley, & Price, 2002). In fact, this would demonstrate a major conceptual point in the history of the study of placebo, that is, there is not one "placebo effect," there are many (see below). The study of placebo mechanism has elegantly brought this to light.

6. PLACEBO "EFFECTS": THE STUDY OF MECHANISMS

As outlined in this paper, there have essentially been three broad aspects related to placebo in medicine. The first has been placebo use in the clinical trial context, where placebo controls have been employed, and more recently response to placebo in these groups studied. The second is the use of placebo (or extrapolation of aspects of placebo use) in routine clinical practice. A third and critical element over time has been the dedicated study of placebo mechanisms. This has made the most substantial change in the field and has arguably progressed it to its current state today.

At a mechanistic level, placebo effects were first studied in some elegant experiments involving classical conditioning in animals (Ader & Cohen, 1975; Herrnstein, 1962). These studies were able to demonstrate conditioned placebo effects, and this work was soon replicated in humans (Laska & Sunshine, 1973). This was remarkable proof of a psychological mechanism (learning principles) mediating responses to placebo. In 1978, a landmark trial was conducted by Levine and colleagues (Levine, Gordon, & Fields, 1978). Noting the title "The mechanism of placebo analgesia," investigators were able to demonstrate significant placebo effects in post-operative pain. This was a double-blinded study, where patients were given a placebo 2h after surgery. Patients were then given a further drug identical to the placebo (but it contained the opioid antagonist naloxone, which in this case should block the action of endogenous opioids). Those who responded to placebo became worse with naloxone, and those who did not respond reported little change. This was an important demonstration that placebo analgesia can be mediated by the endogenous opioid system. There were several studies conducted between 1978 and 1985 supporting the role of opioids in placebo analgesia (for a review, see Benedetti, 2008; Finniss et al., 2010; Price, Finniss, & Benedetti, 2008). Interestingly in some of these experiments, the response to placebo was not always completely blocked by naloxone, demonstrating mechanisms that are independent of the endogenous opioids system. In retrospect this was a critical time point in the conceptualization that placebo was not a single effect, rather there were multiple effects. At a mechanism level, there was a clear opioid-mediated placebo effect, and an effect that was either partially mediated by endogenous opioids (by virtue of some reversibility by naloxone) or independent of the opioid systems. One of these additional mechanisms has been shown to be mediated by endogenous cannabinoids (Benedetti, Amanzio, Rosato, & Blanchard, 2011).

This time period was also important from a methodological and conceptual viewpoint. In two of these studies, investigators used a different scientific paradigm. In these studies, experimental pain was created in volunteers. Placebo injections were administered, and a second injection 40min later (Grevert, Albert, & Goldstein, 1983). However, unlike the study by Levine and colleagues, the second injection was given surreptitiously in a so called "hidden" manner. This way a drug could be given without the patient being aware of its administration. In this paradigm, patients are connected to intravenous infusions. In one condition, the "open" condition, the patient receives an injection with the clinician in the room. In the "hidden" condition, a computer delivers the same dose of the drug but the patient and the clinician are blinded to when it is given. A hidden injection removes as much of the psychosocial component of drug treatment and therefore demonstrates only the pharmacology of the drug (Benedetti et al., 2003; Colloca, Lopiano, Lanotte, & Benedetti, 2004). In this study and a subsequent study, hidden naloxone was able to completely or partially block placebo analgesia, again supporting the role of an opioid-mediated mechanism in a robust and somewhat different manner (Levine & Gordon, 1984). Interestingly, this study design was most useful at a conceptual level in subsequent years. A purely biomedical or pharmacological paradigm would dictate that there should be no difference between "open" and "hidden" administration, for the overall drug effect should be due to the pharmacological properties of the drug, and this effect should be independent of the context around the administration, as long as the dose and route of administration are the same (Finniss & Benedetti, 2005). Benedetti et al. repeated and extended this trial methodology when looking at both the opioid system and the chemical cholecystokinin (CCK) (which is pro-nociceptive, thereby increasing pain) (Benedetti, Amanzio, & Maggi, 1995). Benedetti studied the effect of a CCK blocker with the hypothesis that this may be a good analgesic drug. Interestingly, a cholecystokinin blocker (proglumide) was an effective analgesic when given in a standard manner. However, when it was given as a hidden infusion, it was not effective at all, demonstrating that the drug alone had no inherent analgesic property (Fig. 1). In fact, the open administration of proglumide was more effective than placebo, implying that this drug may act with other endogenous chemicals to elicit its action. This is a remarkable demonstration that links back to the work of Houston and those before him, where one could consider that the result of a treatment may be due to the drug itself combined with the "placebo" or "ritual" or "doctor–patient" component of the treatment (Houston, 1938). If giving a placebo induces endogenous chemical release (such as an opioid), it is entirely possible that

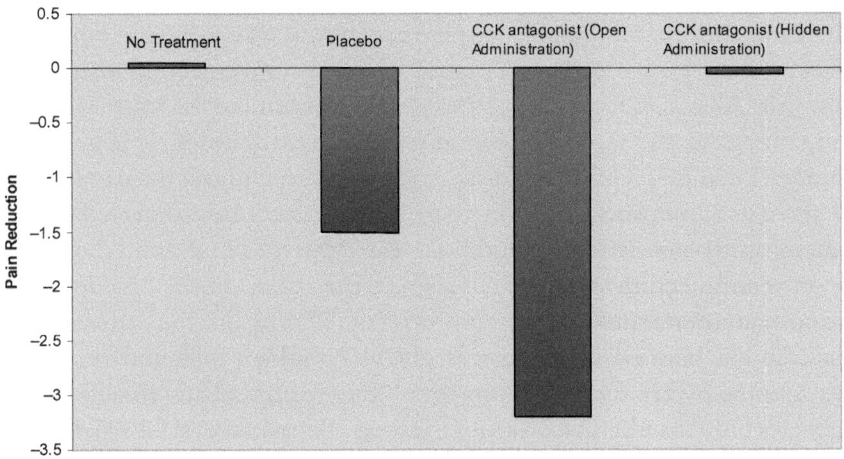

Fig. 1 Potentiation of placebo effects with open–hidden administration of a CCK antagonist. *Finniss, D.G., & Benedetti. (2009). Cousins & Bridenbaugh's neural blockade in clinical anesthesia and pain medicine. 4th ed. LWW with permission (originally adapted from Colloca & Benedetti, 2005).*

these chemicals can be *additive* or *synergistic* with an exogenous drug, such as was the case with proglumide (Enck, Bingel, Schedlowski, & Rief, 2013).

The open–hidden paradigm is a powerful reinforcer that the overall outcome of a treatment is not only due to the treatment itself, but the context in which it is given. This is a potentially challenging concept for biomedicine, although as mentioned this is something that has been raised for over 100 years. Benedetti and his group were able to show more recently that significant differences exist between standard "open" administration and "hidden" administration of common analgesic drugs (opioids, non-steroidal anti-inflammatory drugs),other drugs such as benzodiazepines (Benedetti et al., 2003; Finniss & Benedetti, 2007) and surgical procedure such as deep brain stimulation (Colloca et al., 2004). Despite the same route of administration and same dose of a drug, the ritual of drug delivery was "adding" to the drug effect. This is conceptualized as the placebo component of therapy, even though no placebo is given. In this instance, it is the health care encounter that initiates and maintains placebo effects.

Further work into mechanisms and determinants of placebo effects has occurred over the past 30 years, and there has been a substantial rise in this work. Well-designed studies now clearly demonstrate many discrete neurobiological and psychological mechanisms underpinning the many different placebo effects that exist (Keller, Akintola, & Colloca, 2018; Pecina & Zubieta, 2018; Zion & Crum, 2018) A large number of studies have been done in the experimental setting (such as experimental pain), but more recently work has expanded into various clinical populations, such as pain, depression, Parkinson's disease (Colloca et al., 2004; de la Fuente-Fernandez et al., 2001) and asthma (Wechsler et al., 2011).

Modern studies have supported the initial work and thoughts of authors mentioned in this article, such as Houston and Cullen (Houston, 1938; Kerr et al., 2008). For example, the role of expectancy as a key psychological determinant is clear (Price et al., 2008). Two unique studies have added to our understanding of the doctor–patient interaction. The first was a landmark study performed by Gracely and colleagues in 1985. In this study, patients with dental pain (after third molar extraction) were randomized into two groups (Gracely, Dubner, Deeter, & Wolskee, 1985). All patients were told that they would receive either a drug which would improve their pain (an opioid drug), do nothing (a placebo), or make their pain worse (naloxone). In group one, the clinicians were informed that there was no possibility of giving an opioid (a rationale about drug supply was given), but that strict double blinding must be kept in place. In group two, the clinicians were told that the issue was rectified and there was a reasonable chance that the drug was the opioid. When analyzing responses to placebo, patients in group one actually became worse on average despite a placebo being given. In group two, there was an average placebo effect. In fact, this study was aimed at the clinician's effect on the patients' response, and with strict double blinding in place, the authors concluded that subtle behavioral changes may have existed between the groups, based on the information given to clinician. This would underscore the importance of clinicians' words and behaviors, and supports the early accounts of the importance of this relationship.

The second unique study was conducted by Kaptchuk and colleagues (Kaptchuk et al., 2008). The Gracely's study looked at clinician behavior as a determinant of outcome (Gracely et al., 1985). Kaptchuk and colleagues extended this work by assessing more components of the therapeutic relationship. This study involved two experimental groups and a control group. The experimental groups all received a program of placebo acupuncture (there is an illusion of the needle penetrating the skin but it does not actually do so). In one group, the clinician interacted with the patient in a clear manner, explaining that he or she had read the patient's file, and would deliver treatment which would be of benefit to the patient; however, no further discussion would take place during the therapy, as it was a controlled trial. In the other group there was an attempt to augment the doctor–patient relationship, involving verbal and non-verbal strategies. The "augmented" placebo group had better outcomes when more focus was placed on the doctor–patient relationship, despite both groups receiving placebo (Kaptchuk et al., 2008). This was an elegant dissection of the health care encounter, supporting many previous assertions as to the importance of this interaction. This was also a convincing demonstration that the "power" of placebo was due to the psychosocial context around the patient (rather than the sham needle itself). Taken together, the study of mechanisms (whether at a neurochemical level or psychosocial level) has dramatically advanced the knowledge of the topic and has given support to the field of placebo as a relevant and important area of research.

7. TOWARD A MODERN RE-CONCEPTUALIZATION AND APPLICATION OF PLACEBO EFFECTS

Learning from the rich history in the field has permitted great progress in understanding the mind–brain–body interaction and the way in which this is routinely harnessed at the bedside. It is clear that placebo effects are valid psychobiological processes that are target- (or system-) specific, operating across many different clinical conditions. A large percentage of this work has been performed in the field of pain and in the experimental or laboratory setting (Atlas & Wager, 2014; Colloca, Klinger, Flor, & Bingel, 2013; Hashmi, 2018; Meissner et al., 2011). This is important on two accounts. Firstly, this challenges the pejorative use of the word "placebo" by demonstrating that either a placebo treatment or the placebo component of routine care (where no placebo is given) has the ability to initiate robust and specific biological events that can be large in magnitude (Price et al., 2008). The latter has been referred to as "placebo-rapport" (Sussex, 2018). This contradicts some of the pejorative constructs, such as something being "just" or "merely" a placebo or the idea that giving a placebo is akin to doing nothing (Sussex, 2018). This analysis has demonstrated a tension that has existed for over 200 years, whereby on one hand placebo is a useful and powerful tool (Egbert et al., 1964; Houston, 1938) and on the other it is a sub-optimal alternative that is either *non-specific*, a result of the imagination (Kaptchuk et al., 2009), or primarily useful only for subjective symptoms (Hróbjartsson & Gøtzsche, 2001). Secondly, there is a challenge to explore these mechanisms in larger clinical populations and across the duration of different conditions. This will add to the knowledge of the longevity and clinical significance of placebo effects.

Placebo effects are inherent in most aspects of health care, as most health care encounters involve interaction between the patient and their environment (of which the interaction with the clinician is paramount). Fundamental to the progression of this field is acceptance of the concept of what a placebo is and *how* placebo effects are generated. This analysis has outlined an evolution in the conceptual understanding of placebo, where it was initially an "inert" placebo administration (either as a deceptive treatment or in the setting of a trial) that often resulted in positive responses. The concept was extended to encompass aspects of the health care encounter other than the actual placebo, such as the role of the clinician (Houston, 1938) and the extrapolation of what placebo administration may be representing by enhancing the therapeutic encounter such as "positive placebo action" (Egbert et al., 1964) and "augmented placebo" (Kaptchuk et al., 2008). When a placebo is given, it is the psychosocial context around the patient at the time of administration that generates the placebo effect(s). The elements of that psychosocial context are relatively complex, and the language variable, e.g., health care encounter, therapeutic ritual, doctor–patient interaction, context effects (Moerman & Jonas, 2002; Sussex, 2018). Conceptually, the result of a treatment is due to the effect of the intervention itself and the effect of the health care encounter or psychosocial context (by placebo mechanisms). There are two important aspects for consideration. Firstly, one does not need to give a placebo to generate a placebo effect (Fig. 2). Secondly, as illustrated in this review, a large amount of research has occurred in situations that do not resemble clinical practice, and therefore there is a challenge to continue to build this research in a clinically meaningful manner. On the one hand, placebo treatments are generally not considered ethical in routine care, and responses in these trials (where deception is a part of the experiment) may not be generalizable to the clinical environment (see ensuing discussion). On the other hand, clinical trials do not resemble clinical practice at both a construct level and an outcomes level (where placebo effects are more robust when placebo administration is given in the same manner as a routine treatment would be given) (Vase et al., 2002).

One of the most challenging aspects in the application of this work has been the ethical issue of deception, and this review has outlined some of the debate on this issue since the word appeared in medical literature (Miller & Colloca, 2009). As discussed, the modern re-conceptualization of placebo effects, supported by robust mechanistic studies, demonstrates that one can harness placebo mechanisms during routine care, and no placebo is given. This is an important alternative to deceptive administration, whereby

KEY ASPECTS OF A MODERN RE-CONCEPTUALISATION OF PLACEBO

Placebo Effects are valid psychobiological effects that operate across many conditions

There is not one Placebo Effect, there are many, and they can be target specific

Placebo Effects are generated and maintained by the Psychosocial Context around a patient

Placebo Effects can be harnessed by either non-deceptive administration of placebo or by augmentation of the therapeutic encounter (where no placebo is given)

Deceptive administration of placebo "to please" in replacement of a more suitable treatment is unethical.

Non-deceptive administration of either a placebo or treatments where a significant component of efficacy is that of placebo may represent progressive and ethical application to practice. The therapeutic intent must be clear and this should be articulated to the patient in an open manner

Fig. 2 Key aspects of a modern re-conceptualization of placebo.

patients are given placebo treatments with the belief that they are receiving something different (an active drug for example). This form of placebo administration in clinical practice has been deemed to be unethical (Miller et al., 2005). Furthermore, recent evidence suggests that deceptive administration may not actually be necessary, with novel studies assessing placebo administration when patients are aware that they are receiving a placebo (Charlesworth et al., 2017).

These recent studies have demonstrated the potential utility of "non-blinded" or "honest" placebos (Colloca & Howick, 2018). Two such examples in Irritable Bowel Syndrome and Chronic Low Back Pain followed a similar methodology and were able to show marked reductions in symptoms (such as pain and disability) following administration of placebo tablets when patients were aware of their content (Carvalho et al., 2016; Kaptchuk et al., 2010). These effects occurred over several weeks of administration, demonstrating potential longevity. At the center of these novel experiments was the way in which the placebo was presented. Patients were aware of the placebo and its content, however the idea of placebo being "powerful" was reinforced. In fact, the conceptual advance in these studies was that placebo treatments were presented as something which was likely to benefit patients through mind–brain physiology, that is, there was a "real" physiological mechanism. This is in keeping with contemporary mechanism literature, in that knowledge about placebo mechanisms was delivered in an open and honest way. For example, patients were reminded of the principles of conditioning (that is that they body can respond to the act of taking the pills, similar to "Pavlov and the dogs") and that taking the pills as prescribed was important. This was a remarkable demonstration that placebo mechanisms may be accessed in a non-deceptive manner. Interestingly, at least two components may have been of significance in these trials. Firstly, there is the health care encounter with clinicians which occurred at several time points. Secondly, there was a self-managed component where patients would undertake the important "ritual" or "act" of swallowing the tablets, regularly as prescribed. The relative contributions of the "act" of administration (either clinician or patient administered) and the actual encounter itself remain important future areas of research, particularly when assessing longevity or maintenance of placebo effects.

This review has demonstrated the additional issue of therapeutic intent when prescribing a placebo treatment. This has been evident dating back to early use of placebo in the medical setting, with different themes since this time. There is intent as it relates to the therapeutic goal or mechanism.

There has been intent based on pleasing which is a psychological mechanism, reflecting the origin of the word. Another therapeutic goal has been use of placebo to modulate symptoms of disease rather than a disease process itself. For example, this has been discussed as a possibility when no other curative treatment is possible for example Flint's writings about the usefulness of placebo for symptomatic management in the early 1800s (de Craen et al., 1999), or when a condition is believed to have a substantial psychological component (Ready, 1920). Most recently, this was demonstrated in placebo administration for asthma, where significant symptomatic improvement was seen following use of a placebo and albuterol inhaler, whereas lung function (representing airway physiology) was only significantly altered with albuterol (Wechsler et al., 2011). This study provides a clear demonstration of the ability of placebo mechanisms to improve symptomatology without necessarily affecting the underlying disease process. Conversely, recent mechanistic work has demonstrated changes in pathophysiology in certain conditions such as Parkinson's disease (Benedetti et al., 2004) and in pain (Hashmi, 2018).

The field of pain is particularly interesting when we consider therapeutic intent, as placebo administration may be deemed to be both a symptomatic treatment (for example modulation of nociception by endogenous opioid release in the setting of an acute fracture) or as a disease modifier in the setting of chronic or persisting pain (where there is substantial pathophysiology in nociceptive processing, particularly in the central nervous system) (Gebhart & Bielefeldt, 2016; Tracey, 2008; Woller, Eddinger, Corr, & Yaksh, 2017). In the latter, placebo effects have been shown to alter neuronal activity at both the cortical (Petrovic, Kalso, Petersson, & Ingvar, 2002), subcortical (Benedetti et al., 2004), limbic (Wager, Scott, & Zubieta, 2007), and spinal cord levels (Eippert, Finsterbusch, Bingel, & Buchel, 2009), representing endogenous neuromodulation of nociception. With a modern conceptualization of persisting pain as a disease entity in its own right (Siddall & Cousins, 2004), the notion that placebo may be both a symptomatic and a disease modifying treatment seems a valid one. Importantly, these are concepts which need further empirical validation, particularly with mechanistic studies in clinical populations. However, the theme of therapeutic intent is not new, and the progression in ideas around either pleasing a patient, modifying symptoms, or disease itself has existed for over 200 years, with the concept of disease modification only a recent addition. Further research will assist in at least two ways. Firstly, studies of placebo mechanisms in clinical populations will provide clarity about these mechanisms over time. This will assist in delineating between symptomatic improvement and disease modification. Importantly these two concepts are not mutually exclusive despite their portrayal in this way over time, and integrated use of placebo with other routine treatments may be an optimal way of targeting both processes. Some studies have demonstrated this approach, particularly its utility in a drug reduction setting (Buckley, Sizemore, & Charlton, 1986; Pollo et al., 2001; Ralphs, Williams, Richardson, Pither, & Nicholas, 1994). Secondly, clarity about *how* a placebo may work in a given condition can facilitate open discussion and disclosure between clinician and patient as to why a treatment may be prescribed. This is potentially useful with the new "open labeled" and "dose-extending" (Colloca, Enck, & DeGrazia, 2016) prescription of placebos (these studies have demonstrated disclosure of therapeutic intent and the mechanisms behind the treatment), and when recommending treatments where a significant component of the treatment may be due to placebo mechanisms rather than the treatment itself (e.g., when a treatment is marginally superior to placebo but both are superior to no treatment) (Finniss et al., 2010).

Clarity of therapeutic intent and ensuing open discussion with a patient provides a progressive use of placebo (or treatments where a significant component is due to placebo effects). This eliminates the need for deception, and this tension over time has been outlined in this review. Several studies have been conducted which demonstrate varied prescription of placebos in routine general and internal medicine practice (Fässler, Gnädinger, Rosemann, & Biller-Andorno, 2009; Sherman & Hickner, 2008; Tilburt, Emanuel, Kaptchuk, Curlin, & Miller, 2008). These prescriptions are typically not traditional placebos (for example a sugar tablet), but are often treatments which are given when there is belief that they are not "active" for that particular condition (in this case they are treatment simulations aimed at eliciting placebo effects). The classic example is an antibiotic prescription for a viral illness. The intent of the clinicians is often variable, with one consistent aspect being the wish to "please" the patient (Linde, Fässler, & Meissner, 2011). This is a theme which has remained present for 200 years. A more progressive approach, based on new conceptualizations of placebo, is that these treatments are being given with the therapeutic intent of activating placebo mechanisms. Furthermore, the notion of "pleasing" as an inferior alternative to a valid therapy can be replaced with genuine intent to access very specific psychobiological responses. However, such prescriptions need to be balanced with both open disclosures of intent and appreciation of cost, side effect, and greater issues (such as antibiotic resistance for example). This need for balance is not new and has been discussed over many years (Houston, 1938; Ready, 1920).

A final consideration is the placebo groups in clinical trials. Clinical trial methodology is beyond the scope of this paper. But two aspects should be emphasized which link current concepts and their historical origins. Firstly, the psychosocial context in the placebo arm of a trial is not the same as clinical care. A patient does not meet a clinician, obtain a diagnosis and commence treatment which they are told may or may not be a placebo. Secondly, it is entirely possible that the rapidly growing field of research and evolving views on placebo in fact mean that the "placebo" arm is more than a neutral control, and receiving a placebo may actually be more akin to receiving a treatment (if one considers the marked responses to unblinded placebo administration) (Bennett, 2018; Sussex, 2018). Both these aspects are important in considering responses in the placebo arms of trials and in future clinical trial design.

8. CONCLUSIONS

The history of placebo is very informative, as it spans generations and has paralleled many biomedical discoveries, the emergence of evidence-based health care, and major advances in scientific method. This history demonstrates key themes surrounding placebo use in clinical care and in clinical trials. Placebo effects are valid psychobiological effects that operate across many conditions. It is the therapeutic encounter or psychosocial context around a patient that triggers these effects. Analysis reveals that there has been an appreciation that placebo effects can be harnessed in two different ways. Firstly, administration of a placebo treatment (either a traditional placebo or a treatment simulation where there is no inherent biological action of the treatment being given). This approach has resulted in constant tension between both the intent of therapy (that is, to please or to provide a valid alternative or addition to therapy) and its deceptive nature. Novel "open" or "honest" placebo prescription has provided a new avenue for exploration of placebo administration without deception. Secondly, there has been an anthropological appreciation that it is the context around the patient that makes the placebo *work* rather than the actual placebo itself. This appreciation has been clarified in more recent times, underscoring the principles that a placebo does not need to be given to generate a placebo effect. Therefore, placebo effects can be harnessed by understanding and manipulating the components of the health care encounter. It also challenges a lengthy history whereby placebo was often given *instead* of a treatment, rather placebo effects should be harnessed in combination with routine clinical care. This field underscores the complexity of the health care encounter and the many disciplines required for collaborative work in researching it. Future research will clarify the utility of placebo mechanisms in clinical practice, with evidence already demonstrating positive findings across many different diseases and suggesting placebo effects remain as an important component of routine health care.

ACKNOWLEDGMENT

The author would like to thank Roland Sussex for his helpful comments on this manuscript.

REFERENCES

Ader, R., & Cohen, N. (1975). Behaviourally conditioned immunosuppression. *Psychosomatic Medicine*, *37*, 333–340.

Al-Lamee, R., Thompson, D., Dehbi, H., Sen, S., Tang, K., & Davies, J. (2018). Percutaneous coronary intervention in stable angina (ORBITA): A double-blind, randomised controlled trial. *Lancet (London, England)*, *391*(10115), 31–40.

Aronson, J. (1999). When I use a word ...: Please, please me. *BMJ (Clinical Research ed.)*, *318*, 716.

Atlas, L. Y., & Wager, T. D. (2014). A meta-analysis of brain mechanisms of placebo analgesia: Consistent findings and unanswered questions. *Handbook of Experimental Pharmacology*, *225*, 37–69.

Beecher, H. K. (1955). The powerful placebo. *Journal of the American Medical Association*, *159*, 1602–1606.

Beecher, H. K. (1959). *Measurement of subjective responses: Quantitative effects of drugs.* New York: Oxford University Press.

Benedetti, F. (2008). Mechanisms of placebo and placebo-related effects across diseases and treatments. *Annual Review of Pharmacology and Toxicology*, *48*, 33–60.

Benedetti, F., Amanzio, M., & Maggi, G. (1995). Potentiation of placebo analgesia by proglumide. *Lancet (London, England)*, *346*(8984), 1231.

Benedetti, F., Amanzio, M., Rosato, R., & Blanchard, C. (2011). Nonopioid placebo analgesia is mediated by CB1 cannabinoid receptors. *Nature Medicine*, *17*(10), 1228–1230.

Benedetti, F., Colloca, L., Torre, E., Lanotte, M., Melcarne, A., Pesare, M., et al. (2004). Placebo-responsive Parkinson patients show decreased activity in single neurons of subthalamic nucleus. *Nature Neuroscience*, 7(6), 587–588.

Benedetti, F., Giuliano, M., Lopiano, L., Lanotte, M., Rainero, I., Vighetti, S., et al. (2003). Open versus hidden medical treatments: The patient's knowledge about a therapy affects the therapy outcome. *Prevention & Treatment*, *6*. Article 1.

Bennett, G. J. (2018). Does the word "placebo" evoke a placebo response? *Pain*. https://doi.org/10.1097/j.pain.0000000000001269 (accessed 24/7/18).

Booth, C. (2005). The rod of Aesculapious: John Haygarth (1740–1827) and Perkins' metallic tractors. *Journal of Medical Biography*, *13*(3), 155–161.

Brody, H. (1982). The lie that heals: The ethics of giving placebos. *Annals of Internal Medicine*, *97*(1), 112–118.

Buchbinder, R., Osborne, R. H., Ebeling, P. R., Wark, J. D., Mitchell, P., Wriedt, C., et al. (2009). A randomized trial of vertebroplasty for painful osteoporotic vertebral fractures. *New England Journal of Medicine*, *361*(6), 557–568. https://doi.org/10.1056/NEJMoa0900429.

Buchbinder, R., Osborne, R. H., & Kallmes, D. (2009). Vertebroplasty appears no better than placebo for painful osteoporotic spinal fractures, and has potential to cause harm. *Medical Journal of Australia*, *191*(9), 476–477.

Buckley, P. F., Sizemore, W. A., & Charlton, J. E. (1986). Medication management in patients with chronic non-malignant pain. A review of the use of a drug withdrawal protocol. *Pain*, *26*(2), 153–165.

Carlill, H. (1918). Hysterical sleeping attacks: Treatment by gross suggestion, reinforced by the operation of trephining, with noses on behaviour in hypnotic sleep. *Lancet*, *194*(5025), 1128–1131.

Carvalho, C., Caetano, J. M., Cunha, L., Rebouta, P., Kaptchuk, T., & Kirsch, I. (2016). Open-label placebo treatment in chronic low back pain: A randomized controlled trial. *Pain*, *157*(12), 2766–2772.

Charlesworth, J. E., Petkovic, G., Kelley, J. M., Hunter, M., Onakpoya, I., Roberts, N., et al. (2017). Effects of placebos without deception compared with no treatment: A systematic review and meta-analysis. *Journal of Evidence-Based Medicine*, *10*(2), 97–107.

Cobb, L. A., Thomas, G. I., Dillard, D. H., Marendino, K. A., & Bruce, R. A. (1959). An evaluation of internal mammary artery ligation by a double blind technique. *New England Journal of Medicine*, *260*, 1115–1118.

Cockayne, E. A., & Ward, E. S. (1920). Letter to the editors. *Lancet*, *195*(5030), 221.

Colloca, L., Enck, P., & DeGrazia, D. (2016). Relieving pain using dose-extending placebos: A scoping review. *Pain*, *157*(8), 1590–1598.

Colloca, L., & Howick, J. (2018). Placebos without deception: Outcomes, mechanisms and ethics. *International Review of Neurobiology*, *138*, 219–240.

Colloca, L., Klinger, R., Flor, H., & Bingel, U. (2013). Placebo analgesia: Psychological and neurobiological mechanisms. *Pain*, *154*(4), 511–514.

Colloca, L., Lopiano, L., Lanotte, M., & Benedetti, F. (2004). Overt versus covert treatment for pain, anxiety, and Parkinson's disease. *Lancet Neurology*, *3*, 679–684.

de Craen, A. J., Kaptchuk, T. J., Tijssen, J. G., & Kleijnen, J. (1999). Placebos and placebo effects in medicine: Historical overview. *Journal of the Royal Society of Medicine*, *92*(10), 511–515.

de la Fuente-Fernandez, R., Ruth, T. J., Sossi, V., Schulzer, M., Calne, D. B., & Stoessl, A. J. (2001). Expectation and dopamine release: Mechanism of the placebo effect in Parkinson's disease. *Science*, *293*(5532), 1164–1166.

Dimond, E. G., Kittle, C. F., & Crockett, J. E. (1960). Comparison of internal mammary artery ligation and sham operation for angina pectoris. *American Journal of Cardiology*, *5*, 483–486.

Egbert, L. D., Battit, G. E., Welch, C. E., & Bartlett, M. K. (1964). Reduction in postoperative pain by encouragement and instruction of patients—A study of doctor-patient rapport. *New England Journal of Medicine*, *270*(16), 825–827.

Eippert, F., Finsterbusch, J., Bingel, U., & Buchel, C. (2009). Activation of the opioidergic descending pain control system underlies placebo analgesia. *Neuron*, *63*(4), 533–543.

Enck, P., Bingel, U., Schedlowski, M., & Rief, W. (2013). The placebo response in medicine: Minimize, maximize or personalize? *Nature Reviews. Drug Discovery*, *12*(3), 191–204.

Engel, G. L. (1977). The need for a new medical model: A challenge for biomedicine. *Science*, *196*(4286), 129–136.

Fässler, M., Gnädinger, M., Rosemann, T., & Biller-Andorno, N. (2009). Use of placebo interventions among Swiss primary care providers. *BMC Health Services Research*, *9*(144), 1–9.

Finniss, D. G. (2013). Historical aspects of placebo analgesia. In L. Colloca, M. A. Flaten, & K. Meissner (Eds.), *Placebo and pain: From bench to bedside* (pp. 1–8). London: Academic Press.

Finniss, D. G., & Benedetti, F. (2005). Mechanisms of the placebo response and their impact on clinical trials and clinical practice. *Pain*, *114*, 3–6.

Finniss, D., & Benedetti, F. (2007). The placebo response—Clinical implications. *IASP Clinical Updates*, *XV*(1).

Finniss, D. G., Kaptchuk, T. J., Miller, F. G., & Benedetti, F. (2010). Biological, clinical, and ethical advances of placebo effects. *Lancet (London, England)*, *375*(9715), 686–695.

Garrison, F. H. (1921). *An introduction to the history of medicine* (3rd ed.). Philadelphia: Saunders.

Gebhart, G. F., & Bielefeldt, K. (2016). Physiology of visceral pain. *Comprehensive Physiology*, *6*(4), 1609–1633. https://doi.org/10.1002/cphy.c150049.

Gold, H., Kwit, N. T., & Otto, H. (1937). The Xanthines (Theobromine and Aminophylline) in the treatment of cardiac pain. *Journal of the American Medical Association*, *108*, 2173–2179.

Gracely, R. H., Dubner, R., Deeter, W. D., & Wolskee, P. J. (1985). Clinician's expectations influence placebo analgesia. *Lancet (London, England)*, *325*(8419), 43.

Greiner, T. H., Gold, H., Cattell, M., Travell, J., Bakst, H., Rinzler, S. J. H., et al. (1950). A method for the evaluation of effects of drugs on cardiac pain in patients with angina of effort: A study of Khellin (Visammin). *American Journal of Medicine*, *9*, 143–155.

Grevert, P., Albert, L. H., & Goldstein, A. (1983). Partial antagonism of placebo analgesia by naloxone. *Pain*, *16*, 129–143.

Hart, C. (1999). The mysterious placebo effect. *Modern Drug Discovery*, *2*(4), 30–40.

Hashmi, J. A. (2018). Placebo effect: Theory, mechanisms and teleological roots. *International Review of Neurobiology*, *139*, 233–253.

Hediger, E. M., & Gold, H. (1935). U.S.P ether from large drums and ether from cans labeled 'For anaesthesia'. *Journal of the American Medical Association*, *104*, 2244–2248.

Herrnstein, R. (1962). Placebo effect in the rat. *Science*, *138*, 677–678.

Hewlett, A. W. (1913). Clinical effects of 'natural' and 'synthetic' sodium salicylate. *Journal of the American Medical Association*, *61*, 319–321.

Houston, W. R. (1938). The doctor himself as a therapeutic agent. *Annals of Internal Medicine*, *11*(8), 1416–1425.

Hróbjartsson, A., & Gøtzsche, P. C. (2001). Is the placebo powerless? An analysis of clinical trials comparing placebo with no treatment. *New England Journal of Medicine*, *344*(21), 1594–1602.

Hróbjartsson, A., & Gøtzsche, P. C. (2004). Is the placebo effect powerless? Update of a systematic review with 52 new randomized trials comparing placebo with no treatment. *Journal of Internal Medicine*, *256*, 91–100. http://www.jameslindlibrary.org/haygarth-j-1800/.

Jacobs, B. (2000). Biblical origins of placebo. *Journal of the Royal Society of Medicine*, *93*(4), 213–214.

Kaptchuk, T. (1998a). Intentional ignorance: A history of blind assessment and placebo controls in medicine. *Bulletin of the History of Medicine*, *72*, 389–433.

Kaptchuk, T. J. (1998b). Powerful placebo: The dark side of the randomised controlled trial. *Lancet (London, England)*, *351*(9117), 1722–1725.

Kaptchuk, T., Friedlander, E., Kelley, J. M., Sanchez, M. N., Kokkotou, E., Singer, J. P., et al. (2010). Placebos without deception: A randomized controlled trial in irritable bowel syndrome. *PLoS One*, *5*(12), 15591.

Kaptchuk, T. J., Kelley, J. M., Conboy, L. A., Davis, R. B., Kerr, C. E., Jacobson, E. E., et al. (2008). Components of placebo effect: Randomised controlled trial in patients with irritable bowel syndrome. *BMJ (Clinical Research ed.)*, *336*(7651), 999–1003.

Kaptchuk, T. J., Kerr, C. E., & Zanger, A. (2009). Placebo controls, exorcisms and the devil. *Lancet (London, England)*, *374*(9697), 1234–1235.

Keller, A., Akintola, T., & Colloca, L. (2018). Placebo analgesia in rodents: Current and future research. *International Review of Neurobiology*, *138*, 1–15.

Kerr, C. E., Milne, I., & Kaptchuk, T. J. (2008). William Cullen and a missing mind–body link in the early history of placebos. *Journal of the Royal Society of Medicine*, *101*(2), 89–92.

Lasagna, L. C. (1986). The placebo effect. *The Journal of Allergy and Clinical Immunology*, *78*, 161–165.

Lasagna, L. C., Mosteller, F., von Felsinger, J. M., & Beecher, H. K. (1954). The study of the placebo response. *The American Journal of Medicine*, *16*(6), 770–779.

Laska, E., & Sunshine, A. (1973). Anticipation of analgesia: A placebo effect. *Headache*, *1*, 1–11.

Levine, J. D., & Gordon, N. C. (1984). Influence of the method of drug administration on analgesic response. *Nature*, *312*(5996), 755–756.

Levine, J. D., Gordon, N. C., & Fields, H. L. (1978). The mechanism of placebo analgesia. *Lancet*, *2*(8091), 654–657.

Linde, K., Fässler, M., & Meissner, K. (2011). Placebo interventions, placebo effects and clinical practice. *Philosophical Transactions of the Royal Society of London. Series B, Biological Sciences*, *366*(1572), 1905–1912.

Meissner, K., Bingel, U., Colloca, L., Wager, T. D., Watson, A., & Flaten, M. A. (2011). The placebo effect: Advances from different methodological approaches. *Journal of Neuroscience*, *31*(45), 16117–16124.

Miller, F. G., & Colloca, L. (2009). The legitimacy of placebo treatments in clinical practice: Evidence and ethics. *The American Journal of Bioethics*, *9*(12), 39–47.

Miller, F. G., Wendler, D., Swartzman, L. C., Miller, F. G., Wendler, D., & Swartzman, L. C. (2005). Deception in research on the placebo effect. *PLoS Medicine/Public Library of Science*, *2*(9), e262.

Moerman, D. (1997). Physiology and symbols: The anthropological implications of the placebo effect. In L. Romanucci-Ross, D. E. Moerman, & L. R. Tancredi (Eds.), *The anthropology of medicine: From culture to method* (3rd ed.). Westport, CT: Bergin & Garvey.

Moerman, D. (2002). *Meaning, medicine and the placebo effect*. Cambridge University Press.

Moerman, D. E., & Jonas, W. B. (2002). Deconstructing the placebo effect and finding the meaning response. *Annals of Internal Medicine*, *136*(6), 471–476.

Moseley, B. J., O'Malley, K., Petersen, N. J., Menke, T. J., Brody, B. A., Kuykendall, D. H., et al. (2002). A controlled trial of arthroscopic surgery for osteoarthritis of the knee. *New England Journal of Medicine*, *347*, 81–88.

Pecina, M., & Zubieta, J. K. (2018). Expectancy modulation of opioid neurotransmission. *International Review of Neurobiology*, *138*, 17–37.

Petrovic, P., Kalso, E., Petersson, K. M., & Ingvar, M. (2002). Placebo and opioid analgesia—Imaging a shared neuronal network. *Science*, *295*(5560), 1737–1740.

Pollo, A., Amanzio, M., Arslanian, A., Casadio, C., Maggi, G., & Benedetti, F. (2001). Response expectancies in placebo analgesia and their clinical relevance. *Pain*, *93*(1), 77–84.

Price, D. D., Finniss, D. G., & Benedetti, F. (2008). A comprehensive review of the placebo effect: Recent advances and current thought. *Annual Review of Psychology*, *59*, 565–590.

Ralphs, J. A., Williams, A. C., Richardson, P. H., Pither, C. E., & Nicholas, M. K. (1994). Opiate reduction in chronic pain patients: A comparison of patient-controlled reduction and staff controlled cocktail methods. *Pain*, *56*(3), 279–288.

Ready, F. (1920). The justifyability of therapeutic lying. *The Lancet, 195*(5031), 289.
Risse, G. (1986). *Hospital life in enlightenment Scotland.* Cambridge: Cambridge University Press.
Shapiro, A. (1959). The placebo effect in the history of medical treatment. Implications for psychiatry. *American Journal of Psychiatry, 116*, 298–304.
Shapiro, A. (1964). A historic and heuristic definition of the placebo. *Psychiatry, 27*, 52–58.
Shapiro, A., & Shapiro, E. (1997). The placebo: Is it much ado about nothing. In A. Harrington (Ed.), *The placebo effect—An interdisciplinary exploration.* USA: Harvard University Press.
Sherman, R., & Hickner, J. (2008). Academic physicians use placebos in clinical practice and believe in the mind–body connection. *Journal of General Internal Medicine, 23*(1), 7–10.
Siddall, P. J., & Cousins, M. J. (2004). Persistent pain as a disease entity: Implications for clinical management. *Anesthesia and Analgesia, 99*(2), 510–520.
Sollmann, T. (1917). The crucial test of therapeutic evidence. *Journal of the American Medical Association, 1817*(7), 1439–1442.
Sussex, R. (2018). Describing placebo phenomena in medicine: A linguistic approach. *International Review of Neurobiology, 139*, 49–83.
Tilburt, J. C., Emanuel, E. J., Kaptchuk, T. J., Curlin, F. A., & Miller, F. G. (2008). Prescribing "placebo treatments": Results of national survey of US internists and rheumatologists. *BMJ (Clinical Research ed.), 337*, a1938.
Tracey, I. (2008). Imaging pain. *British Journal of Anaesthesia, 101*(1), 32–39. https://doi.org/10.1093/bja/aen102.
Turner, J. A., Deyo, R. A., Loeser, J. D., Von Korff, M., & Fordyce, W. E. (1994). The importance of placebo effects in pain treatment and research. *JAMA, 271*(20), 1609–1614.
Vase, L., Riley, J. L., 3rd, & Price, D. D. (2002). A comparison of placebo effects in clinical analgesic trials versus studies of placebo analgesia. *Pain, 99*(3), 443–452.
Wager, T. D., Scott, D. J., & Zubieta, J.-K. (2007). Placebo effects on human mu-opioid activity during pain. *Proceedings of the National Academy of Sciences of the United States of America, 104*(26), 11056–11061.
Wechsler, M. E., Kelley, J. M., Boyd, I. O. E., Dutile, B. S., Marigowda, M. B., Kirsch, I., et al. (2011). Active albuterol or placebo, sham acupuncture, or no intervention in asthma. *New England Journal of Medicine, 365*, 119–126.
Wolfe, H. G. (1946). Use of placebos in therapy. *New York State Journal of Medicine, 46*, 1718–1727.
Woller, S. A., Eddinger, K. A., Corr, M., & Yaksh, T. L. (2017). An overview of pathways encoding nociception. *Clinical and Experimental Rheumatology, 35 Suppl 107*(5), 40–46.
Zion, S. R., & Crum, A. J. (2018). Mindsets matter: A new framework for harnessing the placebo effect in modern medicine. *International Review of Neurobiology, 138*, 137–160.

CHAPTER TWO

Lumping and Splitting: Toward a Taxonomy of Placebo and Related Effects

John M. Kelley[1]
Distinguished Professor of Psychology, Endicott College, Beverly, MA, United States
Deputy Director, Program in Placebo Studies, Harvard Medical School, Boston, MA, United States
[1]Corresponding author: e-mail address: jkelley@endicott.edu

Contents

Abstract

The placebo effect is closely related to many other constructs, including most prominently, conditioning and expectancy, but also natural history, regression to the mean, priming, mindset, context effects, the meaning response, specific and non-specific clinical effects, placebo-related effects, the patient–clinician relationship, and the common factors in psychotherapy. How are these various terms related to one another? To what degree do they overlap, and to what degree do they diverge? To form a better theoretical understanding of these constructs and to foster improved empirical research, is it better to lump these terms together in some fashion? Or will progress best be served by maintaining the splits between the constructs? Or would it perhaps be most effective to employ some mixture of lumping and splitting? I will address these and related questions with two major goals: (1) to delineate and clarify the relationship between these

International Review of Neurobiology, Volume 139
ISSN 0074-7742
https://doi.org/10.1016/bs.irn.2018.07.011

various terms and (2) to suggest some possible re-alignments in the way in which we conceptualize the relationships among these constructs that might prove useful in fostering research on placebo and related effects. In addition, clarifying the interconnections between the placebo effect and other related terms has the potential to spark innovative cross-fertilizations between related areas of research.

1. INTRODUCTION

In the spring of 2017, the Society for Interdisciplinary Placebo Studies held its first official conference in the beautiful and historic city of Leiden, in the Netherlands. Prior to the start of the formal conference proceedings, a small group of about 30 scholars were invited to attend a one-day, preconference meeting focused on potential clinical applications of the placebo effect. The attendees were an international collection of distinguished experts in placebo studies who also had expertise in medicine, clinical psychology, and psychiatry; and I expected that we would quickly dive into a lively discussion of the ethics, feasibility, and logistics associated with attempts to harness the placebo effect in clinical settings. Instead, and somewhat to my surprise, we spent a considerable portion of the morning session discussing the *definition* of the placebo effect. Moreover, despite a lengthy discussion, no clear consensus was reached, and the moderator of the meeting eventually urged us to put aside the definitional issues and move on to attend to the nominal focus of the meeting which was the potential clinical applications of the placebo effect, not its definition. Implicit in her appeal was the notion that although our group might not be able to come to consensus on the definition of the placebo effect, we each had a sufficiently reasonable and adequate working definition of the term that we could proceed forward to the question of how we might harness the placebo effect to benefit patients.

What struck me most about this incident was how ironic it was that a gathering of world experts in placebo studies, scholars who had collectively written thousands of peer-reviewed research articles and scores of books on the topic, could not agree on a common definition of the placebo effect. Houston, we have a problem!

In addition, the moderator's advice to the group that we should put aside the issue of how to define the placebo effect put me in mind of US Supreme Court justice Potter Stewart and his famous comment about the difficulty in defining pornography. In a concurring opinion written in 1964, Justice

Stewart allowed that he might never be able to formally define obscenity, but nevertheless he boldly asserted that "I know it when I see it." Using this standard, Justice Stewart decided that the motion picture at the heart of the case was not obscene, and therefore that the right of the cinema owner to screen the movie was a matter of free speech that should be protected by the First Amendment to the US constitution. The case, however, was famously divisive, and the lack of any clear definition of obscenity prevented the Court from providing unambiguous guidance to lower courts regarding the sorts of artistic expression that should or should not be protected by the First Amendment. As I hope to demonstrate in this article, satisfactorily defining the placebo effect in a clear and unambiguous way is at least as difficult as defining obscenity was for the Supreme Court.

Aside from wrestling with the definition of the placebo effect, the main goal of this article is to examine how the placebo effect is related to many other constructs, including most prominently, conditioning and expectancy, but also natural history, regression to the mean, priming, mindset, context effects, the meaning response, specific and non-specific clinical effects, placebo-related effects, the patient–clinician relationship, and the common factors in psychotherapy. How are these various terms related to one another? To what degree do they overlap, and to what degree do they diverge? To form a better theoretical understanding of these constructs and to foster improved empirical research, is it better to lump these terms together in some fashion? Or will progress best be served by maintaining the splits between the constructs? Or would it perhaps be most effective to employ some mixture of lumping and splitting? In this article, I address these and related questions with two major goals: (1) to delineate and clarify the relationship between these various terms, and (2) to suggest some possible re-alignments in the way in which we conceptualize the relationships among these constructs that might prove useful in fostering research on placebo and related effects. In addition, clarifying the interconnections between the placebo effect and other related terms has the potential to spark innovative cross-fertilizations between related areas of research.

I am especially interested in attempting to construct a useful taxonomy of these various constructs, with particular attention paid to the utility in either lumping some terms together to the degree that they are fundamentally similar; or considering whether, in other cases, the splits between the terms are necessary and useful. The subtitle of this article, *Toward a Taxonomy of Placebo and Related Effects*, emphasizes that what I have to say is provisional and primarily meant to stimulate thought and debate on the topic.

2. BACKGROUND

Before turning to a description of the proposed taxonomy itself, I begin with a few brief sections devoted to background issues, including a definition of taxonomy and a description of the fundamental tasks of lumping and splitting that necessarily underlie any taxonomic effort.

Taxonomy (from the ancient Greek *taxis*, meaning "arrangement" or "order," and *nomia*, meaning "method") is the systematic classification of items into a set of categories based on some measure of similarity, with the categories often arranged hierarchically. For example, in biology, organisms are classified in a hierarchical taxonomy that ranges from species, to genus, to family, and so on, each of which denotes an increasingly inclusive category. In biology, taxonomic systems were originally based on morphological or behavioral similarity (i.e., phenetics, which focuses on phenotypes), but more recently and perhaps more effectively, they have been based on genetic relatedness (i.e., cladistics), which reflects evolutionary history and focuses on genotypes. So, for example, by morphology and behavior alone, one might be tempted to place wombats and groundhogs into the same or similar taxons because they look quite similar, and they are both medium-sized herbivores that feed on grasses and dig extensive underground tunnels. However, since wombats are marsupials and groundhogs are placentals, they are actually much more distant from each other in terms of evolutionary history and genetic make-up than their superficial similarities might suggest. Indeed, despite their obvious differences in morphology and behavior, groundhogs and dolphins are more similar to one another both genetically and in terms of evolutionary history, than groundhogs are to wombats.[a]

Taxonomic systems necessarily involve two tasks: grouping similar items together and delineating boundaries between groups of items. These opposing tasks have sometimes been referred to collectively as *lumping and splitting*. Befitting biology's status as the science that most extensively employs a taxonomic approach, one of the earliest uses of this term (or something similar)

[a] The simple wombat/groundhog example used here is meant to clarify the difference between the phenetics and cladistics approaches to taxonomy. However, it should be noted that phenetics does not only examine easily observable, external characteristics, but it also considers morphological and behavioral differences that are more difficult to observe. For example, because marsupials differ substantially from placentals, especially in their reproductive organs, a more sophisticated phenetics approach would place the two species rather far apart from one another, in a fashion similar to that of the cladistics approach.

is found in a letter that Charles Darwin wrote to J.D. Hooker in 1857: "It is good to have hair-splitters & lumpers" (Darwin, 1857). The eminent American paleontologist G.G. Simpson perfectly captured the inherent tension between these two contradictory tasks when he wrote, tongue in cheek, that, "splitters make very small units – their opponents say that if they can tell two animals apart, they place them in different genera, and if they cannot tell them apart, they place them in different species. Lumpers make large units – their opponents say that if a carnivore is neither a dog nor a bear, they call it a cat" (Simpson, 1945, p. 23).

Interestingly, and perhaps ironically, the term taxonomy has itself been subject to some lumping and splitting. As noted above, taxonomy is the general term for the systematic classification of objects into a set of categories based on some measure of similarity, with the categories often arranged hierarchically. Nosology (from the ancient Greek *nosos*, meaning "disease," and *logia*, meaning "study of") is a subset of taxonomy and refers specifically to the classification of diseases within medicine. It is unclear whether splitting the concept of taxonomy by introducing nosology as a subset obscures or clarifies. Indeed, instead of nosology, why not use the term medical taxonomy? In this case, it seems that the introduction of another 50-cent word (nosology) might violate the classic, yet ironic adage that good writers should eschew obfuscation. Moreover, since biology does not have a special term for biological taxonomy, why should medicine?

3. THE PLACEBO EFFECT IS AN OXYMORON

Before we even consider constructing a taxonomy of the phenomena associated with the placebo effect, a fundamental problem must first be acknowledged: the oxymoronic nature of the term. By definition, placebos are inert and contain no active ingredients, and therefore, in and of themselves, placebos can have no effects whatsoever. Thus, the placebo effect is the effect of something that has no effect. In other words, the term "the placebo effect" is an oxymoron. This point has been made many times before (e.g., Moerman, 2003), but much of the difficulty in defining the placebo effect and thinking clearly about how it relates to other constructs stems from this initial confusion.

Given the inadequacies of the term, several alternatives to the placebo effect have been proposed, including most prominently response expectancy (Kirsch, 1985), the meaning response (Moerman, 2003; Moerman & Jonas, 2002), and context effects (Di Blasi, Harkness, Ernst, Georgiu, & Kleijnen,

2001), all of which point away from the placebo itself, and toward the context within which treatment takes place and how that context affects the psychology of the patient. For a variety of reasons that will be discussed in more detail below, none of these alternatives to the placebo effect has taken root and for the foreseeable future we may need to resign ourselves to continuing to struggle with a less than optimal term.

If we insist on keeping the term, how should the placebo effect be defined? Clearly, the placebo effect is not the effect of the placebo itself, but rather, it is the effect that occurs *when* a placebo is given. Although I have argued above that the placebo effect is fundamentally an oxymoron, and although there are certainly differences of opinion between experts about the appropriate boundaries of the definition (i.e., what sorts of phenomena should "count" as being subsumed by the term), there is nevertheless reasonable consensus regarding the core or prototypical definition (Evers et al., 2018). Simply put, the placebo effect is the improvement that occurs when a patient receives a treatment that is inert (most typically a placebo pill). It is important to note that though patient improvement has most often been measured subjectively, there is now abundant evidence that the placebo effect, though initiated via psychological mechanisms, can involve objectively measurable biochemical changes (Benedetti, 2008). Nevertheless, since placebo treatment is by definition inert (but see the concepts of *impure placebo* and the *broad definition of the placebo effect* below), any beneficial effects that emerge are necessarily initiated by psychological mechanisms.

4. TOWARD A TAXONOMY OF PLACEBO AND RELATED EFFECTS

The diagram in Fig. 1 illustrates the proposed taxonomy of placebo and related effects. After this brief initial paragraph orienting the reader to the figure, a more detailed discussion of the various components of the taxonomy is contained in the sub-sections that follow. The large black oval on the right side of the figure represents the *treatment domain*, which lumps together the typical situations in which placebo effects have been studied. Prototypically, the treatment domain involves patients suffering from a disorder who are given a treatment that is prescribed by a clinician. However, as I have defined it, the treatment domain also includes analog experiments in which the "patients" might be healthy volunteers, the "disorder" might be experimentally induced pain, and the "clinician" might be a researcher with no clinical credentials. In addition, this domain also includes self-treatment where there may not be a clinician, as well as treatments

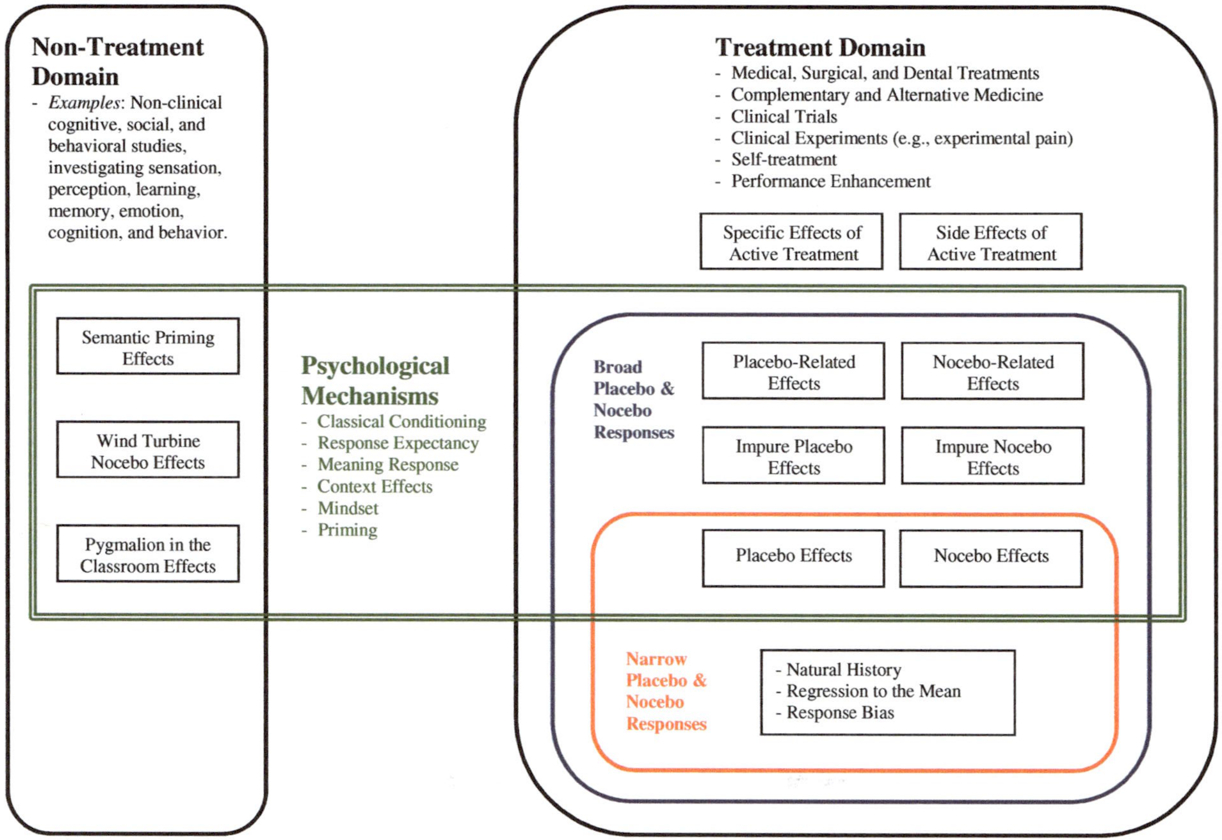

Fig. 1 Diagram of the proposed taxonomy of placebo and related effects.

aimed at enhancing performance where there are no patients (e.g., doping to increase endurance in cycling, or transcranial magnetic stimulation to enhance cognition). The large black oval on the left represents the *non-treatment domain* in which effects that seem similar to placebo effects sometimes occur, but the context is clearly not healthcare, nor any close analog to healthcare. This domain will be more fully examined in later sections below. The red oval at the lower right represents *narrow placebo and nocebo responses*, which refer to the totality of positive or negative responses to inert treatments, part of which is attributable to a variety of artifacts (i.e., natural history, regression to the mean, and response bias, which are in the black rectangular box at the bottom right), and part of which is attributable to true placebo or nocebo effects, which are represented by the two small black rectangles at the top of the red oval. Similarly, the blue oval represents *broad placebo and nocebo responses*, which refer to the totality of positive or negative responses to treatments that might be placebos, but that might also be impure placebos, or even active treatments—the responses in question, however, are independent from any direct physiological action of the treatment (i.e., the specific effects or side effects of treatment, which are represented by the small black rectangles at the top right of the diagram). Once again, a distinction is made between the totality of response (some of which is due to the artifacts at the bottom right of the figure) and the actual placebo and nocebo effects, which are represented by the six black rectangular boxes at the top half of the blue oval. Finally, the green rectangle encloses the *psychological mechanisms* that underlie many of the effects that we refer to as placebo or nocebo effects. Importantly, however, these very same mechanisms seem also to be operative in a very large number of effects in the non-treatment domain on the left side of the diagram. Thus, the green rectangle lumps together a set of effects that are prototypically considered placebo effects or placebo-related effects, with a set of effects that are not often thought of as related to placebo effects.

4.1 Placebo Effects vs. Nocebo Effects

In the middle section of the right-hand oval are three pairs of boxes, which refer to various placebo and nocebo effects. A nocebo effect (from the Latin *nocebo* "I shall harm") is simply a negative effect that occurs after administration of a placebo (from the Latin *placebo* "I shall please"). According to the Oxford English Dictionary (OED), the first printed reference to nocebo as a negative placebo effect (as opposed to its literal meaning in Latin) was in in the early 1960s (Kennedy, 1961); however, the term has become much more common since 2000.[b] The introduction of this term raises the question of whether splitting nocebo from placebo is necessary. If the mechanisms that underlie placebo and nocebo effects are substantially different, then this split may make sense; if, however, the mechanisms do not differ, then lumping the two terms together might be better. Stewart-Williams and Podd (2004) summarize a number of additional reasons to question the utility of splitting up the concepts of placebo and nocebo effects. On the other hand, a strong argument for keeping the two separate terms is the simple fact that placebo and nocebo effects correspond so closely to beneficial treatment effects and negative side effects, respectively. Despite these various considerations, for much of the rest of this article I will describe only placebo effects and placebo responses, lumping those terms together with their nocebo counterparts so as to simplify the process of explicating the rest of the taxonomy.

4.2 Placebo Response vs. Placebo Effect

An important distinction has often been made between the *placebo response* and the *placebo effect*. This distinction is perhaps most easily understood by way of analogy to randomized controlled trials testing medication efficacy. In these trials, the *drug response* is the sum total of improvement (or worsening) that a patient shows in response to the treatment. Of course, some fraction of that total is not due to the physiological effects of the drug, but rather to the response that would have occurred had the patient been given an inert placebo. Thus, we have a drug response—the sum total of symptom change—and a *drug effect*—the fraction of improvement that is attributable to the specific physiological effects of the medication.

The very same analysis can be applied to placebos. The *placebo response* is the sum total of patients' responses to the administration of the placebo. But some fraction of that response would probably have occurred in the absence of any treatment at all and is due to a variety of artifacts, such as regression to the mean and the natural history of the disorder (see the artifacts section below for additional details). Thus, the *placebo effect* is the fraction of patient improvement (or worsening) that remains after controlling for the various artifacts (Ernst & Resch, 1995).

[b] A Google Scholar search for the word "nocebo" on March 2, 2018 yielded nearly 20 times as many academic citations from 2000 to 2018 as compared to the entire 20th century—11,400 vs. 593 citations, respectively.

4.3 Artifacts

The small box at the bottom right of the figure identifies factors that can make it *appear* that a particular treatment (or a placebo) caused a change in symptoms, when, in fact, the changes are actually due to a variety of artifacts. *Natural history* refers to the typical waxing and waning of symptoms over time. Since patients often seek treatment when their symptoms are particularly bothersome, they sometimes show improvement that is simply due to the natural history of the disorder. *Regression to the mean* is a well-known statistical artifact (Campbell & Kenny, 1999) that is associated with random error in measurements and the use of entry criteria for study participation (e.g., participants can only qualify for a study if their symptom levels exceed a particular cut-off score). This selection process nearly always leads to a sample of participants with measurement errors that contribute to upwardly biased severity scores. Therefore, when they are later retested, they will, on average, regress toward the mean (i.e., get lower severity scores), even if there are, in fact, no treatment or placebo effects whatsoever. Finally, *response bias* is the tendency for study participants to provide answers that they think will please the clinician or investigator. Space constraints in the box preclude the inclusion of the many other factors that could produce artifactual improvement or worsening, such as changes in life circumstances (e.g., alterations in diet, exercise, social support, stress management, sleeping patterns), or the independent use of ancillary psychological or physical treatments. The important point here is that any putative treatment or placebo effects must be above and beyond these various artifacts.

4.4 Narrow vs. Broad Placebo Responses

Another important dichotomy is the distinction between narrow and broad placebo responses (Kaptchuk, 2002). The red oval in Fig. 1 denotes the narrow placebo response, and the larger blue oval denotes the broad placebo response. A *narrow placebo response* refers to symptom changes that occur when a patient is given an inert treatment—a true placebo. Although the *broad placebo response* includes, as a subset, the effects due to inert placebos, its defining feature is that it also includes situations in which the treatment is *not* inert, but the effects are due to factors other than the active ingredients of the treatment. Examples of these sorts of effects are described in sections below.

4.5 Impure Placebos

An *impure placebo* is a treatment that includes an active ingredient, but the active ingredient does not have a direct, physiological effect on symptoms of the targeted disorder. An example is the use of *active placebos* in RCTs. Active placeboes include ingredients that induce side effects that are similar to the side effects that occur with the medication that is being tested in the trial. The purpose is to decrease the likelihood that the patient or clinician breaks the blind by inferring treatment assignment from the presence or absence of side effects. In antidepressant trials, the use of active placebos has been shown to decrease the separation between the experimental and control groups, suggesting that active placebos can have effects that go beyond those attributable to an inactive placebo (Moncrieff, Wessely, & Hardy, 2004).

A second type of an impure placebo is the common medical strategy known as "start low, go slow," which is meant to minimize side effects by starting pharmaceutical treatment with a very low dose, and then slowly titrating the dose upward (e.g., Brent & Gibbons, 2014). This strategy is especially useful in patients who tend to experience many side effects in response to medications, either because of a high degree of physiological sensitivity, or due to a nocebo effect. The initial dose in these situations can be so low that it is sub-therapeutic and unlikely to provide any clinical benefit to the patient. In this case, if the patient responds, the response is not due to the active ingredients but rather to an impure placebo effect.

A third type of an impure placebo would be the prescription of a relatively benign treatment (e.g., vitamins) that the doctor thinks probably won't be of any particular help in treating the condition, but which he or she thinks will please the patient without causing any harm. Thus, in theory, any effects that occur are due to psychological factors as opposed to the inherent properties of the treatment.

Finally, a fourth type of impure placebo occurs when a physician prescribes an active medication at a therapeutic dose, but the medication is one that cannot have an impact on the targeted disorder. For example, there is very strong evidence that antibiotics are over-prescribed for a variety of conditions, including especially the treatment of the common cold (Pouwels, Dolk, Smith, Robotham, & Smieszek, 2018). Since colds are caused by viruses, antibiotics cannot have any beneficial effect; and indeed, in this much less benign situation, the treatment could actually produce negative effects such as contributing to antibiotic resistance. Despite the potential for negative outcomes, it is also possible for the inappropriate prescription of antibiotics to provide psychological benefits to the patient (e.g., easing anxiety about the condition), which would be an impure placebo effect. There have been a number of surveys of clinicians that indicate that the prescription of impure placebos is very common and certainly outnumbers the prescription of inert placebos (Fässler, Meissner, Schneider, & Linde, 2010).

4.6 Placebo-Related Effects

Placebo-related effects form the final category subsumed under the *broad placebo responses* umbrella. Placebo-related effects occur when the effects of active treatments are either enhanced or diminished by psychological factors. A classic example is the open–hidden paradigm developed by Benedetti and his colleagues (2003). In this paradigm, clinical or experimental pain is treated with an effective analgesic such as morphine through an IV line. When the patient is aware that she is receiving treatment because the clinician identifies the medication and administers it in front her (the open condition), the efficacy of the medication is enhanced. In contrast, when a computer administers the medication surreptitiously and the patient is unaware that he has received it (the hidden condition), the efficacy of the medication is diminished. Since no placebo has been delivered, this phenomenon does not fit into the narrow definition of the placebo effect. On the other hand, since the psychological mechanisms that likely mediate this phenomenon (e.g., expectancy and conditioning) are probably the same as those that mediate the narrow placebo effect, it makes sense to lump this phenomenon with the other effects that comprise the broad definition of the placebo response.

4.7 Specific Effects of Active Treatment vs. Non-Specific Effects

The specific effects (and side effects) of active treatment are depicted at the top right of Fig. 1. Specific effects are direct, physiological responses that can be attributed to the medication or procedure. Importantly, the existence and magnitude of specific effects can only be determined when all the lower level effects in the figure are controlled for (i.e., the broad and narrow placebo effects, as well as the artifacts). Especially in the pharmacological literature, the lower level effects have sometimes been referred to collectively as *non-specific effects*. The problem with this term is that once these effects are studied in their own right and their mechanisms begin to be determined, we have an oddly oxymoronic situation whereby we are identifying the specific determinants of non-specific effects (Kirsch, 1999). This is likely the reason that the term non-specific effects has not caught on as an alternative to the placebo effect.

A possible alternative to non-specific effects is *the common factors*, which refers to the factors that are common across all forms of psychotherapy (Wampold & Imel, 2015). These factors are the essential features of a good therapeutic alliance, such as warmth, empathy, reassurance, and fostering positive expectations. In many ways, the common factors of psychotherapy closely resemble what has often been termed *good bedside manner* in medical treatment (Kelley et al., 2009). Although the common factors are likely to be an important contributor to placebo effects, their tight association with psychotherapy (as opposed to medicine in general) probably explains why the term has only infrequently been proposed as an alternate term for the placebo effect.

5. PSYCHOLOGICAL MECHANISMS

The green box in Fig. 1 identifies a number of psychological mechanisms that have been proposed as driving placebo effects. Since placebos are physiologically inert, the effects that are attributed to placebo treatment are necessarily initiated via psychological factors, hence the emphasis here on psychological mechanisms of placebo effects. Nevertheless, it's important to note that these initiating psychological mechanisms have clearly been shown to produce genuine psychobiological changes that are downstream mechanisms of the placebo effect. As will be discussed in the sections that follow, in many ways, these psychological mechanisms are in competition with one another as contenders for an alternative to replace the problematic term, the placebo effect. It is particularly noteworthy that this box extends outside of the treatment domain and therefore lumps together the various placebo effects from the treatment domain that were discussed in Section 5, with a wide variety of other phenomena from the non-treatment domain, a few examples of which are shown on the left-hand side of the figure. The relationship between the various placebo effects and these other, non-treatment phenomena will be discussed in the sections below.

Classical conditioning is one of the psychological mechanisms that can drive placebo effects. Conditioning involves the repeated pairing of an initially neutral stimulus with an unconditioned stimulus, such that the initially neutral stimulus eventually becomes a conditioned stimulus that can reliably evoke a conditioned response. In the well-known Pavlovian example, the unconditioned stimulus is the sight of food, which reliably evokes an unconditioned response, salivation. If the unconditioned stimulus is then repeatedly paired with the sound of a ringing bell (the initially neutral stimulus), eventually the sound of the bell will become a conditioned stimulus, which can then, by itself, reliably evoke the conditioned response, salivation (Pavlov, 1927).

This basic formulation of classical conditioning suggests a purely behavioral or mechanical mechanism, unmediated by cognitive functioning on the part of the organism. Contrary to this view, however, it has long been known that conditioning *is* mediated by cognitive functions, and that it is the informational value that the organism extracts from the temporal pairing of the unconditioned stimulus with the initially neutral stimulus that accounts for the phenomenon (e.g., Rescorla, 1988). In other words, even though conditioning does not require elaborate cognitive processing (e.g., Mpitsos and Davis (1973) showed that even a sea slug can be conditioned), it does require that the organism have a sufficiently complex nervous system that is capable of computing at least a rudimentary form of expectancy based on the repeated pairing of the initially neutral stimulus and the unconditioned stimulus. Thus, according to this view, conditioning procedures that lead to placebo effects are mediated by learning processes.

Along with conditioning, response expectancy is one of the two most frequently invoked mechanisms to explain placebo effects; and often, expectancy and conditioning are pitted against one another as rival explanations for placebo effects (Kirsch, 1985). In common parlance, expectancy is typically thought of as conscious and verbalizable; whereas conditioning, perhaps because it was first explored in animal models, was initially thought of as a purely behavioral phenomenon, unmediated by conscious expectations. However, as noted above, conditioning effects are clearly mediated by cognitive mechanisms (though they can often be quite low-level forms of cognition), and so long as one defines expectancy in a sufficiently broad way (i.e., not limiting the concept to conscious, verbalizable expectations), one can argue that all conditioning paradigms produce some form of expectancy in the organism, which then leads to the various placebo effects.

Thus, if expectancy is defined as including relatively low-level cognitive processes, then the putative hard boundary between conditioning and expectancy can be erased. Indeed, under this view, expectancy can be considered as a mediator between conditioning procedures and placebo effects. However, conditioning and expectancy cannot simply be equated with one another. Conditioning might be best thought of as a *procedure* that can lead to conscious, and non-conscious expectancies (Jensen et al., 2012), that then in turn lead to physiological changes that produce the placebo effect. However, conditioning is clearly not the only way in which expectancies can be formed. For example, verbal suggestions alone can produce positive and negative expectancies that can lead to placebo and nocebo effects, even for some autonomic nervous system functions (Meissner, 2011).

Although classical conditioning and response expectancy are the most commonly cited mechanisms of placebo effects, the anthropologist, Daniel Moerman, has advocated for the term *meaning response*, which he asserts arises from "the understandings about illness and treatment that doctors, patients, communities, and cultures may share in their varying ways" (Moerman, 2003, para 3). Furthermore, Moerman makes clear that the meaning response can involve tacit knowledge—indeed Moerman prefers *meaning response* over *response expectancy* precisely because, as he defines the term, it does not require that the meaning be explicit or verbalizable. Nevertheless, the term is limited in that it does not seem able to include the placebo effects that animals show unless we stipulate that animals are also meaning-making organisms, which seems implausible.

Di Blasi and her colleagues have suggested the use of the term *context effects* as an alternative to the placebo effect (Di Blasi et al., 2001). This relatively neutral term is intended to include all of the contextual factors that are associated with healthcare treatment, including treatment characteristics (e.g., pill color and shape), patient characteristics (e.g., beliefs about the illness, personality traits), clinician characteristics (e.g., status, gender) and actions (e.g., reassurance, empathy), and the treatment setting (e.g., prestige of the hospital, room décor). The major disadvantage of this term is that it can easily be misunderstood as referring *only* to the literal context (i.e., the physical space) as opposed to the more general meaning that includes the entire psychosocial context within which treatment takes place.

An interesting psychological mechanism to explain some placebo effects is the concept of *mindset*. Crum and Langer (2007) provided hotel room attendants with verbal and written descriptions indicating that their everyday work routine exceeded the US Surgeon General's guidelines for the amount of daily exercise needed to promote good health. As compared to a control group of room attendants who did not receive the intervention, the experimental group showed improvements in a number of objective health outcomes, including weight, blood pressure, body fat, waist-to-hip ratio, and body mass index. Since the room attendants did not report any increase in exercise or decrease in caloric intake, Crum and Langer interpreted their results as indicating that the change in mindset of the participants potentiated the beneficial health outcomes.

Despite the lack of any evidence to suggest that the room attendants had changed their health-related behaviors, it is still possible that the observed improvements in objective measures of health might have been caused by behavioral changes that the attendants did not report—either because the changes were subtle and the attendants were unaware of them, or because of reporting bias. In a follow-up study, Crum and her colleagues provided data supporting their contention that mindset can directly influence health-related physiology (Crum, Corbin, Brownell, & Salovey, 2011). Using a within-subjects design, the experimenters had participants consume a milkshake on two occasions, 1 week apart. On both occasions, the same 380 cal milkshake was consumed. However, on one occasion the shake was falsely labeled as a "Sensible Shake" that contained only 140 cal; and on the other occasion, the shake was falsely labeled as an "Indulgent Shake" that contained 620 cal. The investigators showed that the divergent labeling produced a shift in mindset regarding the healthfulness of the shake, which in turn had direct physiological effects. In particular, the "indulgent" labeling was associated with a significantly steeper decline in the appetite-regulating hormone ghrelin as compared to the "sensible" labeling. Taken together, these two studies have interesting potential implications for improving healthcare by changing individuals' mindsets.

6. LUMPING AND SPLITTING REDUX

Much of this article has focused on various splits between opposing terms (e.g., placebo vs. nocebo effects; placebo vs. placebo-related effects; and broad vs. narrow placebo effects). Although the process of splitting by drawing clear distinctions between terms is useful to clarify their meanings, it may also be helpful to look for analogies between phenomena that are not frequently lumped together. Although it has not yet been heavily emphasized, I have purposely placed the psychological mechanisms of placebo effects in between the treatment domain and the non-treatment domain. My purpose is to suggest visually that these psychological mechanisms are by no means confined to the treatment domain.

For example, the mindset examples described above were rather distant from the prototypical placebo study in that the participants were volunteers (not patients) and they did not suffer from any specific health problems, nor were they being given any form of treatment from a healthcare provider. Furthermore, the mindset concept has also been applied in educational psychology. Dweck and her colleagues have described the differential effect on children's learning depending on whether they were encouraged to take a growth mindset or a fixed mindset regarding intelligence (Mangels, Butterfield, Lamb, Good, & Dweck, 2006). Those who had a growth mindset were more likely to perceive difficult academic situations as opportunities to challenge themselves, whereas those who had a fixed mindset perceived these same situations as threats. Although Dweck's work stands outside of the treatment domain, it has clear connections to placebo studies and it may be profitable to those who study the placebo effect to look for ways to integrate the two research domains.

Priming effects have been also studied extensively in cognitive and social psychology, but the implications of this work for placebo studies has only occasionally been considered (e.g., Sliwinski & Elkins, 2013). A simple example of semantic priming in reading fluency is that word recognition is faster when the target word ("dog") is preceded by a word that is semantically related ("cat") as compared to an unrelated word ("hat") (McNamara, 2005). Although priming has often been studied in the laboratory using relatively simple stimuli, it has also been studied using more elaborate stimuli in the field. Researchers have applied the concept of priming in investigations of the reported negative health effects associated with the low frequency noise produced by wind farm turbines. Chapman, Joshi, and Fry (2014) suggested that the distribution of health complaints could best be explained by what they termed nocebo priming—news media reports and social network posts describing negative health effects attributed to the noise generated by wind turbines. In a follow-on study, Crichton and Petrie (2015) demonstrated experimentally that self-reports of ill effects following media warnings about the purported health risks caused by low frequency windfarm noise could be reduced by countering the negative priming with an intervention that explained how nocebo responses could lead to symptoms of ill health.

I would like to offer one final analogous connection that attempts to lump together an effect in the non-treatment domain with one in the treatment domain. The classic *Pygmalion in the Classroom* experiment showed that teachers' expectations significantly affected the subsequent academic performance of their students (Rosenthal & Jacobson, 1968; Rosenthal & Rubin, 1978). In particular, after testing the students, the experimenters provided feedback to teachers about the likelihood that each of their students would show academic gains over the coming academic year. Despite the fact that the feedback was actually bogus and randomly determined, the teachers' expectations affected the subsequent academic performance of their students, even though the students and their parents were never provided with any feedback whatsoever.

The Pygmalion in the Classroom study is in many ways quite similar to the concept of *placebo by proxy* (Grelotti & Kaptchuk, 2011). Placebo by proxy is the notion that placebo and nocebo effects can be generated not only via psychological mechanisms originating in the patient, but also by psychological processes originating in the patient's relatives or caregivers (Chapman et al., 1985). Thus, for example, the positive (or negative) expectations held by a parent might influence how his or her child reacts to treatment.

These two phenomena seem quite similar to me, and lumping them together might prove fruitful to researchers, particularly those in placebo studies. This is true for two important reasons. First, the concept of placebo by proxy has existed for less than 10 years, and relatively few studies of the phenomenon have been undertaken. In contrast, there have been literally hundreds of studies of the Pygmalion effect, so there is a rich theoretical and empirical literature to draw upon for inspiration and ideas. Second, the Pygmalion effect can itself by lumped in with a much larger literature on *interpersonal expectancy effects*, with an even larger theoretical and empirical literature from which to draw inspiration and ideas (Rosenthal, 1994). Certainly, lumping can be overdone, and there is, of course, great value in carefully splitting concepts so as to clarify their precise meanings; nevertheless, lumping analogous research fields together has enormous potential to generate new insights and to foster novel research collaborations. We should all strive to be both lumpers and splitters, taking care to balance the respective strengths and weaknesses of these two approaches to taxonomy.

REFERENCES

Benedetti, F. (2008). Mechanisms of placebo and placebo-related effects across diseases and treatments. *Annual Review of Pharmacology and Toxicology*, *48*(2), 33–60.

Benedetti, F., Maggi, G., Lopiano, L., Lanotte, M., Rainero, I., Vighetti, S., et al. (2003). Open versus hidden medical treatments: The patient's knowledge about a therapy affects the therapy outcome. *Prevention and Treatment*, *6*(1), 1a.

Brent, D. A., & Gibbons, R. (2014). Initial dose of antidepressant and suicidal behavior in youth: Start low, go slow. *JAMA Internal Medicine*, *174*(6), 909–911.

Campbell, D. T., & Kenny, D. A. (1999). *A primer on regression artifacts*. New York, NY: Guilford Press.

Chapman, C. R., Casey, K. L., Dubner, R., Foley, K. M., Gracely, R. H., & Reading, A. E. (1985). Pain measurement: An overview. *Pain*, *22*(1), 1–31.

Chapman, S., Joshi, K., & Fry, L. (2014). Fomenting sickness: Nocebo priming of residents about expected wind turbine health harms. *Frontiers in Public Health*, *2*, 279.

Crichton, F., & Petrie, K. J. (2015). Health complaints and wind turbines: The efficacy of explaining the nocebo response to reduce symptom reporting. *Environmental Research, 140*, 449–455.

Crum, A. J., Corbin, W., Brownell, K., & Salovey, P. (2011). Mind over milkshakes: Mindsets, not actual nutrients, determine ghrelin response. *Health Psychology, 30*(4), 424–429.

Crum, A. J., & Langer, E. J. (2007). Mind-set matters: Exercise and the placebo effect. *Psychological Science, 18*(2), 165–171.

Darwin, C. (1857). Letter no. 2130, August 1. Darwin Correspondence Project. Retrieved from http://www.darwinproject.ac.uk/DCP-LETT-2130.

Di Blasi, Z., Harkness, E., Ernst, E., Georgiu, A., & Kleijnen, J. (2001). Influence of context effects on health outcomes: A systematic review. *Lancet, 357*, 757–762.

Ernst, E., & Resch, K. (1995). Concept of true and perceived placebo effects. *British Medical Journal, 311*(7004), 551.

Evers, A. W. M., Colloca, L., Blease, C., Annoni, M., Atlas, L. Y., Benedetti, F., et al. (2018). Implications of placebo and nocebo effects for clinical practice: Expert consensus. *Psychotherapy and Psychosomatics*, 1–7. https://doi.org/10.1159/000490354.

Fässler, M., Meissner, K., Schneider, A., & Linde, K. (2010). Frequency and circumstances of placebo use in clinical practice—A systematic review. *BMC Medicine, 8*, 15.

Grelotti, D. J., & Kaptchuk, T. J. (2011). Placebo by proxy. *The BMJ, 343*, d4345.

Jensen, K. B., Kaptchuk, T., Kirsch, I., Raiceka, J., Lindstromd, K. M., Bernae, C., et al. (2012). Nonconscious activation of placebo and nocebo pain responses. *Proceedings of the National Academy of Sciences, 109*, 15959–15964.

Kaptchuk, T. J. (2002). The placebo effect in alternative medicine: Can the performance of a healing ritual have clinical significance? *Annals of Internal Medicine, 136*(11), 817–825.

Kelley, J. M., Lembo, A. J., Ablon, J. S., Villanueva, J. J., Conboy, L. A., Levy, R., et al. (2009). Patient and practitioner influences on the placebo effect in irritable bowel syndrome. *Psychosomatic Medicine, 71*(7), 789–797.

Kennedy, W. P. (1961). The nocebo reaction. *Medical World, 95*, 203–205.

Kirsch, I. (1985). Response expectancy as a determinant of experience and behavior. *American Psychologist, 40*(11), 1189–1202.

Kirsch, I. (1999). Specifying nonspecifics: Psychological mechanisms of placebo effects. In A. Harrington (Ed.), *The placebo effect: An intedisciplinary exploration* (pp. 166–186). Cambridge, MA: Harvard University Press.

Mangels, J. A., Butterfield, B., Lamb, J., Good, C., & Dweck, C. S. (2006). Why do beliefs about intelligence influence learning success? A social cognitive neuroscience model. *Social Cognitive and Affective Neuroscience, 1*(2), 75–86.

McNamara, T. P. (2005). *Semantic priming: Perspectives from memory and word recognition*. Hove, UK: Psychology Press.

Meissner, K. (2011). The placebo effect and the autonomic nervous system: Evidence for an intimate relationship. *Philosophical Transactions of the Royal Society B: Biological Sciences, 366*(1572), 1808–1817.

Moerman, D. E. (2003). "Placebo" versus "meaning": The case for a change in our use of language. *Prevention and Treatment, 6*(1), 7c.

Moerman, D. E., & Jonas, W. B. (2002). Deconstructing the placebo effect and finding the meaning response. *Annals of Internal Medicine, 136*, 471–476.

Moncrieff, J., Wessely, S., & Hardy, R. (2004). *Active placebos versus antidepressants for depression*. The Cochrane Library.

Mpitsos, G. J., & Davis, W. J. (1973). Learning: Classical and avoidance conditioning in the mollusk Pleurobranchaea. *Science, 180*(4083), 317–320.

Pavlov, I. P. (1927). *Conditioned reflexes: An investigation of the physiological activity of the cerebral cortex*. London: Oxford University Press.

Pouwels, K. B., Dolk, F. C. K., Smith, D. R. M., Robotham, J. V., & Smieszek, T. (2018). Actual versus 'ideal' antibiotic prescribing for common conditions in English primary care. *Journal of Antimicrobial Chemotherapy, 73*(Suppl. 2), 19–26. https://doi.org/10.1093/jac/dkx502.

Rescorla, R. A. (1988). Pavlovian conditioning: It's not what you think it is. *American Psychologist, 43*(3), 151–160.

Rosenthal, R. (1994). Interpersonal expectancy effects: A 30-year perspective. *Current Directions in Psychological Science, 3*(6), 176–179.

Rosenthal, R., & Jacobson, L. (1968). *Pygmalion in the classroom: Teacher expectation and pupil's intellectual ability*. New York, NY: Holt, Rinehart and Winston.

Rosenthal, R., & Rubin, D. B. (1978). Interpersonal expectancy effects: The first 345 studies. *Behavioral and Brain Sciences, 1*(03), 377–386. https://doi.org/10.1017/S0140525X00075506.

Simpson, G. G. (1945). The principles of classification and a classification of mammals. *Bulletin of the American Museum of Natural History, 85*, 1–350.

Sliwinski, J., & Elkins, G. R. (2013). Enhancing placebo effects: Insights from social psychology. *American Journal of Clinical Hypnosis, 55*(3), 236–248.

Stewart-Williams, S., & Podd, J. (2004). The placebo effect: Dissolving the expectancy versus conditioning debate. *Psychological Bulletin, 130*(2), 324–340.

Wampold, B. E., & Imel, Z. E. (2015). *The great psychotherapy debate: The evidence for what makes psychotherapy work* (2nd ed.). New York: Routledge.

CHAPTER THREE

Describing Placebo Phenomena in Medicine: A Linguistic Approach

Roland Sussex[1]
School of Languages and Cultures & Institute for Teaching and Learning Innovation, The University of Queensland, St Lucia, QLD, Australia
[1]Corresponding author: e-mail address: r.sussex@uq.edu.au

Contents

Abstract

The present paper reviews how language and communication are used to describe placebo phenomena in medicine, with particular reference at key points to the question of pain. Selected aspects of placebo language are submitted to lexical, grammatical and semantic analysis. We distinguish three uses of *placebo*, with three semantic components: "pleasing," inert medication or procedure, and what is often called "deception." The three uses of *placebo* are:

PLACEBO-TREATMENT, which combines all three semantic components: "pleasing" patients with inert treatments, and possibly with "deception."

International Review of Neurobiology, Volume 139
ISSN 0074-7742
https://doi.org/10.1016/bs.irn.2018.07.032

PLACEBO-TRIAL, which refers to the closely defined inert control in Randomized Controlled Trials.

PLACEBO-RAPPORT, which covers all aspects of the health care encounter, excluding therapeutic interventions, but including psychosocial factors which can have a beneficial effect on the patient.

Three key conclusions emerge from this analysis. First, the "deception" connotation of *placebo* found in PLACEBO-TREATMENT can contaminate the use of PLACEBO-TRIAL and especially PLACEBO-RAPPORT. Second, the enlarged "pleasing" framework of PLACEBO-RAPPORT, which includes the benefits of quality appropriate and helpful medical care, forces us to ask whether PLACEBO-TRIAL in Randomized Controlled Trials has in fact been a neutral baseline control. Biopsychosocial and other factors in PLACEBO-RAPPORT may have influenced the outcomes of many trials in medicine and pharmacology. And third, extending PLACEBO-RAPPORT to embrace all aspects of clinical care (except specific medication or procedural interventions) and psychosocial factors of the health care encounter engages patient beliefs, mindsets and interpersonal dynamics, as well as values and contextual and cultural issues. These take PLACEBO-RAPPORT substantially beyond what is currently seen as its identity and modes of operation.

1. INTRODUCTION

There are five meanings of the term *placebo* which can be recognized in English, though apparently no major general dictionary of English includes them all. Two uses are historical and now obsolete. The other three uses of the word *placebo* in contemporary English, which we call PLACEBO-TREATMENT, PLACEBO-TRIAL and PLACEBO-RAPPORT, are all medical, though I can find no medical dictionary which goes beyond the equivalent of PLACEBO-TREATMENT and PLACEBO-TRIAL. We analyze their identity and use in language and conversation, selectively utilizing online digitized corpora of English texts. The three uses of PLACEBO rely on three semantic components: "pleasing" the patient, inert medication or procedure, and "deception" (Justman, 2013). We discuss these issues with selective reference to the question of pain (Turner, Deyo, Loeser, von Korff, & Fordyce, 1994). This article grew out of a lecture "Language, communication, culture and placebo (with special reference to pain)" which I gave at the "Harnessing Placebo Mechanisms to Improve Clinical Outcomes" conference, organized on November 17–18 2017 in Sydney by the Pain Management Research Centre and the IASP Placebo Special Interest Group.

2. THE HISTORICAL *PLACEBOS*

The modern English noun *placebo* derives from the Latin verb *placere* "to please," in the first person singular future indicative: "I shall please." The same grammatical analysis applies to *nocebo* "I shall harm" from the verb *nocere* "to harm." In modern medical practice *placebo* and *nocebo* are treated as virtual antonyms (benefit ~ harm), though this is not true of the Latin verbs. Such a repurposing of Latin verbs is known elsewhere in English. The English noun *exit*, for instance, derives from the Latin verb form *exit* "he goes out."

The historical uses of *placebo* are no longer found in current English, and have been amply covered elsewhere (Finniss, 2018; Finniss, Kaptchuk, Miller, & Benedetti, 2010; Shapiro, 1968). Most contemporary dictionaries ignore them, or include them only for etymological reasons as part of the historical record.

Historically *placebo* entered English around the 8c as part of the Latin Catholic service for the dead (Psalms 116:9). It is quite probably based on a mistranslation of the Hebrew Bible. This sense of *placebo* is no longer current in English, and few would recognize it. It is included in historical dictionaries of English, like the *Oxford English Dictionary*, where it is listed first to capture its chronological order of entry into the language.

Over time the term *placebo* came to mean people who turned up to funerals without being part of the genuine mourners. This insincere element was echoed in Chaucer's character Placebo, the name of a subservient sycophant in The Merchant's Tale in his *Canterbury Tales* (1387–1400). During this period *placebo*, in the *Oxford English Dictionary* definition, meant "sycophant, flatterer or parasite." This meaning was still current in 1780, not long before the first medical use of *placebo*.

The following analysis of the three medical uses of *placebo* leaves aside these historical uses of *placebo* because they are no longer part of English.

3. THE MEDICAL *PLACEBOS*: DEFINITIONS

We distinguish three modern meanings of *placebo*. They involve three meaning components—"pleasing" the patient, inert treatment, and "deception"—which are recorded in most of the major current dictionaries. The three meanings are combined in different ways in the medical uses of *placebo*. In this context "pleasing" broadly means either creating a positive atmosphere for the patient to receive treatment, or working with the patient's expectations to help the patient reach a positive attitude toward to the symptoms and the prescribed treatment. In relation to medications and procedures (which we will group together as "treatments"), "inert" means of no known therapeutic value. "Deception" relates to real or imputed lack of openness on the part of the prescribing health care professional.

3.1 Placebo-treatment

Chronologically the first medical use, PLACEBO-TREATMENT relates to inert things (medicines) or actions (procedures, interventions), treatments which are believed to have no proven therapeutic value. The concept of a placebo had earlier been used in tests of religious faith, and then in tests of the authenticity of mesmerism in 1784. An important landmark in the specifically medical use of the word *placebo* dates from 1811 (Shapiro, 1964) in Quincy's *Lexicon-medicum*, where it is defined as "any medicine adapted more to please than benefit the patient" (*Oxford English Dictionary*). Together with the much later *nocebo* (1961), the major dictionaries state that *placebo* refers to treatments which the patient believes will have a beneficial (or in the case of *nocebo*, negative) effect (Colloca & Benedetti, 2007; Richter, Eck, Straube, Miltner, & Weiss, 2010), even though the physician knows that there is no therapeutic potential in the treatment itself. The *Oxford English Dictionary* gives this definition of *placebo*, our PLACEBO-TREATMENT:

> *4. Med. A drug, medicine, therapy, etc., prescribed more for the psychological benefit to the patient [...] being given treatment than for any direct physiological effect; esp. one with no specific therapeutic effect on a patient's condition, but believed by the patient to be therapeutic (and sometimes therefore effective).*

This definition is representative of mainstream dictionaries of English, though some of them include definitions which are less forgiving about the clinician's intentions in administering a placebo:

> *a substance having no pharmacological effect but given merely to satisfy a patient who supposes it to be a medicine.*
>
> **Dictionary.com; *www.dictionary.com***
>
> *a harmless, unmedicated preparation given as a medicine merely to humor a patient.*
>
> **Wiktionary: *https://www.wiktionary.org/***
>
> *A prescription with no pharmacological activity given to a patient to humor or satisfy the desire for medical treatment.*
>
> **Wordnik: *https://www.wordnik.com/***

In these definitions there is no mention of positive therapeutic outcomes.

These definitions imply a measure of what in the literature is most often termed "deception." "Deception" was certainly once true of sham "doctors" purveying snake-oil and other such "remedies" whose quackery was described with the term *placebo*. With modern prescribing practitioners the "deception" component rather involves allowing or encouraging the patient to believe that the treatment is indeed therapeutic when the practitioner believes it to be inert: the treatment may indeed have serendipitous beneficial therapeutic results, but beyond its known proven therapeutic effect.

We take up the issue of "deception" in the discussion of semantics (below). These negative connotations of *placebo* can then contaminate other senses of *placebo*. This inference may be quite unfair to the conscientious clinician, who may simply be hoping to encourage a positive mindset in the patient (Zion & Crum, 2018), together with expectations of a beneficial outcome. In the patient's mind, things done or provided by the doctor are seen as inherently therapeutically beneficial. These are the conditions under which PLACEBO-TREATMENT operates.

We can therefore *reshape* the *Oxford's* definition of the medical PLACEBO-TREATMENT to combine all three semantic components:

1. goal in the mind of the health care professional: to "please" the patient
2. inert treatment
3. the health care professional encourages the patient to believe that the treatment is therapeutic (deception)

From the point of view of the clinician, in PLACEBO-TREATMENT factors 1 and 2 are present, and factor 3 may be present, overtly or covertly. If factor 3 is present in the mind of the patient, then the conditions for a placebo no longer hold. And even then there is some controversy about whether placebos actually work (Hróbjartsson & Gøtzsche, 2001).

3.2 PLACEBO-TRIAL

PLACEBO-TRIAL is the most closely defined and restricted of the three uses of *placebo*. It applies the "inert, neutral" medical sense of PLACEBO-TREATMENT to the testing of treatments in Randomized Controlled Trials (RCTs), which started to become popular around 1955. PLACEBO-TRIAL, like PLACEBO-TREATMENT, refers to a treatment, a medication or procedure. Any idea of "pleasing" is secondary or assumed to be irrelevant. In Shapiro's definition this placebo is "a substance or procedure that is without specific activity for the condition being evaluated" (1968, p. 682). This definition is repeated, with relatively minor variations, in the main dictionaries of English. Subjects do not know whether it is an effective treatment or not.

One group receives the inert placebo. The other group is given the genuine treatment. Both groups are statistically selected on a randomized basis. In double-blind tests neither the patient nor the administering health care professional knows who is receiving the inert or the genuine treatment. A favorable result for the genuine treatment can result in its moving forward to additional tests, or in its being approved for commercial release by bodies like the FDA in America. RCTs are regarded, and commonly described as, the "gold standard" for testing drugs and medical procedures (Brody, 2012; Farrar, Troxel, Stott, Duncombe, & Jensen, 2008; Jamshidian, Hubbard, & Jewell, 2014; Kaptchuk, 1998; Linde et al., 2007;Lund, Näslund, & Lundeberg, 2009; Serres, Skowronski, Wu, & Ambrose, 2013; White, Lewith, Hopwood, & Prescott, 2003). PLACEBO-TRIAL is more concerned with what does **not** happen in the placebo group.

In addition, PLACEBO-TRIAL has undergone a grammatical shift. Whereas PLACEBO-TREATMENT is represented by a regular countable common noun, PLACEBO-TRIAL is commonly, though not always, expressed by an uncountable abstract noun:

> The results showed that medication x was better than placebo
> (*less frequently* a placebo)

which aligns it with other abstracts involving medical procedures of various kinds:

> [...] better than surgery/acupuncture/hospitalization

We discuss this feature below (Grammatical factors).

There is a growing literature which questions aspects of this experimental role of *placebo* (note: not usually *a placebo*), and so the way in which the word *placebo* is to be used and understood. The first of these has to do with ethics, following a landmark paper by Rothman and Michels (1994), and see Annoni (2018), Cohen et al. (2018), and Colloca and Howick (2018). In administering a placebo the health care practitioner needs to believe that they are acting ethically, which is also particularly important for PLACEBO-RAPPORT if the patient is to believe in the efficacy of the treatment. Rothman and Michels point to the Declaration of Helsinki, progressively refined by the World Medical Association:

> *In any medical study every patient—including those of a control group, if any—should be assured of the best proven diagnostic and therapeutic method.*
>
> ***Declaration of Helsinki IV (1989)***

A placebo, then, may be administered as part of an RCT if there is no currently known best method, and administering the placebo is part of determining what such a best method might be. But some trials based on the PLACEBO-TRIAL methodology have not held to this ethical principle, and so are not consistent with the Hippocratic Oath. The Helsinki Declaration also rules out the purely speculative prescription of inert treatments by the practitioner, whose responsibility it is to be up-to-date with the current best procedures. But there are further ethical issues associated with the withholding of effective treatment from patients with symptoms needing attention. And there is a problem of "informed consent." As Rothman and Michels (1994, p. 397) put it,

> *Informed consent is always desirable, but investigators should not put patients in a position in which their health and well-being could be compromised, even if the patients agree. There are several reasons. Despite the best efforts to inform patients, they will rarely if ever be as well informed about their treatment options as their physicians. Moreover, even informed patients may not be disinterested enough to decide rationally whether it is tolerable to be deprived of an accepted treatment. Finally, patients are given the choice of participating in a trial or not, but they are given no choice about which treatments will be studied. It may be more desirable to a patient to be a part of the trial than to decline to participate, but it might have been preferable to be in a different trial that did not have a placebo arm. (1996, p. 397)*

A further line of questioning about outcomes is also relevant to PLACEBO-TRIAL. Colagiuri's (2010) review shows that there are three ways in which even a double-blind RCT can deliver sub-optimal results: if the blinding fails, as it sometimes does; if the participants have strong expectancies, which show up in positive results for **both** treatment and the no-treatment group; and the fact that if participants know that they will be allocated either a placebo or an active treatment, this can lead to weaker treatment responses. To these three we can add regression to the mean (Moerman, 2002), where the patient returns spontaneously to health. And more significantly, there are potential confounding factors in the health care encounter itself, involving mindsets and communication. One key goal of an RCT is to minimize variables, so that outcomes of the trial can be more confidently associated with the non-placebo group. There is now evidence to question whether current RCT designs do in fact achieve this goal. We address these questions below in discussing PLACEBO-RAPPORT.

In one respect, however, the RCT escapes backwash from PLACEBO-TREATMENT: with PLACEBO-TRIAL the "deception" meaning relates rather to the research methodology. Patients agree to take part in the trial, usually knowing that there will be some inert treatments. The goal is to test a medication or procedure, rather than to heal the patient, so any ethical question is rather about the methodology itself. And in a double-blind trial the person administering the placebo does not know whether it is a placebo or not, so there can be no question of active deception of the subject or patient. We take up issues of "deception" under Semantic factors.

3.3 Placebo-rapport

The rise of placebo-rapport coincides chronologically with attempts to divorce medical practice from the "deception" and ethical problems associated with the administering of inert medicines. Part of this had to do with changes in the relationship between doctor and patient, and in particular with the new emphasis:

> *[...] in contemporary medical ethics of individual rights and patient autonomy in the doctor–patient relationship [...], leading to the rejection of many paternalistic assumptions previously thought to justify medical deception.*
>
> ***Brody (1982, p. 112).***

Placebo-rapport was partly anticipated by Shapiro's definition of *placebo effect*:

> *The* placebo effect *is defined as the non-specific psychologic or psychophysiological effect produced by a placebo*
>
> ***Shapiro (1968, p. 682).***

(and see Shapiro, 1964; Shapiro & Struening, 1973; Shapiro, Struening, & Shapiro, 1979). But in placebo-rapport the placebo effect is not produced by a placebo in the older sense of treatment. It is the outcome of actions and mechanisms which are a central part of the health care encounter itself. Placebo-rapport is psychosocial and ubiquitous:

> *Placebo mechanisms are inherent in routine clinical care*
>
> ***Finniss et al. (2010, p. 692)***

Placebo-rapport has in one sense come back to its historical Latin "pleasing" origins. It is wider than placebo-treatment, which is restricted to medications and procedures. Placebo-rapport relates to the elements of the consultation and the treatment, other than therapeutic medications and procedures, which may influence the patient positively and may so help to lead to positive outcomes, psychologically, therapeutically or both. These aspects of placebo-rapport can be summarized:

The sum of the contextual and interpersonal factors in the health care encounter, other than therapeutically validated medications and procedures, which are associated with positive outcomes for the patient. This includes conversation and interpersonal communication, explanation and education about the patient's condition and treatment options, the rapport between practitioner and patient, validation of the patient's condition and concerns, and aspects of health care.

Favorable outcomes from PLACEBO-RAPPORT may not be wholly unplanned or serendipitous, as they are with PLACEBO-TREATMENT, but may be consciously developed in the health care encounter, initially led by the health care professional. To put it another way, PLACEBO-RAPPORT may not involve a purely placebo effect in the sense of PLACEBO-TREATMENT. There is less sense of "deception." PLACEBO-RAPPORT focuses strongly on positive outcomes which go beyond the known explicable effects of authenticated medications and procedures.

It is likely that different elements of "pleasing" will be present in different health care encounters, in different ways, and to different degrees. The above formulation includes "interpersonal" (Necka & Atlas, 2018), which picks up key idea of "social intersubjective space" as proposed by Cohen, Quintner, and van Rysewyk (2018) in their programmatic redefinition of the notion of pain; and the formulation uses "associated," deliberately avoiding explicit causal claims between placebo aspects of the health care encounter and the outcomes which are evident in patients. PLACEBO-RAPPORT elements can interact with medications, treatments, the context and conversation, including the element of "ritual." And it is conversation, and the interpersonal actions which realize conversation, which form the core of the new research domain of PLACEBO-RAPPORT, together with the context, the expectations, values, culture and mindsets which the participants bring to the health care encounter, and the actions of the therapist, together with the patient (Doering, Glombiewski, & Rief, 2018; Incayawar & Todd, 2013; Kazarian & Evans, 1998; Purnell & Paulanka, 1998).

This is not just empathy (Decety & Ickes, 2011; Goubert, Craig, & Buysse, 2011), which is the ability to share and understand the thoughts and feelings of another person. Empathy is directional, from person A to person B. But the goal of the PLACEBO-RAPPORT health care encounter is to create mutual empathy between the practitioner and the patient with respect to the proposed treatment and its outcomes. The more appropriate term is "rapport," which includes the positive shared feelings which arise between the health care professional and the patient, especially in the treatment of chronic pain, where the patient feels that their symptoms and feelings are validated, understood, and appreciated. And it is worth remembering that there is now empirical evidence, biochemical and other, which confirms the changes—psychological and therapeutic—which can take place in the patient in response to the elements of PLACEBO-RAPPORT (Finniss et al., 2010).

In PLACEBO-RAPPORT there is a stronger research concentration on the *placebo effect*, when seen more from the patient's point of view. And the *placebo effect* is seen as arising from *placebo mechanisms*, not just from the patient's positive expectations, nor from the administering of inert treatments, but from a much wider set of factors emerging particularly from the interactions between the health care professional and the patient, including previous experience (Reicherts, Gerdes, Pauli, & Wieser, 2016; Varelmann, Pancaro, Cappiello, & Camann, 2010). This emphasis on outcomes in spite of therapeutics, to adapt a formulation from Harris (2016, Introduction), allows a new focus on benefits (and also negative benefits, in *nocebo* terms) to patients which belongs to a more holistic perspective. Patient outcomes are seen less in terms of purely physical healing, but rather as part of a wider restoration of health, or improvement in health and quality of life.

3.4 Placebo Effect, Placebo Mechanisms

It was PLACEBO-TREATMENT which gave rise to the phrase *placebo effect* (Price, Finniss, & Benedetti, 2008). Strictly speaking, *placebo effect* is self-contradictory, an oxymoron (Sanders, 2017): if a placebo is inert, it cannot by definition have an effect. That is certainly true of PLACEBO-TRIAL, where it is the inertness of the placebo group's treatment which provides the baseline against which new treatments can be tested. But in PLACEBO-TREATMENT, and even more in PLACEBO-RAPPORT, it is the "pleasing" which is associated with the effect. *Placebo effect*, in fact, shows one of the key fracture lines in the use of the term *placebo* across PLACEBO-TREATMENT, PLACEBO-TRIAL and PLACEBO-RAPPORT, and the incompatibility, or at least tension, between the "pleasing" and the "inert" senses. *Placebo effect* is most relevant, and most consistent, with PLACEBO-RAPPORT, which relegates the inert and "deception" senses to the sidelines, and concentrates on the ways in which "pleasing" can be created by the development of rapport in the health care encounter. This focus also allows the investigation of *placebo mechanisms*, which are the actions and strategies in the health care encounter which result in "pleasing" responses, and thence positive therapeutic outcomes, in the patient.

4. LEXICAL, GRAMMATICAL AND SEMANTIC PROPERTIES OF PLACEBO-TREATMENT, PLACEBO-TRIAL AND PLACEBO-RAPPORT

The three uses of the word *placebo* intertwine semantically. There is a line of semantic development from PLACEBO-TREATMENT to PLACEBO-TRIAL, based on the similarity of "inert treatments." And there is a second, and different, line of semantic development from PLACEBO-TREATMENT to PLACEBO-RAPPORT involving "designed to please the patient." Although PLACEBO-RAPPORT has not yet appeared explicitly in mainstream or specialized medical dictionaries, there are growing numbers of examples of this usage in the recent scientific and medical literature.

From a language and usage point of view there are two questions which arise:

- In what respects of form, usage and meaning do the three versions of *placebo* differ?
- Do these differences constitute difficulties for communication, and for the integrity of the various uses of the term *placebo* itself?

We address these questions from the point of view of lexis (vocabulary), grammar, and semantics (meaning).

4.1 Online Resources

The dictionaries fail to provide us with some key aspects of differences in usage between PLACEBO-TREATMENT and PLACEBO-TRIAL, let alone PLACEBO-RAPPORT. Fortunately, Internet-based language corpora give us access to large bodies of English texts where it is possible to search for the word *placebo* and its characteristic usages. As we shall see, however, while it is certainly possible to investigate the occurrence and co-occurrence of the word *placebo* with other words and phrases, it is more difficult to determine certain aspects of its grammatical use, and even more so its semantic properties.

In this article, we make selective use of two general corpora of English:

the iWeb corpus, 14 billion words of English
https://corpus.byu.edu/iweb/.
the NOW corpus, 6 billion words of journalistic English
https://corpus.byu.edu/now/.

There do not seem to be any suitable corpora of specific medical texts in English.

By exploiting the various options available to modify searches, it is possible to search both quantitatively and qualitatively for keywords, phrases consisting of the keywords combined with other words within a certain range (e.g., within three words before and/or after), and (with the NOW corpus) dates. All these searches are on written words, and so concentrate exclusively on the printed form of language.

4.2 Vocabulary Frequency

The online corpora allow us to establish that *placebo* is significantly more common, as a percentage of the total English vocabulary, than it was before about 1955, when RCTs started. Google's NGram Viewer (https://books.google.com/ngrams) corpus, which contains about half a billion words in printed books between 1500 and 2008, reports a sudden rise in *placebo, a placebo* and *placebos* in the late 1950s, with a short (and for now, unexplained) fall around 1980 before climbing again. During this 60+ year period the frequency of *placebo* increased 11-fold from 0.00005% of the recorded vocabulary to 0.00055%. In 1940 *placebo* stood at only 0.000003%.

The larger iWeb corpus, which includes journalism in its 14 billion words, has *placebo* as the 8864th most used word in English, just after *moderator, embryo* and *magistrate,* and just before *displacement, Hispanic* and *majestic.* I would not have anticipated such a high ranking. The most frequent uses of *placebo* are:

placebo	68,454
including	
a placebo	12,464
the placebo	13,514
and there is also	
placebos	2,509

In the 6 billion words NOW corpus of journalistic English, which includes dating information, the raw annual numbers of uses of *placebo* have increased eightfold in the data since 2010, as the corpus has grown. But the overall relative frequency of *placebo* has been steadily declining, from 0.05 words per million in 2010 to 0.01 per million in 2017. This shows a declining share of *placebo* in general English usage.

Without inspecting each example it is not always possible to tell which of PLACEBO-TREATMENT, PLACEBO-TRIAL or PLACEBO-RAPPORT we are dealing with. This is also the case with *placebo* as a prenominal modifier.

4.3 *Placebo* as a Prenominal Modifier

The word *placebo* can be a noun (*placebo, a placebo, the placebo, placebos*). Or it can be a prenominal modifier, in a grammatical construction where it functions like an adjective and modifies a following noun, as in *placebo effect*. Data from the iWeb corpus were analyzed on 29 May 2018. They show that, once we exclude punctuation, prepositions (*in, for*) and conjunctions (*and, or*), it is possible to establish a ranked order of *placebo* combinations with nouns, from the most frequent combinations downward. In the iWeb corpus the most common phrases (collocations) with *placebo* as a pre-nominal modifier, together with the number of citations, are:

placebo group(s):	7,364	consisting of *group* (6554) and *groups* (810)
placebo effect(s):	6,453	consisting of *effect* (5893) and *effects* (560)
placebo treatment(s):	661	consisting of *treatment* (564) and *treatments* (97)
placebo pill(s):	645	consisting of *pill* (265) and *pills* (380)
placebo arm(s)	572	consisting of *arm* (481) and *arms* (91)
placebo control(s):	553	consisting of *control* (421) and *controls* (132)
placebo patient(s):	438	consisting of *patient* (10) and *patients* (428)
placebo response(s):	414	consisting of *response* (349) and *responses* (55)

(*placebo arm* is another term for *placebo group*). Other significant phrases like *placebo surgery* scored only 30. And it is a sign of the recency of the emergence of PLACEBO-RAPPORT, and of its relative quarantining to the research and specialist literature, that the phrase *placebo mechanism*, which is a key component of PLACEBO-RAPPORT, occurs only twice in the 14 billion word iWeb corpus. In contrast, in the professional literature indexed in Google Scholar there are 428 citations of *placebo mechanism*. PLACEBO-RAPPORT has not yet reached public discourse.

The contexts in which these phrases occur are predominantly medical, either professional or in prose connected to medical discourse. It is often difficult to determine which of PLACEBO-TREATMENT, PLACEBO-TRIAL or PLACEBO-RAPPORT they refer to, but they often relate to PLACEBO-TRIAL:

> Because more participants in the placebo group dropped out than the Diclectin group [...]
>
> **Toronto Star, *17 January 2018***

This example, with the pejorative connotation, concerns PLACEBO-TREATMENT:

> Or are we simply paying for a relatively pleasant experience and a dose of the Placebo Effect?
>
> **Daily Edge *(Ireland), 12 April 2018.***

And this one involves "pleasing" and PLACEBO-TREATMENT or PLACEBO-RAPPORT, probably the former:

> Evidence also shows that bigger pills produce a stronger placebo effect than smaller ones.
>
> **National Post, *9 April 2018.***

Placebo effect is more likely to involve PLACEBO-TREATMENT or PLACEBO-RAPPORT, but which? The same is true of *placebo response*, which, when taken out of context, could in principle refer to any of PLACEBO-TREATMENT, PLACEBO-TRIAL or PLACEBO-RAPPORT:

> Currently, placebo response is primarily studied in healthy subjects within controlled experimental settings.
>
> **Free Press Journal, *30 October 2016.***

Placebo surgery could mean either an inert procedure in a trial (PLACEBO-TRIAL), or surgery involving pleasing the patient (PLACEBO-TREATMENT), or could be part of a total health care encounter event (PLACEBO-RAPPORT). Or it could be one of the deceptive PLACEBO-TREATMENT surgeries with no known therapeutic benefit (Beecher, 1961; Harris, 2016; Probst et al., 2016).

This problem involves polysemy, or multiple meanings developing from a single word. It has been created by the expansion and re-specialization of *placebo* from PLACEBO-TREATMENT to PLACEBO-TRIAL and now PLACEBO-RAPPORT, and the co-existence of all three versions of *placebo*, though not to the same extent in all types of texts and uses. And the fact that we cannot confidently identify the sense of *placebo* without looking closely at the context—and sometimes not even then—contributes to the lack of transparency in the term *placebo*.

4.4 Grammatical Factors

Placebo is listed in the mainstream dictionaries as a regular count noun which refers to medications or procedures. This means that it has a singular and a plural (*placebo* and *placebos*), and can occur with numerals and articles:

***the** placebo seems to be working*
***the 3** placebos in this trial performed differently*
*this medication is no better than **a** placebo*

This count noun usage is the only one listed in the examples given in the *Oxford English Dictionary*, which defines *placebo* as "**a** drug, medicine, therapy, etc." (emphasis added).

But in PLACEBO-TRIAL and PLACEBO-RAPPORT the word *placebo* has developed a new grammatical role as an abstract noun. In this role it takes neither articles nor plurals. *Placebo* as an abstract noun has been growing since the middle of the last century, when RCTs started around 1955, with the "inert" sense of PLACEBO-TRIAL:

PLACEBO-TRIAL:
this treatment was better than placebo/surgery/homeopathy/analgesia/
(cf. *better than a placebo/the placebos*)

PLACEBO-RAPPORT is also usually abstract, and in this use is standard only in medicine. The count noun *placebo* is found with both medical texts, especially as PLACEBO-TREATMENT:

the doctor gave her a placebo and she started feeling better the same day

and the metaphorical, non-medical use:

he said he'd be here at 5, but that's just a placebo to stop me worrying

We can track the growth of the abstract use of *placebo* from the numbers of occurrences of *placebos*, the only unambiguously count-usage. Searching for *a placebo* could find *a placebo group*, where *placebo* is a modifier and not a noun. Raw counts will not serve, since both written output and the size of corpus databases, and their coverage, have been growing strongly. So we compare the number of occurrences of *placebos* in relation to the total occurrences of *placebo* + *placebos*, using the NOW database (iWeb does not currently allow searching by date):

Year	*placebo* (a)	*placebos* (b)	total a + b	b/(a + b)
2010	1296	102	1398	0.0729
2017	3402	125	3527	0.0354

This is evidence for the relative decline of the count-noun *placebo* (represented by *placebos*) in journalistic prose when compared to the abstract-noun usage commonly found with PLACEBO-TRIAL and PLACEBO-RAPPORT: it has halved over the period 2010–2017. The raw frequency of *placebo* increased 2.625 times. Remember that over the same period the overall frequency of *placebo* in Google NGram Viewer grew 11-fold. The use of *placebo* in journalistic prose is therefore not matching that of *placebo* in the scientific literature.

4.5 Semantic Factors

In discussing the semantics of *placebo*, we will have the issue of PLACEBO-RAPPORT very much in focus. Compared to PLACEBO-TREATMENT and PLACEBO-TRIAL, its definition is much wider, not yet in any major mainstream dictionary, and fuzzier. There are five topics which we address under the rubric of semantics: "pleasing," inert, the pejorative *placebo*, "deception," and expectation.

4.5.1 Pleasing

The basic meaning of Latin *placebo*, and of English *to please*, is to make another person feel good, happy and satisfied. A patient who is feeling good, happy and satisfied may have reached that position on their own, with positive expectations about the treatment and its effect on the outcome. This positive emotional state may also arise as a result of the actions and demeanor of the health care professional. Or both. A successful placebo achieves its result because of the patient's positive orientation and expectations. It has also been shown that a positive attitude on the part of the health care professional is germane to positive outcomes in the patient (Gracely, Dubner, Deeter, & Wolskee, 1985).

Please is actually something of a misnomer, especially with PLACEBO-TREATMENT. That is why we have so far placed it in inverted commas: "please" and "pleasing." Dictionary definitions of *please* refer to *cause a person to feel happy and satisfied*. But there is a sense in which *please* in *placebo* is more patient-side, and concentrates more on the patient's mood and mindset, rather than on specific actions (other than prescribing the treatment) taken by the practitioner. It is often less a matter of "cause" than of "allow and encourage."

And PLACEBO-RAPPORT moves the focus of being pleased from the patient to the health care encounter, the contributions of both the health care professional and the patient, and their interactions and conversation. There is more talk of *placebo effect*, *placebo mechanisms*, and the processes by which placebo outcomes can be achieved for the good of the patient. The key continuing themes from PLACEBO-TREATMENT are the idea of positive patient mindsets and expectations, independent of the effects of treatments, to create rapport and confidence. The "inert" meaning is left behind with PLACEBO-TRIAL and RCTs. But the "deception" meaning is still potentially operative.

4.5.2 Inert

"Inert" means that a medication or procedure in not effective for the observed symptoms. But it depends who knows or believes what about the inertness:

> A placebo *is defined as any therapy (or that component of any therapy) that is deliberately used for its nonspecific psychologic or psychophysiologic effect, or that is used for its presumed specific effect on a patient, symptom, or illness, but which unknown to therapist and patient is without specific activity for the condition being treated*
>
> ***Shapiro (1968, p. 682).***

For a treatment to be a placebo in the "pleasing" sense, the health care professional must usually believe that it is a placebo. And the patient must believe that it is genuine, otherwise the "pleasing" criterion cannot be fulfilled. Later events, which can prove a placebo to be in fact genuine, or a trusted treatment to be a placebo or even a nocebo, like blood-letting patients, do not change the status of the ±placebo treatment at the time of prescription or administering.

From the point of view of the patient, the patient needs to believe that the medication or procedure is effective. But it is also important that the health care professional believes that the placebo is in the interests of the patient's well-being, whether psychological or therapeutic or both. This is the role of rapport and persuasive talk.

PLACEBO-TRIAL has only the "inert" meaning in its role in RCTs. The "pleasing" component is not engaged on either the physician's or the patient's side. The "deception" component can become involved as an unintended by-product. I also argue below that it could potentially vitiate the results of RCTs (though see Brody, 1982; Kaptchuk et al., 2010 for placebos without "deception").

There is now, with PLACEBO-RAPPORT, the additional factor of the therapeutic impact of the health care encounter, once assumed to be inert, which is starting to show measurable evidence-based results. This once again shifts the definition, and the impact, of factors other than the formerly determined therapeutic value of medications and procedures.

There is also a problem with *inert* and PLACEBO-RAPPORT. *Inert* not only means "lacking effective action," as it does in chemistry. It can also mean "lacking in vigor," and has overtones of laziness or indolence. None of these meanings are welcome in the semantic area of PLACEBO-RAPPORT.

4.5.3 *Pejorative* Placebo

Phrases like *just a placebo*, *merely a placebo*, *a mere placebo* and *only a placebo* suggest non-positive uses of *placebo*. In English such expressions can imply a failure to perform to an expected level: *he's just a beginner* can be read as an excuse for an unsatisfactory performance. Similarly, the frequency of such expressions in the iWeb corpus, with all forms of *placebo*, shows that they are sufficiently numerous to warrant attention in the pejorative sense:

just + *placebo*/*a placebo*/*the placebo*/*placebos* (622)
mere + *placebo*/*a placebo*/*the placebo*/*placebos* (18)
merely + *placebo*/*a placebo*/*the placebo*/*placebos* (46)
only + *placebo*/*a placebo*/*the placebo*/*placebos* (272)

as in

> He recently rushed to give one of his players suffering from cramp a banana, with the idea that the potassium hit would allow him to see through the game. It was probably just a placebo *[...]*
>
> ***(SBS, The World Game, 17 March 2017)***

The *Oxford* notes this sense of *placebo* (also *fig.*), but gives no examples of it.

4.5.4 *Deception*

The fourth meaning of PLACEBO-TREATMENT has to do with "deception." Like *please*, the word *deception* is also partly a misnomer, and we have so far placed it also in inverted commas: "deception." The "deception" connotation is one of the most problematic aspects of *placebo*, since it sits ill with the "please" meaning, especially with PLACEBO-RAPPORT.

Dictionary definitions of *deceive* specify an intention deliberately to cause someone to believe something that is untrue, especially for personal gain. This "deception" sense has been associated with *placebo* since the 13th century, especially in charges against quack doctors and medicines.

But associating the word *deceive* with modern placebos, and specifically PLACEBO-TREATMENT, is too strong. The health care professional may knowingly allow or encourage the patient to believe that an inert treatment is therapeutic. But this is for the patient's benefit (Vance, 2016). This is not "deception" in the malign sense of misinformation for the speaker's profit, nor does it exploit the patient for the physician's benefit, except in the sense that the physician may achieve a good result (and incidentally payment) in spite of the treatment. And as we will see, modern versions of this placebo health care encounter can involve "open-label" placebos, where the patient is aware of the nature of the placebo, which attenuates the notion of "deception" even further. True, there can be negative outcomes and unintentional side effects of placebos, for instance, if the patient is a coeliac and is given a placebo containing gluten, a fact unknown to the practitioner. But a key component of *placebo* is that the doctor will have a benign intention in prescribing it, and the patient will have positive expectancies about its efficacy.

Unfortunately the **word** *placebo* carries with it some less positive baggage.

Doing something just to please another person can be relatively benign: to make them feel better. But there can be a less positive sense of "palliate, placate or please" which implies a lack of sincerity, and acting just to make the other person feel better, or to get them off the speaker's back. So there are echoes, and probable contamination, from the metaphorical use of the non-medical *placebo* to refer to words or actions which are deliberately intended not to be efficacious, something done purely to please or pacify another person (see Wickramasekera, 2001):

> Banning guns is just a placebo, a false sense of security.
>
> ***(CNN, 17 January 2017)***

Using "deception" in connexion with placebos also creates difficulties of frankness, for instance, if the patient asks pointed, detailed questions about what the treatment consists of and how it works. Here the health care professional's justification can be found in this specification from the Declaration of Helsinki IV (1989), which we quoted above:

> *[...] every patient [...] should be assured of the best proven diagnostic and therapeutic method.*

If there is no standard therapeutic "best proven diagnostic and therapeutic method," then a placebo may well fulfill that role. In this case there is an argument that the health care professional is justified in prescribing a placebo, though subject to considerations about discussing the options with the patient. There still remains a question of how the placebo is represented to the subjects, and whether a neutral context can in fact be achieved.

But even the benevolent assumption about *placebo*-prescribing doctors has been challenged. In a 1961 study the American surgeon Beecher claimed that some surgery is done more for psychological than for therapeutic reasons. This argument is taken up in a more thoroughgoing way by Harris in an unsettling book entitled *Surgery, the ultimate placebo* (2016). Harris, a reputable and experienced orthopedic surgeon, presents a detailed argument not only that a great deal of surgery does not have beneficial therapeutic outcomes, but also that practicing surgeons know, or should know, that a substantial proportion of the procedures they do are in fact placebos, with no likely therapeutic benefit. They may possibly improve the patient's sense of well-being, and the sense of well-being of the surgeon too: at least by doing something they may enhance the patient's sense of being cared for. There are further, more cynical, potential interpretations. But so long as the surgeon's main aim is the patient's well-being, their own satisfaction is a by-product. Nonetheless, the element of deception remains, and with it the element of misleading and potential dishonesty.

But in RCTs the subject cannot be told which medication or procedure they are receiving. Here in one sense there may be no ethical issue: if subjects knowingly sign up for a trial, and are told that one group will receive the trial medication or procedure and the other group will not, and they will not know which group is which, then their consent to participate vitiates any claim of deception. The problem, however, may still persist with the **word** *placebo*. Subjects may bring beliefs and mindsets which are detrimental to placebos from previous experience or from the Web: Wikipedia's page on "placebo" uses the words *deception*, *dishonesty* and *sham*. And this could affect the mindset with which they participate in the test, and so potentially the test results as well.

Kaptchuk et al. (2010) aimed to present the placebo concept in a positive light as an "open placebo" by specifying "shown in clinical studies to produce significant improvement" in their instructions to subjects. They call non-deceptive behavior "not concealed" and "not deceptive." In an RCT with Irritable Bowel Syndrome patients, 80 subjects were randomly assigned to two groups, one control group which received no treatment and one group which was given open-label placebo pills:

> *[...] placebo pills made of an inert substance, like sugar pills, that have been shown in clinical studies to produce significant improvement in IBS symptoms through mind-body self-healing processes*
>
> ***Kaptchuk et al. (2010, p. e1559).***

But there is no way of knowing what understanding of *placebo* the subjects had previously formed, or whether any potential fears or negative reactions were neutralized by the phrase with "significant improvement." In other words—and this remains hypothetical at this stage—what the effect was of their previous beliefs about placebo on their participation.

4.5.5 Expectation

Patient expectations are central to the effectiveness of *placebo*, including PLACEBO-RAPPORT. If the patient does not believe that the treatment will have a positive therapeutic effect, then by definition we are not dealing with a placebo, but a neutral situation or possibly even a nocebo (Benedetti, Lanotte, Lopiano, & Colloca, 2007; Lang et al., 2005). It is the creation of such positive expectations which are the business of "placebo mechanisms," and their operation is captured by the phrase *placebo effect* (see Colloca, 2018a).

There is a substantial literature already on the **effect** of patient expectations on therapeutic outcomes, mostly outside the narrower domain of placebos (though see Colagiuri, 2010; Colloca & Benedetti, 2007; Darnall & Colloca, 2018; De Pascalis et al., 2002). Much less research has been devoted to the question of how patient expectations are determined, formed, and the means by which positive expectations are nurtured. And there has been very little on expectations and beliefs from the practitioner's side (Gracely et al., 1985).

The issue of expectations is particularly relevant to cases of pain, with its particular urgency of the patient's condition. Patients hope that the practitioner will provide a single comprehensive effective treatment, and their belief in the medical professions can create positive placebo conditions. The "single silver bullet" expectation can often be met in cases of acute pain, since treating the cause can remove the pain as well. Things are otherwise with chronic pain. Here the patient's expectations may have been damaged by serial encounters with health care professionals who have failed to provide them with relief, and by cycles of hopes raised and dashed. This "previous experience" factor strongly influences the patient's current expectations. These learned expectations can also be positive, as with patients receiving a medication which they have benefited from in the past (Klinger, Stuhlreyer, Schwartz, Schmitz, & Colloca, 2018). Relevant also is the "first meeting experience," a factor found in discussions of PLACEBO-RAPPORT (Finniss et al., 2010, p. 688). Then there is the issue of the patient's knowledge, which is not the same as the information they have collected: online information sources on pain can be misleading, and even patients who have researched their condition extensively may actually understand not very much. They expect that the health care professional will be able to take decisive, directed action to help them, and that treatments will be provided—here the passive verb "be provided" is relevant. Also, by implication, the expectation that the health care professional will "do" certain relevant actions—do "something"—rather than joining with patient in a partnership for the management of the pain, possibly through multiple steps and over a substantial amount of time. The expectation of decisive therapeutic action by the health care professional can be a barrier, not a "placebo mechanism" but a placebo inhibitor. To quote a chronic pain psychiatrist (personal communication), "even doing nothing is doing something"—a lesson which many patients find difficult to accept. If we are thinking of the consultation in terms of a potential "placebo event," then in the context of the placebo event, "doing nothing" is in fact full of placebo meaning potentials which derive from the context and interactions.

Creating positive expectations in others is a well-known goal of discourse in politics, marketing and business, education (though this is probably less widely acknowledged), religion, and many other professional and personal areas of human interaction. It can include propaganda, missionary activities, salesmanship, and effective pedagogy. Health care practitioners know this as caring interaction with their patients. A fair amount is already known about persuasive speech and persuasive behavior in human interactions. What is different about expectations in the context of PLACEBO-RAPPORT is that some of the behavioral and neurobiological consequences can be scientifically investigated, and incorporated into theories, models and therapies (Colloca, 2018a), including factors like the color of the pills (Meissner & Linde, 2018). The detailed investigation of managing placebo mechanisms between health care professionals and patients in the health care encounter is therefore one of the more urgent questions in placebo research. This is centrally related to rapport. Creating expectations to facilitate a placebo effect is more subtle and complicated, as well as collaborative. Once again we come back to the crucial importance of the patient becoming a partner in the management of their condition. And it is problematic to combine it with any notion of deception. A health care professional would have difficulty building placebo trust while at the same time consciously misleading the patient.

4.5.6 PLACEBO-RAPPORT and Terminology

There is therefore a genuine difficulty of cross-contamination from PLACEBO-TREATMENT and PLACEBO-TRIAL to PLACEBO-RAPPORT. Bennett (2018), indeed, advises that the word *placebo* should never be used in front of a patient.

Speaking purely in terms of frequency, the word *placebo*, with nearly 39 million Google hits, is well established. In both its concrete and metaphorical senses it is lively and current. And though *placebo* is not among the most frequently used words in everyday English, it is difficult to see how it could be challenged. Unfortunately, the prevalence of pejorative overtones in phrases like *just a placebo*, which derive from extensions of PLACEBO-TREATMENT, make this expression useful, not just informally but in public life. Such pejorative overtones provide a potential contaminant for the more scientific and technical-medical uses of *placebo*.

PLACEBO-TRIAL, together with the potential problems from its "inert" meaning, is to some extent quarantined behind the firewall of technical language and professional scientific experimentation and practice. Talk about placebos in RCTs is relatively uncommon in public discourse. There is still a possibility of cross-contamination from the pejorative "deception" overtones of PLACEBO-TREATMENT.

PLACEBO-RAPPORT presents the greatest difficulties. The fact that it has established itself within a growing area of intellectual enquiry, with definitions established and accepted by that community, is significant. And there is certainly precedent for the redefinition of terms for specific professional purposes, or for differences between professional usage and general public usage of vocabulary. Outside medical discourse, *tumor*, for instance, is widely understood by default to refer to malignancies. A dental *eruption* is not easily confused with a volcanic one. *Chronic* and *acute* pain have precise meanings in the pain discipline, but are poorly defined and understood by the general public.

But the three *placebo* terms are too close, and in particular PLACEBO-RAPPORT is vulnerable to cross-contamination from everyday use, specifically PLACEBO-TREATMENT. Moerman, in a series of publications (e.g., 2002), has proposed *meaning response* to replace *placebo* and so remove ambiguity. It is difficult not to applaud his intention, but his proposed solution is problematic. *Meaning response* is deliberately high level and not specific. Patients may respond meaningfully in many kind of ways to many kinds of stimuli and inputs in the health care encounter. Other candidates like *expectancy response*, *response expectancy* (Kirsch, 1985), *non-specific effect* and *context effect* (Sanders, 2017) are not persuasive.

5. SOCIAL AND CULTURAL FACTORS AND THE "PLACEBO EVENT"

As we have seen, PLACEBO-TREATMENT concentrates on the inert nature of the placebo, the beliefs and expectations of the health care professional and patient, and the clinical outcomes. PLACEBO-TRIAL focuses on the inert nature of the placebo, and its role in the RCT.

PLACEBO-RAPPORT, as it is presented in the literature so far, refocuses the term *placebo* in the therapeutic context (Leknes et al., 2013). Let us call this the "placebo event," the aspects of the health care encounter which contribute to the *placebo effect*, and incorporate *placebo mechanisms* (Finniss et al., 2010), which operate to create rapport and positive orientation during the placebo event, and then continue after the patient leaves to undertake treatment. The placebo event need not even be spatially contiguous: it could occur without the physical presence of the practitioner, say on telehealth or via software like Skype. In all these contexts PLACEBO-RAPPORT includes the patient's beliefs and mindset, social and psychological dimensions of the placebo event, its mechanisms of operation, the placebo effect, and its psychological, therapeutic and neurobiological outcomes.

These lines of enquiry are consistent with the proposal of the biopsychosocial model as proposed by Engel (1977, 1980) as part of his attempt to broaden psychiatry beyond the bioreductionist model, following the holistic proposals of General Systems Theory (Von Bertalanffy, 1968). And it links with the biopsychosocial theme in definitions of pain, widely accepted among pain specialists since its proposal by Loeser (1982) and Fordyce (1988), and its adaptation by IASP, the International Association for the Study of Pain, in 1979 (Wordsworth, 2018; Gask, 2018). Craig (2009) argued for the incorporation of a stronger emphasis on the social dimension with pain. Main (2016), in a critical historical review of the diagnostic words used in the McGill Pain Questionnaire, proposed that they need to be re-contextualized to incorporate elements of the social context and the speakers. Cohen et al. (2018) take a similar epistemological position in proposing a redefinition of the concept of pain to incorporate "the social 'intersubjective space' in which pain occurs" (and see Goubert et al., 2011).

But in the case of both pain and placebos even the biopsychosocial framework (Turk, 1996) omits crucial factors of values and culture. There are myriad definitions of "culture," but this working definition will suit our purposes:

> *Culture: A set of socially inherited, learnt practices which together underpin the social activity of a group of people and help to define them, and which provide a context in which their interactions find meaning.*

This is particularly true of pain. Pain words can function differently in different cultures, and can mean different things (Diller, 1980; Gaston-Johansson, Albert, Fagan, & Zimmerman, 1990), as I have argued in connection with pain and intercultural communication (Sussex, 2009, 2015, 2018, building especially on earlier work by Zborowski, 1969; and see Power, Bogossian, Sussex, & Strong, 2017). This line of argument takes the statement by Finniss and colleagues that placebo mechanisms are actually present in all health care encounters (Finniss et al., 2010), and develops the contextual implications of "encounters." It is important to note that intercultural communication does not necessarily require a difference of language. It is possible, for instance, for two doctors, one a native speaker of English and the other using English as a second language, to have what is effectively an intra-cultural communication, based on their shared professional expertise. On the other hand, two native speakers of English, say a doctor and a patient with limited education, may have significant difficulties in communicating, and in negotiating and managing the patient's expectations, particularly in an area as fraught and difficult as chronic pain. This constitutes one variety of intercultural communication.

The placebo event is located in a particular, often clinical, setting which itself is pregnant with meaning possibilities for PLACEBO-RAPPORT. In this context, we find interpersonal conversation, collaboration and negotiation, using both language and factors like body language. Cohen et al. (2018) refer to the "limitations of language" in communicating about pain. This is true, but premature, when we have not properly investigated how much language can do in pain communication, and how it does it. The participants bring to the placebo event their existing value systems and expectations, together with a complex set of conceptual categories which constitute their culture. National cultural values, religious beliefs, medical beliefs and knowledge, and previous experiences on the part of both the health care professional and the patient, are all relevant. So also is the notion of "ritual," which is growing in use in the PLACEBO-RAPPORT literature. Ritual is a habitual, repeated format of human behavior. In the placebo event, it will certainly vary widely across cultures, health care professionals, patients, and even individual meetings. Rituals can be shared or single, structured and scripted or partly spontaneous, participatory but not necessarily with equal participation by the participants (Coupland, Robinson, & Coupland, 1994; Heritage & Clayman, 2010; Robinson, 1999, 2003; ten Have, 2006). They are often asymmetrical. But the existence of a structure, and the recognition of that structure, provide a framework which can underpin the placebo event (Robinson, 1999, 2003).

In the case of a placebo event involving chronic pain, for example, the culture of the medical encounter is shifting (Brody, 1982). The goals of the interaction include validation of the pain, understanding, exploration, and diagnoses. In the older model of clinical consultations, the patient was more a passive recipient (the word *patient* derives from the Latin verb *pati* "to suffer"). Now, particularly in the management of chronic pain, the patient can be more of a partner and active stakeholder, who negotiates with the health care professional about shared goals, methods and strategies (Hibbard, Stockard, Mahoney, & Tusler, 2004).

Rapport is central to the placebo effects and the events that trigger them. Without confident, trusting communication, the outcome may be more nocebo than placebo (Benedetti & Amanzio, 2011; Benedetti et al., 2007; Bensing & Verheul, 2010; see Herd et al., 2012 on Parkinson's disease). Placebo effects are closely linked to conversations. And the management of the conversation, which is the business of a branch of linguistics and applied linguistics called Conversation Analysis, is crucial to the outcome. Health care professionals normally occupy the position of power in such conversations, and control key elements like who speaks when and what they say; they ask more questions, and control both the structure of the consultation, including turn taking, and the flow of information. The typical older structure involved a doctor asking questions and a patient responding. The modern consulting room in Western cultures, however, is increasingly characterized by more proactive patients. They may have explored their symptoms on the Web. They have also learnt that they can take a more proactive role, asking unprompted questions and providing unsolicited information more freely than was the case even a generation ago. But this dynamic is not typical of all cultures. Traditionally in Asia the "god-doctor" persists, and patients are still disinclined to move beyond the traditional passive role (Claramita, Nugraheni, van Dalen, & van der Vleuten, 2013; Claramita, Van Dalen, & Van Der Vleuten, 2011; Fielding & Hung, 1996; Ishikawa, Takayama, Yamazaki, Seki, & Katsumata, 2002; Kapoor, Hughes, Baldwin, & Blue, 2003; Ohtaki, Ohtaki, & Fetters, 2003). This may provoke problems in intercultural placebo events and encounters, where, even though a patient may speak good English, they may well be enacting cultural values from their homeland. Such values can include reticence, understatement, a disinclination to give a full and honest report on their pain condition, and much more. And the doctor may or may not be aware of this. Factors like eye contact, touch, body movements, and the management of interpersonal space can differ radically between cultures, and in ways which, if violated, can endanger the establishment and development of rapport. The patient's competence in English does not necessarily mean they are making "English" messages (Sussex, 2018). The same applies to the doctor, who may well have been trained overseas and may come from another cultural background. The ability of both to accommodate and to achieve common ground is something which is familiar in cultural linguistics, but has not so far attracted substantial attention in medical interactions.

In addition, there are many personal factors which affect interpersonal communication, both within and across cultures. They include age, gender (Strong et al., 2009), ethnicity, education level, and socio-economic status, as well as more individual values like confidence and assertiveness, and the degree to which the patient's approach is affected by the context and conduct of the consultation.

Much of this is not overtly known to clinicians, certainly not to patients, and not to medical educators. All, including patients, need educating in cultural and communicative issues in the placebo event. Given this complex context of communication, we need to analyze communication and language use, and their interaction with values and variables, both personal, social and cultural. We can then begin to relate aspects of language and communication in the placebo event to PLACEBO-RAPPORT mechanisms, and to psychological and therapeutic outcomes. And we can investigate the interaction of biological, social, psychological, conversational and interpersonal factors with treatments, how these factors interact to generate rapport, and how these various factors interact with different patients, and even with the same patients on different occasions. This research initiative is fundamentally interdisciplinary (see, e.g., work by the Center for Compassion and Altruism Education and Research at Stanford University; Ruchelli et al., 2014).

It might appear, following Finniss et al. (2010) and others, that if "placebo mechanisms" are present in all health care encounters, or in what we are here calling "placebo events," that we are dealing essentially with a re-badging of the medical consultation. This is a misunderstanding. While "placebo mechanisms" may occur over a whole consultation, the aspects which are focused by "placebo mechanisms" are specifically those which deal with favorable patient outcomes other than therapeutic outcomes which can be ascribed to treatments. What is important is that the factors at work within the health care encounter and a placebo event have been augmented to include contexts and conversations, and the rapport which derives from these conversations contributes critically to driving "placebo mechanisms."

6. CONCLUSION

The medical *placebo*, then, has undergone a major development since its relatively stable emergence as PLACEBO-TREATMENT from the early 19th century. Its three semantic components are: "pleasing" the patient; inert medication or procedure; and the connotation of "deception," both in the relation between the health care professional and the patient, and in the non-medical metaphorical sense of an action cynically done just to appease another person. The pejorative senses associated with PLACEBO-TREATMENT persist, and in a way which simply cannot be either ignored or canceled.

From the grammatical point of view, the word *placebo* has added a new category as an abstract, non-count noun, which now co-exists in the "inert" sense in PLACEBO-TRIAL, and in the "pleasing" sense in PLACEBO-RAPPORT. Especially PLACEBO-TRIAL, but increasingly also PLACEBO-RAPPORT, are now supported by quantitative and qualitative experimental evidence. These two terms are found predominantly in medical discourse, or in language which explicitly aligns with the new abstract usage of *placebo*. PLACEBO-TREATMENT continues in wider current usage.

The expansion of *placebo* has given rise to a number of problems. While the "deception" connotation is most associated with PLACEBO-TREATMENT and the count-noun usage, it is difficult to quarantine the "RCT inert" and "modern pleasing" senses against leakage from the earlier pejorative usage. PLACEBO-TRIAL and especially PLACEBO-RAPPORT are so close to PLACEBO-TREATMENT that some contamination from PLACEBO-TREATMENT is inevitable. This applies not only to the popular usage of *placebo*, but also to its use within medicine. Here we can anticipate that there will be a difficult road for PLACEBO-RAPPORT in particular to establish a clear acceptance outside its current research and laboratory focus, partly because it overlaps with what, for many health care professionals, is their everyday, compassionate, empathetic, considerate practice. It is conceivable that the senses of *placebo* which have been proposed for PLACEBO-RAPPORT will be functional and appropriate within that specialized community of discourse. In other words, specialists (especially academics and laboratory researchers) working in this framework will find it convenient and appropriate to use *placebo* in the sense which they have defined, and to exploit this terminology effectively for their intellectual work.

Clinicians represent a key group who will have to be persuaded. In spite of the positive "pleasing" profile of *placebo* in PLACEBO-RAPPORT, health care professionals may feel that the negative aspects of *placebo* render the word inappropriate because of the association with inert medications and procedures on the one hand, and implications of "deception" and insincerity on the other. The word *placebo*, therefore, will have substantial difficulty in shaking off the pejorative baggage of a thousand years of usage, not the least among health care professionals. Moreover, health care professionals are those who will be crucial in interpreting PLACEBO-RAPPORT to wider audiences, including their patients and a public which is becoming increasingly well-informed, and increasingly curious about, and critical of, the practice of medicine as it affects them.

6.1 Future Directions: PLACEBO-RAPPORT and the RCT

The new, elaborated and scientifically articulated, existence of PLACEBO-RAPPORT paradoxically brings into question the long and substantial research epistemology and methodology of the RCT. These trials are built on the assumption that placebos are inert, and represent a neutral baseline against which new medications and procedures can be tested. But if the context and the conduct of the health care encounter themselves—even the **fact** of a health care encounter, however, minimal—are able to contribute to therapeutic outcomes (leaving aside for the moment psychological outcomes), then what input and effect might these factors have had on a very large number of expensive and highly trusted trials of medications and procedures? These RCTs have long been treated as definitive, the "gold standard" in the evaluation of new medications and procedures (though, as we have seen, there remain questions about their ethical status, and some questions about their methodology). In addition, patients bring to such trials expectations, previous beliefs, and possibly information about placebos which may include the pejorative connotations which are found in the use of the words *deception*, *dishonesty* and *sham* in Wikipedia's entry for "placebo." It is prima facie hard to see how such facts could be confidently factored out of, or excluded from, consideration in evaluating the results of RCTs (see Cohen et al., 2018; Clayman, Smith, Fridberg, & King, 2017).

We appear to be caught between a rock and a hard place. If RCTs are a genuinely inert and reliable experimental baseline, then the psychosocial, contextual, interpersonal and cultural factors of the experimental context, however, minimal, have to be ignored. But if they are not ignored, then how can they be evaluated, and combined especially with quantitative data, in evaluating the outcomes of RCTs? Controlling for these factors in future RCTs will be hard enough. Trying to determine retrospectively what effect they might have had in past trials is a daunting prospect. The only way in which it is possible to neutralize the effects of human beliefs and mindsets, and their reaction to the health care encounter, is to administer the medication or procedure without the patient knowing. And this raises substantial problems for ethical practice.

On the other hand, establishing PLACEBO-RAPPORT as an area for rigorous research opens up substantial new questions. Not only is there already a growing literature on the biochemical patient response to aspects of PLACEBO-RAPPORT from clinical relationships, both psychologically and therapeutically; but there are major new opportunities for collaboration between clinicians, clinical researchers, pharmacologists, neurobiologists, linguists, cultural linguists, sociolinguists, psychologists, and others, to explore which aspects of the health care encounter make which contributions to therapeutic and psychological outcomes in patients. In particular, it is necessary to expand the psychosocial perspective to include cultural factors, values, expectations and individual variables, on the part of **both** the health care professional and the patient. The emerging literature on cultural values in medicine has already given sufficient grounds for us to know that this extension is both necessary and viable.

The next step belongs to the investigation of PLACEBO-RAPPORT. It will be necessary to unpack the various complex elements and dynamics of the health care encounter, to research how these interact with personal and contextual variables, and with both psychological and therapeutic outcomes from the point of view of the patient. And these will have to be analyzed in the context of the cultural and personal values which the health care professional, the patient, and the location and context, all bring to the health care encounter.

We are now at last in a position to be more specific about the nature of PLACEBO-RAPPORT. It is a collaborative, interactive process between a health care professional and patient, led by the health care professional, but with substantial input, reactive as well as proactive, from the patient. Its goal is to exploit a quality health care encounter to create a **positive expectant mindset in the patient**—close to what Kirsch (1985, 2018) called "response expectancy." This "positive expectant mindset" emerges from the health care professional, the health care encounter, the relationship between health care professional and patient, the treatment, and the outcome. If that is what constitutes the "placebo effect," then we can proceed to ask focused questions about the "placebo mechanisms" which will help to bring it about.

Whether or not this program of research and clinical application takes place under the title of PLACEBO-RAPPORT or some other appropriate term or phrase, yet to be determined, we are at the threshold of something new and potentially far-reaching.

ACKNOWLEDGMENTS

I express my warm thanks and appreciation to Luana Colloca, Susan Evans, Laura Hamilton, Frank New and Phuong Tran for their help and advice with this paper.

REFERENCES

Annoni, M. (2018). The ethics of placebo effects in clinical practice and research. In L. Colloca (Ed.), *Neurobiology of the placebo effect, part II, International review of neurobiology: Vol. 139* (pp. 463–484). New York: Academic Press.

Beecher, H. K. (1961). Surgery as placebo: A quantitative study of bias. *JAMA*, *176*(13), 1102–1107. https://doi.org/10.1001/jama.1961.63040260007008.

Benedetti, F., & Amanzio, M. (2011). The placebo response: How words and rituals change the patient's brain. *Patient Education and Counseling*, *84*(3), 413–419. https://doi.org/10.1016/j.pec.2011.04.034.

Benedetti, F., Lanotte, M., Lopiano, L., & Colloca, L. (2007). When words are painful: Unraveling the mechanisms of the nocebo effect. *Neuroscience*, *147*(2), 260–271. https://doi.org/10.1016/j.neuroscience.2007.02.020.

Bennett, G. J. (2018). Does the word "placebo" evoke a placebo response? *Pain*. https://doi.org/10.1097/j.pain.0000000000001269 [Epub ahead of print].

Bensing, J. M., & Verheul, W. (2010). The silent healer: The role of communication in placebo effects. *Patient Education and Counseling*, *80*(3), 293–299. https://doi.org/10.1016/j.pec.2010.05.033.

Brody, H. (1982). The lie that heals: The ethics of giving placebos. *Annals of Internal Medicine*, *97*(1), 112–118.

Brody, T. (2012). *Chapter 7—Placebo arm as part of clinical study design. In T. Brody (Ed.), Clinical trials: Study design, endpoints and biomarkers, drug safety, FDA and ICH guidelines* (pp. 131–141). Boston: Academic Press. https://doi.org/10.1016/B978-0-12-391911-3.00007-4.

Claramita, M., Nugraheni, M. D., van Dalen, J., & van der Vleuten, C. (2013). Doctor–patient communication in Southeast Asia: A different culture? *Advances in Health Sciences Education*, *18*(1), 15–31.

Claramita, M., Van Dalen, J., & Van Der Vleuten, C. P. (2011). Doctors in a Southeast Asian country communicate sub-optimally regardless of patients' educational background. *Patient Education and Counseling*, *85*(3), e169–e174.

Clayman, K., Smith, P., Fridberg, D., & King, A. (2017). The validity of a placebo control in an alcohol challenge paradigm and characteristics of drinkers prone to the placebo effect. *Alcoholism: Clinical and Experimental Research*, *41*(s1), 237A.

Cohen, M., Quintner, J., & van Rysewyk, S. (2018). Reconsidering the international association for the study of pain definition of pain. *Pain Reports*, *3*(2), e634. https://doi.org/10.1097/PR9.0000000000000634.

Colagiuri, B. (2010). Participant expectancies in double-blind randomized placebo-controlled trials: Potential limitations to trial validity. *Clinical Trials*, 7(3), 246–255. https://doi.org/10.1177/1740774510367916.

Colloca, L. (2018a). Responses to the sham treatment vs expectancy effects. *Pain*. https://doi:10.1097/j.pain.0000000000001266 [Epub ahead of print].

Colloca, L., & Benedetti, F. (2007). Nocebo hyperalgesia: How anxiety is turned into pain. *Current Opinion in Anaesthesiology*, *20*(5), 435–439. https://doi.org/10.1097/ACO.0b013e3282b972fb.

Colloca, L., & Howick, J. (2018). Placebos without deception: Outcomes, mechanisms, and ethics. In L. Colloca (Ed.), *Neurobiology of the placebo effect, Part I, International review of neurobiology: Vol. 138.* (pp. 21–240). New York: Academic Press. https://doi.org/10.1016/bs.irn.2018.01.005. (Epub 2018 Apr 4).

Coupland, J., Robinson, J. D., & Coupland, N. (1994). Frame negotiation in doctor-elderly patient consultations. *Discourse and Society*, *5*(1), 89–124.

Craig, K. D. (2009). The social communication model of pain. *Canadian Psychology/Psychologie Canadienne*, *50*(1), 22–32. https://doi.org/10.1037/a0014772.

Darnall, B. D., & Colloca, L. (2018). Optimizing placebo and minimizing nocebo to reduce pain, catastrophizing, and opioid use: A review of the science and an evidence-informed clinical toolkit. In L. Colloca (Ed.), *Neurobiology of the placebo effect, part II, International review of neurobiology: Vol. 139* (pp. 129–157). New York: Academic Press.

De Pascalis, V., Chiaradia, C., & Carotenuto, E. (2002). The contribution of suggestibility and expectation to placebo analgesia phenomenon in an experimental setting. *Pain*, *96*(3), 393–402. https://doi.org/10.1016/S0304-3959(01)00485-7.

Decety, J., & Ickes, W. (Eds.), (2011). *The social neuroscience of empathy.* Cambridge, MA: MIT Press.

Declaration of Helsinki IV. (1989). *World Medical Association, 41st World Medical Assembly, Hong Kong. Retrieved 21 May 2018 from* http://apps.who.int/medicinedocs/en/d/Jh2946e/6.html.

Diller, A. (1980). Cross-cultural pain semantics. *Pain*, *9*, 9–26.

Doering, B. K., Glombiewski, J. A., Rief, W. (Eds.), (2018). Expectation-focused psychotherapy to improve clinical outcomes. In L. Collaca (Ed.), *Neurobiology of the placebo effect, Part I, International review of neurobiology*: *Vol. 138* (pp. 257–270). New York: Academic Press.

Engel, G. L. (1977). The need for a new medical model: A challenge for biomedicine. *Science*, *196*, 129–136.

Engel, G. L. (1980). The clinical application of the biopsychosocial model. *American Journal of Psychiatry*, *137*, 535–544.

Farrar, J. T., Troxel, A. B., Stott, C., Duncombe, P., & Jensen, M. P. (2008). Validity, reliability, and clinical importance of change in a 0–10 numeric rating scale measure of spasticity: A post hoc analysis of a randomized, double-blind, placebo-controlled trial. *Clinical Therapeutics*, *30*(5), 974–985. https://doi.org/10.1016/j.clinthera.2008.05.011.

Fielding, R., & Hung, J. (1996). Preferences for information and involvement in decisions during cancer care among a Hong Kong Chinese population. *Psycho-Oncology*, *5*(4), 321–329.

Finniss, D. G. (2018). Placebo effects: Historical and modern evaluation. In L. Colloca (Ed.), *Neurobiology of the placebo effect, part II, International review of neurobiology: Vol. 139* (pp. 1–27). New York: Academic Press.

Finniss, D. G., Kaptchuk, T. J., Miller, J., & Benedetti, F. (2010). Biological, clinical, and ethical advances of placebo effects. *The Lancet*, *375*(9715), 686–695. Retrieved 16 April 2018, from http://www.thelancet.com/journals/lancet/article/PIIS0140-6736(09)61706-2/abstract.

Fordyce, W. (1988). Pain and suffering: A reappraisal. *American Psychologist*, *43*, 276–283.

Gask, L. (2018). In defence of the biopsychosocial model. *The Lancet. Psychiatry*, *5*, 548–549. https://doi.org/10.1016/S2215-0366(18)30165-2.

Gaston-Johansson, F., Albert, M., Fagan, E., & Zimmerman, L. (1990). Similarities in pain descriptions of four different ethnic-culture groups. *Journal of Pain and Symptom Management*, *5*(2), 94–100. https://doi.org/10.1016/S0885-3924(05)80022-3.

Goubert, L., Craig, K. D., & Buysse, A. (2011). Perceiving others in pain: Experimental and clinical evidence on the role of empathy. In J. Decety & W. Ickes (Eds.), *The social neuroscience of empathy* (pp. 153–165). Cambridge, MA: MIT Press.

Gracely, R. H., Dubner, R., Deeter, W. R., & Wolskee, P. (1985). Clinicians' expectations influence placebo analgesia. *The Lancet*, *325*(8419), 43. https://doi.org/10.1016/S0140-6736(85)90984-5.

Harris, I. (2016). *Surgery, the ultimate placebo*. Sydney: University of New South Wales Press.
Herd, C. P., Tomlinson, C. L., Deane, K. H., Brady, M. C., Smith, C. H., Sackley, C. M., et al. (2012). *Speech and language therapy versus placebo or no intervention for speech problems in Parkinson's disease. In Cochrane Library*: John Wiley & Sons, Ltd. https://doi.org/10.1002/14651858.CD002812.pub2.
Heritage, J., & Clayman, S. (2010). *Talk in action: Interactions, identities, and institutions*. Walden, MA: Wiley-Blackwell.
Hibbard, J. H., Stockard, J., Mahoney, E. R., & Tusler, M. (2004). Development of the Patient Activation Measure (PAM): Conceptualizing and measuring activation in patients and consumers. *Health Services Research, 39*(4p1), 1005–1026. https://doi.org/10.1111/j.1475-6773.2004.00269.x.
Hróbjartsson, A., & Gøtzsche, P. C. (2001). Is the placebo powerless? *New England Journal of Medicine, 344*(21), 1594–1602. https://doi.org/10.1056/NEJM200105243442106.
Incayawar, M., & Todd, K. H. (2013). *Culture, brain, and analgesia: Understanding and managing pain in diverse populations*. Oxford, New York: Oxford University Press.
Ishikawa, H., Takayama, T., Yamazaki, Y., Seki, Y., & Katsumata, N. (2002). Physician–patient communication and patient satisfaction in Japanese cancer consultations. *Social Science & Medicine, 55*(2), 301–311.
Jamshidian, F., Hubbard, A. E., & Jewell, N. P. (2014). Accounting for perception, placebo and unmasking effects in estimating treatment effects in randomised clinical trials. *Statistical Methods In Medical Research, 23*(3), 293–307. https://doi.org/10.1177/0962280211413449.
Justman, S. (2013). Deceit and transparency in placebo research. *The Yale Journal of Biology and Medicine, 86*(3), 323–331.
Kapoor, S., Hughes, P. C., Baldwin, J. R., & Blue, J. (2003). The relationship of individualism–collectivism and self-construals to communication styles in India and the United States. *International Journal of Intercultural Relations, 27*(6), 683–700.
Kaptchuk, T. J. (1998). Intentional ignorance: A history of blind assessment and placebo controls in medicine. *Bulletin of the History of Medicine, 72*(3), 389–433. https://doi.org/10.1353/bhm.1998.0159.
Kaptchuk, T. J., Friedlander, E., Kelley, J. M., Sanchez, M. N., Kokkotou, E., Singer, J. P., et al. (2010). Placebos without deception: A randomized controlled trial in irritable bowel syndrome. *PLoS One, 5*(12), e15591. https://doi.org/10.1371/journal.pone.0015591.
Kazarian, S. S., & Evans, D. R. (1998). Cultural clinical psychology. In S. S. Kazarian & D. R. Evans (Eds.), *Cultural clinical psychology: Theory, research, and practice, 1998* (pp. 3–38). New York: Oxford University Press.
Kirsch, I. (1985). Response expectancy as a determinant of experience and behavior. *American Psychologist, 40*(11), 1189–1202. https://doi.org/10.1037/0003-066X.40.11.1189.
Kirsch, I. (2018). Response expectancy and the placebo effect. In L. Colloca (Ed.), *Neurobiology of the placebo effect, part I, International review of neurobiology: Vol. 138* (pp. 81–93). New York: Academic Press.
Klinger, R., Stuhlreyer, J., Schwartz, M., Schmitz, J., & Colloca, L. (2018). Clinical use of placebo effects in patients with pain disorders. In L. Colloca (Ed.), *Neurobiology of the placebo effect, part II, International review of neurobiology: Vol. 139* (pp. 107–128). New York: Academic Press.
Lang, E. V., Hatsiopoulou, O., Koch, T., Berbaum, K., Lutgendorf, S., Kettenmann, E., et al. (2005). Can words hurt? Patient–provider interactions during invasive procedures. *Pain, 114*(1), 303–309. https://doi.org/10.1016/j.pain.2004.12.028.
Leknes, C., Berna, D., Lee, D., Snyder, D., Biele, D., & Tracey, D. (2013). The importance of context: When relative relief renders pain pleasant. *Pain, 154*(3), 402–410. https://doi.org/10.1016/j.pain.2012.11.018.

Linde, K., Witt, C. M., Streng, A., Weidenhammer, W., Wagenpfeil, S., Brinkhaus, B., ... Melchart, D. (2007). The impact of patient expectations on outcomes in four randomized controlled trials of acupuncture in patients with chronic pain. *Pain, 128*(3), 264–271. https://doi.org/10.1016/j.pain.2006.12.006.

Loeser, J. (1982). Concepts of pain. In M. Stanton-Hicks & R. Boas (Eds.), *Chronic low-back pain* (pp. 145–148). New York: Raven Press.

Lund, I., Näslund, J., & Lundeberg, T. (2009). Minimal acupuncture is not a valid placebo control in randomised controlled trials of acupuncture: A physiologist's perspective. *Chinese Medicine, 4*(1). https://doi.org/10.1186/1749-8546-4-1.

Main, C. J. (2016). Pain assessment in context: A state of the science review of the McGill pain questionnaire 40 years on. *Pain, 157*(7), 1387–1399.

Meissner, K., & Linde, K. (2018). Are blue pills better than green? How treatment features modulate placebo effects. In L. Colloca (Ed.), *Neurobiology of the placebo effect, part II, International review of neurobiology: Vol. 139* (pp. 357–378). New York: Academic Press.

Moerman, D. E. (2002). *Meaning, medicine and the 'placebo effect'*. Cambridge, New York: Cambridge University Press.

Necka, E. A., & Atlas, L. Y. (2018). The role of social and interpersonal factors in placebo analgesia. In L. Colloca (Ed.), *Neurobiology of the placebo effect, Part I, International review of neurobiology: Vol. 138* (pp. 161–180). New York: Academic Press.

Ohtaki, S., Ohtaki, T., & Fetters, M. D. (2003). Doctor–patient communication: A comparison of the USA and Japan. *Family Practice, 20*(3), 276–282.

Power, S., Bogossian, F. E., Sussex, R., & Strong, J. (2017). A critical and interpretive literature review of birthing women's non-elicited pain language. *Women and Birth, 30*(5), e227–e241. https://doi.org/10.1016/j.wombi.2017.02.001.

Price, D. D., Finniss, D. G., & Benedetti, F. (2008). A comprehensive review of the placebo effect: Recent advances and current thought. *Annual Review of Psychology, 59*, 565–590. https://doi.org/10.1146/annurev.psych.59.113006.095941.

Probst, P., Grummich, K., Harnoss, J. C., Hüttner, F. J., Jensen, K., Braun, S., et al. (2016). Placebo-controlled trials in surgery: A systematic review and meta-analysis. *Medicine, 95*(17), e3516. https://doi.org/10.1097/MD.0000000000003516.

Purnell, L. D., & Paulanka, B. J. (1998). *Transcultural health care: A culturally competent approach.* Philadelphia: F.A. Davis.

Reicherts, P., Gerdes, A. B. M., Pauli, P., & Wieser, M. J. (2016). Psychological placebo and nocebo effects on pain rely on expectation and previous experience. *The Journal of Pain, 17*(2), 203–214. https://doi.org/10.1016/j.jpain.2015.10.010.

Richter, M., Eck, J., Straube, T., Miltner, W. H. R., & Weiss, T. (2010). Do words hurt? Brain activation during the processing of pain-related words. *Pain, 148*(2), 198–205. https://doi.org/10.1016/j.pain.2009.08.009.

Robinson, J. (1999). *The organization of action and activity in primary-care, doctor-patient consultations.* Los Angeles, CA: University of California. PhD Dissertation.

Robinson, J. D. (2003). An interactional structure of medical activities during acute visits and its implications for patients' participation. *Health Communication, 15*(1), 27–59.

Rothman, K. J., & Michels, K. B. (1994). The continuing unethical use of placebo controls. *New England Journal of Medicine, 331*(6), 394–398.

Ruchelli, G., Chapin, H., Darnall, B., Seppala, E., Doty, J., & Mackey, S. (2014). Compassion meditation training for people living with chronic pain and their significant others: A pilot study and mixed-methods analysis. *The Journal of Pain, 15*(4 Suppl), S117.

Sanders, L. (2017). Rebranding placebos. *Knowable Magazine*. Retrieved 12 May 2018 from https://www.knowablemagazine.org/article/mind/2017/rebranding-placebos.

Serres, G. D., Skowronski, D. M., Wu, X. W., & Ambrose, C. S. (2013). The test-negative design: validity, accuracy and precision of vaccine efficacy estimates compared to the gold standard of randomised placebo-controlled clinical trials. *Eurosurveillance, 18*(37). https://doi.org/10.2807/1560-7917.ES2013.18.37.20585.

Shapiro, A. K. (1964). A historic and heuristic definition of the placebo. *Psychiatry*, *27*, 52–58.
Shapiro, A. K. (1968). Semantics of the placebo. *Psychiatric Quarterly*, *42*(4), 653–695. https://doi.org/10.1007/BF01564309.
Shapiro, A. K., & Struening, E. (1973). Defensiveness in the definition of placebo. *Comprehensive Psychiatry*, *14*(2), 107–120.
Shapiro, A. K., Struening, E. L., & Shapiro, E. (1979). The reliability and validity of a placebo test. *Journal of Psychiatric Research*, *15*(4), 253–290.
Strong, J., Mathews, T., Sussex, R., New, F., Hoey, S., & Mitchell, G. (2009). Pain language and gender differences when describing a past pain event. *Pain*, *14*, 86–95.
Sussex, R. (2009). The language of pain in applied linguistics. Review article of Chryssoula Lascaratou, *The language of pain*. Amsterdam and Philadelphia: John Benjamins, 2007. *Australian Review of Applied Linguistics*, *32*(1) 6:1–6:14.
Sussex, R. (2015). *How different cultures experience and talk about pain. In The Conversation*. 15 December 2015. https://theconversation.com/how-different-cultures-experience-and-talk-about-pain-49046. Also published in BodyinMind, 1 January 2016. http://www.bodyinmind.org/cultures-pain/.
Sussex, R. (2018). Pain as an issue of intercultural communication. In A. Curtis & R. Sussex (Eds.), *Intercultural communication in Asia: Education, language and values* (pp. 181–204). Berlin and London: Springer Verlag.
ten Have, P. (2006). On the interactive constitution of medical encounters. *Revue Française de Linguistique Appliquée*, *11*(2), 85–98.
Turk, D. C. (1996). Biopsychosocial perspectives on chronic pain. In R. J. Gatchel (Ed.), *Psychological approaches to pain management: A practitioner's handbook* (pp. 3–32). New York, NY: Guildford Press.
Turner, J. A., Deyo, R. A., Loeser, J. D., von Korff, M., & Fordyce, W. E. (1994). The importance of placebo effects in pain treatment and research. *JAMA*, *271*(20), 1609–1614. https://doi.org/10.1001/jama.1994.03510440069036.
Vance, E. (2016). *Suggestible you: The curious science of your brain's ability to deceive, transform, and heal*. National Geographic. Retrieved 26 May 2018 from https://www.amazon.com/Suggestible-You-Curious-Science-Transform/dp/1426217897.
Varelmann, D., Pancaro, C., Cappiello, E. C., & Camann, W. R. (2010). Nocebo-induced hyperalgesia during local anesthetic injection. *Anesthesia and Analgesia*, *110*(3), 868. https://doi.org/10.1213/ANE.0b013e3181cc5727.
Von Bertalanffy, L. (1968). *General system theory: Essays on its foundation and development*. New York: George Brazillier.
White, P., Lewith, G., Hopwood, V., & Prescott, P. (2003). The placebo needle, is it a valid and convincing placebo for use in acupuncture trials? A randomised, single-blind, cross-over pilot trial. *Pain*, *106*(3), 401–409. https://doi.org/10.1016/j.pain.2003.08.013.
Wickramasekera, I. (2001). The placebo efficacy study: Problems with the definition of the placebo and the mechanisms of placebo efficacy. *Advances in Mind-Body Medicine*, *17*(4), 309–312.
Wordsworth, H. (2018). *The history of 'biopsychosocial' pain—A tale of gladiators, war, papal doctrine and a wrestler. Retrieved 8 May 2018 from* https://www.aagbi.org/sites/default/files/WORDSWORTH_anaes_history_prize.pdf.
Zborowski, M. (1969). *People in pain*. San Francisco: Jossey-Bass.
Zion, S. R., & Crum, A. (2018). Mindsets matter: A new framework for harnessing the placebo effect in modern medicine. In L. Colloca (Ed.), *Neurobiology of the placebo effect, Part I, International review of neurobiology: Vol. 138* (pp. 137–160). New York: Academic Press.

FURTHER READING

Colloca, L. (Ed.), (2018b). *Neurobiology of the placebo effect. Part I: Vol. 138* (pp. 1–357). New York: Academic Press.

CHAPTER FOUR

Knowledge Gaps in Placebo Research: With Special Reference to Neurobiology

Paul Enck*[,1], Bjoern Horing[†], Ellen Broelz*, Katja Weimer[‡]
*Department of Internal Medicine VI: Psychosomatic Medicine and Psychotherapy, University Hospital Tübingen, Tübingen, Germany
[†]Institute of Systems Neuroscience, University Medical Center Hamburg-Eppendorf, Hamburg, Germany
[‡]Clinic for Psychosomatic Medicine and Psychotherapy, University Hospital Ulm, Ulm, Germany
[1]Corresponding author: e-mail address: paul.enck@uni-tuebingen.de

Contents

Abstract

This article addresses different ways to identify knowledge gaps in placebo research. Following a short description of the history of placebo research and our contributions, we describe the creation of the Journal of Interdisciplinary Placebo Studies (JIPS) literature database and compare its content (on a meta-level) with two other approaches to survey the current state of the art of placebo research: a bibliometric analysis of a (limited) selection of placebo papers, and the analysis of the content of all abstracts submitted to a 2017 placebo conference. We summarize this discussion with a critical assessment of these approaches and a perspective of future approaches in such placebo "meta"-analyses.

International Review of Neurobiology, Volume 139
ISSN 0074-7742
https://doi.org/10.1016/bs.irn.2018.07.018

1. INTRODUCTION

This article addresses different ways to identify knowledge gaps in placebo research, starting with the description of how—from our perspective—placebo research and our participation in it started, how we created the current literature database, and how one can make use of it. This is followed by a section comparing the database content with two other approaches, namely a bibliometric analysis of a (limited) selection of placebo papers, and an analysis of the content of all abstracts submitted to a 2017 placebo conference. We will close with a critical assessment of these approaches and a perspective of future approaches in such placebo "meta"-analyses.

1.1 How to Start Doing Placebo Research (2003)

During an international medical conference in Tübingen in 2003, the lead author of this chapter (PE) was attending a lecture on placebo effects, and realized that the average knowledge of physicians and even clinical researchers on the mechanisms governing the placebo effect (as seen in placebo arms of randomized controlled trials, RCTs) is rather poor, and is often restricted to what has been taught in medicine, e.g., in pharmacological textbooks, during their medical university training, 20 years ago and longer. With subsequent reviewing of the scientific literature about the placebo effect (see below) another issue became evident: what we and others did in medical subspecialties such as behavioral medicine, psychophysiology, psychoimmunology, stress research, psychotherapy research, and alike, namely investigating psychological mechanisms governing physiological responses, to some degree included phenomena that others had labeled as placebo effects, especially in the tradition of associative learning of physiological responses, and the role of cognitive settings (e.g., expectations) on the outcome of (medical) interventions. In fact, since the mid 1980s we and others had investigated conditioning of nausea and vomiting in the context of anticipatory nausea in cancer chemotherapy, but without explicitly calling it placebo research (Stockhorst, Enck, & Klosterhalfen, 2007). From here it was only a small step to link our actual research to this tradition.

1.2 A Research Discipline Is Born: Placebo Research (1995)

Until the 1990s, dedicated placebo research barely existed, and the term placebo occurred (outside of its mentioning in placebo-controlled RCT) in only a few papers every year across all medical subspecialties (see Fig. 1 in Weimer, Colloca, & Enck, 2015a). While in some of them, e.g., in psychiatry (Loranger, Prout, & White, 1961; Rickels, Lipman, & Raab, 1966), it had been noted quite early that clinical responses among recipients of placebo pills in RCT may yield high improvement, this was not resulting in systematic investigations of the underlying mechanisms, but was rather attributed to patient characteristics, disease history and the design of clinical trials (Loranger et al., 1961). Consequently, attempts were made to improve designs and patient selection to minimize these placebo responses. Whether or not the placebo response was seen as "powerful" in the tradition of Beecher (1955) and Wolf (1959), or as a nuisance necessary to overcome in the interest of better drug therapies, was more a question of belief than an issue of serious investigations. And only with the rise of neurophysiological techniques such as brain imaging this has changed, as it allowed to explore the neurophysiological underpinning of this phenomenon, and placebo research has gained acceptance in many clinical areas — this may be indicated by rising publications in the biomedical literature (see Fig. 2) as well as with increasing popular science books and publications in the mid 1990s. Interestingly, the rise in placebo research is preceded by a rising usage of the word "placebo" in the popular literature, especially in books, much earlier and already in the 1960s (Bennett, 2018), but only in the English speaking world.

1.3 Looking Forward Requires Looking Back

With the decision to join this newly established research agenda, it appeared necessary to us to not only give our own research tradition (learning of physiological responses) a new label (placebo research) but also to look back on what had been achieved and was state of the art of placebo research in 2003. While the historical origins of the term placebo and its use in medicine in RCT and beyond has been well described by others (Jütte, 2013; Kaptchuk, 1998; Shorter, 2011; Wartolowska, Beard, & Carr, 2018), the actual knowledge was distributed across many medical subspecialties and rather sparsely inter-connected, with dominance in the areas of psychiatry, neurology, and internal medicine. Especially pain research was actively investigating placebo analgesia, not the least because of its easier experimental applicability in the laboratory and with healthy volunteers. Up until today, placebo analgesia has remained the best studied condition used to explore the mechanisms behind the placebo effect.

In an attempt to comprehensively screen the published literature for relevant papers focusing on the placebo effect/placebo response, a PubMed search using the single search term "placebo" was conducted, resulting in >100,000 papers in early 2004. The title and abstracts of these papers were reviewed retrospectively (at a frequency of maximally 1000 per day, 7 days a week) by one of us (PE), to identify the approximately 1% of all papers relevant for placebo research. These papers were stored in an *Endnote*-like database, respective PDFs were collected, and made available to the local working group in Tübingen. A few years later, a similar search using the term "nocebo" was added.

At the same time (2004), a *prospective* PubMed search was started that resulted in weekly reports of newly published papers with either of the two terms (on average 200 per week altogether) and again screened for relevance for placebo research; the 1% outcome has increased to about 2% over the years. While the retrospective review lasted for about 1 year, the prospective search is still ongoing. Both resulted in a database of approximately 4000 papers (data papers, meta-analyses, systematic reviews, reviews, commentaries and a limited number of letters) as of mid 2018. Its organization is in the hands of the authors of this paper and follows a simple but effective workflow routine (Fig. 1).

1.4 Making Use of the Database

While the initial plan for the database was merely practical (having local access to the most relevant publications for the Tübingen placebo research team and for new students entering the field), but maybe also reflected a

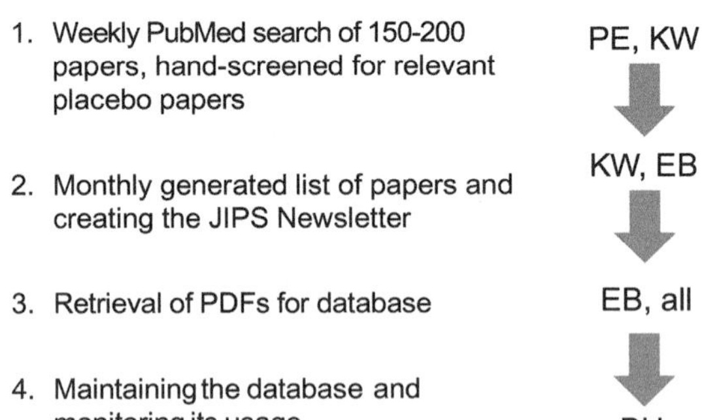

Fig. 1 Workflow for maintenance of the placebo literature database (initials are those of the authors of this article).

compulsive attitude of some of us, it soon became evident that the efforts that were associated with it deserved a wider distribution and recognition.

- A first chance to accomplish this was seen with the foundation and establishing of the Society of Interdisciplinary Placebo Studies (SIPS) in Hermance near Geneva, Switzerland in 2014, until 2018 with approximately 150 registered members. Access to the database (but not to the PDFs of the papers) and distribution of a monthly newsletter of new placebo research papers (Journal of Interdisciplinary Placebo Studies, JIPS) was offered to all members of SIPS on a voluntary basis, and free of charge (see https://jips.online). More subscriptions were achieved after the first SIPS conference held in Leiden, The Netherlands, in 2017. The open access has meanwhile been extended to other interested scientists, e.g., via the Special Interest Group (SIG) *Placebo* of the International Associated for the Study of Pain (IASP). The current number of newsletter subscriptions is around 240.
- Highly specialized literature databases such as ours which have grown to an extent that any individual is not able to overlook anymore, can be analyzed by bibliometric approaches, to identify topics, cluster of topics, scientists, collaborations, networks, medical specialties involved, and journals. It also identifies the origin of a discussion ("footprints") (Park et al., 2017), the most frequent citations ("citation classics") (Chuang & Ho, 2014), forgotten papers ("sleeping beauties") (Zavrsnik, Kokol, Del Torso, & Blazun Vosner, 2016), and other characteristics of the body of knowledge. Bibliometric approaches can also be used to identify gaps in knowledge, raise new topics, and analyze the reception of knowledge (and its change) in the course of time (Eysenbach, 2011; Marashi et al., 2013; Perneger, 2015; van Wesel, 2016). Bibliometry provides standard tools and techniques for analysis and visualization of results. We are currently collaborating with a bibliometric institute and the Humboldt University Berlin, Germany, to establish such an analysis, but are still facing some technical problems to be solved. In the final part of this manuscript, we will discuss a different bibliometric approach to elucidate the changed usage of the term placebo, as addressed recently by Bennett (2018).
- Research projects concentrated on measuring treatment efficacy of specific therapies for specific diseases in medicine, such as by the Cochrane Collaboration, have identified a major problem with current practice of updating through rewriting systematic reviews and meta-analyses with every new RCT published, since the number of RCTs is increasing

every year worldwide, and this questions the value of printed reviews and meta-analyses right from the beginning (Elliott et al., 2017). Out of this dilemma, the idea of "living systematic reviews" was born "to reduce waste in research" (Vandvik, Brignardello-Petersen, & Guyatt, 2016) and "narrow the evidence–practice gap" (Elliott et al., 2014): Once a search algorithm of the scientific literature is established and the basic search has been performed, it may be possible to perform this search on a regular (monthly, weekly, daily, continuous) basis, to add the resulting new RCTs at any stage to the existing collection, re-run the computation of the treatment efficacy (meta-analysis) and provide an update of the resulting evidence to the interested scientific public with the help of electronic online media.

In a series of four papers, the Journal of Clinical Epidemiology has outlined the necessary as well as further optional methodological steps to inaugurate such a strategy (Akl et al., 2017; Elliott et al., 2017), including statistical approaches (Simmonds, Salanti, McKenzie, Elliott, & the Living Systematic Review Network, 2017) and the use of artificial intelligence and machine learning (Thomas et al., 2017). An accompanying editorial (Sutton, 2017) outlines how this could become a method not only for the analysis of treatment trials but for clinical research in general, and in a letter to the editor we have outlined how basic science research may profit from such as strategy as well, with the example of placebo research (Enck, 2018).

1.5 Between Quantitative and Qualitative Research

The present analysis of the placebo literature is our second effort to review the state-of-the-art, to identify knowledge gaps and search for alternative routes of inquiry, after a first one in the past (Enck, Klosterhalfen, & Weimer, 2017). It meanders between qualitative and quantitative analyses techniques, it criticizes a first bibliometric paper in placebo research, and takes a deeper look at all abstracts of the first SIPS meeting 2017, before the backdrop of all placebo papers in our database, to reflect changes and developments.

Neither of these approaches is completely satisfactory in itself, but as an assembly they may provide some insights into the short history of placebo research. Our attempt is pragmatic rather than systematic: Reviewing where the field is coming from and moving to, and providing assistance to establish a research discipline for "harnessing the placebo effect" (Miller & Brody, 2011) at a meta-level, rather than in clinical practice.

To identify knowledge gaps in any area of research, it is required to assess the field as completely as possible. This is even more necessary in fields with a very large body of knowledge, or where contributions are made from very different fields of research, as is the case in placebo research. Only a few tools allow such an assessment—meta-analyses are one example, bibliometric analyses another.

In this article, we are covering placebo research in general, with a focus on its neurobiological aspects. To do so, we approach the field from three different points of view by reviewing (1) a recently (2016) published bibliometric analysis of placebo papers (Boehm, Berger, Ostermann, & Heusser, 2016); (2) all papers in the Journal of Interdisciplinary Placebo Studies (JIPS) database (as of December 2016) (https://jips.online); (3) all abstracts presented at the first Society of Interdisciplinary Placebo Studies (SIPS) conference in Leiden, The Netherlands (https://sipsconference.com) from April 2 to 4, 2017.

All three approaches use sources of different type and complexity and are therefore not directly comparable. The bibliometric analysis is based on 301 papers, selected from open sources such as PubMed following a search strategy vaguely described, but the papers included are neither listed nor made available, therefore remain undisclosed (Boehm et al., 2016). The JIPS database was made available to the scientific community by the authors in 2016. It currently (July 2018) contains nearly 4000 papers pertinent to placebo research and is fully searchable for title, authors, abstract, journals, MESH terms, etc. The 200 SIPS conference abstracts were published as conference material; they include abstracts of oral presentations of invited speakers and state-of-the-art lectures as well as abstracts of oral and poster presentations, with thus far mostly unpublished material.

1.6 The Bibliometric Approach

The authors of this (first) bibliometric analysis of the placebo literature included only 301 papers and based it on a search algorithm not completely transparent: as Boehm et al. (2016) state, "*an already existing and published search strategy was used which was adapted from a protocol published by the Cochrane Collaboration regarding 'The effect of varying practitioner communication on patients' health status and treatment outcomes'*" (Boehm et al., 2016, p. 1), referring to a paper by Verheul et al. (2010) that would include the algorithm in an appendix. However, this paper was withdrawn from publication in 2014 "*because a new protocol is being developed by a new team.*" (http://cochranelibrary-wiley.com/doi/10.1002/14651858.CD008791.pub2/full). This new protocol has not been published so far.

Their search included 61 search terms around a topic closely associated with the placebo response in medicine, namely patient–doctor communication and its relevance for health care outcome (see Verheul et al., 2010, appendix). The search was conducted in November 2011. The selection process and the resulting 301 papers are illustrated in a PRISMA graph; unfortunately, a list of the papers is not provided, e.g., as complementary material. The authors refer to a future paper not yet published.

They present parts of their bibliometric analysis that subdivided the papers into their origin (USA, Europe), year of publication (before and after 2000), medical condition analyzed (healthy, depression, various, hypertension, migraine, irritable bowel syndrome, panic disorder), "main focus" (placebo effects, placebo response, placebo responder, placebo effect/expectation, placebo hypoalgesia, placebo effect/expectancy, and placebo effect/placebo response), journal type (specific condition, pain, psychiatry, psychopharmacology, neurology, general medicine, research), and publication type (meta-analysis, systematic review, randomized controlled trial, clinical controlled trial, uncontrolled trial). Neither of these categories showed any significant difference between Europe- and USA-based papers, and between papers before and after the year 2000, except that the publication type was significantly different between earlier and later published papers, with more meta-analyses being published after 2000.

Surprisingly, the authors provide some rather strong conclusions from their data (Boehm et al., 2016, p. 3f):

1. "*A significant shift has occurred for conditions that have been investigated, the publication and journal type and the main focus of the articles from before to after the year 2000*," with significance only found for the increased number of meta-analyses after 2000.
2. "*Only a few research disciplines cover the field of placebo effects*" (listed are—directly or indirectly—psychiatry, neurology, internal medicine).
3. "*The research field is shrinking as evident by the smaller number of researchers publishing in this field*" (no data provided).
4. "*Increasing concentration of placebo effects [is] being reproduced in experimental settings, namely in randomised controlled trials*" (with an increase from 30% to 39%).

Major concerns can be expressed regarding each of the respective conclusions. Since the included papers are not listed, their selection may or may not include a bias based on their search algorithm. In either case, after contextualizing the findings with the JIPS database (see below), it becomes clear that these 122 and 179 papers before and after 2000, respectively, are a mere fraction of all the papers on the topic: 122 (out of a total of 1057 JIPS-listed papers) published before 2000 constitute just 11.5%, 179 (out of 1303 JIPS-listed papers) published between 2001 and 2011 constitute just 13.5%. Any hard conclusion drawn from this subset is questionable: For example, concerning the distribution of papers from Europe and the United States, one must attest that—at least after 2000—most papers are of European origin.

In an attempt to cover some aspects of the placebo response such as age and sex contributions, we have analyzed a total of 75 meta-analyses across all subspecialties in medicine and have shown (Elsenbruch & Enck, 2015; Enck, Bingel, Schedlowski, & Rief, 2013; Schedlowski, Enck, Rief, & Bingel, 2015; Weimer et al., 2015a, 2013; Weimer, Colloca, & Enck, 2015b) that placebo research has been conducted in *all* areas of medicine even before the year 2000 (see Tables 1–6 in Weimer et al., 2015a), with an increase in the number of publications in the recent decade. Indeed, and with some dominance of psychiatry, neurology and the neurosciences (especially in relation to depression and pain) and various subspecialties of internal medicine.

It is implausible that the steadily increasing number of placebo papers (see Fig. 2) is published by a shrinking community of researchers; this is also evident from the growing number of active delegates participating in placebo-themed conferences held over the years, with >200 active researchers present at the most recent conference in Leiden, The Netherlands. Therefore, the alleged decline in the number of placebo researchers may simply be due to selection bias.

Boehm et al. appear to operate with an incomplete notion of RCTs, as this term usually refers to placebo-controlled trials of novel (or existing) medicines in medical routine and not experimental settings. Every year, roughly 30,000 RCTs are conducted worldwide; however, RCTs usually do *not* provide insight into the size and/or the mechanisms of the placebo response or placebo effect. This is only possible with experimental (and mostly laboratory) studies and the inclusion of a no-treatment control arm; the latter is usually unacceptable under clinical conditions. Indeed, a clear increase in experimental laboratory studies (with patient and/or controls) has been noted over the years.

It is therefore not surprising that the most relevant changes that have occurred in placebo research are not being identified. First and foremost, the implementation of brain imaging technology in experimental placebo research started in 1995 (see below), and marked the beginning of

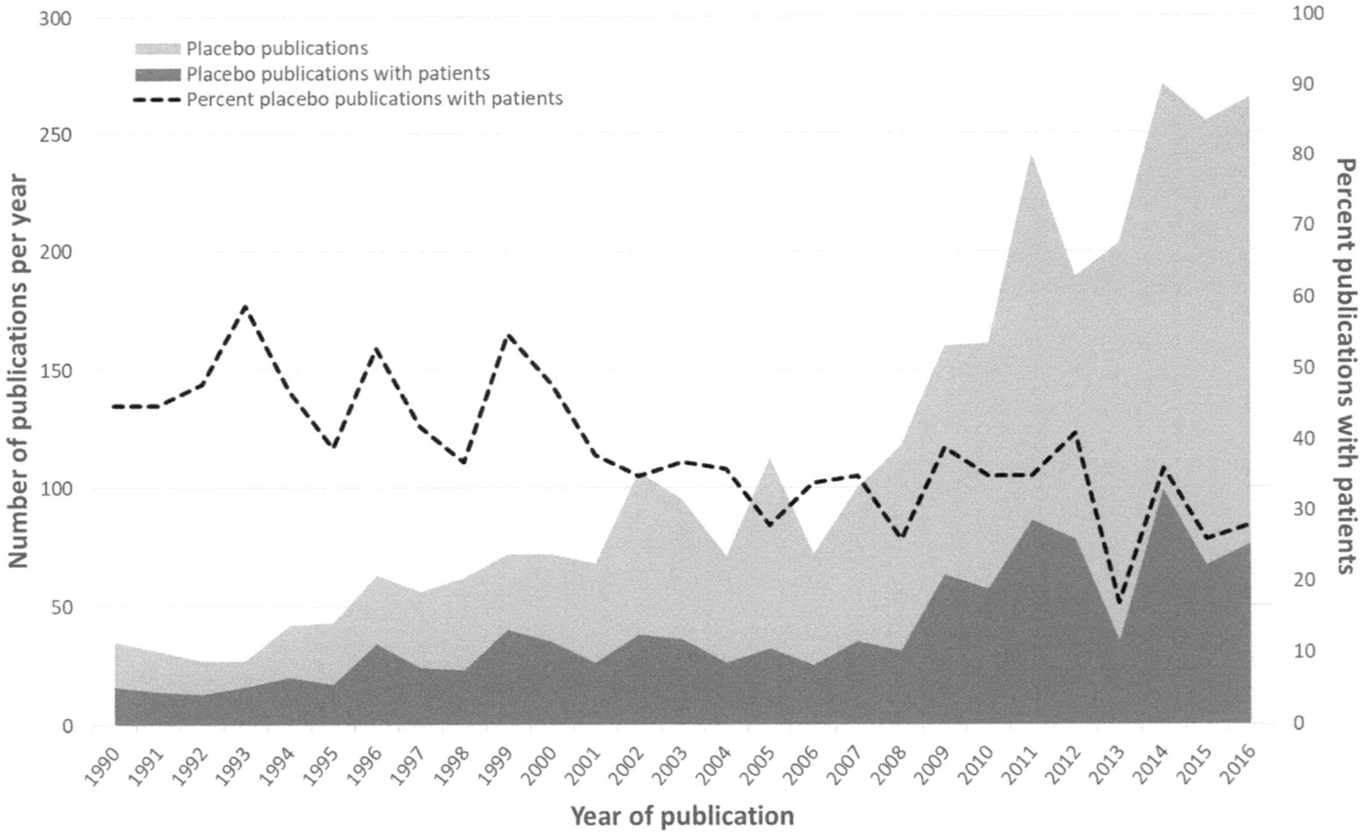

Fig. 2 Absolute numbers (per year) of all placebo studies and studies involving patients (left axis), as well as percentage of studies involving patients in the JIPS database (right axis), at the end of 2016. Note that while absolute numbers are increasing, the relative proportion of patient studies decreased from 50% to 25% over the 15 years covered.

investigation of the neurobiology of the placebo response in health and disease. Among others, this line of research included depression, anxiety, Parkinson's disease, as well as chronic and acute somatic and visceral pain. Further progress is marked by the utilization of human genome analyses (Furmark et al., 2008; Leuchter, McCracken, Hunter, Cook, & Alpert, 2009) that likewise falls into the period covered by the bibliometric analysis but is not being reported. While we acknowledge that the focus of the search algorithm of Boehm et al. (2016) was the effect of doctor communication on patients' health status and treatment outcomes, the overly broad title misleadingly suggests a much more comprehensive approach.

1.7 The JIPS Database

As outlined above, the JIPS database is an attempt to assemble all papers pertinent to placebo research in a database made accessible to all placebo researchers by the authors of this paper. The genesis of the JIPS database has been described elsewhere (Colloca, Enck, & DeGrazia, 2016; Enck, 2018) as well as shortly above. Rather than using a complex search algorithm, e.g., to include experimental studies but to exclude RCTs that only use placebos as control condition (which provides no further insight into placebo mechanisms), the broad search term "placebo" was used on the PubMed meta-database. Searches are supplemented by hand-searched papers from other publications, and other sources. The database currently contains approximately 4000 papers of all kinds.

The most recent use of the database was to conduct systematic reviews of specific areas of the field (Colloca et al., 2016). Taking a broader perspective, we have employed it to identify knowledge gaps in placebo research, e.g., related to food and nutrition trials, sport interventions and physical therapies, and to ethical, legal and cultural aspects of the placebo phenomenon; similarly, we have determined that eHealth and mHealth applications of the placebo response/effects are clearly underrepresented, as compared to their current and future relevance (Enck et al., 2017). In the following, we will focus on its neurobiological content, which is the overarching topic of all contributions to this book.

While the number of articles per year was low and stable until the mid 1990s, the largest relative increase occurred around 1995, with an exponential increase since. This coincides with the first papers including the terms "neurophysiology," "neurobiology," or "imaging." Of these, about 200 have been published until the end of 2016, amounting to around 10% of all placebo papers in this period.

The first studies using brain recording of the placebo effect were EEG studies in 1983 (Irwin & Fink, 1983; Levy & Jankovic, 1983), but it was not until the 1990s when brain imaging technology became the dominant neuroscience tool that initiated the first PET (Schmidt et al., 1996) in 1996, and the first fMRI study in 1999 (Kleinschmidt et al., 1999). From 2000 onward, the number of neurobiological papers increased constantly, up to 24 in 2013 and 2014; the dominant research tool became fMRI (80 papers). Nearly half of the publications were data papers and the other half review-type papers, with occasional meta-analyses. The first study recording from single brain cells in Parkinson's disease was published in 2004 (Benedetti et al., 2004), the first study investigating the genetics of the placebo response was published in 2008 (Furmark et al., 2008). The data shows a moderate decline of the number of papers in 2015 (10 papers) and 2016 (15 papers), but it seems to be too early to posit a trend.

These approximately 200 placebo papers are published in nearly 120 different scientific journals covering many areas of medicine, however with a focus on neuroscience and related areas (*Pain*: 26 papers; *Journal of Neuroscience*: 13 papers; *NeuroImage*: 11 papers; *Science*: 5 papers; various *Nature* journals: 7 papers). A bibliometric count of the authors would yield a long listing, but the ranking of the authors with the largest contribution to the database (for the topics neurobiology/neuroscience/neuroimaging) is led by Fabrizio *Benedetti* (17 papers), Ulrike *Bingel* (14 papers) and Jon-Kar *Zubieta* (12 papers).

Categorizing JIPS-listed papers by content allows us to identify blind spots:

- Of all 3500 papers [as of December 2016], less than half (1586) mention the terms "patient" or "client"; of 210 (out of 3500) including "neurobiology"/"neuroscience"/"neuroimaging" as search terms, about one third (65) also includes the term "patient." Of these 65 papers, 20 included patients, e.g., with irritable bowel syndrome (4 papers), depression (3 papers), low back pain (3 papers), fibromyalgia (2 papers), and Parkinson's disease (2 papers). The remaining papers discuss the clinical relevance for one or more patient groups, but without providing new data.
- This does not reflect completely the representativeness of specific disease topics in the database: Pain is mentioned in 838 papers between 1946 and 2016, depression in 313 (1936–2016), back pain in 123 papers

(1982–2016), Parkinson's disease in 100 papers (1987–2016), schizophrenia in 69 papers (1958–2016), migraine in 59 papers (1987–2016), irritable bowel syndrome in 56 papers (1959–2016), and headache without migraine in 45 papers (1987–2016). Many other clinical conditions are mentioned at least occasionally. However, this cannot distinguish between data and review-type papers and requires a more detailed bibliometric analysis.

- Of the two established mechanisms of the placebo effect, conditioning is mentioned in 284 papers, and expectancy in 197 papers (with partial overlap), indicating that these basic concepts are of rather late origin and occur regularly and with increasing frequency since 1980 (conditioning) and 1996 (expectancy).
- Experimental or clinical intervention techniques frequently used are acupuncture (148 papers), social learning (24 papers), hypnosis (38 papers), and relaxation (18 papers). While systematic reviews and meta-analyses are frequent (192 papers), animal models are scarce in placebo research (126 papers), as are papers mentioning or including children (82 papers), and studies explicitly including the elderly are rare (18 papers).
- The topics of surgery (118 papers), psychotherapy (112 papers), food or dietary intervention (72 papers), manual or physical therapy and sports (60 papers) are underrepresented given their relevance in medicine, presumably due to difficulties in designing control conditions in these fields. While the topic of ethics has received large attention (328 papers), legal issues are rarely discussed (55 papers). Similarly, eHealth and mHealth issues and the role of the internet are almost absent (11 papers).
- A total of 57 papers mention the role of genetics in placebo response, and following the first genetic study (Furmark et al., 2008), a series of studies (see Colagiuri, Schenk, Kessler, Dorsey, & Colloca, 2015; Hall, Loscalzo, & Kaptchuk, 2015) has underscored its relevance in a variety of clinical conditions; this may mark the beginning of a new phase of the neurobiology of placebo response when brain mechanisms are associated with putative genetic markers. Interestingly, twin studies have not yet been utilized except once (Weimer, Mönnikes, Stengel, & Enck, 2017).

While the initial phase of placebo research (i.e., before 2000) may be properly reflected in the bibliometric approach of Boehm et al. (2016), novel developments after the rise of neurophysiological tools and a broader understanding of the neurobiological mechanisms involved in the placebo response were clearly missed. However, the skewed distribution of placebo publications (as evident in Fig. 2) also obscures rather than enlightens the current situation of the discipline for a number of reasons.

Firstly, the number of potentially relevant papers available at any given time vastly exceeds the number of papers citable in a paper (often limited to around 50 in data papers, and between 100 and 200 in review-type papers), especially with electronic databases that are easy to search. It has been noted that, in consequence, more recent papers are being cited earlier, but also show a rapid decreasing trend in being cited (Chuang & Ho, 2014), and older literature is often forgotten—with the exception of citation classics. This is even more surprising given that many journals are only published electronically where space limitations are not an issue any more.

Furthermore, the introduction of open access journals and web-based publications that allow deeper and more thorough validation of papers being cited than "abstract only" citations; as a consequence this has driven the impact (factor) of some journals (Eysenbach, 2006). Therefore, they show a very skewed distribution, which may not completely reflect the relevance of these papers. Open access publishing has an academic, economic and societal impact (Tennant et al., 2016), as has the promotion of science and publications through social networks such as Facebook, Twitter and others (Eysenbach, 2011).

Finally, the short half-life of knowledge is specifically evident in medicine, where nearly all novel ideas are followed by an exponential increase of papers of all types, but usually more reviews (and nowadays more systematic reviews and meta-analyses) than data papers. This is of high relevance for medical decision making in therapy, and has given rise to the pursuit of "living systematic reviews" (Enck, 2018; Sutton, 2017) that seek constant updating rather than occasional re-writing of analyses, as is the case with Cochrane reviews (Elliott et al., 2014).

One of the consequences is that the "footprints" of classical studies (not only in placebo research) get lost (Park et al., 2017), and the search for "sleeping beauties" (Zavrsnik et al., 2016) may become the future, as may be bibliometric approaches that utilize bioinformatics and artificial intelligence technology and tools (Thomas et al., 2017).

1.8 The SIPS 2017 Conference Abstracts

The Society for Interdisciplinary Placebo Studies (SIPS) (http://placebosociety.org) was founded in 2014 during a placebo expert meeting in Hermance near Geneva, Switzerland, by some of the field's eminent researchers and soon became a point of assembly of the placebo research community. It currently lists nearly 150 members. Its first scientific meeting was organized in 2017, and followed individual national and international meetings and conferences over the years since 2000, with an increasing numbers of participants, and including some quite important meeting proceedings and reports (Benedetti, Enck, Frisaldi, & Schedlowski, 2014; Colloca, 2018; Enck, Benedetti, & Schedlowski, 2008).

Because the Leiden meeting attracted the largest group of delegates and submitted abstracts so far ($N=200$), it may best reflect the current state of placebo research. Because the number of abstracts is in the range of annual publications in placebo research, it can be reasonably assumed that the data and approaches presented in these 200 abstracts may anticipate the next 200 or so publications to be published in due course.

We used the text body of these 200 abstracts (excluding authors and affiliations) for a semi-quantitative content analysis using the MAXQDA-12 Plus software (https://www.maxqda.com, VERBI Software, Berlin, Germany) for qualitative and quantitative text analyses. All text segments (abstracts) were coded for their type of presentation (poster, oral, plenary, keynote), whether they report quantitative data (yes, no), whether they deal with placebo or nocebo effects (yes, no), the type of data (review, systematic review, meta-analysis, questionnaire/interview data, experimental data, clinical trial data), the subject of investigation (scientists/physicians, patients, volunteers, animals, theory), the research topic (pain, itch, mood, motor, gastrointestinal/nausea, cognition, rewards, unwanted adverse events, other symptoms/signs), and the related scientific subspecialty (neuroscience, internal medicine, surgery, psychology, ethics, philosophy, other). Due to the global character of some abstracts (e.g., of keynote lectures), abstracts could code for more than one category in each section. Sections (e.g., type of presentation) and codes (subgroups within a section) were predefined by the authors but handled flexibly if inconsistencies became apparent. This agrees with standards of qualitative analyses (Denzin & Lincoln, 2005).

Table 1 lists the distribution of all abstracts rated for all sections listed, separating psychology-related (94 abstracts), physiology-related (24 abstracts), and neuroscience-related abstracts (24 abstracts) from the rest (46 abstracts) for comparison. Psychology-related abstracts are abstracts reporting psychological interventions (expectancy manipulation, conditioning, psychotherapy, etc.) with psychological and behavioral measures, while physiology-related abstracts usually report physiological measures, mostly in addition to psychological measures, and neuroscience-related abstracts

Table 1 Distribution of the 2017 Leiden SIPS Conference Abstracts ($N = 200$) Across Disciplines (Psychology, Physiology, Neuroscience), and Major Medical Topics, Types of Study, Abstracts, and Data (Mutually Exclusive)

Topic		Psychology	Physiology	Neuroscience
	Total	**94**	**24**	**24**
Pain	69	38 (40%)	8 (33%)	10 (41%)
Mood	26	16 (17%)	3 (12%)	2 (8%)
Cognition	5	5 (5%)	0	0
Itch	10	9 (10%)	0	0
Motor function	18	4 (22%)	6 (25%)	4 (17%)
Reviews	47	28 (60%)	4 (17%)	12 (50%)
Systematic reviews	7	0	3 (12%)	0
Meta-analyses	9	2 (2%)	0	2 (8%)
Experiments	100	60 (64%)	15 (62%)	16 (67%)
Clinical trials	21	11 (12%)	4 (17%)	1 (4%)
Quantitative data	168	87 (93%)	24 (100%)	22 (92%)
Patient included	59	27 (29%)	5 (21%)	6 (25%)
Healthy controls	103	62 (66%)	16 (67%)	14 (58%)
Nocebo effects	39	25 (27%)	4 (17%)	5 (21%)

contain brain data, often in addition to physiological and (more often) psychological measures and interventions. The remaining abstracts are attributed to methodology (16 abstracts), philosophy/ethics (11 abstracts), surgery (4 abstracts) and various other disciplines and subspecialties (15 abstracts).

These categories are handled as mutually exclusive for this paper. Further cross-tabulations will be done at a later stage.

This rather superficial analysis reveals several interesting and critical issues, as discussed below.

Clearly, *Pain* dominates as research topic: nearly 40% of all abstracts in the three major categories (but similarly across all 200 abstracts) are devoted to pain and placebo hypoalgesia. This clearly reflects the basic science orientation of placebo research, but may also be due to the availability and reliability of standardized pain and placebo hypoalgesia procedures under experimental conditions.

Furthermore, *Motor functions* as outcome measures of placebo research are on the rise (18 abstracts)—in contrast to their presence in the JIPS database—while neurocognitive functions are still underrepresented (5 abstracts), especially when looking at neuroscience abstracts (nil). The same applies to *food and diet* interventions (4 abstracts).

It is of interest to note that 16 abstracts utilize and/or investigate the more traditional psychological stress concept, in addition to conditioning (52 abstracts) and expectancy manipulations (34 abstracts; 10 abstracts were categorized in both conditioning and expectancy manipulations).

Finally, reviews, systematic reviews and meta-analyses represent nearly 50% of all papers in our JIPS database (see above), but are less prevalent among the SIPS abstracts, where those reporting quantitative data dominate (>90%). Whether all or a major fraction of these papers will get published remains to be shown.

Surprisingly, despite substantial evidence generated by the field, the inclusion of patients—either in experiments or in clinical trials—occurs in less than a third of abstracts, even including meta-analyses of RCTs. Very few animal studies (2 abstracts) supplement basic science research presented at the SIPS conference.

As with the analysis of the JIPS database, neuroscientific techniques other than fMRI are virtually absent, e.g., EEG (4 abstracts), TMS (2 abstracts), PET and NIRS (both none). Similarly, parameters of the autonomic nervous system (such as heart rate variability) are all but non-existent, as no abstract explicitly mentions such measures.

In contrast to the JIPS dataset, 5% of abstracts included children (10 abstracts); however, data from the elderly are missing completely.

1.9 Critical Remarks and Future Direction

Insight into placebo research in toto and not only in specific subdivisions of placebo research (e.g., on the role of conditioning, expectation, personality, genetic contribution, etc.) that may be covered by systematic reviews and meta-analyses, requires methods that are not easily available. Here, we have discussed bibliometric analyses on the one hand, and semi-quantitative or qualitative approaches that we have applied to whole sets of research papers and conference abstracts, to see what they can tell about past achievements and current directions of placebo research. Altogether we feel that such approaches are still underutilized in placebo research, despite their ability for providing important insights: based on our above analyses, we can postulate that placebo research in medicine is focusing too much (and increasingly) on healthy volunteers but is lacking validation of its results in different patient populations, and that the relevance of placebo research results is still to be demonstrated for children and the elderly. We are specifically missing contributions of the "placebo by proxy" concept (Grelotti & Kaptchuk, 2011) beyond a few data in children (Burkart et al., 2017; Liodden, Sandvik, & Norheim, 2015; Whalley & Hyland, 2013) and for acupuncture (Liodden, Pripp, & Norheim, 2018).

Despite the progress achieved in the past decade, placebo research has also not succeeded in demonstrating that it is able to change clinical practice, neither at the level of the individual patient nor at the level of clinical drug testing in RCT. While open-label placebo applications (Charlesworth et al., 2017) have yielded some promising results, their usage in daily medical routine requires testing their efficacy against commonly used drugs, for which protocols may be under way (Ballou et al., 2017); the same holds true for studies using partial reinforcement (does-extension placebos) (Colloca et al., 2016) awaiting the test of its clinical relevance.

With respect to bibliometric analyses, our critical analysis of a rather limited assessment of 300 placebo studies is to contrast with a rather "global" approach taken recently (Bennett, 2018): Among the 500 million words in 5 million books (as accessible in the Google Books Ngram Viewer; https://books.google.com/ngrams), the occurrence of the term "placebo" was counted over time (1950–2008) and for five major languages (English/American, French, German, Italian, and Spanish) in which "placebo" has an identical spelling. While the resulting time lines may not represent the true "public" occurrence of the term (since academic and scholarly publications may be overrepresented in Google Books), the increase in popularization of the term placebo, at least in the English speaking world, may start much earlier (i.e., in the 1960s; see Fig. 1 in Bennett, 2018) than the rise in placebo publications that we have noted for the 1990s (see Fig. 2). For the other four languages, the rise starts around 2000 and thus, follows the increase in placebo publications.

2. SUMMARY AND CONCLUSION

In summary, the apparent lack of innovative research topics, tools and technology that was noted in all three approaches (bibliometrics, the JIPS analysis and the analysis of the SIPS 2017 abstracts) points toward the need for such meta-reflection at least occasionally. As we have pointed out earlier (Enck et al., 2017), the self-limitation of the placebo research community to conventional paradigms may counteract the implicit promise to maximize and harness the placebo response in clinical routine (Colloca & Miller, 2011), to implement its new insights in daily medical practice, and to improve both patient care and drug testing (Enck et al., 2013).

REFERENCES

Akl, E. A., Meerpohl, J. J., Elliott, J., Kahale, L. A., Schunemann, H. J., & the Living Systematic Review Network. (2017). Living systematic reviews: 4. Living guideline recommendations. *Journal of Clinical Epidemiology, 91*, 47–53. https://doi.org/10.1016/j.jclinepi.2017.08.009.

Ballou, S., Kaptchuk, T. J., Hirsch, W., Nee, J., Iturrino, J., Hall, K. T., et al. (2017). Open-label versus double-blind placebo treatment in irritable bowel syndrome: Study protocol for a randomized controlled trial. *Trials, 18*(1), 234. https://doi.org/10.1186/s13063-017-1964-x.

Beecher, H. K. (1955). The powerful placebo. *Journal of the American Medical Association, 159*(17), 1602–1606.

Benedetti, F., Colloca, L., Torre, E., Lanotte, M., Melcarne, A., Pesare, M., et al. (2004). Placebo-responsive Parkinson patients show decreased activity in single neurons of subthalamic nucleus. *Nature Neuroscience*, 7(6), 587–588. https://doi.org/10.1038/nn1250.

Benedetti, F., Enck, P., Frisaldi, E., & Schedlowski, M. (Eds.), (2014). *Placebo (handbook of experimental pharmacology): Vol. 225.* Heidelberg: Springer.

Bennett, G. J. (2018). Does the word "placebo" evoke a placebo response? *Pain*. https://doi.org/10.1097/j.pain.0000000000001269. (Epub ahead of print).

Boehm, K., Berger, B., Ostermann, T., & Heusser, P. (2016). Placebo effects in medicine: A bibliometric analysis. *JRSM Open*, 7(7), 2054270416643890. https://doi.org/10.1177/2054270416643890.

Burkart, T. L., Kraus, A., Koller, B., Natalucci, G., Latal, B., Fauchère, J. C., et al. (2017). Placebo by proxy in neonatal randomized controlled trials: Does it matter? *Children (Basel)*, *4*(6), pii: E43. https://doi.org/10.3390/children4060043.

Charlesworth, J. E. G., Petkovic, G., Kelley, J. M., Hunter, M., Onakpoya, I., Roberts, N., et al. (2017). Effects of placebos without deception compared with no treatment: A systematic review and meta-analysis. *Journal of Evidence-Based Medicine, 10*(2), 97–107. https://doi.org/10.1111/jebm.12251.

Chuang, K. Y., & Ho, Y. S. (2014). A bibliometric analysis on top-cited articles in pain research. *Pain Medicine, 15*(5), 732–744.

Colagiuri, B., Schenk, L. A., Kessler, M. D., Dorsey, S. G., & Colloca, L. (2015). The placebo effect: From concepts to genes. *Neuroscience, 307*, 171–190. https://doi.org/10.1016/j.neuroscience.2015.08.017.

Colloca, L. (Ed.), (2018). *Neurobiology of the placebo effect (part 1).* Amsterdam: Elsevier.

Colloca, L., Enck, P., & DeGrazia, D. (2016). Relieving pain using dose-extending placebos: A scoping review. *Pain, 157*(8), 1590–1598. https://doi.org/10.1097/j.pain.0000000000000566.

Colloca, L., & Miller, F. G. (2011). Harnessing the placebo effect: The need for translational research. *Philosophical Transactions of the Royal Society of London. Series B, Biological Sciences, 366*(1572), 1922–1930. https://doi.org/10.1098/rstb.2010.0399.

Denzin, N. K., & Lincoln, Y. S. (Eds.), (2005). *The sage handbook of qualitative research* (3rd ed.). Thousand Oaks, CA: Sage.

Elliott, J. H., Synnot, A., Turner, T., Simmonds, M., Akl, E. A., McDonald, S., et al. (2017). Living systematic review: 1. Introduction-the why, what, when, and how. *Journal of Clinical Epidemiology*, *91*, 23–30. https://doi.org/10.1016/j.jclinepi.2017.08.010.

Elliott, J. H., Turner, T., Clavisi, O., Thomas, J., Higgins, J. P., Mavergames, C., et al. (2014). Living systematic reviews: An emerging opportunity to narrow the evidence-practice gap. *PLoS Medicine*, *11*(2), e1001603. https://doi.org/10.1371/journal.pmed.1001603.

Elsenbruch, S., & Enck, P. (2015). Placebo effects and their determinants in gastrointestinal disorders. *Nature Reviews. Gastroenterology & Hepatology*, *12*(8), 472–485. https://doi.org/10.1038/nrgastro.2015.117.

Enck, P. (2018). Living systematic reviews, not only for clinical (placebo) research. *Journal of Clinical Epidemiology*, *98*, 152–153. https://doi.org/10.1016/j.jclinepi.2018.01.001.

Enck, P., Benedetti, F., & Schedlowski, M. (2008). New insights into the placebo and nocebo responses. *Neuron*, *59*(2), 195–206. https://doi.org/10.1016/j.neuron.2008.06.030.

Enck, P., Bingel, U., Schedlowski, M., & Rief, W. (2013). The placebo response in medicine: Minimize, maximize or personalize? *Nature Reviews. Drug Discovery*, *12*(3), 191–204. https://doi.org/10.1038/nrd3923.

Enck, P., Klosterhalfen, S., & Weimer, K. (2017). Unsolved, forgotten, and ignored features of the placebo response in medicine. *Clinical Therapeutics*, *39*(3), 458–468. https://doi.org/10.1016/j.clinthera.2016.11.016.

Eysenbach, G. (2006). Citation advantage of open access articles. *PLoS Biology*, *4*(5), e157. https://doi.org/10.1371/journal.pbio.0040157.

Eysenbach, G. (2011). Can tweets predict citations? Metrics of social impact based on twitter and correlation with traditional metrics of scientific impact. *Journal of Medical Internet Research*, *13*(4), e123. https://doi.org/10.2196/jmir.2012.

Furmark, T., Appel, L., Henningsson, S., Ahs, F., Faria, V., Linnman, C., et al. (2008). A link between serotonin-related gene polymorphisms, amygdala activity, and placebo-induced relief from social anxiety. *The Journal of Neuroscience*, *28*(49), 13066–13074. https://doi.org/10.1523/JNEUROSCI.2534-08.2008.

Grelotti, D. J., & Kaptchuk, T. J. (2011). Placebo by proxy. *BMJ*, *343*, d4345. https://doi.org/10.1136/bmj.d4345.

Hall, K. T., Loscalzo, J., & Kaptchuk, T. J. (2015). Genetics and the placebo effect: The placebome. *Trends in Molecular Medicine*, *21*(5), 285–294. https://doi.org/10.1016/j.molmed.2015.02.009.

Irwin, P., & Fink, M. (1983). Familiarization session and placebo control in EEG studies of drug effects. *Neuropsychobiology*, *10*(2–3), 173–177.

Jütte, R. (2013). The early history of the placebo. *Complementary Therapies in Medicine*, *21*(2), 94–97. https://doi.org/10.1016/j.ctim.2012.06.002.

Kaptchuk, T. J. (1998). Intentional ignorance: A history of blind assessment and placebo controls in medicine. *Bulletin of the History of Medicine*, *72*(3), 389–433. https://doi.org/10.1353/bhm.1998.0159.

Kleinschmidt, A., Bruhn, H., Kruger, G., Merboldt, K. D., Stoppe, G., & Frahm, J. (1999). Effects of sedation, stimulation, and placebo on cerebral blood oxygenation: A magnetic resonance neuroimaging study of psychotropic drug action. *NMR in Biomedicine*, *12*(5), 286–292.

Leuchter, A. F., McCracken, J. T., Hunter, A. M., Cook, I. A., & Alpert, J. E. (2009). Monoamine oxidase A and catechol-o-methyltransferase functional polymorphisms and the placebo response in major depressive disorder. *Journal of Clinical Psychopharmacology*, *29*(4), 372–377. https://doi.org/10.1097/JCP.0b013e3181ac4aaf.

Levy, R. S., & Jankovic, J. (1983). Placebo-induced conversion reaction: A neurobehavioral and EEG study of hysterical aphasia, seizure, and coma. *Journal of Abnormal Psychology*, *92*(2), 243–249.

Liodden, I., Pripp, A. H., & Norheim, A. J. (2018). Placebo by proxy expectations toward acupuncture change over time: A survey comparing parental expectations to acupuncture pre- and postoperatively. *BMC Complementary and Alternative Medicine*, *18*(1), 183. https://doi.org/10.1186/s12906-018-2248-z.

Liodden, I., Sandvik, L., & Norheim, A. J. (2015). Placebo by proxy—The influence of parental anxiety and expectancy on postoperative morbidities in children. *Forschende Komplementärmedizin*, *22*(6), 381–387. https://doi.org/10.1159/000442565.

Loranger, A. W., Prout, C. T., & White, M. A. (1961). The placebo effect in psychiatric drug research. *JAMA*, *176*, 920–925.

Marashi, S. A., Hosseini-Nami, S. M., Alishah, K., Hadi, M., Karimi, A., Hosseinian, S., et al. (2013). Impact of Wikipedia on citation trends. *EXCLI Journal*, *12*, 15–19.

Miller, F. G., & Brody, H. (2011). Understanding and harnessing placebo effects: Clearing away the underbrush. *The Journal of Medicine and Philosophy*, *36*(1), 69–78. https://doi.org/10.1093/jmp/jhq061.

Park, K. M., Kim, J. E., Kim, Y., Kim, S. E., Yoon, D. Y., & Bae, J. S. (2017). Searching the footprints of pioneers on neurology: A bibliometric analysis. *European Neurology*, *77*(3–4), 152–161. https://doi.org/10.1159/000455843.

Perneger, T. V. (2015). Online accesses to medical research articles on publication predicted citations up to 15 years later. *Journal of Clinical Epidemiology*, *68*(12), 1440–1445. https://doi.org/10.1016/j.jclinepi.2015.01.024.

Rickels, K., Lipman, R., & Raab, E. (1966). Previous medication, duration of illness and placebo response. *The Journal of Nervous and Mental Disease*, *142*(6), 548–554.

Schedlowski, M., Enck, P., Rief, W., & Bingel, U. (2015). Neuro-bio-behavioral mechanisms of placebo and Nocebo responses: Implications for clinical trials and clinical practice. *Pharmacological Reviews*, *67*(3), 697–730. https://doi.org/10.1124/pr.114.009423.

Schmidt, M. E., Ernst, M., Matochik, J. A., Maisog, J. M., Pan, B. S., Zametkin, A. J., et al. (1996). Cerebral glucose metabolism during pharmacologic studies: Test-retest under placebo conditions. *Journal of Nuclear Medicine*, *37*(7), 1142–1149.

Shorter, E. (2011). A brief history of placebos and clinical trials in psychiatry. *Canadian Journal of Psychiatry*, *56*(4), 193–197.

Simmonds, M., Salanti, G., McKenzie, J., Elliott, J., & the Living Systematic Review Network (2017). Living systematic reviews: 3. Statistical methods for updating meta-analyses. *Journal of Clinical Epidemiology*, *91*, 38–46. https://doi.org/10.1016/j.jclinepi.2017.08.008.

Stockhorst, U., Enck, P., & Klosterhalfen, S. (2007). Role of classical conditioning in learning gastrointestinal symptoms. *World Journal of Gastroenterology*, *13*(25), 3430–3437.

Sutton, A. J. (2017). Not enough I say! Expand the remit of living systematic reviews to inform future research. *Journal of Clinical Epidemiology*, *91*, 54–55. https://doi.org/10.1016/j.jclinepi.2017.08.014.

Tennant, J. P., Waldner, F., Jacques, D. C., Masuzzo, P., Collister, L. B., & Hartgerink, C. H. (2016). The academic, economic and societal impacts of open access: An evidence-based review. *F1000Res*, *5*, 632. https://doi.org/10.12688/f1000research.8460.3.

Thomas, J., Noel-Storr, A., Marshall, I., Wallace, B., McDonald, S., Mavergames, C., et al. (2017). Living systematic reviews: 2. Combining human and machine effort. *Journal of Clinical Epidemiology*, *91*, 31–37. https://doi.org/10.1016/j.jclinepi.2017.08.011.

van Wesel, M. (2016). Evaluation by citation: Trends in publication behavior, evaluation criteria, and the strive for high impact publications. *Science and Engineering Ethics*, *22*(1), 199–225. https://doi.org/10.1007/s11948-015-9638-0.

Vandvik, P. O., Brignardello-Petersen, R., & Guyatt, G. H. (2016). Living cumulative network meta-analysis to reduce waste in research: A paradigmatic shift for systematic reviews? *BMC Medicine*, *14*, 59. https://doi.org/10.1186/s12916-016-0596-4.

Verheul, W., Mistiaen, P., Di Blasi, Z., Kok, L., van Dulmen, S., & Bensing, J. (2010). The effect of varying practitioner communication on patients' health status and treatment outcomes (protocol). *Cochrane Database of Systematic Reviews*, (10), CD008791. https://doi.org/10.1002/14651858.CD008791. (withdrawn).

Wartolowska, K. A., Beard, D. J., & Carr, A. J. (2018). The use of placebos in controlled trials of surgical interventions: A brief history. *Journal of the Royal Society of Medicine*, *111*(5), 177–182. https://doi.org/10.1177/0141076818769833.

Weimer, K., Colloca, L., & Enck, P. (2015a). Age and sex as moderators of the placebo response—An evaluation of systematic reviews and meta-analyses across medicine. *Gerontology*, *61*(2), 97–108. https://doi.org/10.1159/000365248.

Weimer, K., Colloca, L., & Enck, P. (2015b). Placebo effects in psychiatry: Mediators and moderators. *The Lancet Psychiatry*, *2*(3), 246–257. https://doi.org/10.1016/s2215-0366(14)00092-3.

Weimer, K., Gulewitsch, M. D., Schlarb, A. A., Schwille-Kiuntke, J., Klosterhalfen, S., & Enck, P. (2013). Placebo effects in children: A review. *Pediatric Research*, *74*(1), 96–102. https://doi.org/10.1038/pr.2013.66.

Weimer, K., Mönnikes, N., Stengel, A., & Enck, P. (2017). Genetics, shared, or non-shared environment? An experimental twin study on placebo analgesia. *Psychosomatic Medicine*, *79*, A-120.

Whalley, B., & Hyland, M. E. (2013). Placebo by proxy: The effect of parents' beliefs on therapy for children's temper tantrums. *Journal of Behavioral Medicine*, *36*(4), 341–346. https://doi.org/10.1007/s10865-012-9429-x.

Wolf, S. (1959). The pharmacology of placebos. *Pharmacological Reviews*, *11*, 689–704.

Zavrsnik, J., Kokol, P., Del Torso, S., & Blazun Vosner, H. (2016). Citation context and impact of 'sleeping beauties' in paediatric research. *The Journal of International Medical Research*, *44*(6), 1212–1221. https://doi.org/10.1177/0300060516672129.

CHAPTER FIVE

Clinical Use of Placebo Effects in Patients With Pain Disorders

Regine Klinger*[,1], **Julia Stuhlreyer***, **Marie Schwartz***, **Julia Schmitz***, **Luana Colloca**[†,‡,§]

*Department of Anesthesiology, Section Pain Medicine and Pain Psychology, University Medical Center Hamburg-Eppendorf, Hamburg, Germany
[†]Department of Pain Translational Symptom Science, School of Nursing, University of Maryland, Baltimore, MD, United States
[‡]Department of Anesthesiology/Psychiatry, School of Medicine, University of Maryland, Baltimore, MD, United States
[§]Center to Advance Chronic Pain Research, University of Maryland, Baltimore, MD, United States
[1]Corresponding author: e-mail address: r.klinger@uke.de

Contents

Abstract

The analgesic placebo effect is well documented by numerous studies. Many important influencing factors, however, are yet to be discovered. In the arena of placebo effects and clinical implications, expectancies play a central role. Expectancies are shaped by processes of classical and social learning as well as verbal instructions and are strongly related to emotional factors. Expectancies trigger a cascade of endogenous opioids and non-opioids, which alter the experience of pain. For clinical application it is important to know, that placebo research yields ethical possibilities to use placebo effects without deception and without using placebos. Since placebo effects contribute to responses to active analgesics, it is feasible to enhance patients' benefits from pain treatments by increasing the additional placebo effect. There are several possibilities to use the placebo effects via shaping and adapting information about analgesic medication and via associating medication intake with a positive context. A positive patient–clinician communication atmosphere is very important to generate clinically meaningful placebo effects in pain medicine.

International Review of Neurobiology, Volume 139
ISSN 0074-7742
https://doi.org/10.1016/bs.irn.2018.07.015

1. INTRODUCTION

Placebo hypoalgesia is one of the most investigated placebo phenomenon. Many important influential factors have already been discovered with regard to psychological and neurobiological mechanisms (Benedetti, 2006, 2014; Bingel, Colloca, & Vase, 2011; Colloca & Benedetti, 2009; Finniss & Benedetti, 2005; Finniss, Kaptchuk, Miller, & Benedetti, 2010). In the context of pain, the activation of endogenous opioids plays a crucial role—in the case of placebo responders, the cholecystokinin is activated (Colloca, Klinger, Flor, & Bingel, 2013) while in the nonresponders case, it is the dopaminergic system that is de-activated (Scott et al., 2008). These neurobiological processes are triggered by expectancies which are the result of psychological mechanisms and interact closely with emotions and motivations (e.g., anxiety, desire for relief), somatic focus, or cognitions (e.g., attitudes toward the treatment) (Colloca & Grillon, 2014; Finniss & Benedetti, 2005; Geers, Helfer, Weiland, & Kosbab, 2006; Price, Finniss, & Benedetti, 2008; Price et al., 1999; Vase, Petersen, Riley, & Price, 2009). Thus, current research focuses on the clinical relevance of analgesic placebo effects to facilitate better treatment outcomes. To achieve this objective, it is important to make optimal use of placebo effects. Broad evidence exists for the effects of placebo treatments in clinical, neurobiological, and laboratory applications, especially for the area of pain. For clinical applications, it is important to know that placebo research yielded ethical possibilities to use placebo effects without deception and without the use of placebos (c.f. Colloca & Howick, 2018). In other words, the placebo effect can be achieved without inert substances (i.e., placebos) or deception.

2. HOW IS THE ANALGESIC PLACEBO EFFECTS DEFINED IN RESEARCH AND CLINICAL PRACTICE?

A pain reduction after the intake of analgesics does not automatically provide conclusions about the effectiveness of the analgesics. To examine the real effectiveness of a pain killer, the clinical research considers the comparison of a verum and a placebo control group as the gold standard (Bothwell, Greene, Podolsky, & Jones, 2016). In this context, if the verum does not show significant superior results compared to the placebo group, the medication is considered as ineffective even if the symptoms are reduced after medication intake. Such an observed analgesic effect and the related symptom reduction are not automatically due to analgesics. The reduction could be due to other factors as well, such as, natural course of the disease or spontaneous recovery. On the other hand, a treatment with a placebo could also wrongly be interpreted as effective (Kaptchuk, 1998). Consequently, a pain reduction does not automatically indicate the true effectiveness of a placebo; that is, the outcome could result from the natural course of the disease or spontaneous healing. To show the true effect, it is necessary to include a third group consisting of patients who were not treated at all—the so called Natural History (NH) Group. The significant difference in pain reduction between the placebo group and the NH-Group can be interpreted as the real placebo effect (see Fig. 1). Thus, placebo analgesia occurs when alterations in pain perception appear that exceed the specific effect of the pharmacological, psychological, or physical pain treatment (Klinger, Colloca, Bingel, & Flor, 2014).

3. CLINICAL APPLICATION OF ANALGESIC PLACEBOS: ETHICAL TENABLE, USEFUL, AND NECESSARY?

Laboratory studies showed that patients with chronic Irritable Bowel Syndrome (Kaptchuk et al., 2008; Vase, Robinson, Verne, & Price, 2003, 2005), idiopathic and neuropathic pain (Petersen et al., 2014, 2012;

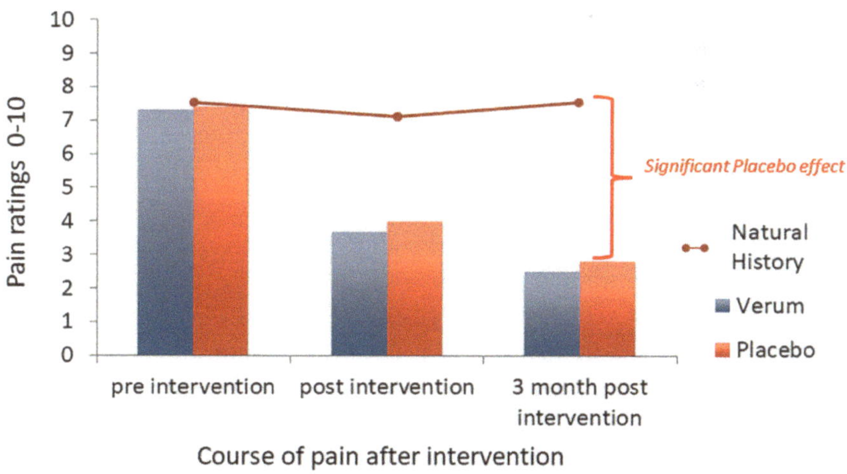

Fig. 1 If a pain reduction is yielded via both, a verum as well as a placebo, this is not a synonym for the existence of a placebo effect: A pain reduction via a placebo does not automatically indicate the true effectiveness of a placebo. The outcome could result from the natural course of the disease or spontaneous healing. A real placebo effect is the significant predominant difference between the level of pain reduction via a placebo and the pain level reported by a Natural History Group (NH).

Vase, Petersen, & Lund, 2014), and low back pain (Carvalho et al., 2016; Hashmi et al., 2012; Klinger, Kothe, Schmitz, Kamping, & Flor, 2017), show robust placebo hypoalgesic responses.

Moreover, analgesic placebos can lead to excellent clinical results for patients with chronic pain (Carvalho et al., 2016; Klinger et al., 2017), idiopathic and neuropathic pain (Petersen et al., 2014, 2012; Vase et al., 2014), migraine (Kam-Hansen et al., 2014), and knee osteoarthritis (Gollub et al., 2018; Tetreault et al., 2016). Therefore, why should analgesic placebos not be used in daily clinical practice? The administration of placebos with the assertion that these are analgesics can lead to difficulties in mid- and/or long term. As soon as patients find out that they were deceived, it is highly likely that they lose trust in their physician. This is because patients usually base their trust mostly on the therapist's credibility. Other contributing factors which create trust are (perceived) competence, compassion, confidentiality, and communication skills (Pearson & Raeke, 2000). At the same time, the concealed application of placebos in the daily clinical practice can greatly tarnish the trust in the relationship between patients and physicians (Miller & Colloca, 2009).

Once the trust relationship between patients and physicians is tarnished, it makes a successful treatment almost impossible.

It seems, however, that there is a very wide range in the application of improper verum medication with the intention of creating placebo effects (Colloca, Jonas, Killen, Miller, & Shurtleff, 2014). Colloca et al. (2014) conducted a cross-cultural, systematic review in which they examined 22 cross-sectional, qualitative, and longitudinal studies. They explored the frequency of placebo use among health care professionals from 12 countries and indicated that the prevalence of placebo use in clinical practice range from 17% to 80% among physicians and from 51% to 100% among nurses. In a sample of 1200 internists and rheumatologists in the United States, the rate of pure placebos used was reported to be 5%. Moreover, 46% of the internists and rheumatologists stated that they recommend a placebo treatment primarily to enhance patient expectancy (Tilburt, Emanuel, Kaptchuk, Curlin, & Miller, 2008). The dissimilar prescription of drugs, which are not medically indicated, is of particular concern, because they can have little actual benefit but potentially serious side effects (Miller & Colloca, 2009). One reason for this medical practice is the pressure of expectancy by the patients to receive medication or a medical prescription. It is possible that physicians receive better ratings when they prescribe medication instead of pointing out the natural healing process (NH). In this context, the prescription of antibiotics for minor illnesses or mild colds, as compared to the prescription of opioids for patients with chronic but not radicular back pain is an important example.

The increased opioid prescription leads to considerable health concerns (Vowles et al., 2015). The United States even declared the opioid crisis a national emergency (Hirschfeld Davis, 2017). Therefore, the legal regulation of placebos, which are not openly labeled as placebos (i.e., hidden placebos), should be reconsidered.

In the United States, drug overdose deaths related to opioids prescribed to treat noncancer chronic pain other than methadone has increased from 2000 to 2014 (Rudd, Aleshire, Zibbell, & Gladden, 2016), with over 47,055 drug overdose deaths. Opioids have identified as "the main drugs associated with overdose deaths" with a recent rate three-times higher than in 2000 (Centers for Disease & Prevention, 2013; Rudd et al., 2016).

4. THE APPLICATION OF ANALGESIC PLACEBO EFFECTS: ETHICAL TENABLE ALTERNATIVES

The placebo research results of the last years demonstrate useful and ethically tenable alternatives to the deceptive use of placebos. The "open-label placebo treatment" use (Carvalho et al., 2016; Kaptchuk et al., 2010), in which patients are openly told that they receive a placebo and are informed about the basic principles of the effectiveness of the placebo effect represents an example of potential use of placebos.

In a randomized controlled trial, Carvalho et al. (2016) investigated whether placebo effects in chronic low back pain could be achieved through open-label placebo (OLP) treatment. They compared the OLP treatment to a group which received a treatment as usual (TAU). They were able to show that OLP elicited greater pain reduction and reduced disability compared to TAU.

Importantly, OLP treatment involves the use of "open medication" (Amanzio, Pollo, Maggi, & Benedetti, 2001; Benedetti, Amanzio, & Maggi, 1995; Colloca, Lopiano, Lanotte, & Benedetti, 2004) and the *additive* placebo effect. The principle of the additive placebo effect is based on the basic concept that the effectiveness all analgesics consist of both, a pharmacological and a psychological (placebo effect) component. The additive placebo effect stands for the psychological component which complements the pharmacological component.

Overtly administered analgesics—whereby patients are fully aware of taking the drug—have a better effect than analgesics given in a covert fashion (e.g., through a computer-controlled infusion pump or as part of a drug cocktail) (Benedetti, 2006; Colloca et al., 2004). The more the administration of a drug is perceived (e.g., sight, smell, taste, touch, and information), the better the placebo effect can be exploited. Consequently, the placebo effect can occur without the use of deception and without the deceptive use of medication with inert substances (placebo), which are misleadingly declared as pharmacological effective. The application of placebo mechanisms can influence patients' expectancies, which results in alterations of pain perception that exceed the specific effect of pharmacological, psychological, or physical pain treatment. An open question becomes: Is the benefit of the OLP equally strong when the patient knows he/she is taking a placebo, as compared to when the patient does not know one of his/her medications is in reality a placebo? In a post-operative setting, acute pain patients were told that they could receive a pain treatment or a placebo (language used from a double-blind clinical trial); told that the pain treatment was an active drug (deceptive administration); or told nothing about the drug (NH-Group) (Pollo et al., 2001). A continuous-drip saline infusion was added to the buprenorphine treatment and the total number of requested doses of buprenorphine over the 3-day treatment was used as outcome (i.e., reduction of buprenorphine request). The requests of buprenorphine decreased in both the deceptive and double-blind groups as compared to the NH-Group with a reduction by 33.8% in the deceptive group and 20.8% in the double-blind group as compared to the NH-Group in which the reduction equaled 16.4% (Pollo et al., 2001). Importantly, a more transparent disclosure promoted a pain improvement and buprenorphine overall intake.

5. UNDERLYING PSYCHOLOGICAL MECHANISMS OF ANALGESIC PLACEBO EFFECTS

The underlying psychological mechanisms of the placebo effect have been intensively researched in the past decades (Enck, Bingel, Schedlowski, & Rief, 2013). At the center of attention of the placebo effect is the patient's expectancy, which plays a crucial role for the possible formation of the placebo effect. The expectancy is mediated and modulated through a variety of psychological mechanisms (see Fig. 2) including learning by instructions, learning by experience, and social observational learning (Colloca, 2014; Colloca & Miller, 2011a, 2011b; Kirsch, 2018).

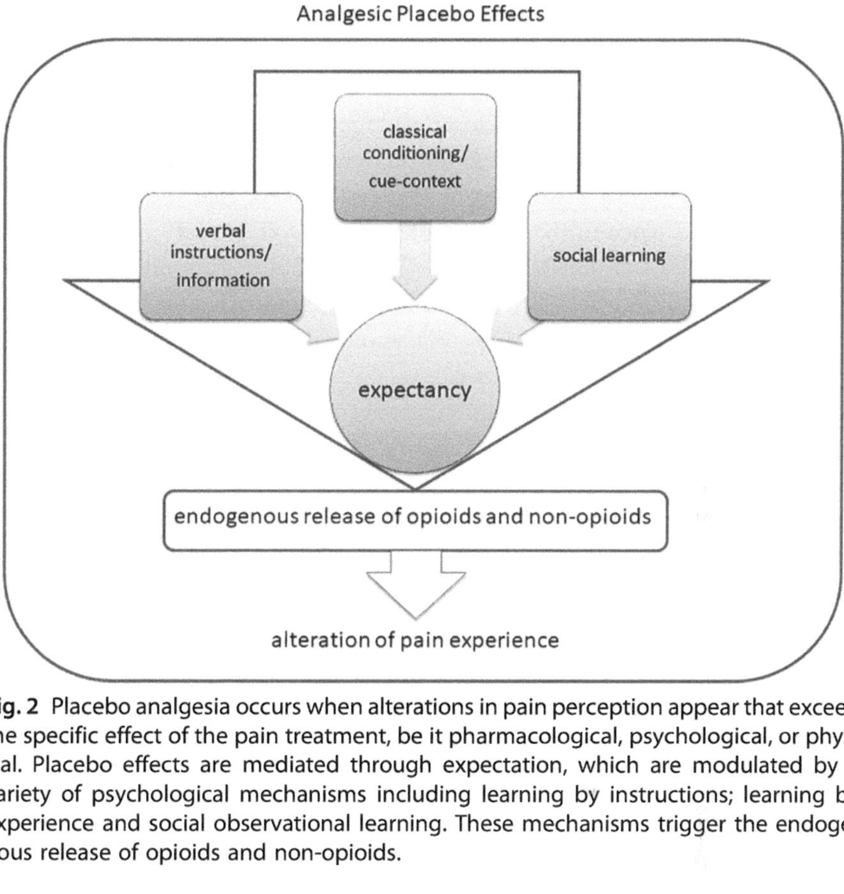

Fig. 2 Placebo analgesia occurs when alterations in pain perception appear that exceed the specific effect of the pain treatment, be it pharmacological, psychological, or physical. Placebo effects are mediated through expectation, which are modulated by a variety of psychological mechanisms including learning by instructions; learning by experience and social observational learning. These mechanisms trigger the endogenous release of opioids and non-opioids.

The placebo effect is achieved through a positive prior experience, the observation of others or/and an instruction that initiates a positive expectancy toward a medication or treatment and that this expectancy initiates the placebo effect (Colloca, 2014; Kirsch, 2018; Kirsch, Lynn, Vigorito, & Miller, 2004). Expectation can be reinforced via a conditioning paradigm, in which the repeated association of a neutral stimulus (e.g., sight, taste of a pill) with the pharmacological effect of the agent leads to a conditioned reaction that is similar to the original response to the pharmacological agent and is then triggered by the placebo (Ader, 1993; Colloca, Enck, & DeGrazia, 2016; Price et al., 2008). Furthermore, according to the social learning model placebo effects can be learned by observation of other patients' treatment effects (Colloca & Benedetti, 2009; Schenk, Krimmel, & Colloca, 2017).

From a neurobiological perspective, the placebo analgesic effect has been linked to the endogenous opioid system. Psychological mechanisms such as verbally induced expectancies, cued and contextual conditioning, and social learning trigger the cascade of endogenous opioids and non-opioids (Colloca et al., 2013). Many functional Magnetic Resonance Imaging and Positron Emission Tomography studies showed, that the analgesic placebo effect takes place in the same brain regions as the "real" analgesic effect (Bingel, Lorenz, Schoell, Weiller, & Buchel, 2006; Eippert et al., 2009; Elsenbruch et al., 2012; Kosslyn, 2005; Petrovic, Kalso, Petersson, & Ingvar, 2002; Wager et al., 2004). Placebo analgesia leads to reduced brain activity in brain areas such as dorsolateral prefrontal cortex, thalamus, insula, periaqueductal area, and anterior cingulate cortex. The placebo analgesia involves the descending pain modulating system. Expectancy plays a central role for the placebo effect. Through the modulation of the expectancy, it is possible to either develop or suppress placebo effects (Colloca et al., 2013). In other words, the patient's expectancy can be systematically modulated by applying these central mechanisms—expectancy is a neural process, it mediates and increases self-efficacy in case of a positive treatment expectancy. Expectancy, for example, can be increased through attention control toward positive aspects of the treatment and their treatment effects. Negative aspects, such as fear of treatment, can be decreased concurrently through the use of positive expectancies (Klinger et al., 2014). However, negative treatment expectancies trigger fear of treatment and reduce the treatment effects (Bingel, Wanigasekera, et al., 2011; Sonthalia et al., 2015). Moreover, there is evidence that the expectancy is influenced by previous experiences (Muller et al., 2016). Though, little research has examined the mechanisms of the expectancy effect yet. In particular, it is unclear which previous experiences are perceived as positive or negative. Furthermore, the rules for the interaction between previous experiences and current manipulation of the experiences to attain a positive treatment expectancy are not known yet. These insights would be highly clinically relevant, because they would enable a decisive influence achieving better treatment result.

6. REQUIREMENTS FOR A CLINICAL APPLICATION OF PLACEBO EFFECTS

The clinical application of placebo effects in a mechanistic-driven medicine has to be differentiated from approaches, which are not evidence based, such as anthroposophical medicine or alternative medicine.

"Evidence-based Medicine" (Cochrane, 1972) states that intuitions, unsystematic clinical experiences, and pathophysiologic rationales are not sufficient grounds for clinical decision making and stresses the importance of examining the evidence from clinical research (Guyatt, 1992). Therefore, the clinical application of placebo effects has to fulfill certain requirements:

1. Placebo research results must be transferable to patients.

   Most studies are conducted with healthy controls and it is important to show, that the placebo response in healthy people is the same as in patients.

2. Placebo effects must be applied deliberately.

   It is important to understand the underlying mechanisms of placebo effects and to translate them into clinical application.

6.1 Translation of Placebo Research Results to Patients With Pain

Only a few studies which investigate analgesic placebo effects deal with patients. The majority of studies, however, are conducted with healthy controls instead. The following studies which were conducted with pain patients outline the evidence for analgesic placebo effects:

First, Vase et al. (2005) examined patients suffering from irritable bowel syndrome. They were able to show that the effect of Lidocaine during a painful rectal examination could also be yielded through a placebo. Hence, Lidocaine and placebo significantly reduced pain levels compared to a NH-Group.

Second, Kaptchuk et al. (2008) investigated practitioners' influences on the placebo effect on patients with irritable bowel syndrome. They compared the effects of two placebo acupuncture treatments, which were realized by using so called Streitberger needles, who disappear in the handle (tube) when located on the skin. One was a placebo acupuncture with personal interactions, and the other was a placebo acupuncture limited to neutral, standard hospital interactions. Both groups were compared to a waiting list group. One very interesting result was, that both placebo acupuncture treatments led to symptom improvements that were beyond the NH-Group (waiting list). The positive therapeutic relationship increased further the effects of placebo acupuncture in comparison to a treatment, in which the practitioners minimized the interaction with patients. Hence, this study shows that placebo effects not only help healthy participants but also patients with chronic pain.

Third, Kam-Hansen et al. (2014) examined placebo effects in patients suffering from migraine headaches. In a sample of 66 patients, they showed that altered placebos and drug labeling changes the outcome of migraine attacks. They compared three different instructions for the analgesic (10 mg rizatriptan) and a corresponding placebo. The participants were either assigned to the "rizatriptan" group, "placebo" group, or "rizatriptan or placebo" group. Interestingly, the placebo treatment mislabeled as 10 mg rizatriptan reduced headache severity as effectively as rizatriptan mislabeled as placebo, and the OLP treatment was superior to no treatment. This study shows that patients' expectancy is highly influenced by instructions and plays a significant role in efficacy of pain medication and pain reducing placebos.

Fourth, Klinger et al. (2017) were able to reduce experimental and clinical pain and to improve the patients' physical capacity in a sample of 48 patients suffering from chronic back pain. These results were generated by using a placebo tincture combined with the instruction that the tincture is a "highly effective opioid" and provided additional conditioning. The pain reduction was of high clinical relevance, as the highly chronic patients had a reduction on average from 5.0 on a numerical rating scale (0 = no pain, 10 = worst imaginable pain) to about 2.5.

Despite the fact that there is evidence for analgesic placebo effects in patients, it is arguable that placebo effects are the same in patients as in healthy controls. There are many differences due to patient's characteristics (e.g., treatment history and illness beliefs). However, very important distinguishing features are the different expectancies resulting from the different previous experiences with treatment, medication, and the healing context (Muller et al., 2016). Studies which compare the placebo effect in healthy people and patients are required to determine whether the placebo response is the same in both samples. Only one study compared the patients' placebo response with the responses of healthy participants so far, concluding that patients and healthy people differ in the maintenance of the placebo effect. Patients only maintained the placebo effect when they experienced a conditioning procedure (Klinger, Soost, Flor, & Worm, 2007).

6.1.1 Do Patients Have Different Expectancies Than Healthy Individuals?

Peerdeman et al. (2016) showed in their meta-analysis that pain in patients could be reduced by modulation of expectancy with medium to large effects for experimental and acute procedural pain and small effects for chronic pain. Expectancies are believed to be a core mechanism of placebo effects

(Enck, Benedetti, & Schedlowski, 2008; Price et al., 2008). Patients, for example with acute or chronic pain, have in contrast to healthy individuals' different expectancies toward a treatment (Constantino, Arnkoff, Glass, Ametrano, & Smith, 2011). Patients with chronic pain have a long history of treatments and therefore entirely different previous experiences with medical treatment. Their expectancies toward the effectiveness of an analgesics may be different than the expectancy of generally healthy people who want to take medication because of minor acute pain.

Treatment expectancies depend on prediction of treatment results (Klinger et al., 2014). This prediction plays an important role for the patient. Patients want to know from the clinicians which treatment, with corresponding effect, will be carried out to relieve their pain. For example, prior to a surgery, most patients want to know how intense the post-operative pain will be. For this purpose, they formulate a prediction from the treatment team on which they can base their specific expectancies on. Moreover, patients observe the course of their diseases and treatments in relation to their expectancies and observe if the clinician's prognosis comes true. Perceived discrepancies seem to alter the perception of pain (Wiech, 2016). If patients expect that they will be pain free after the surgery, low pain levels will be perceived stronger and evaluated as too much. If the physician explains to the patients prior the surgery, that a low level of pain is normal after the surgery, their pain tolerance will be greater and their pain experience lower. The interaction between the prognosis and "true" pain reduction seems to be of great importance for the placebo effect and influences positively the course of treatment. Gonzalez Sáenz de Trejada et al. (2010) showed this in a study which included 881 patients with a knee- and coxarthrosis who received a total joint replacement surgery. The patients, whose own previous high expectancies were fulfilled, had better results in their health-related quality of life three to 12 months after the surgery (Health related Quality of Life/HRQoL (Ethgen, Bruyere, Richy, Dardennes, & Reginster, 2004); Short-Form-36-Questionnaire/SF-12, (Bullinger, 1995); Western Ontario and McMaster Universities Osteoarthritis Index/WOMAC, (Bellamy, Buchanan, Goldsmith, Campbell, & Stitt, 1988)) than patients with unmet expectancies. Consequently, health-care providers should assist their patients in developing realistic expectancies.

Another study investigated the placebo effect on patients with atopic dermatitis in comparison with healthy individuals (Klinger et al., 2007). This study showed that the patients reacted much stronger if their own prognosis, or more specifically their invalidated prediction, became reality. Furthermore, this study found higher initial placebo responses in patients than for healthy controls. However, patients could only maintain placebo responses when they had received a positive reinforcement (learning experiences) subsequently. When the patients did not receive a learning experience, the placebo effect decreased significantly. However, this response did not occur in healthy controls. A possible interpretation of this phenomenon is that the patients are more attentive to their body sensations and expect more from the medication. The lack of expected pain relief (invalidation of their prognoses and wish) led to a sense of disappointment which seemed to cancel out their placebo effect. Therefore, it is clinically relevant to keep in mind that patients are very sensitive to what clinicians say about their pain reducing medication. The pain reducing medication has to be explained realistically in order to avoid a disappointment response. The difference between patients and healthy controls is essential for the clinical practice, because it means that patients react differently—they are more sensitive to the described treatment manipulations than healthy controls and consider positive predictions made by clinicians more critically. Consequently, results from the placebo research on non-patients can not necessarily be transferred to patients. That means for the patient–physician relationship that physicians should carefully choose their words and statements in regards to treatment prognosis. Physicians should tailor the predicted treatment results individually to the patient and to the experiential realistic course and should avoid empty promises.

6.1.2 Patients' Expectancies and Their Impact on Treatment

Patients' expectancies are important predictors of the outcome of analgesic treatments, as demonstrated predominantly in research on placebo effects. In their systematic review, Peerdeman et al. (2016) demonstrated that patients' pain can be relieved and functioning can be improved by increasing expectancy. In their study on joint arthroplasty Gandhi, Davey, and Mahomed (2009) found a relationship between prior expectancies (expectancies patients have preoperative) and surgical outcome related to their functioning. Patients with high prior expectancies had significant better results in pain reduction and functioning than patients with low prior expectancies.

Furthermore, Davidge et al. (2009) investigated the relationship between prior expectancies and function-related surgical outcome (e.g., disability, functioning, and participation) and the health related quality of life in patients with extremity soft tissue sarcoma. The prior expectancies refer to the following three areas: (1) length of recovery, (2) complications, and (3) difficulties with performing daily activities.

Prior expectancies regarding performance of daily activities proved to be a significant predictor for disability ($P=0.0016$) and the ability to reintegrate into their life's roles ($P=0.044$) 1 year later. However, prior expectancies did not predict health related quality of life.

Nevertheless, Engel, Hamilton, Potter, and Zautra (2004) could show that prior expectancies and self-efficacy had a positive influence on the length of recovery period after a total knee replacement surgery. In this study, the surveyed prior expectancies focused on the probability of recovery and the expected change in quality of life (i.e., mental and physical health, pain, and functioning). Consequently, prior expectancies, as well as self-efficacy, predicted physical and mental health outcomes.

Lutz et al. (1999) found a relationship between prior expectancies (i.e., expected recovery time and expected state of health after the surgery, preferring the surgery over waiting whether sciatica improved over time on its own) and the surgical outcome (i.e., satisfaction with the surgery, change in symptoms and overall functioning) in patients with sciatica after surgery. Patients who expected a shorter recovery period were more satisfied with the treatment outcome after 12 months (odds ratio: 2.2; 5% margin of error) than patients who expected a longer recovery period. The comparison between these two groups did not elicit a significant difference in overall functioning and changes in symptoms.

Patients' treatment expectancies result from treatment history and modulate the magnitude of the placebo effect. Therefore, clinical research of analgesic placebo effect should focus on clinical pain and requires investigation models for chronic pain rather than experimental pain. Most studies conducted on patients use experimental or acute pain (Peerdeman et al., 2016; Vase et al., 2009).

6.2 Deliberate Application of Placebo Effects: Making Use of Placebo Effects in Clinical Practice

Clinical work must be a comprehensible process—random positive results are a pleasant benefit but are not usable and therefore not helpful for clinical practice. Accordingly, patient care relies on reliable and predictable results.

Research on placebo effects offers a variety of possibilities to use placebo effects in ethical acceptable ways. One key finding of previous research is that the pharmacological effect of pain medication can be enhanced by the additive effect of analgesic placebo effect (Amanzio et al., 2001; Benedetti et al., 1995; Colloca et al., 2004).

The application of placebo effects in clinical settings does not mean that pain medication should be substituted with placebos. It rather stands for increasing the effectiveness of pain medication by amplifying their inherent placebo component. Every effective pain medication has a pharmacological active component and a psychological (placebo) component. In clinical practice, it is possible to boost an analgesic therapy by supporting it with the placebo component. The principles are to modulate expectancies by a targeted use of verbal instructions, cues, associations, and social learning models in the healing context of pain treatment. According to Amanzio et al. (2001), Benedetti et al. (1995), and Colloca et al. (2004) the open application of medication is an ethical way of using placebo effects in a clinical context. In this approach, physicians and other providers of pain treatment draw patients' attention to the positive aspects of analgesics or other treatments. They give supporting information about the medication, take care of the patients, and try to establish a positive patient–clinician atmosphere.

6.2.1 Modulation of Expectancy via Verbal Instruction

A basic requirement for a successful treatment is to give verbal supporting information about the advantages (positive action) of pain treatments.

The conversation in which the advantages of pain treatments are discussed should follow the following guidelines:

1. Emphasize the positive side of pain medication:
   - Do not overestimate the side effects, if there is a need to inform the patient about the side effects, balance the proportion of negative and positive information and let the patient focus on the positive facts.
   - Explain the mechanisms of the drug action as easy as possible to ensure that the patient understands what is said.
2. Explain the course of the drug action:
   - Explain when effects of pain medication will start being felt by the patient, and prevent any distraction in this critical phase.
3. Avoid unrealistic promises:
   - Explain the realistic effects of the drug which the patient can then validate.
4. Reactivate positive associations, relativate negative associations:
   - Let the patients speak about previous or current negative and positive aspects related to their use of pain medication. Let them associate positive and relativate negative memories.

Modulation of expectancies by different information can completely alter the effectiveness of pain treatment. Bingel, Wanigasekera, et al. (2011)

assessed the effects of remifentanil on constant heat pain under three experimental conditions in a within-subject design: (1) no expectation of analgesia (hidden analgesic infusion), (2) expectation of a positive analgesic effect (instruction that the infusion will reduce pain), and (3) negative expectation (infusion is still running but instruction that the infusion is turned off). Positive treatment expectancy enhanced the analgesic benefit of remifentanil, whereas negative treatment expectancy abolished remifentanil analgesia. These results underline that negative communication during treatment can significantly reduce pharmacological effectiveness.

6.2.2 Modulation of Expectancy via Cues and Context Associations

There are several possibilities to strengthen the positive attention to pain medication in order to allow the patient to experience the effect of pain medication in a positive context:

- Advise patients to pay attention when taking pain medication: sight, smell, taste, and feeling. Encourage them to take time and concentrate on the intake of pain medication to create positive moments to strengthen the association between the conditions and positive analgesic effects from the treatment.
- Advise patients to take pain treatments in a positive internal and external situation/context (association with positive conditions), (e.g., in association with relaxation or positive, pain coping oriented thoughts, or a pleasant situation like a warm perfumed bath).
- Encourage the patient to associate pain medication with other pain reducing techniques (e.g., relaxation or other psychological pain interventions).
- Reactivate previous experiences with pain medication: support the patients to recall positive memories or—in case of negative experiences—to relativize their negative memories.

Respectively, in some clinical situations, it is beneficial to suppress context factors such as association with negative physiological or psychological states (Ader, 2003). For example, patients who undergo a physical withdrawal from medication (e.g., back pain patients with iatrogenic opioid misuse) will experience negative symptoms. These withdrawal symptoms will probably be associated with being withdrawn and are very closely connected with their negative expectancies associated with reduction of the quantity of medication. To cancel out these associations, it is possible to hide the amount of medication, (e.g., in a colored liquid), so that the patients are unaware of how much medication they receive. This way the cognitive and emotional mediated withdrawal phenomenon symptoms can be suppressed. There is also evidence, that this approach is a very successful way to inhibit the development of nocebo nausea (Colagiuri & Zachariae, 2010; Quinn & Colagiuri, 2015).

6.2.3 Open Label Placebos

There is encouraging evidence that OLPs in the treatment of patients with different pain disorders is effective (Carvalho et al., 2016; Kam-Hansen et al., 2014; Kaptchuk et al., 2010). Patients are informed about the fact that they will receive placebos instead of a pharmacological product and are given information about how the placebo effect induces analgesia. This is an ethical way of using placebos in clinical practice (Blease, Colloca, & Kaptchuk, 2016; Colloca & Howick, 2018).

Another useful clinical application is using OLPs to substitute pain treatments (e.g., when a patient undergoes a medication withdrawal). Such procedures can minimize symptoms of an withdraw phenomenon.

Future research should concentrate on strengthening the experience of self-efficacy associated with the intake of a placebo, so that patients can learn to rely on themselves to reduce pain rather than being dependent on a placebo. An OLP could be helpful for the transition from professional pain treatment into self-management of pain by using pain coping strategies learned in multidisciplinary pain centers which also include psychological pain treatments.

6.2.4 Modulation of Expectancy via Social Learning

The social context plays a crucial role in shaping negative or positive expectancies in patients. There are several situations which provide potential risks for negative or positive modulations of expectancies via observation of other patients:

- The waiting area
- Shared rooms on a ward
- Group therapy context
- Patients' observation of other patients' course of disease and treatment on TV, the internet, in lay press and newspapers etc.

Clinicians should be aware of these unsolicited sources of observation and social learning. They should ask what the patient has heard about pain treatments from others and seek positive or negative statements. If negative hearsay is reported, the differences between patients' histories could be discussed. Whereas, if positive experiences with pain treatments were exchanged, the effects of pain medication could be used as a vicarious model.

6.2.5 Positive Patient–Clinician Communication During Treatment

A positive patient–clinician relationship—characterized by trust, warmth, and empathy—is essential to make optimal use of placebo effects for the therapeutic efficacy of a treatment. Every pain physician, psychologist or physiotherapist is responsible to arrange a confiding and confident interpersonal patient–clinician relationship. One very important element for a positive atmosphere is the therapeutic alliance, which involves the bond between the therapist and the patient as well as agreement about the therapeutic goals and tasks (Hatcher & Barends, 2006; Horvath & Bedi, 2002). Del Re, Fluckiger, Horvath, Symonds, and Wampold (2012) showed that the therapists' development of stronger alliances with their patients results in a better outcome at the end of the therapy. Indicators of a strong therapeutic alliance include the patient's perception of feeling respected and cared for (Epstein & Street, 2007). Patient's perception of such a relationships can promote emotional well-being (Salkeld, Solomon, Short, & Butow, 2004; Thorne, Hislop, Kuo, & Armstrong, 2006). The clinician–patient alliance will become stronger when clinicians are optimally informative and show empathy with the patient's circumstances, when patients have an opportunity to express their concerns, and when the patient receives consistent messages and coordinated care from the clinical team. Clinician–patient communication which ensures that the patients feel accepted, validated, hopeful, worthy, reassured, and comforted can enhance the therapeutic outcome (Hack, Degner, Parker, & Team, 2005).

Moreover, a trusting and successful clinician–patient communication is also influenced by a shared understanding. Patients should have an understanding about illness, risks, and benefits of various treatment options in order to make informed decisions about medical treatment (e.g., to adopt analgesic medication). Clinicians need to understand patient's values, preferences, and beliefs about health (Street, Makoul, Arora, & Epstein, 2009). Both, clinicians and patients, need skills to present their perspective. A successful clinician–patient communication enables an effective management of information and can increase the patient's commitment to treatment (DiMatteo et al., 1993).

REFERENCES

Ader, R. (1993). Conditioned responses in pharmacotherapy research. *Psychological Medicine*, *23*(2), 297–299. https://doi.org/10.1017/S0033291700028361.

Ader, R. (2003). Psychosocial factors and the therapeutic response to drugs. *Prevention & Treatment*, *6*(1), Article ID 2c. https://doi.org/10.1037/1522-3736.6.1.62c.

Amanzio, M., Pollo, A., Maggi, G., & Benedetti, F. (2001). Response variability to analgesics: A role for non-specific activation of endogenous opioids. *Pain*, *90*(3), 205–215. https://doi.org/10.1016/S0304-3959(00)00486-3.

Bellamy, N., Buchanan, W. W., Goldsmith, C. H., Campbell, J., & Stitt, L. W. (1988). Validation study of WOMAC: A health status instrument for measuring clinically important patient relevant outcomes to antirheumatic drug therapy in patients with osteoarthritis of the hip or knee. *The Journal of Rheumatology*, *15*(12), 1833–1840.

Benedetti, F. (2006). Placebo analgesia. *Neurological Sciences*, *27*(2), 100–102. https://doi.org/10.1007/s10072-006-0580-4.

Benedetti, F. (2014). Placebo effects. In *Understanding the mechanisms in health and disease*. Oxford: Oxford University Press.

Benedetti, F., Amanzio, M., & Maggi, G. (1995). Potentiation of placebo analgesia by proglumide. *Lancet*, *346*(8984), 1231. https://doi.org/10.1016/S0140-6736(95)92938-X.

Bingel, U., Colloca, L., & Vase, L. (2011). Mechanisms and clinical implications of the placebo effect: Is there a potential for the elderly? A mini-review. *Gerontology*, *57*(4), 354–363. https://doi.org/10.1159/000322090.

Bingel, U., Lorenz, J., Schoell, E., Weiller, C., & Buchel, C. (2006). Mechanisms of placebo analgesia: rACC recruitment of a subcortical antinociceptive network. *Pain*, *120*(1–2), 8–15. https://doi.org/10.1016/j.pain.2005.08.027.

Bingel, U., Wanigasekera, V., Wiech, K., Ni Mhuircheartaigh, R., Lee, M. C., Ploner, M., et al. (2011). The effect of treatment expectation on drug efficacy: Imaging the analgesic benefit of the opioid remifentanil. *Science Translational Medicine*, *3*(70), 1–9. https://doi.org/10.1126/scitranslmed.3001244.

Blease, C., Colloca, L., & Kaptchuk, T. J. (2016). Are open-label placebos ethical? Informed consent and ethical equivocations. *Bioethics*, *30*(6), 407–414. https://doi.org/10.1111/bioe.12245.

Bothwell, L. E., Greene, J. A., Podolsky, S. H., & Jones, D. S. (2016). Assessing the gold standard—Lessons from the history of RCTs. *The New England Journal of Medicine*, *374*(22), 2175–2181. https://doi.org/10.1056/NEJMms1604593.

Bullinger, M. (1995). German translation and psychometric testing of the SF-36 health survey: Preliminary results from the IQOLA project. *Social Science & Medicine*, *41*(10), 1359–1366.

Carvalho, C., Caetano, J. M., Cunha, L., Rebouta, P., Kaptchuk, T. J., & Kirsch, I. (2016). Open-label placebo treatment in chronic low back pain: A randomized controlled trial. *Pain*, *157*(12), 2766–2772. https://doi.org/10.1097/j.pain.0000000000000700.

Centers for Disease Control and Prevention. (2013). Vital signs: Overdoses of prescription opioid pain relievers and other drugs among women—United States, 1999–2010. *MMWR Morbidity and Mortality Weekly Report*, *62*(26), 537–542.

Cochrane, A. L. (1972). *Effectiveness and efficiency: Random reflections on health services*. London: Nuffield Provincial Hospitals Trust London.

Colagiuri, B., & Zachariae, R. (2010). Patient expectancy and post-chemotherapy nausea: A meta-analysis. *Annals of Behavioral Medicine*, *40*(1), 3–14. https://doi.org/10.1007/s12160-010-9186-4.

Colloca, L. (2014). Placebo, nocebo, and learning mechanisms. *Handbook of Experimental Pharmacology*, *225*, 17–35. https://doi.org/10.1007/978-3-662-44519-8_2.

Colloca, L., & Benedetti, F. (2009). Placebo analgesia induced by social observational learning. *Pain*, *144*(1–2), 28–34. https://doi.org/10.1016/j.pain.2009.01.033.

Colloca, L., Enck, P., & DeGrazia, D. (2016). Relieving pain using dose-extending placebos: A scoping review. *Pain*, *157*(8), 1590–1598. https://doi.org/10.1097/j.pain.0000000000000566.

Colloca, L., & Grillon, C. (2014). Understanding placebo and nocebo responses for pain management. *Current Pain and Headache Reports*, *18*(6), 419. https://doi.org/10.1007/s11916-014-0419-2.

Colloca, L., & Howick, J. (2018). Placebos without deception: Outcomes, mechanisms, and ethics. *International Review of Neurobiology, 138*, 219–240. https://doi.org/10.1016/bs.irn.2018.01.005.

Colloca, L., Jonas, W. B., Killen, J., Miller, F. G., & Shurtleff, D. (2014). Reevaluating the placebo effect in medical practice. *Zeitschrift für Psychologie, 222*(3), 124–127. https://doi.org/10.1027/2151-2604/a000177.

Colloca, L., Klinger, R., Flor, H., & Bingel, U. (2013). Placebo analgesia: Psychological and neurobiological mechanisms. *Pain, 154*(4), 511–514. https://doi.org/10.1016/j.pain.2013.02.002.

Colloca, L., Lopiano, L., Lanotte, M., & Benedetti, F. (2004). Overt versus covert treatment for pain, anxiety, and Parkinson's disease. *Lancet Neurology, 3*(11), 679–684. https://doi.org/10.1016/S1474-4422(04)00908-1.

Colloca, L., & Miller, F. G. (2011a). How placebo responses are formed: A learning perspective. *Philosophical Transactions of the Royal Society of London. Series B, Biological Sciences, 366*(1572), 1859–1869. https://doi.org/10.1098/rstb.2010.0398.

Colloca, L., & Miller, F. G. (2011b). Role of expectations in health. *Current Opinion in Psychiatry, 24*(2), 149–155. https://doi.org/10.1097/YCO.0b013e328343803b.

Constantino, M. J., Arnkoff, D. B., Glass, C. R., Ametrano, R. M., & Smith, J. Z. (2011). Expectations. *Journal of Clinical Psychology, 67*(2), 184–192.

Davidge, K., Bell, R., Ferguson, P., Turcotte, R., Wunder, J., & Davis, A. M. (2009). Patient expectations for surgical outcome in extremity soft tissue sarcoma. *Journal of Surgical Oncology, 100*(5), 375–381. https://doi.org/10.1002/jso.21301.

Del Re, A. C., Fluckiger, C., Horvath, A. O., Symonds, D., & Wampold, B. E. (2012). Therapist effects in the therapeutic alliance-outcome relationship: A restricted-maximum likelihood meta-analysis. *Clinical Psychology Review, 32*(7), 642–649. https://doi.org/10.1016/j.cpr.2012.07.002.

DiMatteo, M. R., Sherbourne, C. D., Hays, R. D., Ordway, L., Kravitz, R. L., McGlynn, E. A., et al. (1993). Physicians' characteristics influence patients' adherence to medical treatment: Results from the medical outcomes study. *Health Psychology, 12*(2), 93–102. https://doi.org/10.1037/0278-6133.12.2.93.

Eippert, F., Bingel, U., Schoell, E. D., Yacubian, J., Klinger, R., Lorenz, J., et al. (2009). Activation of the opioidergic descending pain control system underlies placebo analgesia. *Neuron, 63*(4), 533–543. https://doi.org/10.1016/j.neuron.2009.07.014.

Elsenbruch, S., Kotsis, V., Benson, S., Rosenberger, C., Reidick, D., Schedlowski, M., et al. (2012). Neural mechanisms mediating the effects of expectation in visceral placebo analgesia: An fMRI study in healthy placebo responders and nonresponders. *Pain, 153*(2), 382–390. https://doi.org/10.1016/j.pain.2011.10.036.

Enck, P., Benedetti, F., & Schedlowski, M. (2008). New insights into the placebo and nocebo responses. *Neuron, 59*(2), 195–206. https://doi.org/10.1016/j.neuron.2008.06.030.

Enck, P., Bingel, U., Schedlowski, M., & Rief, W. (2013). The placebo response in medicine: Minimize, maximize or personalize? *Nature Reviews. Drug Discovery, 12*(3), 191–204. https://doi.org/10.1038/nrd3923.

Engel, C., Hamilton, N. A., Potter, P. T., & Zautra, A. J. (2004). Impact of two types of expectancy on recovery from total knee replacement surgery (TKR) in adults with osteoarthritis. *Behavioral Medicine, 30*(3), 113–123. https://doi.org/10.3200/BMED.30.3.113-123.

Epstein, R., & Street, R. L. (2007). *Patient-centered communication in cancer care: Promoting healing and reducing suffering*. Bethesda, MD: National Cancer Institute, US Department of Health and Human Services. National Institutes of Health.

Ethgen, O., Bruyere, O., Richy, F., Dardennes, C., & Reginster, J.-Y. (2004). Health-related quality of life in total hip and total knee arthroplasty: A qualitative and systematic review of the literature. *JBJS, 86*(5), 963–974.

Finniss, D. G., & Benedetti, F. (2005). Mechanisms of the placebo response and their impact on clinical trials and clinical practice. *Pain, 114*(1–2), 3–6. https://doi.org/10.1016/j.pain.2004.12.012.

Finniss, D. G., Kaptchuk, T. J., Miller, F., & Benedetti, F. (2010). Biological, clinical, and ethical advances of placebo effects. *Lancet, 375*(9715), 686–695. https://doi.org/10.1016/S0140-6736(09)61706-2.

Gandhi, R., Davey, J. R., & Mahomed, N. (2009). Patient expectations predict greater pain relief with joint arthroplasty. *The Journal of Arthroplasty, 24*(5), 716–721. https://doi.org/10.1016/j.arth.2008.05.016.

Geers, A. L., Helfer, S. G., Weiland, P. E., & Kosbab, K. (2006). Expectations and placebo response: A laboratory investigation into the role of somatic focus. *Journal of Behavioral Medicine, 29*(2), 171–178. https://doi.org/10.1007/s10865-005-9040-5.

Gollub, R. L., Kirsch, I., Maleki, N., Wasan, A. D., Edwards, R. R., Tu, Y., et al. (2018). A functional neuroimaging study of expectancy effects on pain response in patients with knee osteoarthritis. *The Journal of Pain, 19*(5), 515–527. https://doi.org/10.1016/j.jpain.2017.12.260.

Gonzalez Sáenz de Trejada, M., Escobar, A., Herrera, C., García, L., Aizpuru, F., & Sarasqueta, C. (2010). Patient expectations and health-related quality of life outcomes following total joint replacement. *Value in Health, 13*(4), 447–454.

Guyatt, G. (1992). Evidence-based medicine. *JAMA, 268*(17), 2420. https://doi.org/10.1001/jama.1992.03490170092032.

Hack, T. F., Degner, L. F., Parker, P. A., & Team, S. C. (2005). The communication goals and needs of cancer patients: A review. *Psychooncology, 14*(10), 831–845, discussion 846–837. https://doi.org/10.1002/pon.949.

Hashmi, J. A., Baria, A. T., Baliki, M. N., Huang, L., Schnitzer, T. J., & Apkarian, A. V. (2012). Brain networks predicting placebo analgesia in a clinical trial for chronic back pain. *Pain, 153*(12), 2393–2402. https://doi.org/10.1016/j.pain.2012.08.008.

Hatcher, R., & Barends, A. (2006). Thinking about the alliance in practice. *Psychotherapy, Therapy, Research, Practice, Training, 41*, 7–10.

Hirschfeld Davis, J. (2017). Trump declares opioid crisis a 'health emergency' requests no funds news paper. *The New York Times* Retrieved from https://nyti.ms/2iBk6Uc.

Horvath, A. O., & Bedi, R. P. (2002). The alliance. In J. C. Norcross (Ed.), *Psychotherapy relationships that work* (pp. 37–69). New York: Oxford University Press.

Kam-Hansen, S., Jakubowski, M., Kelley, J. M., Kirsch, I., Hoaglin, D. C., Kaptchuk, T. J., et al. (2014). Altered placebo and drug labeling changes the outcome of episodic migraine attacks. *Science Translational Medicine, 6*, 218ra5. https://doi.org/10.1126/scitranslmed.3006175.

Kaptchuk, T. J. (1998). Powerful placebo: The dark side of the randomised controlled trial. *Lancet, 351*(9117), 1722–1725. https://doi.org/10.1016/S0140-6736(97)10111-8.

Kaptchuk, T. J., Friedlander, E., Kelley, J. M., Sanchez, M. N., Kokkotou, E., Singer, J. P., et al. (2010). Placebos without deception: A randomized controlled trial in irritable bowel syndrome. *PLoS One, 5*(12), e15591. https://doi.org/10.1371/journal.pone.0015591.

Kaptchuk, T. J., Kelley, J. M., Conboy, L. A., Davis, R. B., Kerr, C. E., Jacobson, E. E., et al. (2008). Components of placebo effect: Randomised controlled trial in patients with irritable bowel syndrome. *BMJ, 336*(7651), 999–1003. https://doi.org/10.1136/bmj.39524.439618.25.

Kirsch, I. (2018). Response expectancy and the placebo effect. *International Review of Neurobiology, 138*, 81–93. https://doi.org/10.1016/bs.irn.2018.01.003.

Kirsch, I., Lynn, S. J., Vigorito, M., & Miller, R. R. (2004). The role of cognition in classical and operant conditioning. *Journal of Clinical Psychology, 60*(4), 369–392. https://doi.org/10.1002/jclp.10251.

Klinger, R., Colloca, L., Bingel, U., & Flor, H. (2014). Placebo analgesia: Clinical applications. *Pain, 155*(6), 1055–1058. https://doi.org/10.1016/j.pain.2013.12.007.

Klinger, R., Kothe, R., Schmitz, J., Kamping, S., & Flor, H. (2017). Placebo effects of a sham opioid solution: A randomized controlled study in patients with chronic low back pain. *Pain, 158*(10), 1893–1902. https://doi.org/10.1097/j.pain.0000000000000977.

Klinger, R., Soost, S., Flor, H., & Worm, M. (2007). Classical conditioning and expectancy in placebo hypoalgesia: A randomized controlled study in patients with atopic dermatitis and persons with healthy skin. *Pain, 128*(1–2), 31–39. https://doi.org/10.1016/j.pain.2006.08.025.

Kosslyn, S. M. (2005). Mental images and the brain. *Cognitive Neuropsychology, 22*(3), 333–347. https://doi.org/10.1080/02643290442000130.

Lutz, G. K., Butzlaff, M. E., Atlas, S. J., Keller, R. B., Singer, D. E., & Deyo, R. A. (1999). The relation between expectations and outcomes in surgery for sciatica. *Journal of General Internal Medicine, 14*(12), 740–744. https://doi.org/10.1046/j.1525-1497.1999.10417.x.

Miller, F. G., & Colloca, L. (2009). The legitimacy of placebo treatments in clinical practice: Evidence and ethics. *The American Journal of Bioethics, 9*(12), 39–47. https://doi.org/10.1080/15265160903316263.

Muller, M., Kamping, S., Benrath, J., Skowronek, H., Schmitz, J., Klinger, R., et al. (2016). Treatment history and placebo responses to experimental and clinical pain in chronic pain patients. *European Journal of Pain, 20*(9), 1530–1541. https://doi.org/10.1002/ejp.877.

Pearson, S. D., & Raeke, L. H. (2000). Patients' trust in physicians: Many theories, few measures, and little data. *Journal of General Internal Medicine, 15*(7), 509–513. https://doi.org/10.1046/j.1525-1497.2000.11002.x.

Peerdeman, K. J., van Laarhoven, A. I., Keij, S. M., Vase, L., Rovers, M. M., Peters, M. L., et al. (2016). Relieving patients' pain with expectation interventions: A meta-analysis. *Pain, 157*(6), 1179–1191. https://doi.org/10.1097/j.pain.0000000000000540.

Petersen, G. L., Finnerup, N. B., Grosen, K., Pilegaard, H. K., Tracey, I., Benedetti, F., et al. (2014). Expectations and positive emotional feelings accompany reductions in ongoing and evoked neuropathic pain following placebo interventions. *Pain, 155*(12), 2687–2698. https://doi.org/10.1016/j.pain.2014.09.036.

Petersen, G. L., Finnerup, N. B., Norskov, K. N., Grosen, K., Pilegaard, H. K., Benedetti, F., et al. (2012). Placebo manipulations reduce hyperalgesia in neuropathic pain. *Pain, 153*(6), 1292–1300. https://doi.org/10.1016/j.pain.2012.03.011.

Petrovic, P., Kalso, E., Petersson, K. M., & Ingvar, M. (2002). Placebo and opioid analgesia—Imaging a shared neuronal network. *Science, 295*(5560), 1737–1740. https://doi.org/10.1126/science.1067176.

Pollo, A., Amanzio, M., Arslanian, A., Casadio, C., Maggi, G., & Benedetti, F. (2001). Response expectancies in placebo analgesia and their clinical relevance. *Pain, 93*(1), 77–84.

Price, D. D., Finniss, D. G., & Benedetti, F. (2008). A comprehensive review of the placebo effect: Recent advances and current thought. *Annual Review of Psychology, 59*, 565–590. https://doi.org/10.1146/annurev.psych.59.113006.095941.

Price, D. D., Milling, L. S., Kirsch, I., Duff, A., Montgomery, G. H., & Nicholls, S. S. (1999). An analysis of factors that contribute to the magnitude of placebo analgesia in an experimental paradigm. *Pain, 83*(2), 147–156. https://doi.org/10.1016/S0304-3959(99)00081-0.

Quinn, V. F., & Colagiuri, B. (2015). Placebo interventions for nausea: A systematic review. *Annals of Behavioral Medicine, 49*(3), 449–462. https://doi.org/10.1007/s12160-014-9670-3.

Rudd, R. A., Aleshire, N., Zibbell, J. E., & Gladden, R. M. (2016). Increases in drug and opioid overdose deaths—United States, 2000–2014. *MMWR Morbidity and Mortality Weekly Report, 64*(50–51), 1378–1382. https://doi.org/10.15585/mmwr.mm6450a3.

Salkeld, G., Solomon, M., Short, L., & Butow, P. N. (2004). A matter of trust-patient's views on decision-making in colorectal cancer. *Health Expectations*, 7(2), 104–114. https://doi.org/10.1111/j.1369-7625.2004.00257.x.

Schenk, L. A., Krimmel, S. R., & Colloca, L. (2017). Observe to get pain relief: Current evidence and potential mechanisms of socially learned pain modulation. *Pain*, *158*(11), 2077–2081. https://doi.org/10.1097/j.pain.0000000000000943.

Scott, D. J., Stohler, C. S., Egnatuk, C. M., Wang, H., Koeppe, R. A., & Zubieta, J. K. (2008). Placebo and nocebo effects are defined by opposite opioid and dopaminergic responses. *Archives of General Psychiatry*, *65*(2), 220–231. https://doi.org/10.1001/archgenpsychiatry.2007.34.

Sonthalia, S., Sahaya, K., Arora, R., Singal, A., Srivastava, A., Wadhawan, R., et al. (2015). Nocebo effect in dermatology. *Indian Journal of Dermatology, Venereology and Leprology*, *81*(3), 242–250. https://doi.org/10.4103/0378-6323.155573.

Street, R. L., Jr., Makoul, G., Arora, N. K., & Epstein, R. M. (2009). How does communication heal? Pathways linking clinician-patient communication to health outcomes. *Patient Education and Counseling*, 74(3), 295–301. https://doi.org/10.1016/j.pec.2008.11.015.

Tetreault, P., Mansour, A., Vachon-Presseau, E., Schnitzer, T. J., Apkarian, A. V., & Baliki, M. N. (2016). Brain connectivity predicts placebo response across chronic pain clinical trials. *PLoS Biology*, *14*(10), e1002570. https://doi.org/10.1371/journal.pbio.1002570.

Thorne, S., Hislop, T. G., Kuo, M., & Armstrong, E. A. (2006). Hope and probability: Patient perspectives of the meaning of numerical information in cancer communication. *Qualitative Health Research*, *16*(3), 318–336. https://doi.org/10.1177/1049732305285341.

Tilburt, J. C., Emanuel, E. J., Kaptchuk, T. J., Curlin, F. A., & Miller, F. G. (2008). Prescribing "placebo treatments": Results of national survey of US internists and rheumatologists. *BMJ*, *337*, a1938. https://doi.org/10.1136/bmj.a1938.

Vase, L., Petersen, G. L., & Lund, K. (2014). Placebo effects in idiopathic and neuropathic pain conditions. *Handbook of Experimental Pharmacology*, *225*, 121–136. https://doi.org/10.1007/978-3-662-44519-8_7.

Vase, L., Petersen, G. L., Riley, J. L., 3rd, & Price, D. D. (2009). Factors contributing to large analgesic effects in placebo mechanism studies conducted between 2002 and 2007. *Pain*, *145*(1–2), 36–44. https://doi.org/10.1016/j.pain.2009.04.008.

Vase, L., Robinson, M. E., Verne, G. N., & Price, D. D. (2003). The contributions of suggestion, desire, and expectation to placebo effects in irritable bowel syndrome patients. An empirical investigation. *Pain*, *105*(1–2), 17–25.

Vase, L., Robinson, M. E., Verne, G. N., & Price, D. D. (2005). Increased placebo analgesia over time in irritable bowel syndrome (IBS) patients is associated with desire and expectation but not endogenous opioid mechanisms. *Pain*, *115*(3), 338–347. https://doi.org/10.1016/j.pain.2005.03.014.

Vowles, K. E., McEntee, M. L., Julnes, P. S., Frohe, T., Ney, J. P., & van der Goes, D. N. (2015). Rates of opioid misuse, abuse, and addiction in chronic pain: A systematic review and data synthesis. *Pain*, *156*(4), 569–576. https://doi.org/10.1097/01.j.pain.0000460357.01998.f1.

Wager, T. D., Rilling, J. K., Smith, E. E., Sokolik, A., Casey, K. L., Davidson, R. J., et al. (2004). Placebo-induced changes in fMRI in the anticipation and experience of pain. *Science*, *303*(5661), 1162–1167. https://doi.org/10.1126/science.1093065.

Wiech, K. (2016). Deconstructing the sensation of pain: The influence of cognitive processes on pain perception. *Science*, *354*(6312), 584–587. https://doi.org/10.1126/science.aaf8934.

FURTHER READING

Colloca, L., & Benedetti, F. (2005). Placebos and painkillers: Is mind as real as matter? *Nature Reviews Neuroscience*, *6*(7), 545.

CHAPTER SIX

Optimizing Placebo and Minimizing Nocebo to Reduce Pain, Catastrophizing, and Opioid Use: A Review of the Science and an Evidence-Informed Clinical Toolkit

Beth D. Darnall[*,1], **Luana Colloca**[†,‡,§]
[*]School of Medicine, Department of Anesthesiology, Perioperative and Pain Medicine, Division of Pain Medicine, Psychiatry and Behavioral Sciences (by courtesy), Stanford University, Palo Alto, CA, United States
[†]Department of Pain Translational Symptom Science, School of Nursing, University of Maryland, Baltimore, MD, United States
[‡]Departments of Anesthesiology and Psychiatry, School of Medicine, University of Maryland, Baltimore, MD, United States
[§]Center to Advance Chronic Pain Research, University of Maryland, Baltimore, MD, United States
[1]Corresponding author: e-mail address: bdarnall@stanford.edu

Contents

International Review of Neurobiology, Volume 139
ISSN 0074-7742
https://doi.org/10.1016/bs.irn.2018.07.022

Abstract

Pain, a noxious psychosensory experience, motivates escape behavior to assure protection and survival. Psychological factors alter the experience and trajectory of pain, as well as behavior and treatment response. In the context of pain, the placebo effect (expectation for pain relief) releases endogenous opioids and facilitates analgesia from exogenously administered opioids. Nocebo hyperalgesia (expectation for persistent or worsening pain) opposes endogenous opioid analgesia and patient engagement in prescription opioid tapering. Reductions in nocebo hyperalgesia and pain catastrophizing may enhance descending modulation of pain, mediate adaptive structural brain changes and promote patient engagement in opioid tapering. Interventions that minimize nocebo and optimize placebo may adaptively shape the central nervous system toward pain relief and potentially opioid reduction. Here we provide a critical description of catastrophizing and its impact on pain, placebo and nocebo effects. We also consider the importance of minimizing nocebo and optimizing placebo effects during prescription opioid tapering, and offer a clinical toolkit of resources to accomplish these goals clinically.

1. INTRODUCTION

1.1 The Problem of Pain

Pain is a global health problem with broad negative impacts on physical (Sturgeon, Darnall, Kao, & Mackey, 2015; Sturgeon, Dixon, Darnall, & Mackey, 2015), mental (Bair, Robinson, Katon, & Kroenke, 2003; Feinstein et al., 2017; Ziadni, Sturgeon, & Darnall, 2018), spiritual (Halawa, Al-Diri, McLean, & Darnall, 2015), psychosocial (Karos, Meulders, Goubert, & Vlaeyen, 2018; Karos, Williams, Meulders, & Vlaeyen, 2018; Sturgeon et al., 2016; Ziadni, You, Wilson, & Darnall, 2018), and economic domains (Groenewald, Essner, Wright, Fesinmeyer, & Palermo, 2014; Gustavsson et al., 2012). The 2016 Institute of Medicine report on *Relieving Pain in America* estimated that roughly one-third of the world population is living with ongoing pain of some type (IOM Committee on Advancing Pain Research, 2011). Pain is more costly than diabetes, heart disease, and cancer combined, with combined estimates reaching up to $635 billion each year for medical costs and lost productivity in the United States alone (IOM Committee on Advancing Pain Research, 2011). Effective, scalable, and low-risk pain treatment strategies are urgently needed, particularly in light of calls to reduce opioid prescribing as a pathway to mitigate opioid-related morbidity and mortality in the United States·(CDC, 2016; Hoffman, 2018), Australia, Canada, and elsewhere. Indeed, opioid de-prescribing practices have rapidly taken effect across the United States with scant attention given to the potential patient harms caused by aggressive tapering approaches, including clear nocebo effects (Hoffman, 2018; Langreth, 2017). Treating pain effectively and compassionately—and at lowest-risk—requires careful attention to the psychological dimensions of pain and, when relevant, opioid reduction (Darnall, 2014a, 2014b).

Pain is a psychosensory experience wherein the brain perceives and interprets pain signaling (Darnall, 2018b). Indeed, by definition pain comprises psychological elements (IASP, 1994), thereby suggesting that, in part, analgesia depends on them. The extant literature demonstrates that psychopathology is both an antecedent (Gerrits, van Marwijk, van Oppen, van der Horst, & Penninx, 2015) and a consequence of persistent pain (Archer et al., 2016). Extending the scope beyond formal psychopathology, psychological factors that are known to influence pain and analgesia include cognition (Burns, Glenn, Bruehl, Harden, & Lofland, 2003; Darnall et al., 2017; Salomons, Moayedi, Erpelding, & Davis, 2014; Seminowicz & Davis, 2006; Seminowicz et al., 2013; Ziadni, Sturgeon, et al., 2018), emotion (Burns et al., 2015; McCracken & Keogh, 2009; Vlaeyen, Crombez, & Linton, 2016), appraisal (Ziadni, Sturgeon, et al., 2018), expectations (Atlas et al., 2012; Colloca & Miller, 2011b; Palermo & Drotar, 1996; Wager, Atlas, Leotti, & Rilling, 2011), attention (Kucyi, Salomons, & Davis, 2013; Seminowicz & Davis, 2006), beliefs about pain and its treatment (Carriere, Martel, Kao, Sullivan, & Darnall, 2017; Carriere et al., 2018). Mechanisms of psychological effects on pain and analgesia include behavioral factors (Linton, Flink, & Vlaeyen, 2018; Vlaeyen et al., 2016), conditioning, and neurochemical pathways. Neurally, psychological factors can influence pain and analgesia through descending modulation of pain wherein pain is either facilitated or impeded depending on one's adaptive capacities. As such, low-risk analgesia may be achieved by targeting psychological factors known to amplify pain (Darnall, 2014a, 2014b). While adaptively engaging descending modulation confers in vivo analgesia, longitudinal clinical research in chronic pain has shown that a pattern of engaging descending modulation over a period of weeks is associated with structural changes in the brain that appear to prime the central nervous system for future analgesia, thereby altering the trajectory of pain (Seminowicz et al., 2013). We review the relevance of placebo and nocebo processing in shaping the central nervous system either toward relief or pain exacerbation, and provide an evidence-based clinical toolkit to enhance placebo and pain relief.

1.2 Placebo and Nocebo Are Integral to Pain Experience

Decades of mechanistic research on placebo and nocebo effects is serving to inform the development and integration of placebo optimization strategies into clinical care pathways to treat pain. To understand how placebo and nocebo science may be applied to address the current dual pain and opioid crises, we first review several elemental principles and relevant key research findings. Beginning with nomenclature, placebo and nocebo effects are psychoneurobiological responses that occur in the body as result of positive and negative expectations (Colloca, 2018a, 2018b; Wager & Atlas, 2015). Expectations result in brain events that trigger the release of endogenous neuropeptides and influence behaviors. Placebo effects due to positive expectations have been linked to the release of endogenous opioids for a review, see Eippert et al. (2009) and Pecina and Zubieta (2018) and cannabinoids (Benedetti, Amanzio, Rosato, & Blanchard, 2011). Studies using indirect pharmacological approaches have demonstrated that placebo analgesia is antagonized by the opioid antagonist naloxone, thus, indicating that endogenous opioids crucially involved in placebo analgesic effects. Moreover, pharmacological fMRI and PET studies using an in vivo receptor binding with opioidergic ligands have provided evidence of the anatomical localization of the neuropeptides in the brain (Eippert et al., 2009; Wager, Scott, & Zubieta, 2007; Zubieta et al., 2005). In another evoked pain paradigm, Tor Wager and colleagues illustrated engagement of mu-opioid activity during placebo analgesia (Wager et al., 2007). Participants in the study were told that pills they were given would relieve their pain. Results showed that reports of placebo analgesia following administration of an inert pill was correlated with endogenous release of opioids. As such, the release of endogenous opioids depended on the belief that treatment-related pain relief was imminent.

The cannabinoid receptor 1 (CB1) antagonist SR 141716A (rimonabant) blocks placebo analgesia elicited by placebo given after NSAID ketorolac indicating an involvement of to the release of endogenous cannabinoids (Benedetti et al., 2011). Recently, it has been shown that oxytocin and vasopressin agonists given intranasally enhance behavioral placebo analgesia in men (Kessner, Sprenger, Wrobel, Wiech, & Bingel, 2013) and women (Colloca, Pine, Ernst, Miller, & Grillon, 2016), suggesting that the oxytocinergic and vasopressinergic systems, typically involved in the modulation of social behaviors (Campbell, 2010; Heinrichs & Domes, 2008) can be used as enhancers of placebo analgesic effects. Further research is needed to determine how distinct doses of oxytocin and vasopressin influences outcomes and affect brain mechanisms underlying this potentiation.

On the contrary, nocebo effects have been linked to the release of cholecystokinins that are involved in the modulation of anxiety and hyperalgesia. The block of the CCK A and B receptors with the type A/B receptor antagonist proglumide antagonizes nocebo hyperalgesia (Benedetti, Amanzio, Vighetti, & Asteggiano, 2006).

A noxious stimulus, pain signals threat or harm to an individual, and motivates escape behavior to achieve protection and survival. Viewed in this light, acute or current pain may serve as a direct nocebo reaction. Context information that is derived from one's environment and habitus (i.e., pain itself) interact with an individual's psychological status—consciously or subconsciously—in dynamic fashion. Indeed, pain placebo and nocebo are opposing phenomena existing most likely on a continuum.

Accordingly, optimizing placebo in the clinical setting requires one identify and extinguish any existing nocebo effects. In later sections, we will address the importance and procedures for attempting to extinguish nocebo effects for pain and opioid reduction.

1.3 Conceptualizing Nocebo to Encompass Pain Proper

Expectation for pain elicits a nocebo effect: facilitation of pain, distress and disease. Given that pain triggers a latent appraisal of noxious experience of varying degrees and related components (e.g., sensory, physical, emotional), it could be argued that an individual's positive appraisal regarding their ability to reduce their pain constitutes a viable way to create placebo effects. Whereas, an appraisal that pain will only worsen and there is nothing that can reduce one's pain constitutes a nocebo effect, whether such appraisal is due to poor faith in current treatments or one's ability to effectively self-manage or self-modulate pain. The patient's prior negative experience (memory, learning, priming and conditioning), internal states, and external context cues may interact dynamically to influence her/his brain responses to either inhibit or facilitate pain (Benedetti, Mayberg, Wager, Stohler, & Zubieta, 2005; Di Blasi & Kleijnen, 2003; Mistiaen et al., 2016).

Research suggests that stronger belief in the treatment enhances its effects (Doering, Glombiewski, & Rief, 2018; Kong et al., 2006; Kube, Glombiewski, & Rief, 2018; Seminowicz, 2006). In this case, greater belief in one's ability to self-modulate pain in turns may facilitate the expected analgesic outcome. Whereas, strong belief in a deficiency to self-modulate pain in turn may result in amplified pain.

1.4 Nocebo and Pain Catastrophizing

Accordingly, pain catastrophizing, a psychological construct and potent index of descending modulation of pain, is a particularly useful model for studying placebo and nocebo effects. Pain catastrophizing is a persistent pattern of distressing cognitive and emotional responses to current or anticipated pain. First described by Rosenstiel and Keefe and measured with the catastrophizing subscale of the Coping Skills Questionnaire, pain catastrophizing is comprised of a pattern of rumination on pain and feelings of helplessness about pain (Rosenstiel & Keefe, 1983). This definition of pain catastrophizing was later expanded upon by Michael Sullivan and colleagues to also include a third component: magnification of pain. The expanded definition is assessed with the Pain Catastrophizing Scale (PCS), a lengthier 13-item measure that prompts respondents to reference painful incidents in their life and to rate the degree to which they tend to have various negative thoughts, expectations or emotions about pain when they are experiencing it (Sullivan, 1995). Example items from the PCS include: "I become afraid that the pain will get worse"; "there's nothing I can do to reduce the intensity pain"; and "[my pain] is terrible and I think it's never going to get any better." Each item is rated on a 0–4 scale wherein 0 = "not at all" and 4 = "all the time." The 13 items are summed to arrive at a total PCS score, which reflects an individual's trait disposition toward pain catastrophizing. Pain catastrophizing is a malleable construct that is responsive to both longer course (Cherkin et al., 2016; Seminowicz et al., 2013; Turner et al., 2016) and brief, targeted psychological treatment (Darnall, Sturgeon, Kao, Hah, & Mackey, 2014; Thorn et al., 2007).

Placebo effects have been linked to a distinct series of psychological factors such as dispositional optimism, hypnotic suggestibility, somatic focus, empathy, neuroticism, altruisms, and the locus of ego-reliance (for a review, see Colloca & Grillon, 2014). Conversely, anxiety (Staats, Staats, & Hekmat, 2001), harm avoidance and persistence (Corsi, Emadi Andani, Tinazzi, & Fiorio, 2016) and anxiety sensitivity and physiological suggestibility (Corsi & Colloca, 2017) have been associated with nocebo effects in healthy subjects. In particular, pain catastrophizing has also been associated to nocebo effects in experimental settings in which nocebo manipulations are performed (Corsi & Colloca, 2017; Swider & Babel, 2013; Vogtle, Barke, & Kroner-Herwig, 2013).

A recent study looked at low scores of pain catastrophizing along with expectation for pain relief, anxiety, depression and the personality trait neuroticism and high levels of positive emotions as predictors of placebo responses in randomized, double-blind, placebo-controlled, crossover study with the anticonvulsant oxcarbazepine for treatment of peripheral neuropathic pain (Lund et al., 2017). Multiple regression analyses with pain reduction during placebo treatment as the dependent variable and baseline pain, age, sex, and pain duration revealed high significance ($P < 0.001$), while other psychological variables did not reach statistical significance. Further studies are needed exploring the link between placebo and nocebo effects in patients suffering from chronic and acute pain.

Before we discuss the possibility to target psychotherapy to treat catastrophizing thoughts and pain exacerbation = in extinguishing nocebo, we first review its impacts and importance as a therapeutic target.

1.4.1 Experimental Studies

The extent to which a persistent pattern of catastrophizing might sensitize the nervous system has been a recent topic of interest and investigation. Neuroimaging studies conducted on healthy volunteers have shown that pain catastrophizing is associated with amplified activity in areas of the brain associated with the experience of pain, and this directly correlates with report of increased pain (Seminowicz & Davis, 2006).

Our group conducted a controlled pilot study that tested the psychosensory effects of a pain catastrophizing induction in women with chronic pain (evoked pain catastrophizing), and specifically tested whether secondary hyperalgesia was associated with pain catastrophizing during evoked pain (Taub, Sturgeon, Johnson, Mackey, & Darnall, 2017). Study participants assigned to the catastrophizing condition were guided to imagine their pain worsening and to envision great negative consequence resulting from increased pain. The imagined scenario was designed to allow for a personally relevant construction of worsening pain, reduced function, and "worst case scenario" as it pertained to each individual. Quantitative sensory testing conducted before and after the induction was designed to reveal whether greater pain catastrophizing was associated with increased pain intensity (hyperalgesia) as well as spread of pain to non-painful areas (allodynia). Findings revealed that two quantitative sensory tests (weighted pin and mechanical allodynia) were associated with secondary hyperalgesia for women with greater levels of evoked pain catastrophizing. Replication of this work would confirm central sensitization as a mechanism of pain catastrophizing (Taub et al., 2017). Other research provides supporting evidence of catastrophizing having a mechanistic role in central sensitization, though viewed from the other direction. Salomons and colleagues conducted a randomized controlled experiment to test a brief cognitive behavioral intervention designed to enhance descending modulation of experimental pain could mitigate catastrophizing and pain (Salomons et al., 2014). The intervention included 5-min instructional sessions on cognitive regulation of pain given prior to each of eight sessions of evoked heat pain. The researchers found that the brief cognitive training reduced pain unpleasantness—but not pain intensity—as well as secondary hyperalgesia and pain catastrophizing. The authors stated that their reported association between reduced secondary hyperalgesia and pain catastrophizing suggested that reductions in central sensitization are related to volitional alterations of pain-related cognitions (Salomons et al., 2014).

Longitudinal studies conducted in clinical samples are needed to extend beyond the proximal impacts of nocebo and characterize the mechanisms of pain catastrophizing and placebo and nocebo effects.

1.4.2 Clinical Research

Across different treatment settings spanning outpatient, inpatient, and perioperative environments, pain catastrophizing consistently emerges as one of the strongest predictors of pain treatment outcomes.

A systematic review of pain catastrophizing (total $N = 2269$) found that pain catastrophizing predicted pain intensity and disability, and pain catastrophizing mediated back pain treatment efficacy in outpatients seeking specialty pain care (Wertli et al., 2014). Independent of depression, pain catastrophizing has been associated with an array of negative phenomena including increased affective distress (Picavet, Vlaeyen, & Schouten, 2002; Spinhoven et al., 2004), muscle and joint tenderness (Severeijns, Vlaeyen, van den Hout, & Weber, 2001), muscular tension at rest (Smeets, van Geel, Kester, & Knottnerus, 2007), pain-related disability (Severeijns, Vlaeyen, & van den Hout, 2004; Severeijns et al., 2001), and poor response to various pain treatments including surgery (Abbott, Tyni-Lenne, & Hedlund, 2011; Helmerhorst, Vranceanu, Vrahas, Smith, & Ring, 2014; Jensen, Thomsen, & Hojsted, 2006; Kennedy, Vranceanu, Nunez, & Ring, 2010; Smeets et al., 2007; Spinhoven et al., 2004; Theunissen, Peters, Bruce, Gramke, & Marcus, 2012). Indeed, the perioperative setting is useful for investigating the clinical impacts of pain, nocebo-induced hyperalgesia, particularly with surgery often involving a clear pain stimulus and a defined recovery period. To investigate the impact of pain catastrophizing on post-surgical outcomes, researchers typically measure pain catastrophizing tendencies prior to surgery. Greater pain catastrophizing has been shown to be directly associated with greater use of opioids and pain in the hospital after surgery (Papaioannou et al., 2009; Wright, Hoang, Sofine, Silva, & Schwarzkopf, 2017), longer hospital stay (Wright et al., 2017), delayed recovery from surgery (Roh et al., 2014), and persistent opioid use (Helmerhorst et al., 2014). Pain catastrophizing also harmful in the context of acute pain and even for individuals who are pain-free. For instance, pain catastrophizing is implicated in the persistence of back pain (Picavet et al., 2002) and researchers found that pain catastrophizing accounted for 47% of the variance in the *development* of chronic back pain following an episode of acute back pain (Burton, Tillotson, Main, & Hollis, 1995). Moreover, a cross-sectional observational population study revealed that among all factors investigated, pain catastrophizing best predicted the acquisition of chronic low back pain 1 year after a pain-free baseline (Linton, 2005). Taken together, these data illustrate the detrimental impacts of pain catastrophizing across settings and populations, and suggest that early treatment for pain catastrophizing may serve as prophylaxis for prevention of chronic pain phenotypes.

Given that pain catastrophizing appears to undermine response to medical pain treatments, it is perhaps unsurprising that *reductions in pain catastrophizing are associated with subsequent* improvements in pain and intervention effectiveness. In a prospective study, Burns et al. used cross-lagged analysis to show that early reductions in pain catastrophizing significantly improved later response to multidisciplinary treatment in terms of pain intensity and pain interference (Burns, Glenn, et al., 2003; Burns, Kubilus, Bruehl, Harden, & Lofland, 2003). Neuroimaging studies conducted on individuals with chronic pain reveal that greater pain catastrophizing is associated with volumetric deficits in key brain regions associated with descending modulation of pain (Seminowicz et al., 2013). While it is unknown whether structural deficits were antecedents or consequences of catastrophizing, their co-occurrence suggests a mutually-reinforcing biobehavioral cycle of pain and potential nocebo-induced hyperalgesia (Blasini, Corsi, Klinger, & Colloca, 2017). Indeed, pain catastrophizing in individuals with chronic pain has been shown to associate with altered neural functioning outside of the context of evoked pain or in vivo catastrophizing (Jiang et al., 2016). Greater pain catastrophizing is associated with altered brain functioning in the default mode network—an over-coupling between the central executive network and the amygdala—that attune the brain to pain (Jiang et al., 2016). These findings suggest that pain catastrophizing is associated with neural alterations in individuals with chronic pain that appear to prime their nervous pain signaling systems for future pain.

1.5 Reducing Pain Catastrophizing: Shaping Patient Expectations Toward Pain Relief

Cognitive behavioral therapy for chronic pain (pain-CBT) effectively reduces pain catastrophizing and increases pain self-efficacy; that is, the belief in one's ability to engage in various life activities despite living with ongoing pain (Cherkin et al., 2016; Stewart et al., 2015; Thorn et al., 2007; Williams, Eccleston, & Morley, 2012). Pain-CBT is typically delivered by a trained psychologist either individually or in group classes. Most often, group pain-CBT is a series of 2-h classes (often 8–11 classes) delivered weekly. Classes include social interaction, didactic content, and experiential exercises. Didactic content includes education about pain and the role of psychology in its treatment, thereby establishing the treatment rationale, as well as remoralization around the notion that personal control over pain may be cultivated. Participants learn about the role of cognition in pain perception, emotional experience, and physiological responses. Importantly, participants learn to identify their maladaptive thought patterns that maintain and amplify pain and distress. Often negative thoughts involve catastrophizing, and pain-related negative thoughts (e.g., "My pain is never going to go away," or "I am at the mercy of my pain."). Acquired skills specifically enhance descending modulation of pain. Cognitive restructuring, relaxation training, and positive distraction are adaptive modulatory skills that may effectively interrupt in vivo catastrophizing. Repeated application and thought interruption weakens the negative neural and behavioral patterns. Volitionally calming the nervous system allows for recoding of physiological responses that oppose pain and nocebo (e.g., relaxation, positive expectation for relief and belief in one's ability to self-soothe) and lead to lasting adaptive changes in the nervous system. Seminowicz and colleagues provided strong evidence in this direction. The research group conducted pre-post treatment neuroimaging on patients with chronic pain who underwent an 11-week course of group pain-CBT (Seminowicz et al., 2013). The authors reported that prior to pain-CBT, patients evidenced volumetric deficits in regions of the brain associated with pain control. However, the pre-treatment volumetric deficits were mitigated substantially in the post-treatment scans, and the pre-frontal gray matter brain volume increases were entirely mediated by reductions in pain catastrophizing. The adaptive structural brain changes and reductions in pain catastrophizing correlated directly with decreased pain intensity.

These findings underscore that descending modulation of pain may be applied to directly impact pain perception and cultivated to shape enduring brain changes that confer relief. Placebo and patient engagement in pain-CBT may be optimized by sharing these specific neuroscientific findings for treatment research conducted on clinical samples. Indeed, a central goal of pain-CBT is to enhance descending modulation of pain. Colloca and colleagues' work reveals that a reinforced expectancy (e.g., via conditioning) strongly create large pain modulatory effects (Au Yeung, Colagiuri, Lovibond, & Colloca, 2014; Colloca, Jonas, Killen, Miller, & Shurtleff, 2014; Klinger, Colloca, Bingel, & Flor, 2014). The greater one's expectations for pain to improve (or worsen), the greater pain modulation occurs in the expected direction (Corsi & Colloca, 2017). Shaping placebo to enhance descending modulation and facilitate ongoing engagement with adaptive skills may optimize its clinical manifestation (Klinger, Blasini, Schmitz, & Colloca, 2017). A recent meta-analysis of 27 studies showed that how interventions are presented to patients impacts their pain. Basic information about a treatment can serve to reassure patients that they will have a good response to the treatment—they experience greater analgesia, including individuals with chronic pain (Peerdeman et al., 2016). Providing compelling positive results for scientific studies may boost placebo further, in part by enhancing patient preference and engagement in the treatment. Indeed, a common notion is that psychological treatment for pain is "palliative care," something to be administered when all real medical treatment fails. This common flawed perspective that relegates psychological treatment to "learning how to cope with pain" can severely limit engagement in pain-CBT and undermine placebo. Placebo optimization for pain-CBT includes providing patients with the scientific evidence that psychological treatment extends well beyond "pain coping"; rather, pain-CBT alters the pain experience itself, shapes the trajectory of pain, changes the functioning and the structure of the brain so that the nervous system becomes "primed" for relief.

1.6 The "Actual" Effect of a Treatment: A Mythical Pursuit in Chronic Pain?

To date, research has mainly focused on controlling for placebo effects and minimizing them for the purpose of elucidating the "actual" effect of a therapy in the context of clinical trials. However, the recent research suggests that it is challenging to isolate the so-called "actual" treatment effect in real-world clinical settings (emerging uncertainty principle, see Colloca & Benedetti, 2005) wherein treatments are applied to patients who bring their entire psychology with them—including their cognition, emotion, beliefs and expectations about the treatment they are about to receive—the very factors that profoundly influence pain and treatment response. Furthermore, the success of behavioral treatment such as pain psychology treatment is dependent on patient engagement and a belief that the treatment will benefit them (placebo). As such, the notion of a static pain treatment effect existing in the absence of placebo/nocebo may have questionable value in everyday clinical settings. Pain is an individual experience, as is treatment response, and both interact dynamically with psychological factors. As such, perhaps the most useful clinical pathway is to phenotype patients prior to treatments, identify therapeutic targets for minimizing nocebo and optimizing placebo, and direct resources to enhance pain treatment outcomes (Darnall, 2018a). Owing to its impacts and relationship to nocebo and treatment outcomes, pain catastrophizing remains a primary, high-yield therapeutic target.

1.7 Patient Preference: A Fly in the Ointment

The potential mechanisms of placebo effects are manifold and in addition to neural and pharmacological pathways include behavioral factors that impact engagement and adherence to the treatment regimen, including patient preference for a particular treatment. Patient preference may partially index a belief that the treatment will be beneficial, although various other factors are known to influence patient preferences (Enck, Grundy, & Klosterhalfen, 2012), such as burdens related to actively engaging in a treatment, associated costs, and potential side effects.

Above we briefly touched on the importance of placebo optimization for pain-CBT as a pathway to enhance patient preference and engagement in this clinical treatment pathway. The goal is to provide compelling contextual information that makes the patient *want* the treatment, then boost their engagement in the treatment (behavior change). Combined, this creates a powerful cycle of cognitive, emotional, and behavioral reinforcement related to the treatment. Despite its strong influence on placebo and treatment response, patient preference for treatment type is often ignored in pain research, thereby confounding study findings. Indeed, similar to the placebo studies, in the "gold standard" of clinical trials designs, the randomized controlled clinical trial, patient participant treatment preference is often "subtracted out" as if real-world treatment response does not depend on it. Admittedly, controlling for patient preference in analytic models does inform its predictive value. However, assigning a patient to a treatment group they do not believe will benefit them reduces the likelihood of positive treatment response. In the absence of patient choice, treatment research results likely underestimate true treatment effects (Bingel et al., 2011), particularly in psychological treatment studies that require a high degree of active engagement compared to relatively passive pharmacologic treatments. Recognizing that clinical care does not exist in blinded fashion, research that aims study treatment effects should consider allowing patient choice in the treatment (Enck et al., 2012) whenever possible (equipoise randomized stratification is one statistical method that can be applied for this purpose (Lavori et al., 2001)), as well as include strategies to further enhance preference. The rationale is that the true *available* treatment effect is:

Treatment applied in an engaged patient who chooses the treatment pathway based on a belief that the treatment will be of benefit:

- With treatment = x and placebo = y, the true available treatment effect is x + y.

While the treatment (x) is relatively static, (y) is malleable and dependent on a variety of contextual factors, including the patient-provider dynamics, careful education and patient comprehension about treatment benefits and why they are important to the patient. A new era of patient-centered care and precision medicine stands to improve the effectiveness of various treatments both because the treatment has greater precision and because patient receptivity is enhanced by clear understanding of the personal relevance and importance of the treatment. Placebo optimization strategies may also be applied within the context of any type of analgesic treatment, including psychological treatment, physical therapy, as well as strategies individuals use in their daily lives to self-manage their pain.

1.8 Minimizing Nocebo and Optimizing Placebo for Opioid Reduction

Klinger and colleagues discussed several approaches to optimize placebo response for prescribed analgesic medications (Klinger et al., 2017, 2014). Two examples of strategies designed to enhance positive patient expectations include emphasizing the drug's positive value while minimizing side effects, as well as carefully explaining the drug's mechanisms of actions to the patient (Klinger et al., 2017, 2014). Here, through the lens of prescription opioids, we extend the discussion of placebo optimization beyond the medication to include medication prescribing procedures (e.g., patient engagement), medication titration procedures and specifically symptom management in regards to prescription opioid tapering.

1.9 Avoiding the Nocebo Pitfall of Opioid Tapering

Human physiology fairly rapidly adapts to daily administration of opioids. Over the course of weeks of daily opioid use, most people will develop a degree of physical dependence; that is, if the drug is suddenly withdrawn, noxious symptoms arise (e.g., withdrawal symptoms). Withdrawal symptoms may include increased pain, nausea, anxiety, restlessness, opioid craving, muscle aches, and stomach cramps. While not dangerous, the severity of withdrawal symptoms and related discomfort may range from mild to intolerable. Most patients taking long-term opioids understand withdrawal symptoms through prior experience: they may have accidentally missed a dose of medication, or may have tried to taper or stop opioids and experienced withdrawal symptoms. The experience of opioid withdrawal symptoms may lead patients to encode a false belief that they are unable to taper their opioids and must maintain their current dose. In fact, withdrawal symptoms do not index an individual's capacity to reduce prescription opioid dose; rather, withdrawal symptoms index a need for better tapering methods. Moreover, increased pain is a common opioid withdrawal symptom; however, many patients may encode the false belief that this amplified withdrawal-related pain is their "baseline pain level," thereby leading them to conclude that opioids are the only way to maintain a tolerable level of pain. These common false beliefs are may be powerfully anchored with negative reinforcement in that re-administration of opioids eliminates noxious symptoms, including amplified pain. Unfortunately, current guidelines may be in some clinical cases too aggressive for chronic pain patients who often have been taking opioids for years or decades (Berna, Kulich, & Rathmell, 2015). Aggressive tapers may trigger withdrawal symptoms and unintended nocebo effects, thereby perpetuating the false beliefs that can maintain patients on opioids when they otherwise would have been interested in reducing or stopping opioids if offered a successful pathway forward.

Even in the absence of prior experience with opioid withdrawal symptoms, it is intuitive for patients to assume that their pain will increase in the absence of their pain medication. However, data from opioid tapering studies demonstrate that opioid reduction is more often associated with *reduced* pain when they are tapered the right way (Baron & McDonald, 2006;

Crisostomo et al., 2008; Murphy, Clark, & Banou, 2013). While these studies involved resource-intense methods to achieve opioid cessation (e.g., inpatient interdisciplinary treatment delivered over the course of weeks), intensive and costly treatment may not be requisite. For instance, Darnall and colleagues' findings suggested that patient-centered opioid tapering methods may help community-based outpatients achieve opioid reduction without costly resources, and without increased pain (Darnall et al., 2018). A key aspect of patient-centered opioid tapering methods involves identifying and addressing opioid tapering negative expectations and related-nocebo effects as a pathway to patient engagement in the taper process. A second key aspect is reducing the pace of the taper to allow ample time for physiologic and psychologic adaptive to occur; this serves to obviate nocebo effects, as well as contextually cultivate placebo and a belief that successful tapering is possible.

To set the stage for placebo optimization and patient-centered opioid tapering, we administered an online survey to 1561 patients with chronic pain taking long term opioids to understand their opinions and concerns regarding potential opioid reduction. Surveys were completed by 248 patients (16% response rate). Results were perhaps unsurprising: patients reported that their primary concerns about opioid reduction were increased pain and withdrawal symptoms—negative thoughts and nocebo effects about opioid reduction. Results also revealed that respondents were unaware that opioid reduction could be associated with reduced pain. Seventy percent of patients reported that they would be interested in trying to reduce their opioids if they knew first about the positive results for prior opioid tapering studies (unpublished data). These findings dovetail with work conducted by Darnall, Colloca and others showing that patient concerns and fears about opioid tapering must be addressed first to minimize nocebo effects, empowering positive expectations by optimizing patients' education and patient–clinician communication (Colloca & Finniss, 2012) including amplified pain and poor taper result, and to best ensure patient engagement in the opioid tapering process and clinically-relevant outcomes (Benedetti, Lanotte, Lopiano, & Colloca, 2007; Colasanti, Rabiner, Lingford-Hughes, & Nutt, 2011; Colloca & Benedetti, 2007; Colloca, Klinger, Flor, & Bingel, 2013; Colloca & Miller, 2011a, 2011b; Darnall et al., 2018; Horin, Lee, & Colloca, 2014). Brief education from the prescribing physician can reassure, soften or eliminate negative expectations and enhance patient receptivity and actual analgesic response to the intervention (Benedetti et al., 2007; Colasanti et al., 2011; Colloca & Benedetti, 2007; Colloca & Finniss, 2012;

Colloca et al., 2013), a particularly crucial strategy when opioids are being reduced (Colloca & Miller, 2011b; Darnall et al., 2018; Horin et al., 2014). Indeed, positive patient expectations can enhance response to opioids, reduce pain, and help opioids work better at lower doses.

Assessment of patient expectations and readiness to taper opioids is vitally important for clinical outcomes as well as empirical study on the topic. Scant research exists on prescription opioid tapering, and few studies that have assessed patient expectations prior to the taper. To address this unmet need, our national clinical trial on patient-centered prescription opioid tapering will be carefully assessing patient expectations for opioid tapering (https://www.pcori.org/research-results/2017/comparative-effectiveness-pain-cognitive-behavioral-therapy-and-chronic-pain, 2017). Further, it includes methods to enhance placebo and patient readiness to engage in opioid tapering process. This pragmatic study will allow us to conduct a large scale test of placebo/nocebo on opioid tapering, as well as methods to optimize placebo effects and taper response.

An additional strategy can be employed to challenge patient expectation that they will experience withdrawal symptoms and pain as a consequence of prescription opioid tapering: micro-dose decrements. Anxiety regarding opioid reduction is likely to be highest at the outset of a taper. As such, making tiny reductions in dose can obviate withdrawal symptoms and allow patients to remain comfortable and gain confidence in their ability to reduce their opioids very slowly without experiencing noxious symptoms (Darnall et al., 2018). As such, preventing noxious symptoms at the outset of a taper—and providing sufficient time for physiologic and psychologic adaptation to opioid dose reductions—may minimize attrition and improve taper response (Darnall et al., 2018).

Finally, whenever possible, helping patients have choice and control in the process will best support successful outcomes. As discussed earlier, patient preference *for the intervention*—in this case choosing to reduce opioids because they are convinced of its benefits—enhances treatment outcomes. Going one step further, accounting for patient preferences *during the tapering process* may be equally important. Allowing patients to control over the pace of their taper—pausing their taper or go more slowly when desired—can provide added reassurance during critical time points that are likely to be laden with emotional distress. Continuing the taper process when the patient has confidence in their own readiness signals optimized placebo and increased likelihood for successful outcome.

1.10 Clinical Implications of Placebo and Nocebo Effects and Endogenous Mediated-Opioid Analgesia

As discussed earlier, evidence reveals that placebo/nocebo expectations influence the endogenous release of opioids, suggesting influence on analgesic response to exogenously administered opioids. Bingel et al. investigated the impacts of opioid analgesia on 22 healthy volunteers who were exposed to a heat pain stimulus while simultaneously being administered IV remifentanil in each of three conditions (Bingel et al., 2011). Pain intensity was individually determined to a self-reported moderate intensity, and was applied to participants during in all three conditions. In the first condition, participants were told they were receiving a powerful painkiller during the pain experiment. In the second condition, participants were told they were receiving only saline through the IV, and therefore they would experience the moderate amount of pain that had been individually determine. In the third condition, participants were told they would receive something that would amplify their experience of the heat pain. As such, the researchers only altered participant expectations for pain and relief thusly: (1) placebo, (2) neutral, (3) nocebo. Positive expectations for pain relief due to opioids (opioid placebo) were found to double the analgesic effect of remifentanil relative to the neutral condition. Conversely, nocebo expectations that were induced in the third condition effectively abolished the analgesic effect of remifentanil. Moreover, pain and analgesia findings correlated with functional neuroimaging data supporting modulation of pain processing in the brain based on the condition group. Finally, across the conditions, modulation of anxiety directly aligned with expectations for pain. The findings from this study suggest profound influence of placebo and nocebo effects on exogenously administered opioid analgesia with implications for real-world patients receiving prescription opioids for acute and chronic pain.

To summarize, findings for nocebo/placebo suggest that: (1) treatment beliefs (placebo) are sufficient to release endogenous opioids; (2) opioid analgesia was doubled when coupled with placebo relative to when opioids are administered without knowledge of receipt; (3) nocebo can block analgesia from exogenously administered opioids. Given that placebo/nocebo profoundly influence opioid analgesia it is somewhat surprising that there are no widely used interventions to that target placebo as a pathway to boost opioid analgesia. Moreso, it is clear that opioid reduction nocebo is a timely and urgent issue given its potential iatrogenic harms.

In the United States multiple government agencies and leaders, including the Centers for Disease Control, the U.S. surgeon general, the Institute of

Medicine, and the Department of Health and Human Services have called for reduced opioid prescribing. Such calls have led to local, state, and federal guidelines and policies that recommend or enforce prescribing limits, regardless of patient readiness or willingness, two indices of potential nocebo for opioid reduction (Hoffman, 2018; Langreth, 2017; McCoy, 2018). Overlooking these key patient factors may greatly undermine patient response to opioid tapering because (1) nocebo increases distress and amplifies pain; (2) nocebo opposes opioid analgesia; and (3) forced tapers may contaminate the doctor-patient bond and its positive influence on treatment outcomes; (4) the placebo context of the medical environment can quickly shift to a nocebo context with detrimental effects. Forced tapers have questionable clinical value, amplify patient suffering and may contribute to self-harm and suicide (Demidenko et al., 2017). Compassionate opioid tapering requires attention to patient preference and willingness to taper and applying placebo optimization to cultivate patient engagement, placebo effects, and enhanced outcomes. Outside the context of opioid tapering and considering new or existing opioid prescriptions, such findings underscore the importance of minimizing nocebo effects to potentially prevent risky dose escalation, and optimizing placebo as a pathway to either enhance opioid analgesia or obviate the need for opioids. The question then becomes how might we help patients cultivate placebo effects for improved outcomes?

As outlined in Table 1, we argue that it is possible to minimize nocebo effects and optimize placebo for pain relief and opioid reduction and summarize the clinical strategies as follows.

(1) *Set positive expectations.* Placebo effects and positive expectations for treatments are strengthened when patients are educated about the treatments and their analgesic effects. In the case of pain being the contextual cue, the discussion centers around the potential *placebo cultivation* by facilitating belief in one's ability to shape adaptive brain responses toward pain relief and wellbeing.

(2) *Identify and Extinguish Nocebo.* Assess cognition negative thoughts, expectations and beliefs about pain and opioid use/reduction, self-efficacy to self-manage pain, and treatments.

(3) *Equip individuals with skills to enhance descending modulation of pain and distress reduction.* Enhanced descending modulation of pain dually promotes awareness of pain control and therefore placebo proper. While the Clinical Toolkit (Table 1) is not exclusive to psychologists several of the skillsets are specific to trained pain psychologists, including cognitive behavioral therapy (CBT) for pain.

Table 1 Placebo Optimization Clinical Toolkit

Clinical Goal	Placebo Optimization Toolkit	
	The Clinician Toolkit[b]	**The Patient Toolkit**
Enhance positive expectation for the cultivation of pain relief skills[a]	• Establish the basic treatment rationale by explaining the role of psychology in the experience of pain • Establish psychological treatment and enhanced descending modulation of pain as *primary* pain treatment • Summarize results of psychology treatment research as evidence-informed medicine • Provide education on psychological treatment and self-treatment yielding some of the largest effects • Provide education regarding psychological treatment's ability to "boost" medical treatments for pain • To ensure comprehension, answer questions, ask patients to tell you their understanding of the information you provided	• Acquire a fundamental understanding of the importance of psychoneurobiological processes on pain experience Understand that there are many evidence-informed simple skills that can be learned, that, when used over time, shape brain changes toward the experience of pain relief
Extinguish pain nocebo. Part 1	• Assess nocebo/catastrophizing using the Pain Catastrophizing Scale or other relevant measure • Review the patient's findings with them. (e.g., "Your score tells me that we can help you learn to reduce your distress around pain, and even the pain itself.") • Validate their distress, and connect their distress to being a pain amplifier • Using patient-friendly language, such as "negative pain mindset" in lieu of the term pain catastrophizing may enhance receptivity to the concepts and the treatment plan • Use imagery and narrative: "Nobody want more pain, but having a negative pain mindset is like picking up the can of gasoline and pouring it on a fire. You can learn to put the can of gasoline down, and by doing so it changes your pain in the moment, and steers your nervous away from pain in the future"	• Acquire a personal understanding the relevance of psychoneurobiological processes (e.g., one's score on the measure) • Understand that negative thought patterns amplify pain processing and oppose relief • Receive reassurance that while it's not "all in your head," there is much you can do to impact brain-pain experience • Become re-moralized that while a "negative pain mindset" amplifies pain, mindset is under your control. "I can learn to put the can of gasoline down so that I am not unwittingly contributing to my pain and distress"

Continued

Table 1 Placebo Optimization Clinical Toolkit—cont'd

	Placebo Optimization Toolkit	
Clinical Goal	**The Clinician Toolkit**	**The Patient Toolkit**
Part 2. Entrain descending modulation with frequent application of acquired skills	• Cognitive restructuring (reframing negative/nocebo thoughts) • Positive self-talk; learning to "talk back" to the automatic negative thoughts • Demonstrate the relaxation response and discuss its utility for comfort and control over physiological processes and self-treatment of pain • Mindfulness observing • Learn to identify pain-related distress and apply self-soothing techniques (e.g., relaxation response, mindfulness meditation, diaphragmatic breathing) • Utilize distraction	• Acquire ability to identify negative thoughts and reactions (physical, emotional) and apply adaptive strategies to interrupt pain nocebo: ✓ Cognitive reframing ✓ Positive self-talk ✓ Relaxation response ✓ Mindfulness observing ✓ Distraction ✓ Self-soothing actions • Practice the relaxation response as a self-treatment tool to reduce pain, distress and adaptive conditioning. Positive biofeedback enhances placebo—a belief that one can modulate pain and distress
Enhancements: Behavioral engagement and reinforcements	• Provide visual material, handouts and video-clips • Provide clinical worksheets • Prescribe exercises that cultivate a pain-relief mindset • Review progress in follow-up; acknowledge any challenges that arise, provide supportive encouragement, and highlight small successes • Remind patients that structural brain changes are shown after 11 weeks of skills use • Encourage a focus on skills use and behavior change vs change in pain intensity—which typically follows later	• Review visual material, handouts and video-clips • Complete worksheets and apply information • Set goals • Review progress in follow-up • Adopt an approach to skills use that is not pain contingent but rather focuses on long-range adaptation with an eye to achieving adaptive structural changes of the nervous system

Extinguish opioid reduction nocebo	• "Tell me your concerns about reducing opioids." Listen and address their fears • Discuss the data for patient-centered opioid tapering • Review the physiology of opioid reduction and how very slow tapering will allow for comfortable adaptation • Prescribers may use adjuvant medications to address discomfort and optimize placebo with a non-opioid medication • Withdrawal symptoms are just a sign that the taper is going too fast. Remind them you have a plan to prevent withdrawals, and you will work with them to adjust the taper if any discomfort arises • Keep the process very simple, avoid making any other changes during an opioid taper to obviate confounding, patient anxiety, and negative effects • Maintain very small dose reduction for the first month (see Darnall et al., 2018) • Partner with your patient. As much as possible, allow them to feel and be in control (e.g., allow them to go slower or pause the taper) • Follow-ups every 3 weeks for the first few months for close monitoring, to address any discomfort or concerns quickly, and to solidify therapeutic trust and placebo • Provide access to descending modulatory skills	• Understand the science behind endogenous pain modulation and opioid tapering: most patients experience the same or less pain when opioids are tapering the right way—very slowly so that brain and body have time to adjust • Encode that withdrawals are not harmful; they are uncomfortable and mostly preventable. My doctor will help me stay comfortable and will track me closely to make sure I'm doing ok • My doctor and I created a plan that helps me be in control. I can pause my taper if I need to during a difficult time • Understand that there's much I can do during my taper to help manage my pain (see above Patient Toolkit for the clinical goal, "Entrain descending modulation with frequent application of acquired skills" • Additional targeted reading and skills application specific to opioid reduction may be useful (Darnall, 2014a, 2014b, 2016)

[a]Enhancing placebo may dually optimize patient preference for the treatment.
[b]To assure comprehension, use simple lay language to explain complex concepts. Ask patients to explain their understanding of the concepts you describe to (1) facilitate learning through verbal recall; and allow you to (2) positively reinforce their accurate comprehension, (3) correct any misunderstandings, and (4) identify and address any concerns they raise.

2. CONCLUSION

Historically, pain-related placebo and nocebo effects have been viewed as psychological responses to external contextual information, often involving aspects of treatment. Considering the role of the patient in self-modulating, self-managing, and self-treating chronic pain, we argue for an expanded therapeutic exploitation of placebo and nocebo effects to include strategies immediately feeding back to either amplify the analgesic experience or diminish the pain experience. Pain catastrophizing illustrates the concept of pain amplification and related nocebo effects, with supporting experimental and clinical data suggesting that it may contribute inhibiting descending pain modulation. Therefore, it is necessary to face the burden of pain and the epidemic of opioids with novel approaches including psychological interventions to manage catastrophizing thoughts and other psychological factors known to amplify pain and undermine pain treatment outcomes. Clinical toolkits are needed, and we have provided a resource that may facilitate these goals of patient-centered pain management and successful tapering of prescription opioids by minimizing and optimizing placebo effects and descending pain control.

ACKNOWLEDGMENTS

We acknowledge support from NCCIH 1R01AT008561-01A1 (BDD) and NIDCR 1R01DE025946 (LC) the Patient-Centered Outcomes Research Institute® (PCORI®; 1610-37007) (BDD and LC). The views presented in this publication are solely the responsibility of the authors and do not necessarily represent the views of the Patient-Centered Outcomes Research Institute® (PCORI®), its Board of Governors or Methodology Committee.

REFERENCES

Abbott, A. D., Tyni-Lenne, R., & Hedlund, R. (2011). Leg pain and psychological variables predict outcome 2–3 years after lumbar fusion surgery. *European Spine Journal, 20*(10), 1626–1634. https://doi.org/10.1007/s00586-011-1709-6.

Archer, K. R., Heins, S. E., Abraham, C. M., Obremskey, W. T., Wegener, S. T., & Castillo, R. C. (2016). Clinical significance of pain at hospital discharge following traumatic orthopedic injury: General health, depression, and PTSD outcomes at 1 year. *The Clinical Journal of Pain, 32*(3), 196–202. https://doi.org/10.1097/AJP.0000000000000246.

Atlas, L. Y., Whittington, R. A., Lindquist, M. A., Wielgosz, J., Sonty, N., & Wager, T. D. (2012). Dissociable influences of opiates and expectations on pain. *The Journal of Neuroscience, 32*(23), 8053–8064. https://doi.org/10.1523/JNEUROSCI.0383-12.2012.

Au Yeung, S. T., Colagiuri, B., Lovibond, P. F., & Colloca, L. (2014). Partial reinforcement, extinction, and placebo analgesia. *Pain*, *155*(6), 1110–1117. https://doi.org/10.1016/j.pain.2014.02.022.

Bair, M. J., Robinson, R. L., Katon, W., & Kroenke, K. (2003). Depression and pain comorbidity: A literature review. *Archives of Internal Medicine*, *163*(20), 2433–2445.

Baron, M. J., & McDonald, P. W. (2006). Significant pain reduction in chronic pain patients after detoxification from high-dose opioids. *Journal of Opioid Management*, *2*(5), 277–282.

Benedetti, F., Amanzio, M., Rosato, R., & Blanchard, C. (2011). Nonopioid placebo analgesia is mediated by CB1 cannabinoid receptors. *Nature Medicine*, *17*(10), 1228–1230. https://doi.org/10.1038/nm.2435.

Benedetti, F., Amanzio, M., Vighetti, S., & Asteggiano, G. (2006). The biochemical and neuroendocrine bases of the hyperalgesic nocebo effect. *The Journal of Neuroscience*, *26*(46), 12014–12022. https://doi.org/10.1523/JNEUROSCI.2947-06.2006.

Benedetti, F., Lanotte, M., Lopiano, L., & Colloca, L. (2007). When words are painful: Unraveling the mechanisms of the nocebo effect. *Neuroscience*, *147*(2), 260–271. https://doi.org/10.1016/j.neuroscience.2007.02.020.

Benedetti, F., Mayberg, H. S., Wager, T. D., Stohler, C. S., & Zubieta, J. K. (2005). Neurobiological mechanisms of the placebo effect. *The Journal of Neuroscience*, *25*(45), 10390–10402. https://doi.org/10.1523/JNEUROSCI.3458-05.2005.

Berna, C., Kulich, R. J., & Rathmell, J. P. (2015). Tapering long-term opioid therapy in chronic noncancer pain: Evidence and recommendations for everyday practice. *Mayo Clinic Proceedings*, *90*(6), 828–842. https://doi.org/10.1016/j.mayocp.2015.04.003.

Bingel, U., Wanigasekera, V., Wiech, K., Ni Mhuircheartaigh, R., Lee, M. C., Ploner, M., et al. (2011). The effect of treatment expectation on drug efficacy: Imaging the analgesic benefit of the opioid remifentanil. *Science Translational Medicine*, *3*(70), 70ra14. https://doi.org/10.1126/scitranslmed.3001244.

Blasini, M., Corsi, N., Klinger, R., & Colloca, L. (2017). Nocebo and pain: An overview of the psychoneurobiological mechanisms. *Pain Reports*, *2*(2). https://doi.org/10.1097/PR9.0000000000000585.

Burns, J. W., Gerhart, J. I., Bruehl, S., Peterson, K. M., Smith, D. A., Porter, L. S., et al. (2015). Anger arousal and behavioral anger regulation in everyday life among patients with chronic low back pain: Relationships to patient pain and function. *Health Psychology*, *34*(5), 547–555. https://doi.org/10.1037/hea0000091.

Burns, J. W., Glenn, B., Bruehl, S., Harden, R. N., & Lofland, K. (2003). Cognitive factors influence outcome following multidisciplinary chronic pain treatment: A replication and extension of a cross-lagged panel analysis. *Behaviour Research and Therapy*, *41*(10), 1163–1182.

Burns, J. W., Kubilus, A., Bruehl, S., Harden, R. N., & Lofland, K. (2003). Do changes in cognitive factors influence outcome following multidisciplinary treatment for chronic pain? A cross-lagged panel analysis. *Journal of Consulting and Clinical Psychology*, *71*(1), 81–91.

Burton, A. K., Tillotson, K. M., Main, C. J., & Hollis, S. (1995). Psychosocial predictors of outcome in acute and subchronic low back trouble. *Spine*, *20*(6), 722–728.

Campbell, A. (2010). Oxytocin and human social behavior. *Personality and Social Psychology Review*, *14*, 281–295. https://doi.org/10.1177/1088868310363594.

Carriere, J. S., Martel, M. O., Kao, M. C., Sullivan, M. J., & Darnall, B. D. (2017). Pain behavior mediates the relationship between perceived injustice and opioid prescription for chronic pain: A collaborative health outcomes information registry study. *Journal of Pain Research*, *10*, 557–566. https://doi.org/10.2147/JPR.S128184.

Carriere, J. S., Sturgeon, J. A., Yakobov, E., Kao, M. C., Mackey, S. C., & Darnall, B. D. (2018). The impact of perceived injustice on pain-related outcomes: A combined model examining the mediating roles of pain acceptance and anger in a chronic pain sample. *The Clinical Journal of Pain*, *34*, 739–747. https://doi.org/10.1097/AJP.0000000000000602.

CDC, C. f. D. C. (2016). https://www.cdc.gov/drugoverdose/prescribing/patients.html.

Cherkin, D. C., Sherman, K. J., Balderson, B. H., Cook, A. J., Anderson, M. L., Hawkes, R. J., et al. (2016). Effect of mindfulness-based stress reduction vs cognitive behavioral therapy or usual care on back pain and functional limitations in adults with chronic low back pain: A randomized clinical trial. *JAMA*, *315*(12), 1240–1249. https://doi.org/10.1001/jama.2016.2323.

Colasanti, A., Rabiner, E. A., Lingford-Hughes, A., & Nutt, D. J. (2011). Opioids and anxiety. *Journal of Psychopharmacology*, *25*(11), 1415–1433. https://doi.org/10.1177/0269881110367726.

Colloca, L. (2018a). Neurobiology of the placebo effect—Part I. In L. Colloca (Ed.), *International review of neurobiology: Vol. 138*. Oxford University Press.

Colloca, L. (2018b). Preface: The fascinating mechanisms and implications of the placebo effect. *International Review of Neurobiology*, *138*, xv–xx. https://doi.org/10.1016/S0074-7742(18)30027-8.

Colloca, L., & Benedetti, F. (2005). Placebos and painkillers: Is mind as real as matter? *Nature Reviews. Neuroscience*, *6*(7), 545–552. https://doi.org/10.1038/nrn1705.

Colloca, L., & Benedetti, F. (2007). Nocebo hyperalgesia: How anxiety is turned into pain. *Current Opinion in Anaesthesiology*, *20*(5), 435–439. https://doi.org/10.1097/ACO.0b013e3282b972fb.

Colloca, L., & Finniss, D. (2012). Nocebo effects, patient-clinician communication, and therapeutic outcomes. *JAMA*, *307*(6), 567–568. https://doi.org/10.1001/jama.2012.115.

Colloca, L., & Grillon, C. (2014). Understanding placebo and nocebo responses for pain management. *Current Pain and Headache Reports*, *18*(6), 419. https://doi.org/10.1007/s11916-014-0419-2.

Colloca, L., Jonas, W. B., Killen, J., Miller, F. G., & Shurtleff, D. (2014). Reevaluating the placebo effect in medical practice. *Zeitschrift für Psychologie*, *222*(3), 124–127. https://doi.org/10.1027/2151-2604/a000177.

Colloca, L., Klinger, R., Flor, H., & Bingel, U. (2013). Placebo analgesia: Psychological and neurobiological mechanisms. *Pain*, *154*(4), 511–514. https://doi.org/10.1016/j.pain.2013.02.002.

Colloca, L., & Miller, F. G. (2011a). The nocebo effect and its relevance for clinical practice. *Psychosomatic Medicine*, *73*(7), 598–603. https://doi.org/10.1097/PSY.0b013e3182294a50.

Colloca, L., & Miller, F. G. (2011b). Role of expectations in health. *Current Opinion in Psychiatry*, *24*(2), 149–155. https://doi.org/10.1097/YCO.0b013e328343803b.

Colloca, L., Pine, D. S., Ernst, M., Miller, F. G., & Grillon, C. (2016). Vasopressin boosts placebo analgesic effects in women: A randomized trial. *Biological Psychiatry*, *79*(10), 794–802. https://doi.org/10.1016/j.biopsych.2015.07.019.

Corsi, N., & Colloca, L. (2017). Placebo and nocebo effects: The advantage of measuring expectations and psychological factors. *Frontiers in Psychology*, *8*, 308. https://doi.org/10.3389/fpsyg.2017.00308.

Corsi, N., Emadi Andani, M., Tinazzi, M., & Fiorio, M. (2016). Changes in perception of treatment efficacy are associated to the magnitude of the nocebo effect and to personality traits. *Scientific Reports*, *6*, 30671. https://doi.org/10.1038/srep30671.

Crisostomo, R. A., Schmidt, J. E., Hooten, W. M., Kerkvliet, J. L., Townsend, C. O., & Bruce, B. K. (2008). Withdrawal of analgesic medication for chronic low-back pain patients: Improvement in outcomes of multidisciplinary rehabilitation regardless of surgical history. *American Journal of Physical Medicine & Rehabilitation*, *87*(7), 527–536.

Darnall, B. D. (2014a). *Less pain, fewer pills: Avoid the dangers of prescription opioids and gain control over chronic pain*. Boulder, CO: Bull Publishing.

Darnall, B. D. (2014b). Minimize opioids by optimizing pain psychology. *Pain Management*, *4*(4), 251–253. https://doi.org/10.2217/pmt.14.18.

Darnall, B. D. (2016). *The opioid-free pain relief kit.* Boulder, CO: Bull Publishing.

Darnall, B. (2018a). To treat pain, study people in all their complexity. *Nature, 557*(7703), 7. https://doi.org/10.1038/d41586-018-04994-5.

Darnall, B. D. (2018b). *Psychological treatment for patients with chronic pain* (176 pp.). Washington, D.C: American Psychological Association Press.

Darnall, B. D., Sturgeon, J. A., Cook, K. F., Taub, C. J., Roy, A., Burns, J. W., et al. (2017). Development and validation of a daily pain catastrophizing scale. *The Journal of Pain, 18*(9), 1139–1149. https://doi.org/10.1016/j.jpain.2017.05.003.

Darnall, B. D., Sturgeon, J. A., Kao, M. C., Hah, J. M., & Mackey, S. C. (2014). From catastrophizing to recovery: A pilot study of a single-session treatment for pain catastrophizing. *Journal of Pain Research*, 7, 219–226. https://doi.org/10.2147/JPR.S62329.

Darnall, B. D., Ziadni, M. S., Stieg, R. L., Mackey, I. G., Kao, M. C., & Flood, P. (2018). Patient-centered prescription opioid tapering in community outpatients with chronic pain. *JAMA Internal Medicine, 178*(5), 707–708. https://doi.org/10.1001/jamainternmed.2017.8709.

Demidenko, M. I., Dobscha, S. K., Morasco, B. J., Meath, T. H. A., Ilgen, M. A., & Lovejoy, T. I. (2017). Suicidal ideation and suicidal self-directed violence following clinician-initiated prescription opioid discontinuation among long-term opioid users. *General Hospital Psychiatry, 47*, 29–35. https://doi.org/10.1016/j.genhosppsych.2017.04.011.

Di Blasi, Z., & Kleijnen, J. (2003). Context effects. Powerful therapies or methodological bias? *Evaluation & the Health Professions, 26*(2), 166–179. https://doi.org/10.1177/0163278703026002003.

Doering, B. K., Glombiewski, J. A., & Rief, W. (2018). Expectation-focused psychotherapy to improve clinical outcomes. *International Review of Neurobiology, 138*, 257–270. https://doi.org/10.1016/bs.irn.2018.02.004.

Eippert, F., Bingel, U., Schoell, E. D., Yacubian, J., Klinger, R., Lorenz, J., et al. (2009). Activation of the opioidergic descending pain control system underlies placebo analgesia. *Neuron, 63*(4), 533–543. https://doi.org/10.1016/j.neuron.2009.07.014.

Enck, P., Grundy, D., & Klosterhalfen, S. (2012). A novel placebo-controlled clinical study design without ethical concerns—The free choice paradigm. *Medical Hypotheses, 79*(6), 880–882. https://doi.org/10.1016/j.mehy.2012.09.017.

Feinstein, A. B., Sturgeon, J. A., Darnall, B. D., Dunn, A. L., Rico, T., Kao, M. C., et al. (2017). The effect of pain catastrophizing on outcomes: A developmental perspective across children, adolescents, and young adults with chronic pain. *The Journal of Pain, 18*(2), 144–154. https://doi.org/10.1016/j.jpain.2016.10.009.

Gerrits, M. M., van Marwijk, H. W., van Oppen, P., van der Horst, H., & Penninx, B. W. (2015). Longitudinal association between pain, and depression and anxiety over four years. *Journal of Psychosomatic Research, 78*(1), 64–70. https://doi.org/10.1016/j.jpsychores.2014.10.011.

Groenewald, C. B., Essner, B. S., Wright, D., Fesinmeyer, M. D., & Palermo, T. M. (2014). The economic costs of chronic pain among a cohort of treatment-seeking adolescents in the United States. *The Journal of Pain, 15*(9), 925–933. https://doi.org/10.1016/j.jpain.2014.06.002.

Gustavsson, A., Bjorkman, J., Ljungcrantz, C., Rhodin, A., Rivano-Fischer, M., Sjolund, K. F., et al. (2012). Socio-economic burden of patients with a diagnosis related to chronic pain—Register data of 840,000 Swedish patients. *European Journal of Pain, 16*(2), 289–299. https://doi.org/10.1016/j.ejpain.2011.07.006.

Halawa, O., Al-Diri, I., McLean, S.A., Darnall, B.D. (2015). When I'm in pain, I feel god is testing me: Differences in middle eastern and US Cancer Pain Experiences. BAOJ Palliative Medicine, Vol. 1:001. https://bioaccent.org/palliative-medicine/palliative-medicine01.php.

Heinrichs, M., & Domes, G. (2008). Neuropeptides and social behaviour: Effects of oxytocin and vasopressin in humans. *Progress in Brain Research, 170*, 337–350. S0079-6123(08)00428-7 [pii]https://doi.org/10.1016/S0079-6123(08)00428-7.

Helmerhorst, G. T., Vranceanu, A. M., Vrahas, M., Smith, M., & Ring, D. (2014). Risk factors for continued opioid use one to two months after surgery for musculoskeletal trauma. *The Journal of Bone and Joint Surgery. American Volume, 96*(6), 495–499. https://doi.org/10.2106/JBJS.L.01406.

Hoffman, J. (2018). *Medicare is cracking down on opioids. Doctors fear pain patients will suffer.* The New York Times.

Horin, A. P., Lee, K. M., & Colloca, L. (2014). Placebo effects in therapeutic outcomes. *Current Clinical Pharmacology, 9*(2), 116–122.

IASP. (1994). Part III: Pain terms, a current list with definitions and notes on usage. In H. M. A. N. Bogduk (Ed.), *Classification of chronic pain* (2nd ed., pp. 209–214). Seattle: IASP Press.

IOM Committee on Advancing Pain Research, Care, and Education. (2011). *Relieving pain in America: A blueprint for transforming prevention, care, education, and research.* Retrieved from.

Jensen, M. K., Thomsen, A. B., & Hojsted, J. (2006). 10-year follow-up of chronic non-malignant pain patients: Opioid use, health related quality of life and health care utilization. *European Journal of Pain, 10*(5), 423–433.

Jiang, Y., Oathes, D., Hush, J., Darnall, B., Charvat, M., Mackey, S., et al. (2016). Perturbed connectivity of the amygdala and its subregions with the central executive and default mode networks in chronic pain. *Pain, 157*(9), 1970–1978. https://doi.org/10.1097/j.pain.0000000000000606.

Karos, K., Meulders, A., Goubert, L., & Vlaeyen, J. W. S. (2018). The influence of social threat on pain, aggression, and empathy in women. *The Journal of Pain, 19*(3), 291–300. https://doi.org/10.1016/j.jpain.2017.11.003.

Karos, K., Williams, A. C. C., Meulders, A., & Vlaeyen, J. W. S. (2018). Pain as a threat to the social self: A motivational account. *Pain.* [Epub ahead of print]. https://doi.org/10.1097/j.pain.0000000000001257.

Kennedy, S. A., Vranceanu, A. M., Nunez, F., & Ring, D. (2010). Association between psychosocial factors and pain in patients with trigger finger. *Journal of Hand and Microsurgery, 2*(1), 18–23. https://doi.org/10.1007/s12593-010-0009-4.

Kessner, S., Sprenger, C., Wrobel, N., Wiech, K., & Bingel, U. (2013). Effect of oxytocin on placebo analgesia: A randomized study. *JAMA, 310*(16), 1733–1735. https://doi.org/10.1001/jama.2013.277446.

Klinger, R., Blasini, M., Schmitz, J., & Colloca, L. (2017). Nocebo effects in clinical studies: Hints for pain therapy. *Pain Reports, 2*(2), pii: e586. https://doi.org/10.1097/PR9.0000000000000586.

Klinger, R., Colloca, L., Bingel, U., & Flor, H. (2014). Placebo analgesia: Clinical applications. *Pain, 155*(6), 1055–1058. https://doi.org/10.1016/j.pain.2013.12.007.

Kong, J., Gollub, R. L., Rosman, I. S., Webb, J. M., Vangel, M. G., Kirsch, I., et al. (2006). Brain activity associated with expectancy-enhanced placebo analgesia as measured by functional magnetic resonance imaging. *The Journal of Neuroscience, 26*(2), 381–388. https://doi.org/10.1523/JNEUROSCI.3556-05.2006.

Kube, T., Glombiewski, J. A., & Rief, W. (2018). Using different expectation mechanisms to optimize treatment of patients with medical conditions—A systematic review. *Psychosomatic Medicine, 80*(6), 535–543. https://doi.org/10.1097/PSY.0000000000000596.

Kucyi, A., Salomons, T. V., & Davis, K. D. (2013). Mind wandering away from pain dynamically engages antinociceptive and default mode brain networks. *Proceedings of the National Academy of Sciences of the United States of America, 110*(46), 18692–18697. https://doi.org/10.1073/pnas.1312902110.

Langreth, R. (2017). *Millions of patients face pain and opioid withdrawal as* Bloomberg Business. https://www.bloomberg.com.

Lavori, P. W., Rush, A. J., Wisniewski, S. R., Alpert, J., Fava, M., Kupfer, D. J., et al. (2001). Strengthening clinical effectiveness trials: Equipoise-stratified randomization. *Biological Psychiatry*, *50*(10), 792–801.

Linton, S. J. (2005). Do psychological factors increase the risk for back pain in the general population in both a cross-sectional and prospective analysis? *European Journal of Pain*, *9*(4), 355–361.

Linton, S. J., Flink, I. K., & Vlaeyen, J. W. S. (2018). Understanding the etiology of chronic pain from a psychological perspective. *Physical Therapy*, *98*(5), 315–324. https://doi.org/10.1093/ptj/pzy027.

Lund, K., Demant, D. T., Vase, L., Sindrup, S. H., Jensen, T. S., & Finnerup, N. B. (2017). Placebo responses in patients with peripheral neuropathic pain. *Scandinavian Journal of Pain*, *5*(3), 210. https://doi.org/10.1016/j.sjpain.2014.05.019.

McCoy, T. (2018). *Unintended consequences: Inside the fallout of America's crackdown on opioids.* The Washington Post.

McCracken, L. M., & Keogh, E. (2009). Acceptance, mindfulness, and values-based action may counteract fear and avoidance of emotions in chronic pain: An analysis of anxiety sensitivity. *The Journal of Pain*, *10*(4), 408–415. https://doi.org/10.1016/j.jpain.2008.09.015.

Mistiaen, P., van Osch, M., van Vliet, L., Howick, J., Bishop, F. L., Di Blasi, Z., et al. (2016). The effect of patient-practitioner communication on pain: A systematic review. *European Journal of Pain*, *20*(5), 675–688. https://doi.org/10.1002/ejp.797.

Murphy, J. L., Clark, M. E., & Banou, E. (2013). Opioid cessation and multidimensional outcomes after interdisciplinary chronic pain treatment. *The Clinical Journal of Pain*, *29*(2), 109–117. https://doi.org/10.1097/AJP.0b013e3182579935.

Palermo, T. M., & Drotar, D. (1996). Prediction of children's postoperative pain: The role of presurgical expectations and anticipatory emotions. *Journal of Pediatric Psychology*, *21*(5), 683–698.

Papaioannou, M., Skapinakis, P., Damigos, D., Mavreas, V., Broumas, G., & Palgimesi, A. (2009). The role of catastrophizing in the prediction of postoperative pain. *Pain Medicine*, *10*(8), 1452–1459. https://doi.org/10.1111/j.1526-4637.2009.00730.x.

PCORI Award. (2017). https://www.pcori.org/research-results/2017/comparative-effectiveness-pain-cognitive-behavioral-therapy-and-chronic-pain.

Pecina, M., & Zubieta, J. K. (2018). Expectancy modulation of opioid neurotransmission. *International Review of Neurobiology*, *138*, 17–37. https://doi.org/10.1016/bs.irn.2018.02.003.

Peerdeman, K. J., van Laarhoven, A. I., Keij, S. M., Vase, L., Rovers, M. M., Peters, M. L., et al. (2016). Relieving patients' pain with expectation interventions: A meta-analysis. *Pain*, *157*(6), 1179–1191. https://doi.org/10.1097/j.pain.0000000000000540.

Picavet, H. S., Vlaeyen, J. W., & Schouten, J. S. (2002). Pain catastrophizing and kinesiophobia: Predictors of chronic low back pain. *American Journal of Epidemiology*, *156*(11), 1028–1034.

Roh, Y. H., Lee, B. K., Noh, J. H., Oh, J. H., Gong, H. S., & Baek, G. H. (2014). Effect of anxiety and catastrophic pain ideation on early recovery after surgery for distal radius fractures. *The Journal of hand surgery*, *39*(11), 2258–2264. e2252. https://doi.org/10.1016/j.jhsa.2014.08.007.

Rosenstiel, A. K., & Keefe, F. J. (1983). The use of coping strategies in chronic low back pain patients: Relationship to patient characteristics and current adjustment. *Pain*, *17*(1), 33–44.

Salomons, T. V., Moayedi, M., Erpelding, N., & Davis, K. D. (2014). A brief cognitive-behavioural intervention for pain reduces secondary hyperalgesia. *Pain*, *155*(8), 1446–1452. https://doi.org/10.1016/j.pain.2014.02.012.

Seminowicz, D. A. (2006). Believe in your placebo. *The Journal of Neuroscience*, *26*(17), 4453–4454. https://doi.org/10.1523/JNEUROSCI.0789-06.2006.
Seminowicz, D. A., & Davis, K. D. (2006). Cortical responses to pain in healthy individuals depends on pain catastrophizing. *Pain*, *120*(3), 297–306.
Seminowicz, D. A., Shpaner, M., Keaser, M. L., Krauthamer, G. M., Mantegna, J., Dumas, J. A., et al. (2013). Cognitive-behavioral therapy increases prefrontal cortex gray matter in patients with chronic pain. *The Journal of Pain*, *14*(12), 1573–1584. https://doi.org/10.1016/j.jpain.2013.07.020.
Severeijns, R., Vlaeyen, J. W., & van den Hout, M. A. (2004). Do we need a communal coping model of pain catastrophizing? An alternative explanation. *Pain*, *111*(3), 226–229.
Severeijns, R., Vlaeyen, J. W., van den Hout, M. A., & Weber, W. E. (2001). Pain catastrophizing predicts pain intensity, disability, and psychological distress independent of the level of physical impairment. *The Clinical Journal of Pain*, *17*(2), 165–172.
Smeets, R. J. E. M., van Geel, A. C., Kester, A. D., & Knottnerus, J. A. (2007). Physical capacity tasks in chronic low back pain: What is the contributing role of cardiovascular capacity, pain and psychological factors? *Disability and Rehabilitation*, *29*(7), 577–586.
Spinhoven, P., Ter Kuile, M., Kole-Snijders, A. M., Hutten Mansfeld, M., Den Ouden, D. J., & Vlaeyen, J. W. (2004). Catastrophizing and internal pain control as mediators of outcome in the multidisciplinary treatment of chronic low back pain. *European Journal of Pain*, *8*(3), 211–219.
Staats, P. S., Staats, A., & Hekmat, H. (2001). The additive impact of anxiety and a placebo on pain. *Pain Medicine*, *2*(4), 267–279. https://doi.org/10.1046/j.1526-4637.2001.01046.x.
Stewart, M. O., Karlin, B. E., Murphy, J. L., Raffa, S. D., Miller, S. A., McKellar, J., et al. (2015). National dissemination of cognitive-behavioral therapy for chronic pain in veterans: Therapist and patient-level outcomes. *The Clinical Journal of Pain*, *31*(8), 722–729. https://doi.org/10.1097/AJP.0000000000000151.
Sturgeon, J. A., Carriere, J. S., Kao, M. J., Rico, T., Darnall, B. D., & Mackey, S. C. (2016). Social disruption mediates the relationship between perceived injustice and anger in chronic pain: A collaborative health outcomes information registry study. *Annals of Behavioral Medicine*, *50*(6), 802–812. https://doi.org/10.1007/s12160-016-9808-6.
Sturgeon, J. A., Darnall, B. D., Kao, M. C., & Mackey, S. C. (2015). Physical and psychological correlates of fatigue and physical function: A collaborative health outcomes information registry (CHOIR) study. *The Journal of Pain*, *16*(3), 291–298. e291. https://doi.org/10.1016/j.jpain.2014.12.004.
Sturgeon, J. A., Dixon, E. A., Darnall, B. D., & Mackey, S. C. (2015). Contributions of physical function and satisfaction with social roles to emotional distress in chronic pain: A collaborative health outcomes information registry (CHOIR) study. *Pain*, *156*(12), 2627–2633. https://doi.org/10.1097/j.pain.0000000000000313.
Sullivan, M. J. L. (1995). The pain catastrophizing scale: Development and validation. *Psychological Assessment*, 7(4), 524–532.
Swider, K., & Babel, P. (2013). The effect of the sex of a model on nocebo hyperalgesia induced by social observational learning. *Pain*, *154*(8), 1312–1317. https://doi.org/10.1016/j.pain.2013.04.001.
Taub, C. J., Sturgeon, J. A., Johnson, K. A., Mackey, S. C., & Darnall, B. D. (2017). Effects of a pain catastrophizing induction on sensory testing in women with chronic low back pain: A pilot study. *Pain Research & Management*, *2017*, 7892494. https://doi.org/10.1155/2017/7892494.
Theunissen, M., Peters, M. L., Bruce, J., Gramke, H. F., & Marcus, M. A. (2012). Preoperative anxiety and catastrophizing: A systematic review and meta-analysis of the association with chronic postsurgical pain. *The Clinical Journal of Pain*, *28*(9), 819–841. https://doi.org/10.1097/AJP.0b013e31824549d6.

Thorn, B. E., Pence, L. B., Ward, L. C., Kilgo, G., Clements, K. L., Cross, T. H., et al. (2007). A randomized clinical trial of targeted cognitive behavioral treatment to reduce catastrophizing in chronic headache sufferers. *The Journal of Pain*, *8*(12), 938–949. https://doi.org/10.1016/j.jpain.2007.06.010.

Turner, J. A., Anderson, M. L., Balderson, B. H., Cook, A. J., Sherman, K. J., & Cherkin, D. C. (2016). Mindfulness-based stress reduction and cognitive behavioral therapy for chronic low back pain: Similar effects on mindfulness, catastrophizing, self-efficacy, and acceptance in a randomized controlled trial. *Pain*, *157*(11), 2434–2444. https://doi.org/10.1097/j.pain.0000000000000635.

Vlaeyen, J. W., Crombez, G., & Linton, S. J. (2016). The fear-avoidance model of pain. *Pain*, *157*(8), 1588–1589. https://doi.org/10.1097/j.pain.0000000000000574.

Vogtle, E., Barke, A., & Kroner-Herwig, B. (2013). Nocebo hyperalgesia induced by social observational learning. *Pain*, *154*(8), 1427–1433. https://doi.org/10.1016/j.pain.2013.04.041.

Wager, T. D., & Atlas, L. Y. (2015). The neuroscience of placebo effects: Connecting context, learning and health. *Nature Reviews Neuroscience*, *16*(7), 403–418. https://doi.org/10.1038/nrn3976.

Wager, T. D., Atlas, L. Y., Leotti, L. A., & Rilling, J. K. (2011). Predicting individual differences in placebo analgesia: Contributions of brain activity during anticipation and pain experience. *The Journal of Neuroscience*, *31*(2), 439–452. https://doi.org/10.1523/JNEUROSCI.3420-10.2011.

Wager, T. D., Scott, D. J., & Zubieta, J. K. (2007). Placebo effects on human mu-opioid activity during pain. *Proceedings of the National Academy of Sciences of the United States of America*, *104*(26), 11056–11061. https://doi.org/10.1073/pnas.0702413104.

Wertli, M. M., Burgstaller, J. M., Weiser, S., Steurer, J., Kofmehl, R., & Held, U. (2014). The influence of catastrophizing on treatment outcome in patients with non-specific low back pain: A systematic review. *Spine*, *39*(3), 263–273. https://doi.org/10.1097/BRS.0000000000000110. Review.

Williams, A. C., Eccleston, C., & Morley, S. (2012). Psychological therapies for the management of chronic pain (excluding headache) in adults. *Cochrane Database of Systematic Reviews*, *11*, CD007407. https://doi.org/10.1002/14651858.CD007407.pub3.

Wright, D., Hoang, M., Sofine, A., Silva, J. P., & Schwarzkopf, R. (2017). Pain catastrophizing as a predictor for postoperative pain and opiate consumption in total joint arthroplasty patients. *Archives of Orthopaedic and Trauma Surgery*, *137*(12), 1623–1629. https://doi.org/10.1007/s00402-017-2812-x.

Ziadni, M. S., Sturgeon, J. A., & Darnall, B. D. (2018). The relationship between negative metacognitive thoughts, pain catastrophizing and adjustment to chronic pain. *European Journal of Pain*, *22*(4), 756–762. https://doi.org/10.1002/ejp.1160.

Ziadni, M., You, D. S., Wilson, A. C., & Darnall, B. D. (2018). CARE scale—7: Development and preliminary validation of a measure to assess factors impacting self-care in chronic pain. *The Clinical Journal of Pain*. [Epub ahead of print]. https://doi.org/10.1097/AJP.0000000000000606.

Zubieta, J. K., Bueller, J. A., Jackson, L. R., Scott, D. J., Xu, Y., Koeppe, R. A., et al. (2005). Placebo effects mediated by endogenous opioid activity on mu-opioid receptors. *The Journal of Neuroscience*, *25*(34), 7754–7762. https://doi.org/10.1523/JNEUROSCI.0439-05.2005.

CHAPTER SEVEN

Choose Your Words Wisely: The Impact of Message Framing on Patients' Responses to Treatment Advice

Paul Glare*,[1], Ilona Fridman[†], Claire E. Ashton-James*
*Pain Management Research Institute, The University of Sydney, Sydney, NSW, Australia
[†]Fuqua School of Business, Duke University, Durham, NC, United States
[1]Corresponding author: e-mail address: paul.glare@sydney.edu.au

Contents

Abstract

Other people's words can have a powerful influence on how we interpret our environment, what we expect and experience, what we value, how we feel, what we choose, and how we behave. Placebo (and nocebo) effects are a dramatic example of this. The way in which healthcare professionals discuss, describe, and inform patients about the characteristic effects of a given disease and it prevention, diagnosis and treatment influence patients' feelings and expectations which in turn affects their psychobiological responses to, and subjective experiences and outcomes of the disease and its treatment effects.

The effect of clinicians' words on patients' responses to treatments and procedures, both active and inert or sham is nothing less than incredible. The way in which information about treatment effects is delivered to patients can even reverse the clinically proven effects of an active treatment, or increase the adverse effects of it.

In this chapter, we begin by presenting the data on the impact of message framing on affect and expectations of health care in experimental situations followed by the

International Review of Neurobiology, Volume 139
ISSN 0074-7742
https://doi.org/10.1016/bs.irn.2018.07.020

evidence that indicates how various patient, disease and clinician related factors modify framing effects in the clinic. Finally we discuss how framing effects affect clinical practice. They can be leveraged to enhance placebo effects and minimize nocebo effects. They can provide strategies to assist shared-decision making in the face of complex uncertainty.

Going forward, automation of health care and artificial intelligence may change the delivery of health care but patients will continue to be humans seeking health gains while avoiding health losses and how the information is presented will always be susceptible to framing effects.

1. INTRODUCTION

Other people's words can have a powerful influence on how we interpret our environment, what we expect and experience, what we value, how we feel, what we choose, and how we behave. Placebo (and nocebo) effects are a clear example of this: The way in which healthcare professionals discuss, describe, and inform patients about the characteristic effects of a given treatment influences patients' feelings and expectations about that treatment, psychobiological responses to the treatment, and the subjective experience of treatment effects. The effect of clinicians' words on patients' responses to active (and inert) treatments and procedures is nothing less than incredible. Active treatments are more effective when described to patients in a way that reduces patient anxiety and increases patient expectations of positive outcomes (Koban, Jepma, Geuter, & Wager, 2017). When administered *without* instruction or explanation from a doctor or nurse, commonly administered analgesic medications for post-operative pain—buprenorphine, tramadol, ketorolac, and metamizol—were found to be 50% less effective (Amanzio, Pollo, Maggi, & Benedetti, 2001). Even inactive treatments—inert substances or procedures (placebos)—can provide effective pain relief when administered to patients by a doctor who describes the treatment as a powerful analgesic ("the pain will subside in a few minutes") (Levine, Gordon, Smith, & Fields, 1981). Indeed, the words that are used to convey treatment effects are so influential that they can even reverse the clinically proven effects of active treatments. For instance, researchers found that the application of a topical analgesic (Emla) increased, rather than decreased, pain sensitivity when it was administered by an experimenter who informed participants that "increased pain sensitivity" was a possible side effect of the treatment (Aslaksen, Zwarg, Eilertsen, Gorecka, & Bjørkedal, 2015).

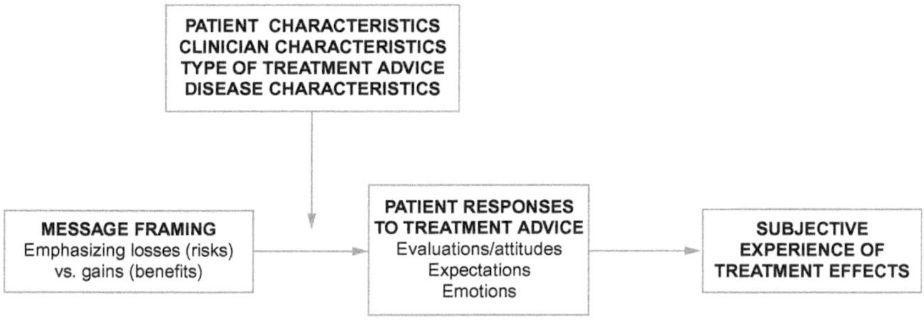

Fig. 1 A conceptual model of the processes by which message framing may influence patients' subjective experience of treatment effects.

The forgoing examples of placebo (and nocebo) effects serve to demonstrate the potential for message framing—the way in which information regarding treatment is relayed to patients—to influence patients' affective responses to treatment advice, evaluations and expectations of treatments, psychobiological responses to treatments, and subjective experience of treatment effects (see Fig. 1).

The most common facet of any patient–provider discussion about treatment alternatives is the presentation of information regarding the evidence-based risks and benefits of treatment. One of the most important clinical implications of research into placebo effects is the potential for the discussion of intended treatment effects (benefits) to increase patient expectations of a positive treatment outcome, and consequently improve patients' subjective experience of treatment effects. Despite the potential for the delivery of treatment information to enhance patients' responses to treatment, there is a dearth of research into *how* to deliver treatment information in a way that minimizes patient anxiety and enhances patient preferences for, expectations of, and affective responses to treatment recommendations.

The goals of this article are to review evidence-based strategies for delivering treatment information in a manner that optimizes patients' affective responses to and evaluations of treatment advice. To begin, we discuss the impact of message framing on patients' evaluations of and affective responses to treatment advice (Section 2). It is important to take a critical approach to the application of message framing research to clinical practice. Therefore, Section 3 of this chapter explores potential moderators or boundary conditions on the effectiveness of message framing strategies in practice, including clinician characteristics, patient characteristics, and disease characteristics. To conclude, we apply the evidence-based principles of message framing to complex treatment scenarios, and consider the implications of message framing strategies for enhancing placebo effects (Section 4).

2. MESSAGE FRAMING EFFECTS

Nearly all health related information, including treatment advice, can be construed in terms of benefits and costs. For example, a clinician who is advising a patient on chronic opioid therapy for chronic pain to taper off opioid medications could either emphasize the benefits of opioid tapering (e.g., "If you gradually reduce your use of opioid medications, you will have more clarity of mind, be more able to engage with others in social interaction, and have a better overall quality of life"), or the costs of not tapering off opioids (e.g., "If you do not taper off these medications, you may experience worsening side effects and increased risk of morbidity and mortality"). Hence the majority of research into the communication of treatment recommendations has examined the relative efficacy of framing expected outcomes in terms of benefits (gains) or costs (losses). For example, does it matter whether a clinician describes the benefits of condom use to a teenager or the costs of unprotected sex? Are patients more likely to adhere to a recommended diet plan if the benefits of diet adherence are emphasized (better physical functioning, improved self-esteem, reduced risk of heart disease) over the costs of non-adherence (weight gain, physical fatigue, increased risk of heart disease)? Keeping in mind that placebo responses to treatments are facilitated by patients' evaluations, expectations, and affective responses to treatment advice, is there any "placebo advantage" to emphasizing the potential gains associated with following a treatment recommendation rather than emphasizing the losses avoided by following the recommended treatment?

The evaluation of treatment alternatives frequently involves substantial uncertainty and risk. For instance, a patient who is being advised to reduce their reliance on opioid medications for pain management cannot be certain whether or how much their pain may increase if they taper off opioids, nor the extent to which they will have a better quality of life when opioid-free. Prospect theory (Kahneman & Tversky, 1979) postulates that if people are presented with two choices—one posing little risk and one posing some higher degree of risk—people's preference for (or evaluation of) one alternative over the other will depend on the manner in which the choices are framed. In general, people will be more willing to accept higher risks to avoid or prevent losses than to achieve gains, and conversely, are less willing to accept risk to secure gains. For instance, people tend to express a preference for an offer of $800 for certain over an offer of having an 85% chance of receiving $1000, and prefer an 85% chance of losing $1000 (a 15% chance of avoiding loss entirely) over a certain loss of $800. In short, people are risk averse when prospective outcomes are framed in terms of gains, and have less negative attitudes toward risk (are risk-seeking) when faced with the prospect of avoiding certain losses.

Rothman and Salovey (1997) applied this reasoning to investigate the impact of loss vs. gain framing on responses to health messages—including treatment advice. In line with prospect theory, Rothman and Salovey suggested that *gain-framed health messages* should be more persuasive than loss-framed health messages for treatment advice or health behaviors that are perceived to carry only *minimal risk* and effort (e.g., applying sunscreen, exercising regularly). For health behaviors with a *higher perceived risk* (i.e., potential side effects, or negative outcomes), *loss-framed health messages* should be more persuasive. In general, researchers have found this to be the case: people are more likely to respond more "favorably" to loss-framed healthcare advice when the risks, costs, or losses associated with *not* taking the advice are perceived to be greater than the risks, costs, or losses associated with taking the advice (Gallagher & Updegraff, 2011).

The concept of "perceived risk" is critical to understanding the variable effects of message framing in the delivery of healthcare advice. The subjective experience of "risk" or "threat" that is associated with a given health behavior does not necessarily follow from the health behavior itself, but may be elicited by the perceived threat of the disease associated with the health behavior (Salovey & Wegener, 2003). For example, there is a certain level of "objective" risk associated with liver biopsy. However, the subjective experience of risk or threat associated with a liver biopsy will be greater for an individual who believes that there is a high probability that the liver biopsy will reveal cancer. Demonstrating this, Gallagher, Updegraff, Rothman, and Sims (2011) found that among women who perceive themselves to have high susceptibility to breast cancer, a loss-framed message led to significantly greater rates of subsequent screening than a gain-framed message. Interestingly, in this particular study, framing had no effect on screening among women who perceived themselves as being low in susceptibility to breast cancer. A similar pattern of was revealed in a study of women responding to HIV-testing advice (Apanovitch, McCarthy, & Salovey, 2003). When HIV-testing advice was given using gain framing, it was only effective in promoting HIV testing in people who were *certain* that the test would not detect HIV. Loss-framed messages were relatively more effective with participants who were *uncertain* about the possible results of HIV testing. Hence, in order for message framing to have its intended effects on patient evaluations, expectations, emotions, and behavioral intentions, clinicians must first evaluate the level of risk or threat associated with the treatment or health behavior advice one gives before framing it in terms of losses or gains.

2.1 Treatment Expectations

Message framing effects on treatment expectations has primarily been studied within the context of presenting patients with probabilities of treatment benefits and probabilities of treatment costs, examining the impact of *probability framing* on treatment expectations and outcomes. This research highlights the importance of *the way in which the probability of gains* (and losses or side effects) *are described* for patient expectations and treatment outcomes. O'Connor, Pennie, and Dales (1996) seminal research examined the impact of the way in which gains are framed and losses are framed on expectations of side effects, the experience of side effects, and consequent work absenteeism. Two hundred and ninety-two participants received information about influenza and all participants received information about the benefits of vaccination (gain framing) as well as information about potential side effects of vaccination. Participants were randomly assigned to one of two probability framing conditions in which the probability of gains and side effects were framed "positively," emphasizing potential for gains and avoidance of side effects, or "negatively," emphasizing the potential for no gains and the presence of side effects. The "positive" framing of gains and side effects described the percentage of vaccinated individuals who remain influenza-free (70%) and free of vaccine side effects (60%). The negative framed described the chance of acquiring influenza despite vaccination (30%) and the chance of experiencing vaccine size effects (40%). Hence, this was not a typical study of framing effects, in that the researchers investigated variations upon the construction of gain frames (positive vs. negative presentation of probabilities) rather than comparing gain frames with equivalent loss frames.

All messages presented to participants were variations on gain framing. Hence, there were no significant differences in the percentage of participants who chose to receive vaccines in response to gain frames that were constructed "positively" vs. "negatively." There was, however a significant impact of the way in which gain frames were worded on participants' expectations of treatment benefit and expectations of side effects. The group assigned to the "positive" framing of gains had significantly higher expectations that they would remain free of influenza and free of side effects. Furthermore, the group who received "positive" framing of side effects reported a lower incidence of both systemic and local side effects described as well as lower work absenteeism.

Taber and Aspinwall (2015) is one example of research in which treatment expectations, operationalized as perceived treatment efficacy, has been explored as a mediator of the relationship between message framing (loss vs. gain) and behavioral intentions. The researchers randomly assigned undergraduate psychology students to one of four experimental conditions in which they watched a video of a genetic counselor providing them with information that they had either a low (15%) or high (75%) genetic risk of melanoma. Participants were then provided with recommendations to use sunscreen that were either framed in terms of loss (risk of developing melanoma may increase by 15% if you don't use sunscreen) or gain (risk of developing melanoma may decrease by 15% if you use sunscreen), before being asked to indicate the extent to which they perceived sunscreen use to be protective against melanoma, and their intentions to use sunscreen. Consistent with Rothman and Salovey's (1997) model of framing effects, Taber and Aspinwall found that when undergraduates believed that their risk of developing melanoma was *low* (15%), gain-framed recommendations to use sunscreen ("Risk may decrease by 15%) resulted in greater intentions to use sunscreen. On the other hand, when undergraduates believed that their risk of developing melanoma was *high* (76%), loss-framed healthcare advice ("avoid increasing risk of developing melanoma by 15%") was associated with greater intentions to use sunscreen. Importantly, when undergraduates believed that their risk of developing melanoma was high (76%), loss-framed healthcare advice had a significant impact on participants' treatment expectations, as measured by participants' ratings of the perceived efficacy of sunscreen use for preventing melanoma.

The forgoing studies, while limited in number and replication, are suggestive that message framing has the potential to improve patients' treatment expectations and subjective experience of treatment outcomes (see Fig. 1). These findings may provide the impetus for researchers to explore the impact of message framing on treatment expectancies and outcomes in future research.

2.2 Affective Responses

There are sound theoretical grounds for expecting message framing (specifically loss vs. gain framing) to influence message recipients' affective states. According to contemporary theories of emotion, positive and negative affective states are elicited by appraisals of the relationship between one's current goals and the environment. These appraisals yield positive affect when an individual perceives that the environment is congruent with their goals, and yields negative affect when the environment is incongruent with their goals (Frijda, 1986; Lazarus, 1991). As Shen and Dillard (2007) point out, gain-framed messages promoting preventative health behaviors (e.g., using sun protection) focus on beneficial or "advantageous" outcomes, and in this respect, gain-framed messages essentially emphasize congruence between the message recipient's goals (health promotion) and the current environmental cues (information promoting the use of sunscreen). On the other hand, loss-framed messages focus on costs or disadvantages and in a sense emphasize incongruence between one's goals (heath promotion) and the current environment. For this reason, it has often been hypothesized that, generally speaking, gain-framed messages will elicit more positive affect and loss-framed messages will elicit more negative affect.

Relatively little research has been conducted to empirically test this hypothesis, but the available research, while inconsistent, is generally supportive. For example, Millar and Millar (2000) measured the impact of message framing and issue involvement on intentions to engage in safe driving behavior, agreement with messages promoting safe driving behavior, and affective responses to message framing. Participants who were randomly assigned to read either a gain-framed message ("by driving at the recommended speed, the driver will gain time needed for braking in an emergency") or a loss-framed message ("by not driving at the recommended speed, the driver will lose time needed for braking in an emergency") were asked to indicated how sad or happy, anxious or relaxed, and nervous or calm the message made them feel on a series of 9-point Likert scales. Participants' responses to the three affect scales were summed to create a single index of positive or negative affect, and analyses of the interaction between issue involvement and message framing revealed only a main effect of message framing on affective response. Namely, gain-framed messages were associated with more positive affective reactions than were loss messages.

These results were subsequently replicated by Schneider et al. (2001), Cox, Cox, and Zimet (2006), and Shen and Dillard (2007) among others.

At the same time, however, there have been several "failed attempts" to find affective consequences of message framing (see for example, Banks et al., 1995; Detweiler, Bedell, Salovey, Pronin, & Rothman, 1999; Meyerowitz & Chaiken, 1987). There are several possible reasons for this. In part, it is may be an issue of measurement error incurred by the explicit (self-report) of emotions. Affective processes are subtle, and not always consciously recognized (Winkielman & Berridge, 2004). Indeed, when emotions are implicit, or experienced without conscious reflection or awareness, they are more likely to influence judgment and behavior since reflection upon one's emotional state elicits emotion-regulation strategies prevents affective influence or bias (Lieberman et al., 2007). Hence, if affective processes are instigated by message framing, it is likely to be an implicit process that will not be captured by self-reports. Evidence for the impact of implicit affect and emotions on judgment and decision making has relied upon neuroscientific measures (Bechara & Damasio, 2005), psychophysiological measures (Smith, Cacioppo, Larsen, & Chartrand, 2003), and implicit social cognitive measures such as spontaneous thought-listing tasks, which are often used to assess participants' unprompted reactions to experimental stimuli (Petty & Cacioppo, 1986).

Taber and Aspinwall (2015, described above) utilized a thought-listing task in part to assess participants emotional responses to loss- and gain-framed messages promoting sunscreen use. Two independent raters blind to experimental condition coded participants' statements for perceived control in addition to negative emotions and cognitions. The researchers found that when fair-skinned participants (a cue to high risk) were given low-risk genetic counseling (15% chance of developing melanoma) in combination with gain framing, they spontaneously listed thoughts indicating hope, optimism for disease prevention. These results are by no means conclusive, but demonstrate the utility of using more indirect measures of emotions when evaluating the potential impact of message framing on affective responses to treatment information.

It is also possible that affective responses to message framing are indirectly captured by measures of attitudes and expressed behavioral intentions. Indeed, how one "feels" about issues, persons objects or events (their affective response) contributes in large part to one's evaluations of these issues (Zanna & Rempel, 1988). Indeed, people often infer their attitudes from their feelings (see for example, the "somatic marker hypothesis" (Bechara & Damasio, 2005) and the affect-as-information model (Schwarz & Clore, 1996)). In their systematic review of research into message framing effects on evaluations, intentions, and behaviors, Gallagher and Updegraff (2011) point out that the majority of work in this area has measured participants' evaluations or attitudes to health care advice (including behavioral intentions) rather than actual behavior. Hence, to the extent that evaluations of treatment advice, expressed preferences for treatment alternatives, and behavioral intentions reflect, in part, people's affective responses to loss- or gain-framed messages, there is arguably a great deal of evidence in support of the impact of message framing on affective responses to treatment advice.

Arguably, research into the impact of message framing on the subjective experience of regulatory fit, described as a sense of "feeling right" (or a feeling of correctness or importance) that comes from the recognition of a match between a person's motivational orientation (risk-seeking or risk-avoiding) (Cesario, Grant, & Higgins, 2004) provides further support for the notion that message framing has the potential to modulate affective states. Although regulatory fit theory argues that the experience of "feeling right" is more than simply positive affect, and evidence suggests that the feeling of regulatory fit is indeed independent of self-reported positive mood (Higgins, Idson, Freitas, Spiegel, & Molden, 2003), it is difficult to argue that the notion of a "feeling of rightness," which directly transfers to and informs a person's evaluation of messages (Cesario, Higgins, & Scholer, 2008) in a manner consistent with the "somatic marker" hypothesis and affective-as-information models mentioned above, is not (at least in part) an affective response.

Notwithstanding the presence of several null findings regarding affective responses to message framing present in the literature, more recent research has revived the notion that framing effects on behavioral intentions and attitudes may be mediated by affect. Taking a more nuanced approach to the role of affect in message framing effects, van't Riet, Ruiter, Werrij, Candel, and de Vries (2010) argued gain-framed information motivates behavior through positive affect, while loss-framed information motivates behavior through negative affect. Hence, both gain- and loss-framed messages exert their influence on behavior through affect, but through quite different "substrates" (positive vs. negative affect).

van't Riet et al. (2010) tested their hypothesis in an experiment in which participants were provided with information about reducing salt intake which either emphasized the advantages of little salt (gain framing) or the dangers of too much salt (loss framing). Four items assessed positive affective reactions to the information ("How happy, relieved, optimistic, and content did the information make you feel?") and four items assessed negative affective reactions to the information ("How sad, afraid, tense, and uncomfortable did the information make you feel?"). Results of analyses confirmed the authors' hypotheses that gain-framed messages evoked more positive affect than loss-framed messages, and loss-framed messages elicited more negative affect than gain-framed messages. Interestingly, however, there was no effect of framing on behavioral intentions in this study.

Taken together, these results together with the results of earlier studies suggest that message framing can influence patients' experience of positive or negative affect, although there is as yet very little understanding of the conditions under which message framing is most likely to elicit a positive or negative affective response. Given the importance of affective responses to treatment advice for not only patient decision making, but the formation of patients' treatment expectancies and subjective experience of treatment effects, we believe this is a critical area for future research.

3. THE COMPLEXITIES MESSAGE FRAMING EFFECTS

The studies reviewed in Section 2 highlighted the potential for message framing to influence patient preferences, expectations, and affective responses to treatment advice. Dozens of experiments have presented data consistent that Rothman and Salovey's (1997) seminal review in which the central tenets of Prospect Theory are applied to predict the outcomes of loss and gain framing in the context of delivering healthcare advice. In general, health promoting messages which emphasize achieving gains are more effective in facilitating health-affirming behavior, whereas loss-framed messages were more effective in facilitating illness-detecting behaviors (Rothman, Martino, Bedell, Detweiler, & Salovey, 1999), with perceived risk or feelings of uncertainty or threat being a key moderator of these relationships (e.g., Taber & Aspinwall, 2015). A decade later however, a meta-analyses of message framing effects in health communication have found only weak effects on individuals' behavior in terms of both prevention and detection (O'Keefe & Jensen, 2007). These inconsistencies across studies indicate that there may be a range of contextual factors related to the message, the messenger, the message receiver, and even the environment that could modulating the effects of framing.

3.1 Inconsistency of Framing Effect in Clinical Communications

Framing a health communication may help a physician to elicit a health enhancing behavior in a patient, but may also serve to undermine good quality decision making. Consequently, the potential importance of framing effects in medical decision making has been recognized over the years by the editors of high-impact mainstream medical journals such as New England Journal of Medicine (McNeil, Pauker, Sox, & Tversky, 1982), Lancet (Bobbio, Demichelis, & Giustetto, 1994) and Journal of the American Medical Association (Verma, Razak, & Detsky, 2014). Despite this interest, there is currently only a thin evidence base for framing effects in health care, making it hard to make strong recommendations to clinicians and policy makers. Few studies have involved real patients making actual treatment decisions with their treating clinician(s). For example, a recent systematic review of cognitive bias in health care professionals found evidence of a framing effect operating in only 3 of the 20 studies included (Saposnik, Redelmeier, Ruff, & Tobler, 2016). An systematic review in 2001 purporting to focus on framing effect located 24 studies, but in fact it addressed manipulation of information in the context of clinical risk communication more broadly, i.e., not limited to framing that was logically equivalent (Edwards, Elwyn, Covey, Matthews, & Pill, 2001).

A 2011 Cochrane review of framing effects focused on attribute framing and goal framing across the three main types of health information messages: prevention, detection, treatment (Akl et al., 2011). They identified 35 studies (most of which were published between 2000 and 2006), involving 16,342 patients. The 35 studies comprised 12 studies of *attribute framing* (e.g., "the chance of surviving vs. dying from this disease is …") and 23 studies of *goal framing* (e.g., "if you quit smoking, then the chance of living longer vs. dying sooner is …"). The studies reported four different effects of framing: 1. understanding the message; 2. perception of effectiveness of the message; 3. persuasiveness of the message; and 4. behavior change after receiving the message.

Health messages were judged more persuasive when the attribute was framed positively (e.g., "quitting smoking saves lives") rather than negatively ("not quitting smoking causes deaths"). In contrast, goal framing had the opposite effect: a negatively-framed message emphasizing losses ("if you don't quit smoking, you will die sooner") was rated as more persuasive than a comparable positively-framed message emphasizing gains ("if you quit smoking, you'll live longer"). However, gain/loss framing does not have much effect on health behavior.

In more detail, the 12 studies of attribute framing included one which addressed the effect on understanding the message, and found that participants better understood a message about cancer screening when it was framed negatively than positively; 2 studies on perceived effectiveness (cancer screening and harm from vaccination), and found positively-framed messages may lead to more positive perception of effectiveness than negatively-framed messages; 8 studies on persuasiveness of various prevention, screening and treatment messages but found little or no difference in persuasiveness; 1 study addressing behavior (flu vaccine) found no effect of positive/negative attribute framing. Regarding the 39 studies of goal framing, they can be summarized as follows: No study assessed the effect on understanding the message; loss messages led to a more positive perception of effectiveness of screening programs (5 programs) than gain messages, but not prevention programs (9 studies); persuasiveness of framing as gain/loss was evaluated in 6 studies of screening messages, 13 studies of prevention messages and 3 studies of treatment messages (BP meds; common surgical procedures), and found loss messages may have been more persuasive for treatment messages, but gain/loss framing did not affect screening or prevention messages; gain/loss framing made little or no difference in behavior (16 studies). The review concluded that contrary to commonly held beliefs, the available evidence (albeit of low to moderate quality) suggests that both attribute and goal framing may have little if any consistent effect on health consumers' behavior. However, the heterogeneity of the results, as described here, suggests the possibility of a framing effect of health messaging under specific conditions.

Subsequently, another review primarily targeted studies published in the 2005–12 came to the same conclusion as the Cochrane review, namely that unlike in the general cognitive psychology literature, evidence for a framing effect is inconsistent in the health care literature, that it may apply in specific circumstances, but seems to have little impact on behaviors such as compliance with health care recommendations (Gong et al., 2013).

3.2 Moderators (Direction and Magnitude) of Message Framing Effects

While there is currently a lack of strong evidence regarding which variables influence the effect of framing in a health care setting, some information is known about characteristics of the messenger (the clinician), the recipient (the patient) and the milieu (societal context) which modulate the effect of message framing (Gong et al., 2013).

3.2.1 *Patient Characteristics*

3.2.1.1 Disease Context

Framing effects which are consistently reported in experimental studies may not be so reliably found in clinical studies because the context of the decision is different (Edwards et al., 2001). In applying framing concepts to the health domain, Rothman and Salovey hypothesized that the framing effect varies with for the three main categories of health care intervention messages—prevention of disease, early detection of disease, and treatment of established disease—such that gain frames and loss frames have differential effects (Rothman & Salovey, 1997). Basing their hypothesis on what prospect theory would predict, they proposed that gain frames would be more effective for disease prevention and treatment interventions, because these are perceived as "non-risky" in the sense they are preventing disease or returning health to normal. On the other hand, detection behaviors, such as cancer screening, are perceived as "risky," at least in the short-term, because they may discover early signs of a health problem and loss frames should be more persuasive for these types of risk-seeking behaviors. Subsequently, research has confirmed differences in framing effects depending on the kind of health message, but they have not always been consistent with the original hypothesis (Akl et al., 2011). Modulating factors in this situation may include: 1. recency of illness experience; 2. severity of current illness; and 3. significance and seriousness of outcome concerned.

As an example, in the chemotherapy study by O'Connor (1989), cancer patients' preferences for the new but more toxic chemotherapy were significantly stronger than those of healthy volunteers.

3.2.1.2 Level of Perceived Personal Risk and Benefit

Related to disease context is the patient's level of personal risk. Varying the degree of the risk has been shown to alter the size of the framing effect (Fagley & Miller, 1990) and in the case of life-limiting illness can appear to even contradict prospect theory's principle that people prefer to avoid losses over achieving gains. A powerful example of this moderator at work was demonstrated in a study of 100 outpatients of the family practice, hypertension, and cardiology services at a large Canadian university hospital regarding preferences for lipid lowering drugs with limited benefits for improving survival (Hux & Naylor, 1995a). When the group data were presented (average increase in survival of only 4 months), receiving no treatment was preferred to taking the medicine. But when the chances of benefitting were disaggregated and stratified (i.e., 6% will gain 4 years extra life,

6% will gain 3 extra years and 4% will gain 2 years, while 84% will gain no increase in survival), the preference switched valence to accepting treatment (Yusuf et al., 1994). This finding was interpreted to indicate that compared to epidemics like Tversky and Kahneman's (1981) Asian disease conundrum, optimism bias trumps prospect theory when patients are thinking about themselves rather than a population (Hux & Naylor, 1995b), and this bias extends to the treating physician (Hux, Levinton, & Naylor, 1994).

The study of hypothetical preferences for a new, toxic chemotherapy (O'Connor, 1989) over a currently available non-toxic one included in its design pegging 5-year survival at various probability levels, and framing survival as a gain or loss (living/dying). Both the cancer patients' and the healthy controls' choices were significantly influenced by the probability percentage that was presented. Preferences for the toxic treatment were less preferred when the chance of survival were dropped below 50% (O'Connor, 1989). This weakening of preference was enhanced in subjects who were given the outcome was framed negatively (expressed as chance of dying greater than 50%). The authors interpreted these findings as indicating that a "dying mode"-type value system is stimulated by a negative frame or probability level below 0.5, in which quality of life becomes more salient in decision making than quantity of life.

3.2.1.3 Demographic Factors

Personal characteristics that include age, sex and education could influence how individuals perceive a message. For instance, adolescents have been shown to reverse frame, preferring risky gains and sure losses (Reyna, Wilhelms, McCormick, & Weldon, 2015), while women have been shown to be more susceptible to framing effect than men (Fagley & Miller, 1990).

Patients' education level also plays an important role in clinical communications. Interpretation of numeric data, however framed, requires the decider to have adequate numeracy skills, and framing effects have been shown to be more pronounced in people with poor probability reasoning skills (Peters et al., 2006).

Many patient don't understand risk or to recall discussing it (Lloyd, 2001). A study of the effect of risk framing on preferences for using lipid lowering agents to prevent heart disease (with a large relative risk reduction but a small absolute risk reduction) found that participants with a scientific background, who were more numerate or had more years of education were more likely to decide not to take statins (Carling et al., 2008).

3.2.1.4 Personal Goal Orientation

While making decisions, individuals have a natural tendency to focus on gains or losses (Higgins, 1997). For instance, in a medical context, some people are more concerned about what gains they could receive (promotion focus). When choosing a treatment, they want to know, "Could treatment X improve my current state?" Other individuals are concerned about safety and preventing losses (prevention focus). When choosing a treatment, they want to know, "Could treatment X make my state worse?"

Whether people approach a decision with a promotion focus or a prevention focus depends on their personal and situational characteristics. The environment in which individuals grew up influences whether individuals more often focus on promotion or prevention concerns. If one's social environment emphasized accomplishments and growth, this person tends to focus on improving their current states and achieving gains (promotion focus). In contrast, if one's social environment emphasized safety and obligation, this person tends to focus on securing their current states and avoiding losses (prevention focus) (Higgins & Silberman, 1998).

The characteristics of a particular situation could also influence and elicit individuals' promotion or prevention focus (Higgins et al., 2001). If an individual faces a decision about a current injury, the situation by itself could induce individuals' promotion focus. The individual's main concern likely would be about how treatment could help him or her facilitate (promote) healing. In contrast, a decision about an elective surgery could induce a prevention focus. In this case, the individual would likely consider how treatment could help him or her prevent worsening the current state.

Independent from whether regulatory focus results from personal or situational characteristics, regulatory focus influences what information individuals choose to attend to, how they process and interpret information, and what information they evaluate as being valuable and worth acting upon (Lanaj, Chang, & Johnson, 2012). The same information framed differently could enhance motivation to follow a medical recommendation for some people but reduce the motivation of others (Motyka et al., 2014).

Many studies have demonstrated advantages of matching the message and individuals' promotion or prevention focus in health care. For example, researchers distributed among participants materials developed by Cancer Research UK that intentionally triggered worry and fear to facilitate motivation for cancer-screening tests. Individuals who are concerned about safety (promotion focused) were more motivated to engage in cancer-screening behavior than those who focus on achievements and gains (prevention focused) (Uskul, Keller, & Oyserman, 2008). Other researchers found that smoking cessation information was more effective for adolescents if adolescents with a promotion focus received the message that emphasized benefits of abstaining from smoking. For adolescents with a prevention focus, the message was more effective if it emphasized how abstaining from smoking could prevent losses (Kim, 2006).

At times, not only matching the message to individual goal orientations, but also mismatching could be helpful. In health communications, patients' strong positive or negative attitudes could bias patients' decisions away from statistical likelihoods. Mismatching individuals' promotion/prevention focus and message framing might reduce patients' engagement and motivate them to stop and think before making a choice that supports their initial strong attitude.

For instance, patients with advanced cancer read a hypothetical scenario describing a patient with lung cancer who faced a decision: to continue cancer treatment, or stop treatment and choose hospice care. Promotion and prevention foci of participants were experimentally controlled. Participants then learned that a physician recommended choosing hospice care. The recommendation was randomly framed, emphasizing either how hospice could help one achieve gains, or how hospice could help one avoid losses. This randomization match or mismatched participants' promotion or prevention foci. The participants who initially strongly disliked a hospice option before the advice was provided became less negative toward this option if the recommendation did not match their promotion or prevention focus (Fridman et al., 2018). This line of the research illustrated that personal goal orientation should be considered when framing health messages.

3.2.1.5 Other Personal Differences

Other personal differences such personality, preferences, and cognitive biases can affect decision making processes and the response to framing strategies. Personality attributes such as optimism, overconfidence, and risk-taking propensity affect decision making. Positive framing may be more persuasive than negative framing when patients are willing to take a risky option (as in the study of surgery vs. radiotherapy for operable lung cancer where surgery trumped radiotherapy when outcome presented as years gained despite risk of perioperative death).

Individuals have also different preferences for the way they like probabilistic information to be presented and discussed with them. A systematic review on this topic recommended that professionals should possess and understand a range of complementary formats for presenting risk data, e.g., a range of numerical data (ARR, RRR and NNT) as well as graphical representations, like bar graphs and 100-person infographics. The review recommended that clinicians should use them when discussing treatment choices with patients (Edwards & Elwyn, 2001).

Phenomena such as the representative heuristic (e.g., "I have a healthy diet and exercise, so I won't get bowel cancer") or the availability heuristic ("I don't want analgesia because of what I saw about overdoses in *Time* magazine") may affect patient's perception of risk for a disease, not only moderating framing effects but generally leading them to make decisions that are no in their best interests (Rice, 2013). A full consideration of the broad topic of behavioral economics and clinical decision making is beyond the scope of this chapter.

3.2.2 Clinician Characteristics

Clinicians are not immune to these "cognitive failures" either (Croskerry, 2013; Roman & Asch, 2014; Saposnik et al., 2016), and this includes being susceptible to framing. This susceptibility appears to be independent of independent of the physician's age, sex, speciality and experience (Bobbio et al., 1994). For example, a survey consisting of short vignettes covering four variations of framing effect was undertaken in a representative sample of US internists (Bui, Krieger, & Blumenthal-Barby, 2015). The survey found that physicians' recommendations were significantly different when the predicted probability of the outcome was framed as mortality vs. survival (breast cancer treatment vignette) and as frequency vs. percentage (cervical cancer-screening vignette).

3.2.3 Circumstantial Characteristics: Effect of Societal Factors

Often in a health context, pictures of healthy, active, fit people are used to emphasize the "gains" of a healthy life style, because the marketing world believe messaging should adhere to the positive. There appears to be a growing awareness within marketing of the need for health messaging that's rooted in a more negative approach. Loss-framed messages emphasizing the negative outcomes of an unhealthy life style, such placing as a picture of a gangrenous foot on a pack of cigarettes, may be more impactful on contemporary health care consumers who may require a more in-depth education on the pros and cons of what it is being recommended and the health care costs they will be up for (Smith, 2014).

In the world of marketing, there is an unresolved issue around gain and loss framing, statistics and quality data as the influence they have on the decisions that patients and consumers make about health choices. Consistent with the findings of Rothman and Salovey (1997), marketers find that some conversations focus on the positive, like improved survival statistics, and sometimes they are negative, such as the risks of hidden or undiagnosed conditions, and that while most of the current research focuses on whether to present a message in print or video; how the message is framed may be crucial (Smith, 2014).

4. CLINICAL IMPLICATIONS AND APPLICATIONS

Given that modulating factors result in a variable and inconsistent impact of message framing on health decision making and behaviors, some examples of situations in which framing does appear to be clinically relevant are now presented.

4.1 Using Framing to Induce Placebo Effects, While Minimizing Nocebo Effects

As discussed at the beginning of this article, only one part of the effect of a pharmaceutical or other therapeutic intervention is produced by the intervention itself. This is especially so when the intervention is aimed at relieving symptoms, such as pain. The rest of the therapeutic effect, commonly referred to as the "placebo effect," is achieved via complex physiological events (e.g., activation of opioid and descending pain inhibition pathways) occurring in the central and peripheral nervous systems in response to certain contextual factors and psychological processes that result from conditioning and expectations (Colloca, Klinger, Flor, & Bingel, 2013).

Inducing and enhancing this placebo effect may optimize the efficacy of the intervention, and positive attribute framing of the intervention should be one way of doing this. At the same, health care professionals should take care to minimize manipulation of health consumers' decisions by "over-framing" (Edwards & Elwyn, 2001).

Beyond the mathematical psychology of prospect theory and the analysis of the evidence base for evidence as to whether or not the "framing effect" impacts on clinical decisions by patients and providers presented in the preceding sections of this chapter, studies of placebo and nocebo effects indicate that the affective psychology surrounding the delivery of health care is powerful and can modulate the outcome of a therapeutic intervention.

As per the hidden vs. open paradigm (Colloca, Lopiano, Lanotte, & Benedetti, 2004), framing morphine as a powerful analgesic that requires double checking and recording in a register before being injected by a trained nurse is likely to produce a greater change in pain scores than an over-the-counter analgesic, even when the relative potency of the two medicines is taken into account. Likewise patients may report the generic brand of a drug is less effective or produces different side effects to the same medicine dispensed under its trademark protected name (Faasse, Cundy, Gamble, & Petrie, 2013).

The compelling preclinical research defining the psychological and neurobiological mechanisms of placebo analgesia translate across to clinical practice. A Cochrane review of studies up until 2008 found evidence for a placebo effect contributing to the outcome of treatments of pain and nausea, and possibly for asthma and phobia (Hrobjartsson & Gotzsche, 2010), although there were several caveats—larger placebo effects were associated with physical placebo interventions (e.g., sham acupuncture), patient-reported outcomes, small trials, and trials with the explicit purpose of studying placebo. Larger effects of placebo were also found in trials that did not inform patients about the possible placebo intervention.

That review needs updating, as a number well-designed studies have been subsequently published that could alter its conclusions. One such study aimed to investigate was a randomized clinical trial which sought to determine whether placebo effects in chronic low back pain could be harnessed ethically by adding an open-label placebo (Ludolph & Schulz, 2015) pill to treatment as usual (TAU) (Carvalho et al., 2016). Ninety-seven participants with persistent low back pain for more than 3 months' duration and diagnosed by a board-certified pain specialist were randomized to receive OLP or continue TAU. Participants randomized to receive the OLP were informed by the trial nurse that although the pills were chemically inert some patients seem to benefit from taking them. Eighty-three completed the 3-week trial. Pain severity was assessed on three 0- to 10-point Numeric Rating Scales, scoring maximum pain, minimum pain, and usual pain, and a composite outcome, total pain score. Back pain-related dysfunction was the other primary outcome. Compared to TAU, OLP elicited greater pain reduction on each of the pain scores, achieving moderate- to-large, clinically significant effect sizes, by 1.5 points on the composite pain score in the OLP group. Pain intensity was reduced, but not by a clinically significant amount (0.21 points) in the TAU group. OLP also reduced disability, with a large, clinically significant effect size of 2.9 points while TAU did not (no change). On switching to OLP, the TAU group showed significant reductions in both pain (1.5 points) and disability (3.4 points). Of note, many participants asked where they could obtain an on-going supply of the OLP pills on completing the trial. These findings suggest that OLP pills presented in a positive context may be helpful in chronic low back pain.

This study indicates that health care professionals can use attribute and goal framing of their interventions to enhance the placebo effect to improve the outcomes of the care they provide. This is achieved as a result of conditioning and by enhancing the expectations concerning the effect of a treatment, and both of these factors can be modified by the health care provider. Conditioning occurs in classic Pavlovian terms whereby verbal and other cues recall previous episodes of pain relief (Colloca & Benedetti, 2006). Examples include (a) verbal induction, by pairing placebo treatments with reductions in pain intensity, enhancing expectations for future analgesia from placebo; (b) pharmacological conditioning, via repeated administration of active drug then replacing it with an inert agent, or by inducing some side effects, e.g., dry mouth with a small dose of atropine, to recall active analgesics with anticholinergic side effects, and (c) the development of anticipatory effects, e.g., nausea and vomiting from chemotherapy. Expectations are learnt from three sources: personal experience (including conditioned responses), instruction, and social learning including watching another being administered an effective analgesic (Colloca & Benedetti, 2009). Expectations can also be "unlearned" therefore indicating that framing strategies may have a role in reversing negative expectations and changing behaviors and outcomes.

Nocebo effects occur when information and expectancies about adverse effects of an intervention produce negative outcomes, often with significant consequences (Faasse & Petrie, 2013). The problem of anticipatory nausea (ANV) before chemotherapy, which can be severe enough to interfere with the cancer treatment schedule, is a prime example of this (Kamen et al., 2014). Patients obtain information from many sources, and it is frequently negative information when it comes via the news media or internet (Faasse & Petrie, 2013). Medical professionals need to be aware that their own negative beliefs about a treatment can contribute to patients' expectations of a bad outcome. In the case of ANV, oncologists and chemotherapy nurses professionals can use attribute and goal framing to advise cancer patients that chemotherapy-induced nausea is now a manageable problem that is less common than it used to be, and should not prevent them "staying the course" with treatment.

The impact of nocebo effects on health extends beyond the side effects of intervention to a poor health state generally. The so-called "infrasound syndrome" describes the adverse health effects attributed to wind turbines on wind farms. These have been blamed for various symptoms including tinnitus, ear pressure, vertigo, dizziness, sleep disturbance, headaches, nausea, visual blurring, irritability, memory and concentration problems, panic episodes, tachycardia, and body vibrations, forcing some nearby residents to abandon their homes. However, a study exposing volunteers to simulated infrasound (albeit short-term) while being influenced into states of high or low expectancy of negative effects from infrasound found that exposure to the sound had no statistically significant effect on symptoms but the participants' prior concerns about its ill effects did have a significant effect, supporting a nocebo effect hypothesis for the infrasound syndrome (Tonin, Brett, & Colagiuri, 2016).

4.2 Influencing Wisely

4.2.1 Pain Management: Framing and Harnessing Placebo Effects

The development of evidence-based medicine combined with the rise of the patient-centered care movement has undoubtedly improved the quality of health care in the past two decades. Practicing in this manner, physicians are expected provide recommendations and guidance based on their appraisal of the medical research literature and their understanding of the patient's values but then ultimately allow patients final authority over major decisions (Quill & Holloway, 2012). As Quill and Holloway (2012) have pointed out, actualizing this process is not simple. As well as abdicating their decision-making responsibilities by putting too much of the burden on patients in the interest of "patient autonomy," health care providers have been ill-equipped to contribute to the patient's decision process more fully because they are untrained in the psychology of choosing, the irrationality of human decisions, and the impact of specific thought process such as framing effects. Consequently patients may decide to pursue options that are inconsistent with their expressed preferences or deemed low-value by initiatives such as *Choosing Wisely* (Roman & Asch, 2014).

For patients with persistent pain on chronic opioid therapy, a physician practicing evidence-based patient-centered care nowadays will be recommending discontinuation of this treatment on the grounds that long term opioids are neither effective (Krebs et al., 2018) nor safe (Gomes, Mamdani, Dhalla, Paterson, & Juurlink, 2011; Vowles et al., 2015). These can be very difficult discussions and communication skills training to assist physicians in negotiating disagreements about pain management has been posited as the solution (Henry, Bell, Fenton, & Kravitz, 2018). However, teaching clinicians to frame their advice more effectively will also be important (see also, Darnall & Colloca, 2018).

There has been little research on message framing and pain. It has been reported that physicians often perceive pain patients as "narcotic seeking" while patients report physicians do not appreciate their plight or take their pain seriously (Henry et al., 2018). Certainly, opioid analgesics are being attribute-framed negatively in the news media, bestowed with labels such as "crisis" and "epidemic" and many physicians are confused about whether or not to prescribe them.

In the absence of data, the etymology of pain shows that it began with a negative connotation, derived from the Latin word *poena* (punishment), but suffering pain also has had a positive, redemptive aspect. During most of the 20th century, pain was seen to be a biomedical problem by most patients and physicians, a symptom of an injury or disease that should be diagnosed accurately so that it can be fixed or controlled. Toward the end of the 20th century, Fordyce argued the case for reframing most chronic pain as a behavioral problem rather than a biomedical one (Fordyce, 1982).

A study of experimental pain found that attribute framing of pain as having a positive or negative effect influenced the pain experience and this was mediated by both endogenous opioids and cannabinoids (Benedetti, Thoen, Blanchard, Vighetti, & Arduino, 2013). These findings may have a profound impact on clinical practice. For example, post-operative pain, which means healing, can be perceived as less unpleasant than cancer pain, which means death. Therefore, the behavioral and/or pharmacological manipulation of the meaning of pain can represent an effective approach to pain management. A clinical study of post-thoracotomy pain with an opioid plus a saline infusion found that patients receiving saline misrepresented as an analgesic drug had better pain relief that those where it was presented as having a 50:50 chance of being an adjuvant analgesic, or when it was correctly presented as fluid for hydration purposes (Pollo et al., 2001). These results indicate that different verbal instructions about certain and uncertain expectations of analgesia produce different placebo analgesic effects, which in turn trigger a dramatic change of behavior leading to a significant reduction of opioid intake.

Framing is also important in harnessing the placebo effect associated with the administration of pain treatments and reducing any nocebo effects associated with them. Several old studies reported the benefits of framing pain as normal in the perioperative period on reducing perioperative anxiety (Egbert, Battit, Turndorf, & Beecher, 1963) with a 40% reduction in opioid requirements post-operatively, from 38 mg to 26 mg on Day 1 and 29 mg to 13 mg on Day 2 (Egbert, Battit, Welch, & Bartlett, 1964). Clearly, the effect of administering opioids is influenced the context which drives expectancy. This context includes how it is delivered (by trustworthy staff behaving like professional experts), the explanation given (e.g., "an injection is more powerful than a tablet"), and the setting (e.g., university hospital vs. community hospital), and contemporary anesthesiologists learned from Egbert et al. (1964), Beecher (1955) and others to harness the "placebo" component of the therapeutic ritual of the acute pain round.

Expectations are also important in managing chronic pain. Patients with chronic primary pain such as mechanical low back pain, for which no specific treatments are available, often have low expectations for treatment when they attend a pain clinic. This is because of their personal experience of not benefiting from biomedical treatments, what they have been told by surgeons and what they have learned from family and significant others in their social networks ("it's all in your head"). It is important to help them to move away from a biomedical model that is applicable to acute post-trauma pain (surgery, analgesia, rest and recuperation) but not applicable to their problem. They need to be convinced to focus more on coping with the adverse consequences of being in chronic pain, such as physical inactivity, deconditioning and disability, distress and depression, damaged relationships and over-reliance on medications. Despite their low initial expectations, many patients can benefit from treatment if the pain physician reframes their pain as a cognitive and behavioral problem rather than a biomedical one and leads them to embrace and use strategies that are effective for reducing pain-related disability and distress (Kamper et al., 2015; Nicholas & Blyth, 2016) rather than those that are not (Chou et al., 2009).

4.2.2 Making Difficult Treatment Decisions at the End of Life: Focus on Gain or Losses

One of the challenges to eliciting patients' values and preferences about treating a disease is the complexity. Often there are multiple treatment options which carry different risks, burdens and benefits, none of which are guaranteed to eventuate or not. This conundrum gets to the behavioral underpinnings of choice, and how choices are made in the face of uncertainty.

Principles that distinguish between framing outcomes in terms of benefits (gains) or losses could be leveraged in solving difficult health care problems where the stakes are high, there is more than one option, and the various stakeholders evaluate the risks and benefits of each option differently (Verma et al., 2014). Verma et al. (2014) used an example of an elderly patient with advanced lung cancer who had to deal with the difficult decision of whether to pursue more chemotherapy (responders have an average survival of 15 months but with the risk of life-shortening toxicity) or not (average survival of 8 months, but no risk of life-shortening toxicity). In this example, the oncologist recommended chemotherapy and the family initially agreed with the recommendation. But the family physician, familiar with the principles of prospect theory, opined against chemotherapy recognizing three decision-making biases that were leading the patient and family not to be acting in his true best interests:

- *Oversimplifying the choice by only focusing on the key differences between the options* (important gains vs. losses)—in this case, focusing on the potential gain in survival from chemotherapy while ignoring the losses, (time at hospital having treatment; risk of hospitalization with toxicity) and the delays in other decisions (e.g., advance care planning) with risk of futile end of life care.
- *Loss aversion from not having chemotherapy:* Framing chemotherapy as the loss of 7 extra months was aversive despite the potential for the loss of quality of life from toxicity. Despite the framing hospice as a gain, with 8 quality months, chemotherapy is selected.
- *Placing distorted values on the options*: people tend to be overly optimistic about the chances of success (Quill & Holloway, 2012), overweight short-term gains, and discount future losses (Roman & Asch, 2014).

The family physician, on the other hand, was able to posit a less sanguine view of chemotherapy. She saw the certainty of 8 good months on hospice a better bet than 15 poor months suffering through chemotherapy and delayed hospice referral. By reframing the data on chemotherapy outcomes (emphasizing the losses of this option), the family physician made a stronger case for hospice care. The scenario concluded with the patient reconsidering his options from this reframed perspective, enabling him to find the best match between his preferences and the options available.

4.2.3 Individual Goal Orientation and Message Tailoring

Careful tailoring message and individual goal orientation on gain/losses could be useful in clinical communications. The question is how a clinician could evaluate whether a patient focuses on gain or losses during an encounter?

There are a few methods for understanding patients' goal orientation and tailoring message to their focus. First, clinician could simply observe whether patients' questions are more about safety of a treatment (prevention focus) or about efficacy of a treatment (promotion focus). At times, the observation of patients questions might not lead to a simple answer as patients might be concern about both benefits and losses associated with the treatment. In this case, a clinician could frame own questions focusing patients more on benefits or losses in past or present (Kanze, Huang, Conley, & Higgins, 2018). In fact, researchers frequently use regulatory focus induction by asking promotion or prevention focused question in studying health communication (e.g., Hong & Lee, 2008).

Understanding (or inducing) patients' regulatory focus might be beneficial in health communications. It was shown that if researchers matched message framing and participants regulatory focus, participants were more motivated to for healthy eating (Jin, 2012; Tam, Bagozzi, & Spanjol, 2010), physical activities (Gallagher & Updegraff, 2011; Jin, 2010; Martinez, Duncan, Rivers, Latimer, & Salovey, 2013; Pfeffer, 2013), smoking cessation (Zhao & Pechmann, 2007) and vaccination (Gerend & Shepherd, 2007; Jin, 2010; Park, 2012). A recent review has summarized that 26 out of 30 studies which confirm that fitting a health message to individuals' motivational orientation has a substantial enhancing effect on individuals' attitudes and health behaviors (Ludolph & Schulz, 2015).

4.2.4 Use of Communication Tools to Present Data on Uncertain Risks and Benefits

Looking for effective evidence-based communication tools for prostatic specific antigen testing highlighted the challenges for clinicians and consumers in accessing tools that are evidence-based in design as well as content (Trevena, Davey, Barratt, Butow, & Caldwell, 2006). Communication tools in most formats (verbal, written, video, provider-delivered, and computer-based) will increase patients' understanding of probabilistic outcomes and clarifying preferences. They are more likely to do so if they are structured, tailored and/or interactive. Probabilistic information is best represented as event rates (natural frequencies) in relevant groups of people, rather than words, probabilities or summarized as effect measures such as relative risk reduction. Illustrations such as cartoons, graphs (vertical bar charts) or 100 person infographics appear to aid understanding (Paling, 2003). Values clarification exercises may be better than standard utility techniques for eliciting preferences in individual decision making (Trevena et al., 2006).

5. CONCLUSIONS AND FUTURE DIRECTIONS

Going forward, health care is becoming increasingly automated as we move through the 21st century, which at first glance may be seen to reduce the impact of framing effects. Artificial intelligence will provide individuals with personalized outcome data, removing much of the uncertainty about the harms and benefits. And avatar professionals will present these personalized facts in an unbiased, trustworthy way. In this way, many of the problems which currently plague patients and their clinicians when they are working together to make complex health care decisions will be removed.

But the patient will still be a human being with a disease who will continue to seek health gains and avoid health losses. What represents a gain or a loss will continue to be influenced by their values, beliefs, goals, priorities and expectations for care and ultimately, placebo and nocebo effects. These in turn will continue to be the by-product of conditioning and learning, which will continue to be sensitive to framing effects.

REFERENCES

Akl, E. A., Oxman, A. D., Herrin, J., Vist, G. E., Terrenato, I., Sperati, F., et al. (2011). Framing of health information messages. *Cochrane Database of Systematic Reviews, 12*, CD006777. https://doi.org/10.1002/14651858.CD006777.pub2.

Amanzio, M., Pollo, A., Maggi, G., & Benedetti, F. (2001). Response variability to analgesics: A role for non-specific activation of endogenous opioids. *Pain, 90*(3), 205–215.

Apanovitch, A. M., McCarthy, D., & Salovey, P. (2003). Using message framing to motivate HIV testing among low-income, ethnic minority women. *Health Psychology, 22*(1), 60.

Aslaksen, P. M., Zwarg, M. L., Eilertsen, H.-I. H., Gorecka, M. M., & Bjørkedal, E. (2015). Opposite effects of the same drug: Reversal of topical analgesia by nocebo information. *Pain, 156*(1), 39–46.

Banks, S. M., Salovey, P., Greener, S., Rothman, A. J., Moyer, A., Beauvais, J., et al. (1995). The effects of message framing on mammography utilization. *Health Psychology, 14*(2), 178.

Bechara, A., & Damasio, A. R. (2005). The somatic marker hypothesis: A neural theory of economic decision. *Games and Economic Behavior, 52*(2), 336–372.

Beecher, H. K. (1955). The powerful placebo. *Journal of the American Medical Association, 159*(17), 1602–1606.

Benedetti, F., Thoen, W., Blanchard, C., Vighetti, S., & Arduino, C. (2013). Pain as a reward: Changing the meaning of pain from negative to positive co-activates opioid and cannabinoid systems. *Pain, 154*(3), 361–367. https://doi.org/10.1016/j.pain.2012.11.007.

Bobbio, M., Demichelis, B., & Giustetto, G. (1994). Completeness of reporting trial results: Effect on physicians' willingness to prescribe. *Lancet, 343*(8907), 1209–1211.

Bui, T. C., Krieger, H. A., & Blumenthal-Barby, J. S. (2015). Framing effects on Physicians' judgment and decision making. *Psychological Reports, 117*(2), 508–522. https://doi.org/10.2466/13.PR0.117c20z0.

Carling, C., Kristoffersen, D. T., Herrin, J., Treweek, S., Oxman, A. D., Schunemann, H., et al. (2008). How should the impact of different presentations of treatment effects on patient choice be evaluated? A pilot randomized trial. *PLoS One, 3*(11), e3693. https://doi.org/10.1371/journal.pone.0003693.

Carvalho, C., Caetano, J. M., Cunha, L., Rebouta, P., Kaptchuk, T. J., & Kirsch, I. (2016). Open-label placebo treatment in chronic low back pain: A randomized controlled trial. *Pain, 157*(12), 2766–2772. https://doi.org/10.1097/j.pain.0000000000000700.

Cesario, J., Grant, H., & Higgins, E. T. (2004). Regulatory fit and persuasion: Transfer from "feeling right". *Journal of Personality and Social Psychology, 86*(3), 388.

Cesario, J., Higgins, E. T., & Scholer, A. A. (2008). Regulatory fit and persuasion: Basic principles and remaining questions. *Social and Personality Psychology Compass, 2*(1), 444–463.

Chou, R., Loeser, J. D., Owens, D. K., Rosenquist, R. W., Atlas, S. J., Baisden, J., et al. (2009). Interventional therapies, surgery, and interdisciplinary rehabilitation for low back pain: An evidence-based clinical practice guideline from the American Pain Society. *Spine (Phila Pa 1976), 34*(10), 1066–1077. https://doi.org/10.1097/BRS.0b013e3181a1390d.

Colloca, L., & Benedetti, F. (2006). How prior experience shapes placebo analgesia. *Pain, 124*(1–2), 126–133. https://doi.org/10.1016/j.pain.2006.04.005.

Colloca, L., & Benedetti, F. (2009). Placebo analgesia induced by social observational learning. *Pain, 144*(1–2), 28–34. https://doi.org/10.1016/j.pain.2009.01.033.

Colloca, L., Klinger, R., Flor, H., & Bingel, U. (2013). Placebo analgesia: Psychological and neurobiological mechanisms. *Pain, 154*(4), 511–514. https://doi.org/10.1016/j.pain.2013.02.002.

Colloca, L., Lopiano, L., Lanotte, M., & Benedetti, F. (2004). Overt versus covert treatment for pain, anxiety, and Parkinson's disease. *The Lancet. Neurology, 3*(11), 679–684. https://doi.org/10.1016/S1474-4422(04)00908-1.

Cox, A. D., Cox, D., & Zimet, G. (2006). Understanding consumer responses to product risk information. *Journal of Marketing, 70*(1), 79–91.

Croskerry, P. (2013). From mindless to mindful practice—Cognitive bias and clinical decision making. *The New England Journal of Medicine, 368*(26), 2445–2448. https://doi.org/10.1056/NEJMp1303712.

Darnall, B. D., & Colloca, L. (2018). Optimizing placebo and minimizing nocebo to reduce pain, catastrophizing, and opioid use: A review of the science and an evidence-informed clinical toolkit. In L. Colloca (Ed.), *Neurobiology of the placebo effect, part II: Vol. 139* (pp. 129–157). New York: Academic Press.

Detweiler, J. B., Bedell, B. T., Salovey, P., Pronin, E., & Rothman, A. J. (1999). Message framing and sunscreen use: Gain-framed messages motivate beach-goers. *Health Psychology, 18*(2), 189.

Edwards, A., & Elwyn, G. (2001). Understanding risk and lessons for clinical risk communication about treatment preferences. *Quality in Health Care, 10*(Suppl 1), i9–13.

Edwards, A., Elwyn, G., Covey, J., Matthews, E., & Pill, R. (2001). Presenting risk information—A review of the effects of "framing" and other manipulations on patient outcomes. *Journal of Health Communication, 6*(1), 61–82. https://doi.org/10.1080/10810730150501413.

Egbert, L. D., Battit, G., Turndorf, H., & Beecher, H. K. (1963). The value of the preoperative visit by an anesthetist. A study of doctor-patient rapport. *JAMA, 185*(7), 553–555.

Egbert, L. D., Battit, G. E., Welch, C. E., & Bartlett, M. K. (1964). Reduction of postoperative pain by encouragement and instruction of patients. A study of doctor-patient rapport. *The New England Journal of Medicine, 270*, 825–827. https://doi.org/10.1056/NEJM196404162701606.

Faasse, K., Cundy, T., Gamble, G., & Petrie, K. J. (2013). The effect of an apparent change to a branded or generic medication on drug effectiveness and side effects. *Psychosomatic Medicine, 75*(1), 90–96. https://doi.org/10.1097/PSY.0b013e3182738826.

Faasse, K., & Petrie, K. J. (2013). The nocebo effect: Patient expectations and medication side effects. *Postgraduate Medical Journal, 89*(1055), 540–546. https://doi.org/10.1136/postgradmedj-2012-131730.

Fagley, N. S., & Miller, P. M. (1990). The effect of framing on choice: Interactions with risk-taking propensity, cognitive style and sex. *Personality and Social Psychology Bulletin, 16*(3), 496–510.

Fordyce, W. E. (1982). A behavioural perspective on chronic pain. *The British Journal of Clinical Psychology, 21*, 313–320. Pt 4.

Fridman, I., Glare, P. A., Stabler, S. M., Epstein, A. S., Wiesenthal, A., Leblanc, T. W., et al. (2018). Information framing reduces initial negative attitudes in cancer patients' decisions about hospice care. *Journal of Pain and Symptom Management, 55*(6), 1540–1545. https://doi.org/10.1016/j.jpainsymman.2018.02.010.

Frijda, N. H. (1986). *The emotions.* Cambridge University Press.

Gallagher, K. M., & Updegraff, J. A. (2011). When 'fit' leads to fit, and when 'fit' leads to fat: How message framing and intrinsic vs. extrinsic exercise outcomes interact in promoting physical activity. *Psychology & Health, 26*(7), 819–834. https://doi.org/10.1080/08870446.2010.505983.

Gallagher, K. M., Updegraff, J. A., Rothman, A. J., & Sims, L. (2011). Perceived susceptibility to breast cancer moderates the effect of gain-and loss-framed messages on use of screening mammography. *Health Psychology, 30*(2), 145.

Gerend, M. A., & Shepherd, J. E. (2007). Using message framing to promote acceptance of the human papillomavirus vaccine. *Health Psychology, 26*(6), 745–752. https://doi.org/10.1037/0278-6133.26.6.745.

Gomes, T., Mamdani, M. M., Dhalla, I. A., Paterson, J. M., & Juurlink, D. N. (2011). Opioid dose and drug-related mortality in patients with nonmalignant pain. *Archives of Internal Medicine, 171*(7), 686–691. https://doi.org/10.1001/archinternmed.2011.117.

Gong, J., Zhang, Y., Feng, J., Huang, Y., Wei, Y., & Zhang, W. (2013). Influence of framing on medical decision making. *EXCLI Journal, 12*, 20–29.

Gong, J., Zhang, Y., Yang, Z., Huang, Y., Feng, J., & Zhang, W. (2013). The framing effect in medical decision-making: A review of the literature. *Psychology, Health & Medicine, 18*(6), 645–653. https://doi.org/10.1080/13548506.2013.766352.

Henry, S. G., Bell, R. A., Fenton, J. J., & Kravitz, R. L. (2018). Communication about chronic pain and opioids in primary care: Impact on patient and physician visit experience. *Pain, 159*, 371–379. https://doi.org/10.1097/j.pain.0000000000001098.

Higgins, E. T. (1997). Beyond pleasure and pain. *The American Psychologist, 52*(12), 1280–1300.

Higgins, E. T., Friedman, R. S., Harlow, R. E., Idson, L. C., Ayduk, O. N., & Taylor, A. (2001). Achievement orientations from subjective histories of success: Promotion pride versus prevention pride. *European Journal of Social Psychology, 31*(1), 3–23.

Higgins, E. T., Idson, L. C., Freitas, A. L., Spiegel, S., & Molden, D. C. (2003). Transfer of value from fit. *Journal of Personality and Social Psychology, 84*(6), 1140.

Higgins, E. T., & Silberman, I. (1998). Development of regulatory focus: Promotion and prevention as ways of living. In J. Heckhausen & C. S. Dweck (Eds.), *Motivation and self-regulation across the life span* (pp. 78–113). New York: Cambridge University Press.

Hong, J. W., & Lee, A. Y. (2008). Be fit and be strong: Mastering self-regulation through regulatory fit. *Journal of Consumer Research, 34*(5), 682–695.

Hrobjartsson, A., & Gotzsche, P. C. (2010). Placebo interventions for all clinical conditions. *Cochrane Database of Systematic Reviews, 1* CD003974. https://doi.org/10.1002/14651858.CD003974.pub3.

Hux, J. E., Levinton, C. M., & Naylor, C. D. (1994). Prescribing propensity: Influence of life-expectancy gains and drug costs. *Journal of General Internal Medicine, 9*(4), 195–201.

Hux, J. E., & Naylor, C. D. (1995a). Communicating the benefits of chronic preventive therapy: Does the format of efficacy data determine patients' acceptance of treatment? *Medical Decision Making, 15*(2), 152–157. https://doi.org/10.1177/0272989X9501500208.

Hux, J. E., & Naylor, C. D. (1995b). In the eye of the beholder. *Archives of Internal Medicine, 155*(21), 2277–2280.
Jin, S.-A. (2010). The roles of regulatory focus and medical recommendation avatars' trustworthiness in virtual environment-based E-health. *Cyberpsychology, Behavior and Social Networking, 13*(4), 461–466.
Jin, S.-A. (2012). Self-discrepancy and regulatory fit in avatar-based exergames. *Psychological Reports, 111*(3), 697–710.
Kahneman, D., & Tversky, A. (1979). On the interpretation of intuitive probability: A reply to Jonathan Cohen. *Cognition, 7*(4), 409–411.
Kamen, C., Tejani, M. A., Chandwani, K., Janelsins, M., Peoples, A. R., Roscoe, J. A., et al. (2014). Anticipatory nausea and vomiting due to chemotherapy. *European Journal of Pharmacology, 722*, 172–179. https://doi.org/10.1016/j.ejphar.2013.09.071.
Kamper, S. J., Apeldoorn, A. T., Chiarotto, A., Smeets, R. J., Ostelo, R. W., Guzman, J., et al. (2015). Multidisciplinary biopsychosocial rehabilitation for chronic low back pain: Cochrane systematic review and meta-analysis. *BMJ, 350*, h444. https://doi.org/10.1136/bmj.h444.
Kanze, D., Huang, L., Conley, M. A., & Higgins, E. T. (2018). We ask men to win and women not to lose: Closing the gender gap in startup funding. *The Academy of Management Journal, 61*(2), 586–614.
Kim, Y. J. (2006). The role of regulatory focus in message framing in antismoking advertisements for adolescents. *Journal of Advertising, 35*(1), 143–151.
Koban, L., Jepma, M., Geuter, S., & Wager, T. D. (2017). What's in a word? How instructions, suggestions, and social information change pain and emotion. *Neuroscience & Biobehavioral Reviews, 81*, 29–42.
Krebs, E. E., Gravely, A., Nugent, S., Jensen, A. C., DeRonne, B., Goldsmith, E. S., et al. (2018). Effect of opioid vs nonopioid medications on pain-related function in patients with chronic back pain or hip or knee osteoarthritis pain: The SPACE Randomized Clinical Trial. *JAMA, 319*(9), 872–882. https://doi.org/10.1001/jama.2018.0899.
Lanaj, K., Chang, C. H., & Johnson, R. E. (2012). Regulatory focus and work-related outcomes: A review and meta-analysis. *Psychological Bulletin, 138*(5), 998–1034. https://doi.org/10.1037/a0027723.
Lazarus, R. S. (1991). Cognition and motivation in emotion. *American Psychologist, 46*(4), 352.
Levine, J. D., Gordon, N. C., Smith, R., & Fields, H. L. (1981). Analgesic responses to morphine and placebo in individuals with postoperative pain. *Pain, 10*(3), 379–389.
Lieberman, M. D., Eisenberger, N. I., Crockett, M. J., Tom, S. M., Pfeifer, J. H., & Way, B. M. (2007). Putting feelings into words. *Psychological Science, 18*(5), 421–428.
Lloyd, A. J. (2001). The extent of patients' understanding of the risk of treatments. *Quality in Health Care, 10*(Suppl. 1), i14–i18.
Ludolph, R., & Schulz, P. J. (2015). Does regulatory fit lead to more effective health communication? A systematic review. *Social Science & Medicine, 128*, 142–150. https://doi.org/10.1016/j.socscimed.2015.01.021.
Martinez, J. L., Duncan, L. R., Rivers, S. E., Latimer, A. E., & Salovey, P. (2013). Examining the use of message tailoring to promote physical activity among medically underserved adults. *Journal of Health Psychology, 18*(4), 470–476. https://doi.org/10.1177/1359105312445798.
McNeil, B. J., Pauker, S. G., Sox, H. C., Jr., & Tversky, A. (1982). On the elicitation of preferences for alternative therapies. *The New England Journal of Medicine, 306*(21), 1259–1262. https://doi.org/10.1056/NEJM198205273062103.
Meyerowitz, B. E., & Chaiken, S. (1987). The effect of message framing on breast self-examination attitudes, intentions, and behavior. *Journal of Personality and Social Psychology, 52*(3), 500.

Millar, M. G., & Millar, K. U. (2000). Promoting safe driving behaviors: The influence of message framing and issue involvement. *Journal of Applied Social Psychology*, *30*(4), 853–866.

Motyka, S., Grewal, D., Puccinelli, N. M., Roggeveen, A. L., Avnet, T., Daryanto, A., et al. (2014). Regulatory fit: A meta-analytic synthesis. *Journal of Consumer Psychology*, *24*(3), 394–410.

Nicholas, M. K., & Blyth, F. M. (2016). Are self-management strategies effective in chronic pain treatment? *Pain Management*, *6*(1), 75–88. https://doi.org/10.2217/pmt.15.57.

O'Connor, A. M. (1989). Effects of framing and level of probability on patients' preferences for cancer chemotherapy. *Journal of Clinical Epidemiology*, *42*(2), 119–126.

O'Connor, A. M., Pennie, R. A., & Dales, R. E. (1996). Framing effects on expectations, decisions, and side effects experienced: The case of influenza immunization. *Journal of Clinical Epidemiology*, *49*(11), 1271–1276.

O'Keefe, D. J., & Jensen, J. D. (2007). The relative persuasiveness of gain-framed and loss-framed messages for encouraging disease prevention behaviors: A meta-analytic review. *Journal of Health Communication*, *12*(7), 623–644. https://doi.org/10.1080/10810730701615198.

Paling, J. (2003). Strategies to help patients understand risks. *BMJ*, *327*(7417), 745–748. https://doi.org/10.1136/bmj.327.7417.745.

Park, S. Y. (2012). The effects of message framing and risk perceptions for HPV vaccine campaigns: Focus on the role of regulatory fit. *Health Marketing Quarterly*, *29*(4), 283–302. https://doi.org/10.1080/07359683.2012.732847.

Peters, E., Vastfjall, D., Slovic, P., Mertz, C. K., Mazzocco, K., & Dickert, S. (2006). Numeracy and decision making. *Psychological Science*, *17*(5), 407–413. https://doi.org/10.1111/j.1467-9280.2006.01720.x.

Petty, R. E., & Cacioppo, J. T. (1986). The elaboration likelihood model of persuasion. *Communication and persuasion*. New York, NY: Springer, pp. 1–24.

Pfeffer, I. (2013). Regulatory fit messages and physical activity motivation. *Journal of Sport & Exercise Psychology*, *35*(2), 119–131.

Pollo, A., Amanzio, M., Arslanian, A., Casadio, C., Maggi, G., & Benedetti, F. (2001). Response expectancies in placebo analgesia and their clinical relevance. *Pain*, *93*(1), 77–84.

Quill, T. E., & Holloway, R. G. (2012). Evidence, preferences, recommendations—Finding the right balance in patient care. *The New England Journal of Medicine*, *366*(18), 1653–1655. https://doi.org/10.1056/NEJMp1201535.

Reyna, V. F., Wilhelms, E. A., McCormick, M. J., & Weldon, R. B. (2015). Development of risky decision making: Fuzzy-trace theory and neurobiological perspectives. *Child Development Perspectives*, *9*(2), 122–127. https://doi.org/10.1111/cdep.12117.

Rice, T. (2013). The behavioral economics of health and health care. *Annual Review of Public Health*, *34*, 431–447. https://doi.org/10.1146/annurev-publhealth-031912-114353.

Roman, B. R., & Asch, D. A. (2014). Faded promises: The challenge of deadopting low-value care. *Annals of Internal Medicine*, *161*(2), 149–150. https://doi.org/10.7326/M14-0212.

Rothman, A. J., Martino, S. C., Bedell, B. T., Detweiler, J. B., & Salovey, P. (1999). The systematic influence of gain-and loss-framed messages on interest in and use of different types of health behavior. *Personality and Social Psychology Bulletin*, *25*, 1355–1369.

Rothman, A. J., & Salovey, P. (1997). Shaping perceptions to motivate healthy behavior: The role of message framing. *Psychological Bulletin*, *121*(1), 3–19.

Salovey, P., & Wegener, D. T. (2003). Communicating about health: Message framing, persuasion, and health behavior. In J. Suls & K. A. Wallston (Eds.), *Social psychological foundations of health and illness* (pp. 54–81). Malden, MA: Blackwell Publishing Ltd.

Saposnik, G., Redelmeier, D., Ruff, C. C., & Tobler, P. N. (2016). Cognitive biases associated with medical decisions: A systematic review. *BMC Medical Informatics and Decision Making*, *16*(1), 138. https://doi.org/10.1186/s12911-016-0377-1.

Schneider, T. R., Salovey, P., Apanovitch, A. M., Pizarro, J., McCarthy, D., Zullo, J., et al. (2001). The effects of message framing and ethnic targeting on mammography use among low-income women. *Health Psychology*, *20*(4), 256.

Schwarz, N., & Clore, G. L. (1996). Feelings and phenomenal experiences. *Social Psychology: Handbook of Basic Principles*, *2*, 385–407.

Shen, L., & Dillard, J. P. (2007). The influence of behavioral inhibition/approach systems and message framing on the processing of persuasive health messages. *Communication Research*, *34*(4), 433–467.

Smith, S. J. (2014). Framing health care messages: Why interpersonal context matters. *Marketing Health Services e-Newsletter*.

Smith, N. K., Cacioppo, J. T., Larsen, J. T., & Chartrand, T. L. (2003). May I have your attention, please: Electrocortical responses to positive and negative stimuli. *Neuropsychologia*, *41*(2), 171–183.

Taber, J. M., & Aspinwall, L. G. (2015). Framing recommendations to promote prevention behaviors among people at high risk: A simulation study of responses to melanoma genetic test reporting. *Journal of Genetic Counseling*, *24*(5), 771–782.

Tam, L., Bagozzi, R. P., & Spanjol, J. (2010). When planning is not enough: The selfregulatory effect of implementation intentions on changing snacking habits. *Health Psychology*, *29*(3), 284–292.

Tonin, R., Brett, J., & Colagiuri, B. (2016). The effect of infrasound and negative expectations to adverse pathological symptoms from wind farms. *Journal of Low Frequency Noise, Vibration and Active Control*, *35*(1), 77–90.

Trevena, L. J., Davey, H. M., Barratt, A., Butow, P., & Caldwell, P. (2006). A systematic review on communicating with patients about evidence. *Journal of Evaluation in Clinical Practice*, *12*(1), 13–23. https://doi.org/10.1111/j.1365-2753.2005.00596.x.

Tversky, A., & Kahneman, D. (1981). The framing of decisions and the psychology of choice. *Science*, *211*, 453–458.

Uskul, A. K., Keller, J., & Oyserman, D. (2008). Regulatory fit and health behavior. *Psychology & Health*, *23*(3), 327–346. https://doi.org/10.1080/14768320701360385.

van't Riet, J., Ruiter, R. A., Werrij, M. Q., Candel, M. J., & de Vries, H. (2010). Distinct pathways to persuasion: The role of affect in message-framing effects. *European Journal of Social Psychology*, *40*(7), 1261–1276.

Verma, A. A., Razak, F., & Detsky, A. S. (2014). Understanding choice: Why physicians should learn prospect theory. *JAMA*, *311*(6), 571–572. https://doi.org/10.1001/jama.2013.285245.

Vowles, K. E., McEntee, M. L., Julnes, P. S., Frohe, T., Ney, J. P., & van der Goes, D. N. (2015). Rates of opioid misuse, abuse, and addiction in chronic pain: A systematic review and data synthesis. *Pain*, *156*(4), 569–576. https://doi.org/10.1097/01.j.pain.0000460357.01998.f1.

Winkielman, P., & Berridge, K. C. (2004). Unconscious emotion. *Current Directions in Psychological Science*, *13*(3), 120–123.

Yusuf, S., Zucker, D., Peduzzi, P., Fisher, L. D., Takaro, T., Kennedy, J. W., et al. (1994). Effect of coronary artery bypass graft surgery on survival: Overview of 10-year results from randomised trials by the Coronary Artery Bypass Graft Surgery Trialists Collaboration. *Lancet*, *344*(8922), 563–570.

Zanna, M. P., & Rempel, J. K. (1988). Attitudes: A new look at an old concept. In D. Bar-Tal & A. W. Kruglanski (Eds.), *The social psychology of knowledge* (pp. 315–334). New York, NY: Cambridge University Press.

Zhao, G., & Pechmann, C. (2007). The impact of regulatory focus on adolescents' response to antismoking advertising campaigns. *Journal of Marketing Research*, *44*(4), 671–687.

CHAPTER EIGHT

The Therapeutic Value of the Relationship for Placebo Effects and Other Healing Practices

Bruce E. Wampold[1]
Modum Bad Psychiatric Center, Vikersund, Norway
University of Wisconsin—Madison, Madison, WI, United States
[1]Corresponding author: e-mail address: bwampold@wisc.edu

Contents

Abstract

Response to placebos has been documented experimentally for many medical disorders and had been verified in laboratory studies. However, the mechanisms that produce the response to placebos are not well identified, although it appears that the relationship between the medical practitioner and the patient plays an important role. Humans and other eusocial animals evolved to heal in a social context, so it is not surprising that the healing relationship is an important factor in the delivery of health services. Several well-designed studies have examined aspects of the relationship and response to placebo and the evidence is strong that various aspects of the relationship augment response to placebo. There is also meta-analytic evidence that the relationship improves physical and mental health services. Finally, there are three possible ways the relationship augments the response to placebo: combats loneliness, increases expectancies for relief, and coregulates emotions and decreases arousal.

International Review of Neurobiology, Volume 139
ISSN 0074-7742
https://doi.org/10.1016/bs.irn.2018.07.019

Placebos are inert substances or procedures. Although placebos contain nothing that theoretically or empirically should have an effect on humans, experimental evidence has consistently shown that these sham treatments have a demonstrable effect on subjective outcomes (e.g., symptoms) as well as on the physiology of those who are administered the placebo (Ashar, Chang, & Wager, 2017; Benedetti, 2014; Price, Finniss, & Benedetti, 2008; Wampold, Minami, Tierney, Baskin, & Bhati, 2005). Placebo effects have been detected in multiple conditions, including migraines, Parkinson's disease, postoperative pain, dental pain, osteoarthritis, irritable bowel syndrome (IBS), depression, anxiety, respiratory illnesses, urological conditions, menopausal symptoms, among others (Ashar et al., 2017; Benedetti, 2014). The placebos themselves may be sham pills, inoculations, creams, or surgery.

As is the case involving many phenomena, the mechanisms that produce the observations of the effect of placebos are difficult to identify and verify. In general, there is agreement that the effects of placebos "depend on a person's psychological and brain responses to the *treatment context*, which influence appraisals of future well being" (emphasis added, Ashar et al., 2017, p. 73). Moreover, "recent research has revealed that… psychosocial-induced biochemical changes in a patient's brain and body in turn may affect the course of a disease and the response to a therapy" (Benedetti, 2008, p. 33). There appears to be a general agreement that the psychosocial context of placebo is in a general sense the central characteristic of the placebo response. The psychosocial context includes the relationship between the patient and the clinician, the healing rituals, the physical office space (including, e.g., diplomas on the wall), reputational cues, cultural beliefs about healing, and so forth. The focus of this article is on one aspect of the psychosocial context, namely, the relationship between the patient and the clinician.

To fully understand placebos and the research evidence for the magnitude of placebo effects and the mechanisms that produces responses, a differentiation among the concepts placebo response, placebo effect, and placebo-like effects must be made clear (Benedetti, 2008). A placebo response refers to the change in the health status of the patient from before placebo administration to after placebo administration. Of course, this response may be due to factors other than the placebo, including regression toward the mean, the cyclical nature of the distress, expectancies for improvement unrelated to expectancies created by the placebo and its administration, among others. To isolate the placebo effect, change in health status must be compared to the natural history of the distress. The difference between the health status of patients receiving placebos and those receiving no intervention is called the placebo effect. Finally, placebo-like effects are effects produced by the administration of a verum (i.e., a medically recognized treatment) with various psychosocial conditions as compared to the verum administered in the absence of these psychosocial conditions (Benedetti, 2008). As a generic term, the expression "response to placebo" is used. In this chapter evidence for the importance of the interpersonal relationship between the clinician and the patient in producing a response to placebo as a mechanism is derived from studies that examine placebo response, placebo effect, and placebo-like effect. In some of the studies, the participants are patients but in others they are healthy volunteers.

1. DEFINITION OF RELATIONSHIP

For the purpose of this discussion, an interpersonal relationship involves a face-to-face interaction where the clinician (or experimenter) explains the treatment being administered (either verum or placebo). This explanation involves a cognitive component of the relationship, where information is transmitted (Di Blasi, Harkness, Ernst, Georgiou, & Kleijnen, 2001; Kelley, Kraft-Todd, Schapira, Kossowsky, & Riess, 2014). But the relationship also involves an emotional component that involves warmth, caring, empathy, and understanding (Di Blasi et al., 2001; Kelley et al., 2014). The mechanisms and contributions of these two components of the relationship may have different effects, either additive or multiplicative, on response to placebos. Most studies do not differentiate between these two components or focus on only one component, which will make some conclusions tentative. Some studies vary the cognitive or emotional components or compare the presence of these components to a condition where they do not exist. To the extent possible, this review will include experimental studies where participants are randomized to conditions that vary some aspect of the relationship but correlational designs that examine the association of the relative presence of a relationship component with health outcomes will also be mentioned when appropriate.

The article will begin with a discussion of healing practices in social contexts from an evolutionary and anthropological perspective in humans and other eusocial species. Next, the evidence for the importance of relationship factors in producing responses to placebos will be reviewed, providing some detail of the most convincing studies. Finally, theoretical explanations for relationship factors in response to placebos will be proposed (briefly).

2. HEALING IN A SOCIAL CONTEXT

In their discussion of the healing power of placebos, Miller, Colloca, and Kaptchuk (2009) make a distinction between disease and illness. Disease is defined as biological dysfunction and is the focus of Western medicine. On the other hand, illness involves the experience of ill health, which includes symptoms of disease as well as the phenomenological experiences that might be unrelated to any disease. So, one can have a disease but not feel ill (e.g., the person has an asymptomatic disease and is not been made aware of the disease, say by a clinician after testing) and could feel ill without any disease process. One of the objectives of medicine, as well as allied health fields (e.g., psychotherapy) is to reduce suffering. All human civilizations, from the origins of the species to the current day, have had healing practices to relieve suffering (Shapiro & Shapiro, 1997; Wilson, 1978), and these practices involved socially sanctioned healers and healing rituals (Frank & Frank, 1991). Miller et al. (2009), among others, make the claim that the effects of placebos are most pronounced for treating illness than treating disease, although this arguable.

Miller et al. (2009) also make distinctions among three different types of healing: *natural healing*, *technological healing*, and *social healing*. Humans are exposed to many pathogens and traumas during the course of life and survive through various biological mechanisms for healing (e.g., immune functioning, barriers such as skin, blood coagulation, and so forth). Technological healing involves medical and surgical procedures that are available to physicians and other health professionals. Miller et al. defined interpersonal healing as the "context of the clinical encounter and the relationship between a healer and a patient [that] produce therapeutic benefit" (p. 527). The placebo effect, the placebo response, and placebo-like effects are created primarily within interpersonal healing. However, to make sense of the evidence for relationship aspects of the placebo it is important to keep in mind that the three types of healing often work conjointly. A patient with acute appendicitis has a disease but also is feeling ill, as their health condition is discouraging, perhaps because it interferes at a critical time in the patient's career, after several other setbacks. The patient who will undergo an appendectomy receives information about the surgery from the surgeon or a surrogate, who (hopefully) shows warmth, caring, and concern (interpersonal healing), is anesthetized and the appendix is removed (technological healing), after which the sutured incision heals naturally (natural healing), and receives analgesics and other postoperative care given by health professionals (technological and interpersonal healing). Optimal health delivery will make use of the three types of healing in effective ways.

There is a tendency in discussions of healing and in the practice of healing to consider the patient as the object of healing practice, particularly with regard to technological healing. The patient has a disease, and if natural healing is insufficient (or the patient believes it is insufficient), then the patient presents to a physician and technological healing is applied to remediate the pathophysiology. Despite pleas for a biopsychosocial model of medicine, the social aspects of healing typically are considered secondary, if not entirely ignored. In the framework in which placebos are considered, the social aspect of healing is given prominence. This prominence is deserved and indeed there is evidence that eusocial specifies (i.e., those species with the highest level of sociality) have evolved to heal in social contexts, a topic that will be briefly discussed here.

The transmission of pathogens from one organism to another often creates epidemics as disease spreads throughout the population. Thus, it might seem counter-adaptive for organisms to remain in close contact with infected conspecifics. However, in ant colonies when there are individual ants who are suffering from an infection, healthy ants spend time grooming the infected ants, which exposes the healthy ant to the pathogen. However, the transmission is limited, which elicits a low-level infection in the healthy ant, which in turn creates a pathogen-specific immune response. Thus, the social activity provides the healthy ant immunity to the pathogen, a phenomenon that has been labeled as "social immunization" (Konrad et al., 2012). When an infection is present in a honeybee colony, the worker bees will simultaneously fan their wings, raising the temperature of the hive, which provides a defensive reaction against the infection (Cremer & Sixt, 2009). In various insect societies, natural healing mechanisms at the organism level, such as border protection, soma defenses, and germ line defenses, have social healing analogues and evolutionary processes that provide for group fitness (Cremer & Sixt, 2009). The social healing practices do not replace individual healing mechanism, rather they work conjointly: "At the interface between social and individual immunity, several findings indicate that a strong social defense may replace to a certain extent the need for a sophisticated individual immune system" (Cremer & Sixt, 2009, p. 138).

One of the issues in the idea of social healing is recognition. How does an infected insect signal that it is sick and how do healthy ants identify the sick ant? Although there are several mechanisms for signaling and recognition in insects (Cremer & Sixt, 2009), it is more informative to expand the discussion to social healing in humans, beginning with the adaptive features of experiencing pain. Of course, pain has value as it identifies harmful situations, gives priority to escaping the harmful situation immediately, and leads to avoidance of similar harmful situations. However, the most important function of pain may be to elicit assistance from others (Steinkopf, 2016; Williams, 2002). The facial features expressed with pain are the same in all cultures, exist across ages, and are consistent across various sources of pain. It is thought that the facial expression of pain evolved as a means to elicit social assistance for healing and for safety (Williams, 2002). Steinkopf (2015) proposed a signaling theory of symptoms, which explains the evolution of placebo effects that are produced through social means. According to this theory, symptoms such as pain, swelling, and nausea have a signaling component as well as a natural healing component: "Symptoms signal the need for care and treatment to potential helpers. Once help and treatment are granted, the signaling function is fulfilled and the symptoms diminish" (p. 1). The theory notes that prior to modern medicine, extensive social support, comfort, protection, and help were critical.

3. EVIDENCE FOR RELATIONSHIP IN PLACEBO

The evidence for the importance of the interpersonal relationship between clinician and patient in producing responses to placebos is accumulating. The most rigorous studies are presented in some detail to weigh the evidence and to understand the designs used to identify the magnitude of the relationship effect. There are also meta-analyses that are useful to estimate relationship effects.

3.1 Well-Designed Experiments on Relationship Effects With Placebos

The first set of studies to examine use the hidden versus open paradigm. In the hidden condition patients receive medication out of their awareness whereas in the open condition a clinician informs them that they will be receiving the medication and provides some information about the medication. The differences between the two conditions is the relationship with the clinician, so observed outcomes in the two conditions provide an estimate of the effect of relationship (with some caveats to be discussed below).

To illustrate the results of an open-hidden study, an experiment of postoperative pain patients is presented (Amanzio, Pollo, Maggi, & Benedetti, 2001). A total of 278 patients who had received thoracic surgery were assigned to either the open condition or the hidden condition. In the open condition analgesic medications were administered via infusion by a physician who told the patient that the medication was a powerful painkiller. In the hidden condition, the same analgesic medication was administered by a pre-programmed infusion machine without any clinicians in the room. Pain was assessed every 15 min and the outcome analyzed was the dose required to achieve a 50% reduction in pain. The patients in the hidden administration required significantly more medication to achieve a 50% reduction, demonstrating that administering medication in the context of an interpersonal relationship produces greater benefit than administration of the same medication without the relationship. As well, there was significantly less variability in the dose needed in the hidden condition than in the placebo condition, suggesting that the effect of the relationship of the physician with the patient depends on some characteristics of the patient and/or the relationship. The superiority of open administration of medications to hidden administrations has been replicated for several conditions (e.g., pain, anxiety, and Parkinson's disease), suggesting the conclusion that "the decreased effectiveness of hidden treatments indicates that knowledge about a treatment affects outcome and highlights the importance of the patient–provider interaction" (Colloca, Lopiano, Lanotte, & Benedetti, 2013, p. 679).

There are some issues in the interpretation of the open-hidden results. First, the study presented was a study of placebo-like response because no placebo was administered. Nevertheless, the increase in efficacy of an openly administered medication suggests that the mechanisms of the placebo were responsible for the effects of these studies: "We found that the hidden injections were significantly less effective and less variable compared with open injections (in full view of the subject), suggesting that part of the response variability was due to non-specific factors (placebo)" (Amanzio et al., 2001, p. 205). What is not clear is what aspect of the physician/patient interaction was responsible for the benefits of the open administration, as there is a cognitive component (providing information) and an emotional component (giving support and caring), which do not exist in the hidden condition. There are instances where a placebo effect can be obtained by written information (i.e., providing information without a relationship) (e.g., Kam-Hansen et al., 2014) and certainly various medical paraphernalia (e.g., tablets, capsules, syringes, medical equipment, and so forth) have symbolic meaning, which may well induce the expectancies needed for a placebo response (Geers et al., 2018; Moerman & Jones, 2002). In a study of neurotic patients receiving sham treatments for which they either attributed success to their own hard work or to a placebo, expectancies were created intentionally without a relationship so that the patients would attribute benefits to the tasks and not to the clinician (Liberman, 1978). In this study, the sham treatments had beneficial effects without any relationship. The issue is not whether a placebo effect can be induced without a relationship but rather whether the relationship augments the response. Is there something about the clinical relationship that is optimal for creating a response to placebos (and to medications, for that matter)?

The importance of the relationship, particularly the emotional component, is documented by a study of placebos in the treatment of IBS (Kaptchuk et al., 2008). IBS, a prevalent disorder in primary care, has been found to be responsive to placebos. In this study, the placebo used was acupuncture placebo delivered by a validated sham acupuncture device that creates the impression that the needles are piercing the skin when in fact they do not. Patients were assigned to a waiting-list control group (no interaction with an acupuncturist), sham acupuncture twice a week for 3 weeks with limited interaction with the acupuncturist, and sham acupuncture with the same frequency but with an augmented interaction. In the limited interaction, in the first session the acupuncturists introduced themselves, said they had reviewed the patient's information, and indicated that they knew the procedure to be used. The augmented interaction, which lasted about 45 min prior to the first acupuncture procedure, involved what the researchers called "an optimal patient-practitioner relationship" (p. 3). This interaction involved questions about IBS symptoms, other symptoms, the effects of IBS on functioning, and how the patient understood the cause and meaning of IBS. The acupuncturists were trained to be warm and friendly, listen actively, and communicate expertise and experience; as well, the acupuncturist was not allowed to use any specific cognitive or behavioral interventions.

The outcomes measured at the end of the 3-week IBS trial were global improvement, adequate relief, symptom severity, and quality of life. For all outcome measures, the limited sham acupuncture was superior to waiting list and the augmented sham acupuncture was superior to limited sham acupuncture. For example, 3% of the waiting-list patients reported moderate or substantial improvement on the global improvement scale, while 20%

reported the same improvement in the limited sham acupunctures condition, and 37% in the augmented sham acupuncture. According to the authors, "The magnitude of non-specific effects in the augmented arm is not only statistically significant but also clearly clinically significant in the management of irritable bowel syndrome."

The IBS study assessed the placebo response (i.e., response compared to no-treatment) of sham acupuncture with minimal contact with the clinician and sham acupuncture with an "optimal patient–practitioner relationship." This is a study that clearly showed that the relationship with the clinician provided benefits in excess to the expectancies created by administration of a placebo. The augmented interaction was primarily composed of the emotional component of the relationship as no information about the anticipated effects of the acupuncture was included.

A follow-up analysis of the of the IBS study data was conducted to examine the variability among acupuncture practitioners (Kelley et al., 2009). The question was whether some practitioners achieved better outcomes than did other practitioners, which could be considered unlikely given that the practitioners were trained and interactions standardized. Nevertheless, there were significant differences among practitioners in terms of outcomes produced, regardless of acupuncture condition—that is, the most effective practitioner had better outcomes in both conditions (limited and augmented interaction) than the other practitioners, even though their individual outcomes were superior in the augmented condition. Thus, even after training and adherence to the interaction protocols, some practitioners consistently produced better outcome in both conditions.

The results of the analysis of the practitioner effects in the IBS study suggests that some clinicians are more effective than others due to their ability to use the relationship more productively. McKay, Imel, and Wampold (2006) analyzed data from a double-blind clinical trial that compared antidepressant medication to pill placebo (viz., data from the NIMH Treatment for Depression Collaborative Research Program; see Elkin, Parloff, Hadley, & Autry, 1985). According to Elkin et al., psychiatrists were instructed to "provide a generally supportive atmosphere" (p. 311) during clinical management, which involved an initial 45–60 min session and then 20–30 min sessions approximately weekly thereafter. Thus, the conditions were seen as medication (or placebo) "plus minimal supportive therapy" (p. 311); the psychiatrist were not allowed to give advice or strategies for coping with depression. As reported elsewhere, the antidepressant arm produced superior outcomes relative to the placebo arm (Elkin et al., 1989).

McKay et al. found that being assigned to medication or placebo accounted for about 3% of the variability in outcomes. However, psychiatrists accounted for about 9% of the variability in outcomes: The most effective physiatrists had better outcomes administering placebo than the least effective therapists had giving the antidepressant medication! Because this was a double-blind trial, the differences among the therapists in terms of outcomes was due to the quality of the relationship during the initial appointment and/or the brief (viz., 20–30 min) clinical management sessions.

Fuentes et al. (2014) examined the effect of an enhanced therapeutic alliance on pain intensity and pain sensitivity of patients with chronic low back pain. This study was similar in design to the IBS study except a verum and a placebo were given in high and low relationship conditions, resulting in a 2 (verum versus placebo) by 2 (high relationship versus low relationship). Patients received either active interferential current therapy (IFC) or sham IFC. Patients received these two treatments in conjunction with a limited therapeutic alliance or an enhanced therapeutic alliance. In the limited alliance condition, the practitioner introduced him- or herself and explained the purpose of the treatment. In the augmented alliance condition, the practitioner asked about the patient's symptoms, lifestyle, and attributions about the cause of his or her symptoms. This was combined with active listening, warm tone of voice, facilitative nonverbal behaviors (e.g., eye contact), and empathy. In the limited alliance condition, the practitioner left the room during the procedure whereas in the augmented condition the practitioner remained. For both the verum and for the placebo, the augmented alliance conditions produced superior outcomes relative to the limited alliance condition. The authors concluded, "The context in which physical therapy interventions are offered has the potential to dramatically improve therapeutic effects" (p. 477).

The final study to be reviewed attempted to tease out two aspects of the clinical relationship: warmth and competence (Howe, Goyer, & Crum, 2017). Warmth is clearly an emotional component of the relationship and competence is related to the cognitive component of the relationship. Presumably the positive expectancies for placebo are more persuasively induced by a more competent practitioner than by a less competent practitioner.

In this study, the participants were given a physical screening, which involved importantly an "allergy test," as a precursor to a purported study. The allergy test in actuality involved a skin prick with a histamine, which caused an allergic reaction in all participants. The participants were then given a sham placebo cream, which the participants were told would reduce their allergic reaction (n.b., there was also a condition where the participants were told that the cream would aggravate their reaction, but that part of the study is not relevant to the current discussion of placebo rather than nocebo). This procedure was imbedded in four conditions—warmth (high and low) crossed with competence (high and low). High warmth was induced by asking for the participant's name, making eye contact, smiling, sitting close during the exam, and displaying posters with warm images; the low warmth condition had the opposite characteristics. In the high competence condition, the practitioner used a clear, confident tone and made no mistakes in procedures, and as well the room was neat and organized; the low competence condition has the opposite characteristics. The outcomes of the experimental manipulations were assessed by the size of the reaction (viz., wheal diameter). Because the experiment did not contain a no-treatment condition (i.e., a no cream condition), placebo response to the sham cream was not assessed but the relative effectiveness of the placebo in the four conditions (high and low warmth crossed with high and low competence) was assessed. Generally, wheal diameter decreased over time, but the decrease was most rapid and the final wheal diameter was smallest in the condition where the practitioner was both warm and competent. On the other hand, the decrease in wheal diameter was smallest and final wheal diameter largest when the practitioner was low in both warmth and competence. The contribution of warmth and competence appeared to be about equal as the mismatched conditions (high competence/low warmth and low competence/high warmth) produced about equal outcomes that were intermediary to the low/low and high/high conditions. This study was particularly well designed and showed convincing evidence for importance of two components of the relationship on placebo response.

The studies designed to detect the influence of the therapeutic relationship between practitioner and patient have shown produced strong evidence that the relationship augments the response to placebo. In many ways, the relationship is one critical component of the psychosocial context that is considered the essence of the placebo as a phenomenon (Benedetti, 2011, 2014; Benedetti et al., 2003; Colloca et al., 2013; Geers et al., 2018).

There are no systematic reviews of studies that have directly investigated the role of the relationship in response to placebos. However, a recent meta-analysis of randomized clinical trials examined predictors of placebo analgesia response in chronic pain and found that the number of face-to-face visits was related to the size of the placebo effect (Vase et al., 2015), although this was not a meta-analysis of experimental manipulations of the relationship. However, more studies that directly test the effects of the relationship as well as research synthesis of those trials are needed.

3.2 Systematic Reviews

As mentioned, syntheses (e.g., meta-analyses) of research on the effect of relationship on the placebo response do not exist. However, there are systematic reviews of the effects of the relationship on the outcomes of medical and psychological treatments. There are two types of studies reviewed: those that randomize patients to a treatment with higher quality relationship or a lower quality relationship and those that examine the association of the quality of the relationship and outcome in naturalistic settings or within arms of clinical trials. If one views the relationship as a component of the placebo, then the studies in these reviews can be conceived of as investigating the relationship (or said another way, a component of the placebo) as a factor of outcome of medical and psychological treatments. These reviews will be summarized briefly.

At the turn of the century, Di Blasi et al. (2001) completed a systematic review of studies that randomized patients with physical disorders to treatment conditions, one of which involved a contextual intervention involving either cognitive or emotion components of the relationship. A search yielded 19 studies that examined the effect of influencing patients' expectancies about treatment, half of which produced significant findings. None of the studies examined emotional care by itself, but four trials contained conditions with both cognitive and emotional care and three of the four found that providing positive expectancies in an emotionally supportive context improved outcomes.

Recently Kelley et al. (2014) conducted a meta-analysis of studies the examined the effect of relationship on objective or validated subjective measures (e.g., pain ratings) of health. To be included, studies had to systematically manipulate the patient–clinician relationship so that the effects of the relationship on health outcomes could be assessed. Surprisingly, only 13 randomized trials met the inclusion/exclusion criteria. Effect sizes (standardized mean difference, SMD) comparing a more favorable relationship to a less favorable relationship ranged from −0.23 to 0.66, with an aggregate

SMD of 0.11, which was small but statistically significant. The authors made the following conclusion:

> *This systematic review and meta-analysis of RCTs suggests that the patient–clinician relationship has a small, but statistically significant effect on healthcare outcomes. Given that relatively few RCTs met our eligibility criteria, and that the majority of these trials were not specifically designed to test the effect of the patient–clinician relationship on healthcare outcomes, we conclude with a call for more research on this important topic.*
>
> ***(Kelley et al., 2014, p. 1)***

An area where the effect of the relationship on outcome has been studied extensively is psychotherapy (Wampold & Imel, 2015). Psychotherapy is not an ideal model to study the influence of relationship on placebo response because there are some very complex issues with the theoretical status of placebo in psychotherapy and psychotherapy research (Kirsch, Wampold, & Kelley, 2016; Wampold, Frost, & Yulish, 2016). In psychotherapy, the relationship is necessary to conduct the treatment, regardless of the treatment model—in a way, psychotherapy and the relationship are not separable. Simply stated, psychotherapy does not exist without a relationship. Furthermore, the factors that contribute to the benefit of psychotherapy are similar to the factors that have been identified to be important for the response to placebo and it is very difficult to disentangle specific or active ingredients and psychosocial and contextual ingredients in the way that it is done in physical medicine (Wampold & Imel, 2015). Nevertheless, psychotherapy has studied extensively the relationship and its influence on outcome.

The studies of relationship in psychotherapy are predominantly passive, correlating the quality of the relationship assessed during therapy with the outcome of therapy. Meta-analyses of such studies have found that the therapeutic alliance, the real relationship, empathy, positive regard and affirmation, goal collaboration and consensus, congruence and genuineness, real relationship, and creation of positive expectancies are consistently correlated with outcome, with SMD (converted from aggregate correlation coefficients) in the range of 0.40–0.60 (Norcross & Lambert, 2018). The therapeutic alliance, which is a concept that reflects the collaborative work between the therapist and patient, is the most widely studied relationship variable in psychotherapy (Wampold & Imel, 2015; Zilcha-Mano, 2017). Sophisticated designs have been used to probe aspects of the alliance and it appears that the alliance predicts improvement (e.g., decrease in symptoms) on a session-by-session basis and that the therapist contribution to the alliance is critical for therapeutic success (Baldwin, Wampold, & Imel, 2007; Falkenström, Granström, & Holmqvist, 2013).

4. THEORETICAL CONSIDERATIONS

As argued earlier, humans evolved to heal in a social context. It may well be that symptoms function, at least in part, to elicit social support and assistance. This evolutionary process occurred pre-modern medicine. It is reasonable to speculate that social support and assistance were more important before medicine was reliably able to cure many diseases but may no longer be necessary to heal as medicines and surgeries will be more effective and efficient than social support and assistance. If someone is suffering from the consequences of disease, then one can take an Uber/Lyft/Taxi car to the doctor's office and have the disease cured, without invoking social means of healing. Perhaps a social healing module evolved but is no longer adaptive in a modern world. Yet, the evidence that has been presented suggests that social interactions—the relationship between clinician and patient—promotes healing. Thus, there is a need to speculate about how social healing, particularly in a modern world, works. Several possible mechanisms are proposed. They are not mutually exclusive and may work together to produce the effects of relationship in medical practices. It is worth emphasizing that the following (brief) theoretical explanations for the salubrious effects of the relationship in health care settings are speculative.

4.1 Relationship Combats Loneliness

There is convincing evidence that smoking, obesity, lack of exercise, excessive drinking, absence of influenza vaccine, and environmental pollution have deleterious effects on health and increase mortality risk. However, a greater morality risk factor than any of these factors is loneliness (Cacioppo & Cacioppo, 2012; Cacioppo, Capitanio, & Cacioppo, 2014; Holt-Lunstad, Robles, & Sbarra, 2017; Holt-Lunstad, Smith, Baker, Harris, & Stephenson, 2015). It appears that comfort from an intimate other is necessary to maintain health. Moreover, the prevalence of loneliness is great, in excess of 0.40 by some estimates, and is greater than prevalence of obesity, inactivity, and smoking; furthermore, the prevalence of loneliness is increasing (Holt-Lunstad et al., 2017). A warm, caring, and understanding clinician might well provide an antidote for patients who are particularly lonely. A patient might way to themselves, "My doctor really cares about me and my well-being. Whenever, I am feeling distressed or sick, she is there for me and helps me through this tough time." Empathy, as an important factor in the delivery of medical care, has been a focus of researchers for some time (Decety & Fotopoulou, 2015). Certainly, here the discussion is on the emotional aspect of the relationship.

Earlier several studies that varied the warmness, concern, and empathy of practitioners found that these characteristics augmented response to placebos, as discussed earlier. Of interest is the study of IBS, which found that the largest effect was for quality of life (Kaptchuk et al., 2008), which makes sense as loneliness is a general state and not symptom specific.

4.2 Creates Expectancies for Relief

Expectancies are thought to be central to response to placebos (Ashar et al., 2017; Benedetti, 2014; Colloca et al., 2013; Price et al., 2008). Less well understood is how communications about treatments create the expectancies. Certainly, expectancies can be created without face-to-face interactions, as mentioned earlier (e.g., Kam-Hansen et al., 2014; Liberman, 1978). The question is whether face-to-face interpersonal interaction is an efficient and effective means to induce expectancies that a medical treatment will be effective.

There are many ways to acquire expectancies. We have the expectation that inserting a metal object, say a tine of a fork, into an electrical socket will create a painful shock. It is relatively unlikely this was learned by classical conditioning (insertion of the tine followed by a shock, a pairing that generalized to other metal objects) or by vicarious learning (say, by observing an older sibling being shocked). There is no doubt that learning plays a role in producing placebo effects (Colloca & Miller, 2011; Schenk, Krimmel, & Colloca, 2017), but clearly there are other mechanisms involved in creating expectancies. As well, there has been insufficient time since the installation of electric sockets in homes to have evolved a module for fear of such things. We learned not to inset metal objects into electrical sockets the way we learn many things—someone we trust, most likely a parent, told us this was dangerous. That is, the expectation of an outcome (here a negative outcome) was created by a verbal transaction with a trusted person. This is exactly how we come to have expectancies about the consequences of many thoughts, feelings, and actions (Geers et al., 2018; Lieberman, 2013).

There are interesting developments in many fields that support the notion of the social communication in learning and behavior. Indeed, our large brain evolved to manage large networks of social relations

(Dunbar & Shultz, 2007; Lieberman, 2013). Many thoughts, feelings, mental states, preferences, and behaviors spread through social networks (Cacioppo, Fowler, & Christakis, 2009; Christakis & Fowler, 2007, 2009; Fowler & Christakis, 2009, 2010; Rosenquist, Fowler, & Christakis, 2011). In his book on the neuroscience of sociality, Lieberman (2013) states, "Our brains are designed to be influenced by others" (p. 8). That is, patients are wired to believe in the explanations provided by clinicians, particularly if they are perceived to be competent and expert, a conjecture consistent with the result discussed earlier with regard to clinician competence and response to a sham antihistamine cream (Howe et al., 2017). Recently Geers et al. (2018) applied persuasion theory (Petty & Briñol, 2014) to understand how the clinician and the context use social means to produce placebo effects.

4.3 Coregulation of Emotions

One of the factors that moderates response to placebos is the desire to experience less distress (Price et al., 2008). Patients typically present to clinicians when they are concerned about their health and well-being and seek distress reduction. The presence of an intimate other can attenuate arousal and distress through a process that is referred to as coregulation (Butler & Randall, 2013). Coregulation "refers to the process by which relationship partners form a dyadic emotional system involving an oscillating pattern of affective arousal and dampening that dynamically maintains an optimal emotional state" (Butler & Randall, 2013, p. 202). Physiological equilibrium is needed for psychological, physical, and social well-being. The tenet of coregulation is that emotional regulation is not under the control of an individual but is determined within social relationships with intimate others. In a study of martially satisfied women, it was found that holding the hand of their husbands reduced arousal in a stressful situation in comparison to holding the hand of a stranger or not holding anyone's hand—and the more martially satisfied the women were, the greater the effect (Coan, Schaefer, & Davidson, 2006). A key emotion in coregulation is empathy and it appears that touch and empathy can work together to reduce pain (Goldstein, Shamay-Tsoory, Yellinek, & Weissman-Fogel, 2016; Goldstein, Weissman-Fogel, & Shamay-Tsoory, 2017).

5. CONCLUSIONS

Humans evolved to heal in social contexts and there is strong evidence that the relationship augments response to placebos. There are a number of studies that have manipulated aspects of the relationship and found that the relationship is an important aspect of creating a placebo response, although to date there are no meta-analyses related to this question. However, there is meta-analytic evidence, although not particularly strong, that a competent and empathic physician augments the effect of medical procedures. On the other hand, there is strong meta-analytic evidence that the relationship between psychotherapist and patient is associated with better outcomes in psychotherapy.

The mechanisms through which the relationship creates effects is not well known. However, three possibilities exist: (a) the relationship combats loneliness, which is known to be related to pathology, (b) the relationship is an effective means to create expectancies for benefits of treatments (verum and placebos), and (c) the healing relationship between the clinician and the patient involves coregulation of affect, which reduces distress. Further research is needed to identify the features of the optimal clinician/patient interaction and to investigate the mechanisms that produce therapeutic benefits. Nevertheless, the importance of the relationship for the delivery of physical and mental health services is clear and systematic efforts are needed to utilize relationship factors in the delivery of these effective health services.

REFERENCES

Amanzio, M., Pollo, A., Maggi, G., & Benedetti, F. (2001). Response variability to analgesics: A role for non-specific activation of endogenous opioids. *Pain, 90*, 205–211.

Ashar, Y. K., Chang, L. J., & Wager, T. D. (2017). Brain mechanisms of the placebo effect: An affective appraisal account. *Annual Review of Clinical Psychology, 13*, 73–98. https://doi.org/10.1146/annurev-clinpsy-021815-093015.

Baldwin, S. A., Wampold, B. E., & Imel, Z. E. (2007). Untangling the alliance-outcome correlation: Exploring the relative importance of therapist and patient variability in the alliance. *Journal of Consulting and Clinical Psychology, 75*, 842–852.

Benedetti, F. (2008). Mechanisms of placebo and placebo-related effects across diseases and treatments. *Annual Review of Pharmacology and Toxicology, 48*, 33–60. https://doi.org/10.1146/annurev.pharmtox.48.113006.094711.

Benedetti, F. (2011). *The patient's brain: The neuroscience behind the doctor-patient relationship*. New York: Oxford University Press.

Benedetti, F. (2014). *Placebo effects: Understanding the mechanisms in health and disease* (2nd ed.). New York: Oxford University Press.

Benedetti, F., Maggi, G., Lopiano, L., Lanotte, M., Rainero, I., Vighetti, S., et al. (2003). Open versus hidden medical treatments: The patient's knowledge about therapy affects the therapy outcome. *Prevention & Treatment, 6*. Article 1.

Butler, E. A., & Randall, A. K. (2013). Emotional coregulation in close relationships. *Emotion Review, 5*(2), 202–210. https://doi.org/10.1177/1754073912451630.

Cacioppo, S., & Cacioppo, J. T. (2012). Decoding the invisible forces of social connections. *Frontiers in Integrative Neuroscience, 6*. https://doi.org/10.3389/fnint.2012.00051.

Cacioppo, S., Capitanio, J. P., & Cacioppo, J. T. (2014). Toward a neurology of loneliness. *Psychological Bulletin, 140*(6), 1464–1504. https://doi.org/10.1037/a0037618.

Cacioppo, J. T., Fowler, J. H., & Christakis, N. A. (2009). Alone in the crowd: The structure and spread of loneliness in a large social network. *Journal of Personality and Social Psychology, 97*(6), 977–991. https://doi.org/10.1037/a0016076.

Christakis, N. A., & Fowler, J. H. (2007). The spread of obesity in a large social network over 32 years. *The New England Journal of Medicine, 357*(4), 370–379. https://doi.org/10.1056/NEJMsa066082.

Christakis, N. A., & Fowler, J. H. (2009). Connected: The surprising power of our social networks and how they shape our lives—How your friends' friends' friends affect everything you feel, think, and do. New York: Little, Brown & Co.

Coan, J. A., Schaefer, H. S., & Davidson, R. J. (2006). Lending a hand: Social regulation of the neural response to threat. *Psychological Science, 17*(12), 1032–1039. https://doi.org/10.1111/j.1467-9280.2006.01832.x.

Colloca, L., Lopiano, L., Lanotte, M., & Benedetti, F. (2013). Overt versus covert treatment for pain, anxiety, and Parkinson's disease. In F. G. Miller, L. Colloca, R. A. Crouch, & T. J. Kaptchuk (Eds.), *The placebo: A reader* (pp. 220–226). Baltimore, MD: Johns Hopkins University Press. (chapter xx), 327 pages.

Colloca, L., & Miller, F. G. (2011). How placebo responses are formed: A learning perspective. *Philosophical Transactions of the Royal Society B: Biological Sciences, 366*(1572), 1859–1869. https://doi.org/10.1098/rstb.2010.0398.

Cremer, S., & Sixt, M. (2009). Analogies in the evolution of individual and social immunity. *Philosophical Transactions of the Royal Society of London B Biological Sciences, 364*(1513), 129–142. https://doi.org/10.1098/rstb.2008.0166.

Decety, J., & Fotopoulou, A. (2015). Why empathy has a beneficial impact on others in medicine: Unifying theories. *Frontiers in Behavioral Neuroscience, 8*. https://doi.org/10.3389/fnbeh.2014.00457.

Di Blasi, Z., Harkness, E., Ernst, E., Georgiou, A., & Kleijnen, J. (2001). Influence of context effects on health outcomes: A systematic review. *The Lancet, 357*(9258), 757–762.

Dunbar, R. I. M., & Shultz, S. (2007). Evolution in the social brain. *Science, 317*(5843), 1344–1347. https://doi.org/10.1126/science.1145463.

Elkin, I., Parloff, M. B., Hadley, S. W., & Autry, J. H. (1985). NIMH Treatment of Depression Collaborative Research Program: Background and research plan. *Archives of General Psychiatry, 42*, 305–316.

Elkin, I., Shea, T., Watkins, J. T., Imber, S. D., Sotsky, S. M., Collins, J. F., et al. (1989). National Institute of Mental Health treatment of depression collaborative research program: General effectiveness of treatments. *Archives of General Psychiatry, 46*, 971–982.

Falkenström, F., Granström, F., & Holmqvist, R. (2013). Therapeutic alliance predicts symptomatic improvement session by session. *Journal of Counseling Psychology, 60*(3), 317–328. https://doi.org/10.1037/a0032258.

Fowler, J. H., & Christakis, N. A. (2009). Dynamic spread of happiness in a large social network: Longitudinal analysis over 20 years in the Framingham Heart Study. *BMJ (Clinical Research Ed.), 338*(7685), 1–13. https://doi.org/10.1136/bmj.b1.

Fowler, J. H., & Christakis, N. A. (2010). Cooperative behavior cascades in human social networks. *Proceedings of the National Academy of Sciences of the United States of America, 107*(12), 5334–5338. https://doi.org/10.1073/pnas.0913149107.

Frank, J. D., & Frank, J. B. (1991). *Persuasion and healing: A comparative study of psychotherapy* (3rd ed.). Baltimore: Johns Hopkins University Press.

Fuentes, J., Armijo-Olivo, S., Funabashi, M., Miciak, M., Dick, B., Warren, S., et al. (2014). Enhanced therapeutic alliance modulates pain intensity and muscle pain sensitivity in patients with chronic low back pain: An experimental controlled study. *Physical Therapy, 94*(4), 477–489. https://doi.org/10.2522/ptj.20130118.

Geers, A. L., Brinol, P., Vogel, E. A., Aspiras, O., Caplandies, F. C., & Petty, R. E. (2018). The application of persuasion theory to placebo effects. *International Review of Neurobiology, 138*, 113–136. https://doi.org/10.1016/bs.irn.2018.01.004.

Goldstein, P., Shamay-Tsoory, S. G., Yellinek, S., & Weissman-Fogel, I. (2016). Empathy predicts an experimental pain reduction during touch. *The Journal of Pain, 17*(10), 1049–1057. https://doi.org/10.1016/j.jpain.2016.06.007.

Goldstein, P., Weissman-Fogel, I., & Shamay-Tsoory, S. G. (2017). The role of touch in regulating inter-partner physiological coupling during empathy for pain. *Scientific Reports, 7*. https://doi.org/10.1038/s41598-017-03627-7.

Holt-Lunstad, J., Robles, T. F., & Sbarra, D. A. (2017). Advancing social connection as a public health priority in the United States. *American Psychologist, 72*(6), 517–530. https://doi.org/10.1037/amp0000103.

Holt-Lunstad, J., Smith, T. B., Baker, M., Harris, T., & Stephenson, D. (2015). Loneliness and social isolation as risk factors for mortality: A meta-analytic review. *Perspectives on Psychological Science, 10*(2), 227–237. https://doi.org/10.1177/1745691614568352.

Howe, L. C., Goyer, J. P., & Crum, A. J. (2017). Harnessing the placebo effect: Exploring the influence of physician characteristics on placebo response. *Health Psychology, 36*(11), 1074–1082. https://doi.org/10.1037/hea0000499.

Kam-Hansen, S., Jakubowski, M., Kelley, J. M., Kirsch, I., Hoaglin, D. C., Kaptchuk, T. J., et al. (2014). Altered placebo and drug labeling changes the outcome of episodic migraine attacks. *Science Translational Medicine, 6*(218), 7. https://doi.org/10.1126/scitranslmed.3006175.

Kaptchuk, T. J., Kelley, J. M., Conboy, L. A., Davis, R. B., Kerr, C. E., Jacobson, E. E., et al. (2008). Components of placebo effect: Randomised controlled trial in patients with irritable bowel syndrome. *BMJ (Clinical Research Ed.), 336*(7651), 999–1003. https://doi.org/10.1136/bmj.39524.439618.25.

Kelley, J. M., Kraft-Todd, G., Schapira, L., Kossowsky, J., & Riess, H. (2014). The influence of the patient-clinician relationship on healthcare outcomes: A systematic review and meta-analysis of randomized controlled trials. *PLoS One, 9*(4). https://doi.org/10.1371/journal.pone.0094207.

Kelley, J. M., Lembo, A. J., Ablon, J. S., Villanueva, J. J., Conboy, L. A., Levy, R., et al. (2009). Patient and practitioner influences on the placebo effect in irritable bowel syndrome. *Psychosomatic Medicine, 71*(7), 789–797. https://doi.org/10.1097/PSY.0b013e3181acee12.

Kirsch, I., Wampold, B., & Kelley, J. M. (2016). Controlling for the placebo effect in psychotherapy: Noble quest or tilting at windmills? *Psychology of Consciousness: Theory, Research, and Practice, 3*(2), 121–131. https://doi.org/10.1037/cns0000065.

Konrad, M., Vyleta, M. L., Theis, F. J., Stock, M., Tragust, S., Klatt, M., et al. (2012). Social transfer of pathogenic fungus promotes active immunisation in ant colonies. *PLoS Biology, 10*(4). https://doi.org/10.1371/journal.pbio.1001300.

Liberman, B. L. (1978). The role of mastery in psychotherapy: Maintenance of improvement and prescriptive change. In J. D. Frank, R. Hoehn-Saric, S. D. Imber, B. L. Liberman, & A. R. Stone (Eds.), *Effective ingredients of successful psychotherapy* (pp. 35–72). Baltimore: Johns Hopkins University Press.

Lieberman, M. D. (2013). *Social: Why our brains are wired to connect*. New York: Crown Publishing Group.

McKay, K. M., Imel, Z. E., & Wampold, B. E. (2006). Psychiatrist effects in the psychopharmacological treatment of depression. *Journal of Affective Disorders, 16*, 236–242.

Miller, F. G., Colloca, L., & Kaptchuk, T. J. (2009). The placebo effect: Illness and interpersonal healing. *Perspectives in Biology and Medicine, 52*(4), 518–539.

Moerman, D. E., & Jones, W. B. (2002). Deconstructing the placebo effect and finding the meaning response. *Annals of Internal Medicine, 136*(6), 471–476.

Norcross, J. C., & Lambert, M. J. (Eds.), (2018). *Psychotherapy relationships that work. Vol. 1. Evidence-based therapist contributions* (3rd ed.). New York: Oxford University Press.

Petty, R. E., & Briñol, P. (2014). The elaboration likelihood and metacognitive models of attitudes. In J. W. Sherman, B. Gawronski, & Y. Trope (Eds.), *Dual-process theories of the social mind* (pp. 172–187). New York, NY: Guilford Press.

Price, D. P., Finniss, D. G., & Benedetti, F. (2008). A comprehensive review of the placebo effect: Recent advances and current thought. *Annual Review of Psychology*, *59*, 565–590.

Rosenquist, J. N., Fowler, J. H., & Christakis, N. A. (2011). Social network determinants of depression. *Molecular Psychiatry*, *16*(3), 273–281. https://doi.org/10.1038/mp.2010.13.

Schenk, L. A., Krimmel, S. R., & Colloca, L. (2017). Observe to get pain relief: Current evidence and potential mechanisms of socially learned pain modulation. *Pain*, *158*(11), 2077–2081. https://doi.org/10.1097/j.pain.0000000000000943.

Shapiro, A. K., & Shapiro, E. S. (1997). *The powerful placebo: From ancient priest to modern medicine*. Baltimore: The Johns Hopkins University Press.

Steinkopf, L. (2015). The signaling theory of symptoms: An evolutionary explanation of the placebo effect. *Evolutionary Psychology*, *13*(3), 1–12. https://doi.org/10.1177/1474704915600559.

Steinkopf, L. (2016). An evolutionary perspective on pain communication. *Evolutionary Psychology*, *14*(2), 7. https://doi.org/10.1177/1474704916653964.

Vase, L., Vollert, J., Finnerup, N. B., Miao, X. P., Atkinson, G., Marshall, S., et al. (2015). Predictors of the placebo analgesia response in randomized controlled trials of chronic pain: A meta-analysis of the individual data from nine industrially sponsored trials. *Pain*, *156*(9), 1795–1802. https://doi.org/10.1097/j.pain.0000000000000217.

Wampold, B. E., Frost, N. D., & Yulish, N. E. (2016). Placebo effects in psychotherapy: A flawed concept and a contorted history. *Psychology of Consciousness: Theory, Research, and Practice*, *3*(2), 108–120. https://doi.org/10.1037/cns0000045.

Wampold, B. E., & Imel, Z. E. (2015). *The great psychotherapy debate: The evidence for what makes psychotherapy work* (2nd ed.). New York: Routledge.

Wampold, B. E., Minami, T., Tierney, S. C., Baskin, T. W., & Bhati, K. S. (2005). The placebo is powerful: Estimating placebo effects in medicine and psychotherapy from clinical trials. *Journal of Clinical Psychology*, *61*, 835–854.

Williams, A. C. (2002). Facial expression of pain: An evolutionary account. *Behavioral and Brain Sciences*, *25*(4), 439–488. https://doi.org/10.1017/s0140525x02000080.

Wilson, E. O. (1978). *On human nature*. Cambridge, MA: Harvard University Press.

Zilcha-Mano, S. (2017). Is the alliance really therapeutic? Revisiting this question in light of recent methodological advances. *American Psychologist*, *72*(4), 311–325. https://doi.org/10.1037/a0040435.

CHAPTER NINE

The Role of Patient–Practitioner Relationships in Placebo and Nocebo Phenomena

Maxie Blasini*, Nathalie Peiris†, Thelma Wright†, Luana Colloca†,‡,§,¶,1
*Acupuncture and Chinese Medicine Program, Pacific College of Oriental Medicine, San Diego, CA, United States
†Department of Anesthesiology, School of Medicine, University of Maryland, Baltimore, MD, United States
‡Department of Psychiatry, School of Medicine, University of Maryland, Baltimore, MD, United States
§Department of Pain Translational Symptom Science, School of Nursing, University of Maryland, Baltimore, MD, United States
¶Center to Advance Chronic Pain Research, University of Maryland, Baltimore, MD, United States
[1]Corresponding author: e-mail address: colloca@umaryland.edu

Contents

Abstract

Introduction: Placebo and nocebo effects form part of all therapeutic environments and play a significant role in the effectiveness of treatment outcomes. Patient expectancies drive these phenomena, which can be shaped through contextual factors including verbal suggestions, conditioning, and social observation.

Objectives: This review seeks to identify the biopsychosocial factors of the patient–practitioner interaction that play a role in the development of placebo and nocebo effects, as well as the anthropological elements of the biodynamic process of relating that are meaningful in the development of expectancies.

Methods: We conducted a narrative review of frameworks of the placebo and nocebo effect, including the impact of expectancies and interpersonal relationships in the context of healing and the clinical setting.

Results: Expectancies leading to placebo and nocebo effects can be modified by macro and micro factors, such as culture and society, as well as individual psychobiological traits, respectively. The developmental sociobiological adaptations that form and

International Review of Neurobiology, Volume 139
ISSN 0074-7742
https://doi.org/10.1016/bs.irn.2018.07.033

consolidate mindsets and meaningful contexts play an important role in shaping patient expectancies, as well as patients' conscious and subconscious reactions to signs and actions taking place within the clinical environment. Practitioner characteristics, like empathy, friendliness, and competence, favor the formation of positive expectancies. Caring and warm patient–practitioner interactions can enhance the therapeutic value of clinical encounters when patients' positive expectancies are actively encouraged and engaged.

Conclusion: A patient-centered approach rooted in demonstrating care and empathy can positively enhance a patient's experience within the clinical environment and activate psychosociobiological adaptations associated with the placebo phenomenon. Pain patients could particularly benefit from non-invasive approaches for improving treatment effectiveness and quality-of-life.

1. BACKGROUND

The placebo and nocebo effects have fascinated scholars for decades due to the powerful and significant impact they have on different experimental and health outcomes. Numerous and rigorous research has found that placebo and nocebo effects are associated to multiple eliciting factors, such as verbal suggestions, social observation, conditioning and prior experience, individual personality traits, and genetic variants (Blasini, Corsi, Klinger, & Colloca, 2017; Colagiuri, Schenk, Kessler, Dorsey, & Colloca, 2015; Colloca, 2014; Corsi & Colloca, 2017). These different aspects interact with each other and modulate the *expectancies* that individuals may have regarding a treatment, intervention, or practitioner. If this combination of factors leads to a positive expectancy, it will likely result in a beneficial or improved outcome through the placebo effect, whereas negative expectancies could induce harmful outcomes due to the activation of the nocebo effect. Overall, we can state that these phenomena are associated to cognitive–affective factors that can trigger a top-down neural and biochemical modulation of different physiological processes, thus influencing pain (Benedetti, Pollo, & Colloca, 2007; Colloca & Grillon, 2014; Petersen et al., 2014; Price, Craggs, Nicholas Verne, Perlstein, & Robinson, 2007), fatigue (Shaibani, Frisaldi, & Benedetti, 2017), Parkinson's disease (Frisaldi et al., 2017; Udupa & Fox, 2015), immunity (Tekampe et al., 2017; Wendt, Albring, & Schedlowski, 2014), irritable bowel syndrome (Ballou et al., 2017; Kaptchuk et al., 2010), and the effectiveness of medications (Ader et al., 2010; Bingel et al., 2011; Colloca, Enck, & DeGrazia, 2016), among others. While ethical considerations have always been of concern due to an association of the effectiveness of placebos with deception, newer research has found that placebo effects can occur even when an individual is aware of receiving or having received a placebo intervention (Ballou et al., 2017; Carvalho et al., 2016; Colloca, Pine, Ernst, Miller, & Grillon, 2016; Kaptchuk et al., 2010; Locher et al., 2017).

Although the attention to the placebo and nocebo continues to increase as more is discovered about their intricate origin and nature, these are not phenomena that have only recently appeared in the modern human experience of health and illness. Researchers have suggested that these mechanisms are part of adaptations favored through sociobiological evolution (Colloca, 2014). Some of the earliest reports of placebos date back to the 1500s, when a "trick trial" was done to disprove exorcisms in face of rising Protestant skepticism around these events and the Catholic Church. In this trial, ordinary water was placed in a religious flask and was given to a girl who was deemed to be possessed by the devil; an event which should have caused the girl to go into contortions typically described during exorcisms. However, in other occasions, true holy water was covertly given to the girl, and no effects occurred (Kaptchuk, Kerr, & Zanger, 2009). In 1955, Dr. Henry K. Beecher, an anesthesiologist, published in the Journal of the American Medical Association that placebos accounted for an average of approximately 35% of therapeutic effectiveness across a variety of conditions ranging from headache and nausea, to post-operative pain and anxiety, based on 15 studies that included a total of 1082 patients (Beecher, 1955). In particular, Beecher pointed out how much of a drug's action can be related to the individual processing of "*suffering*" rather than direct effects on the original ailments requiring medication (Beecher, 1955).

Placebo and nocebo effects do not solely need the administration of an inert substance (i.e., a placebo) for them to occur. Contextual factors, which include the patient–practitioner interaction, social observation of others, and environmental cues, can add to the stimuli of these effects. In other words, the occurrence and magnitude of the placebo or nocebo effect can be modulated through the individual positive or negative psychosocial perceptions resulting from the contexts in which they occur (Colloca & Benedetti, 2005; Colloca, Lopiano, Lanotte, & Benedetti, 2004; Di Blasi, Harkness, Ernst, Georgiou, & Kleijnen, 2001; Di Blasi & Reilly, 2005).

Inevitably, the clinical setting is accompanied by a myriad of messages carried through symbols, signs, words, and actions that convey meaning to each patient. These carriers of information lead to the changes in expectancy that can cause placebo and nocebo effects. This article explores important biopsychosocial aspects of these phenomena, namely, through a characterization of anthropological and contextual elements involved in the patient–practitioner relationship. Further, we discuss how to pragmatically approach these aspects within the clinical environment. Finally, we provide questions that can serve as potential avenues of research to overcome challenges arising from the complexity of systemic, sociocultural, and psychobiological factors involved in the context of health and the placebo effect.

2. THE PATIENT–PRACTITIONER INTERACTION: A SOCIOBIOLOGICAL APPROACH

The interaction between a patient and a healthcare provider has the potential to influence clinical outcomes through the continual psychosocial process of relating (Adler, 2007). *Relating* represents a set of complex interactions in which the engaged parties reciprocally effect the behavior, as well as the experience, of the other (Adler, 2007). Given that no psychosocial influence occurs without a precedent and subsequent psychophysiological response, this continual process may also significantly impact factors relevant to clinical and health outcomes. The patient–practitioner interaction can be then described as an emergent, reciprocal, and self-organizing process that creates feedback loops leading to the mutual regulation of "biopsychosocial responses" within the medical context (Adler, 2007).

Adler and colleagues supported the theory that the placebo effect is fundamentally rooted in the evolutionary human necessity for groups and systems (Adler & Hammett, 1973). The complex and progressive evolutionary events that led to human development as a social animal, such as the need for prolonged care of infants postpartum when compared to other mammals and the need for adequate integration and communication of individual physical and external environments, have played a vital role in the ability of humans to form complex psychosocial matrices. These psychosocial aspects of the human experience are intrinsically and essentially connected to proper psychophysical function.

As described by Adler and others, social interactions continue to actively modify neural structures (Greenough, 1992; Kandel, 1979) and circuits (Kandel & Squire, 2000) that impact neurobiological, neuroendocrine, and affective responses within the context of social attachment throughout the lifespan (Adler, 2007; Insel, 1997; Nelso & Panksepp, 1998; Young, Wang, & Insel, 1998). Interpersonal interactions and relationships have the ability to influence and regulate physiological function (Adler, 2007; Hofer, 1981, 1984). An important aspect of this mutual co-regulation is the concept of *affect attunement*. This can be understood as a "secure, affectional bond" (Adler, 2007) leading to a level of autonomic synchronization that promotes feelings of satisfaction and calmness that are similar in nature to what is encountered between a caregiver and an infant (Adler, 2007; Faude, Jones, & Robins, 1996). Affect attunement, which is likely not under complete conscious control due to its instant and continuous characteristics, can play a significant role in creating feelings of empathy in the person who is functioning primarily in the role of caregiving (Adler, 2007; Murray, 1979).

Importantly, empirical evidence suggests that empathy can be considered a neurobiological response, as demonstrated by a study with positron emission tomography (PET) in which empathy of distress was significantly associated with specific neural network activation (Adler, 2007; Shamay-Tsoory et al., 2005).

A recently published study supports the importance of physician characteristics during interpersonal interactions within an experimental medical setting. Investigators found that perceived physician warmth and competence, along with positive expectations of treatment, boosted placebo responses that ameliorated skin allergic reactions resulting from a histamine skin prick with the subsequent application of a cream with no active ingredients (Howe, Goyer, & Crum, 2017). Although these physician characteristics relate more to perceived likeability and credibility rather than empathy itself, they form part of the contextual psychosocial factors that help establish *rapport* with patients (see also Sussex, 2018).

The distress and suffering that an illness or disease can produce in a patient likely result in the activation of the sympathetic nervous system, primarily the hypothalamic–pituitary–adrenal axis (HPA). Anxiety can play a strongly influential role in the process of illness, and thus, this period may be characterized by a vulnerability in which the patient may either intentionally or inadvertently seek relief and psychobiological security by bonding with a caregiver figure, in this case the physician or health provider (Adler, 2007; Benedetti, Lanotte, Lopiano, & Colloca, 2007; Colloca & Benedetti, 2007). This vulnerability could likely increase the opportunity for the application of clinical empathy to create a positive psychobiological enhancement during the clinical encounter, and thus potentially improve both psychological and physiological outcomes. The skillful engagement of emotional resonance and affect attunement by a physician in the form of clinical empathy can help regulate the patient's autonomic arousal and decrease allostatic load (McEwen, 1998), thus promoting physiological homeostasis (Adler, 2007; Carter, 1998; Porges, 1998).

Although the process of empathy can be variable and driven by many subjective factors, the concept of clinical empathy is characterized by certain elements that allow for the consideration of it as a "clinical procedure" (Adler, 2007). Adler presents three primary features that support this idea. First, the engagement of clinical empathy in the context of disease and relief of suffering requires a specific medical interaction, and thus clinical empathy would have a medical indication. Second, the use of clinical empathy would require a set of interpersonal skills that depend upon active "emotional labor" by the physician or health provider (Adler, 2007; Larson & Yao, 2005). Finally, clinical empathy is a purposeful process attempting to attain specific outcomes within the clinical context that will result in the improvement of health through the engagement of psychobiological mechanisms (Adler, 2007).

It is important to note that by deeply immersing into an empathic experience with their patients, physicians may experience a high degree of emotional vulnerability that could potentially lead to emotional imbalance (Brody, 1997). Given that current burn out rates among physicians in the United States are estimated to be over 50% (Shanafelt, Dyrbye, & West, 2017; Shanafelt et al., 2015), it is critical to engage many entities, leaders, medical educators, and organizations within the healthcare system to address this issue that not only affects physicians, but by extension, can also affect the health outcomes of their patients. Building emotional resilience among physicians can counteract the burden that proactive empathy and emotional attunement may cause (Brody, 1997; Coulehan, 1995).

3. INTERPERSONAL HEALING

Research has suggested that physician expectations and beliefs influence their patients' clinical outcomes. A study by Gracely, Dubner, Deeter, and Wolskee (1985) compared postoperative pain after dental surgery in a double-blind setting. Patients were told that they would be administered fentanyl (opioid), naloxone (opioid antagonist), or a placebo. A total of 60 patients were divided into a placebo–naloxone (PN) group and a placebo–naloxone–fentanyl group (PNF). Doctors were told that patients would be receiving one of the substances corresponding to each group (i.e., placebo or naloxone only for PN group; placebo, naloxone or fentanyl for PNF group). Neither patients nor doctors knew which of the substances were given at the time of administration. Results showed that pain levels in the placebo arm of the PNF group were significantly lower when compared to the placebo PN group. The specific results regarding naloxone administration were not discussed (Gracely et al., 1985). This study suggested that doctors' expectations, and likely, subtle behaviors of the clinician administering the drug, might have had an effect in pain levels. These effects could be associated to the way that physicians communicate and express their own personal biases while engaging in their interpersonal rapport, and thus can influence patients' perceptions of the information provided by the physician in terms of treatment, health and disease outcomes, as well as the perceptions of the physician characteristics themselves. Thus, it was concluded that "clinical analgesia depends not only on the physiological action of the treatment administered but also on the expectations of both the patient and clinician" (Gracely et al., 1985).

Miller and colleagues proposed the consideration of placebo effects as a form of *interpersonal healing* (Miller, Colloca, & Kaptchuk, 2009). The authors describe how modern biomedicine, primarily focused in pathophysiological states of disease, has paid much less attention to the concept of illness and how the medical encounter can help in the "relief of suffering," *suffering* being described as the "illness component of disease and injury." Treating an illness can be supported and encouraged through the patient–physician relationship, rather than solely by administering specific treatments (Miller et al., 2009). Although placebo effects include changes in measurements within subjective and objective domains, of which the latter have been confirmed by brain-imaging, biochemical and pharmacological studies (Colloca, Benedetti, & Porro, 2008; Faria, Fredrikson, & Furmark, 2008; Miller et al., 2009), there is little quality evidence linking mitigation of symptoms with actual changes in the pathophysiology of diseases.

The placebo effect can then be considered as a general concept through which multiple mechanisms and pathways implicated in relief and coping can be engaged and established during the patient–clinician interaction as well as during the patient's experience of therapeutic outcomes. These pathways are varied, and can be psychologically related to expectancy mechanisms, and neurobiologically related to the activation of endogenous opioid, dopamine, endocannabinoid, oxytocin, vasopressin and serotonin systems, and the inactivation of cholecystokinin (Benedetti, 2009; Miller et al., 2009).

Importantly, Miller et al. describe certain aspects related to the concept of interpersonal healing that differ from natural and technological healing, the latter two referring to the natural progressions of disease and illness, and biomedical intervention, respectively (Miller et al., 2009). Specifically, it is mentioned how some level of a patient's attention and alertness is required for interpersonal healing and the mechanisms of the placebo effect to occur (Miller et al., 2009). This concept has been demonstrated by Colloca and colleagues, in which a paradigm of open and hidden administration of medication showed improved outcomes in pain, bradykinesia, and anxiety in patients who were aware of receiving an active medication through direct interaction with a health practitioner along with suggestions of benefit, when compared to those patients who received the same pharmacological medication through an automated pump (Colloca et al., 2004). Additionally, shifting a patient's attention away from a focus on noxious or other negative stimuli can help in reducing fear and other factors that influence the placebo and nocebo phenomenon, as well as patient wellbeing, and can thus help mitigate the symptoms of illness (Miller et al., 2009).

4. SIGNS AND MEANING

As demonstrated by the open-hidden paradigm, interactions and communications serve as a vehicle for facilitating both placebo and nocebo effects. Miller and Colloca applied the theory of signs developed by American philosopher Charles Sanders Peirce, named *semiotics*, as a way to bridge the neurobiological, psychological and cultural components of the placebo effect with the concept of meaning (Colloca et al., 2011; Miller & Colloca, 2010). Peirce developed the theory of signs to form a systematic comprehension of the phenomenon of logic, communication, and learning (Peirce, 1940). Within this context, Peirce defined signs as "something which stands to somebody for something" (Peirce, 1940). Peirce's theory described the consequences and influences of signs as a relationship between three main components: (1) the sign *vehicle* "carrying" or conveying an object; (2) the *significance* of the object represented by the sign vehicle; and (3) the *idea* within the mind of the interpreter about the sign vehicle and associated object. In these circumstances, the interpreter consciously and subconsciously interprets the information provided by the sign.

Furthermore, Peirce categorized signs into three primary types: *icons, indexes and symbols*. *Icons* provide meaning to signs through observable likeness between the sign and the represented object, and can include, for example, diagrams and pictures (Miller & Colloca, 2010; Peirce, 1940). Icons convey information through resemblance.

Indexes, on the other hand, represent a direct association between a sign and the object, which also encompass the memories and senses of the interpreter. The use of conditioned and unconditioned stimuli to create conditioned responses, namely, classical conditioning, falls within this category. The recognition of, and response to, an index sign is analogous to the conditioned response brought by a conditioned stimulus (Colloca et al., 2011; Miller & Colloca, 2010).

Finally, Pierce defines *symbols* as signs that associate a general concept by virtue of conventional rules and the conceptualizations and understandings of the interpreter. This can include, for example, verbal communication (Colloca et al., 2011; Miller & Colloca, 2010; Peirce, 1904). Within a clinical context, a symbol can be the verbal language coming from a physician stating "this will help your pain" prior to the injection of a solution, which will be interpreted by the patient as a symbol meaning that pain relief should be expected (Colloca et al., 2011; Miller & Colloca, 2010). As noted by Kirsch (1997) these verbal suggestions and symbols can engage placebo effects through the manipulation of individual expectancies, which include aspects of conditioned responses (Kirsch, 1997; Miller & Colloca, 2010).

Based on this theory, signs give information about objects to an interpreter, and these signs, whether icons, indexes or symbols, or indexes, serve as a contextual vehicle for the placebo effect in the clinical setting. For the patient, the interpreted signs can function as therapeutic agents for responses that have been learned through the continuous processing of what signs, within the clinical setting, convey. This interpretation by the patient can be shaped by both emotion and cognition and can cause both positive or negative effects on the patient's health or outcome, namely, a placebo or a nocebo effect (Colloca & Miller, 2011; Miller & Colloca, 2010). As a result, the semiotics model helps bridge the linguistic, sociocultural and neurobiological elements of the placebo phenomenon within the dynamics of learning, meaning, and contextual factors.

An important aspect in the way signs convey information relates to the meaning that is developed with continuous exposure to these signs. Daniel Moerman, a well-known anthropologist, developed the concept and role of "meaning" within health and illness as a framework for understanding the placebo phenomenon. In fact, Moerman proposed replacing the vernacular of *placebo response*—to *meaning response* (Moerman, 2002), as he believed that changing the terminology was important for mitigating negative historical connotations around the term "placebo" (i.e., the placebo effect being thought of as a "nuisance variable"). His proposal allowed for a focus on what he believed is a key part of the healing and disease process: the *meaning* of an illness. Moerman and Jonas co-authored an article (2002) that defined the meaning response as the "physiologic or psychological effects of meaning in the origins or treatment of illness" (Moerman & Jonas, 2002).

For example, a British study of 835 women frequently using analgesics for headaches evaluated the effects of branding in both aspirin and placebo. Investigators found that aspirin provided in a branded box worked better than its unbranded counterpart, and that the branded placebo worked better than the unbranded placebo, although the former did not surpass the effectiveness of the unbranded aspirin (Branthwaite & Cooper, 1981; Moerman & Jonas, 2002). From a total of 435 headaches that were reported by those receiving branded placebo, a significant 64% reported symptom improvement after only 1 hour following medication administration, while from the 410 headaches reported by the unbranded placebo group, only a 45% reported improvement at the same time point (Branthwaite & Cooper, 1981).

It is important to consider the high variability that can exist in individual *meaning* due to terminology and conceptual differences originating from social and cross-cultural factors. As outlined by Moerman and Jonas (2002), "...1) meaning has biological consequence and 2) meanings vary across cultures, [thus] we can anticipate that biology will differ in different places, not because of genetics but because of these entangled ideas" (Moerman & Jonas, 2002).

Although the individual concept of meaning varies, meaning itself is a fundamental construct shaped by social contexts and processes, perceptions, and identities (Fife, 2005). Contextual meaning goes beyond an individual's self-meaning, which is more closely associated to identity, by encompassing specific circumstantial factors related to an individual's perception of the responses of others in association to changes brought by an event (Fife, 2005). In the context of illness or a treatment, different meanings lead to distinct adaptive responses, as well as varying levels of stress responses. Thus, since these fundamental cognitions shape the individual interpretation and classification of events and have the potential to influence both behavior and physiology, it would be plausible to consider that within a clinical context these cognitive and mentalizing processes may engage psychosocial elements which in turn, engage placebo and nocebo effects. Whether these effects are positive or negative depends upon the context and perspectives associated with circumstantial factors surrounding the individual, the clinical setting, and treatments (Moerman & Jonas, 2002).

Some point to limitations of the meaning framework when it comes to explaining the placebo phenomenon within certain contexts. Specifically, it relies on the perception of psychosocially-driven "symbolic meaning," and thus may potentially fail to account for the observed phenomenon of some conditioned placebo effects that occur independently of conscious perception or meaning (Miller et al., 2009). Further, placebo effects have been experimentally elicited through conditioning in animals such as mice and rats, in which the influence of symbolic meaning is expected to be, for the most part, absent (Keller, Akintola, & Colloca, 2018). In summary, although meaning itself could be understood as a human characterization, it undoubtedly plays a role in the placebo and nocebo phenomenon by influencing learning and shaping expectancies, as meaningful symbols also form part of conditioned stimuli and responses.

5. CONNECTING THE DOTS: A CLINICAL APPROACH

During a period of illness, patients often find themselves in a vulnerable and stressful position. In this case, the simple behavior of going to see a healthcare practitioner creates a context of healing that forms part of the therapeutic process (Adler, 2002). The physician himself/herself becomes a sign that conveys a meaning, whether positive or negative, depending on the context upon which prior and current interactions occur. Establishing a positive therapeutic alliance with patients is crucial during the clinical encounter not only to promote placebo-induced relief of symptoms, but to preserve and enhance treatment efficacy and effectiveness (Fig. 1).

For a patient–practitioner relationship to be therapeutic in itself, it would need to be perceived by patients as a caring, non-judgmental, and supportive interaction that provides relief during a period of high stress (Kornhaber, Walsh, Duff, & Walker, 2016). Characteristics of such alliances include friendliness, empathy, warmth, and genuine involvement and interest. Being purposeful with body language and facial expressions, active listening, showing concern, receptiveness, and responsiveness can strengthen the therapeutic relationship with a patient and set a tone for an empowering alliance. Non-verbal communication is of vital importance during the patient–practitioner interaction, as it is estimated that the meaning of a message can be divided into 7% of content, 38% tone of voice, and 55% body language (Mehrabian, 1971). A recent review found that there is substantial

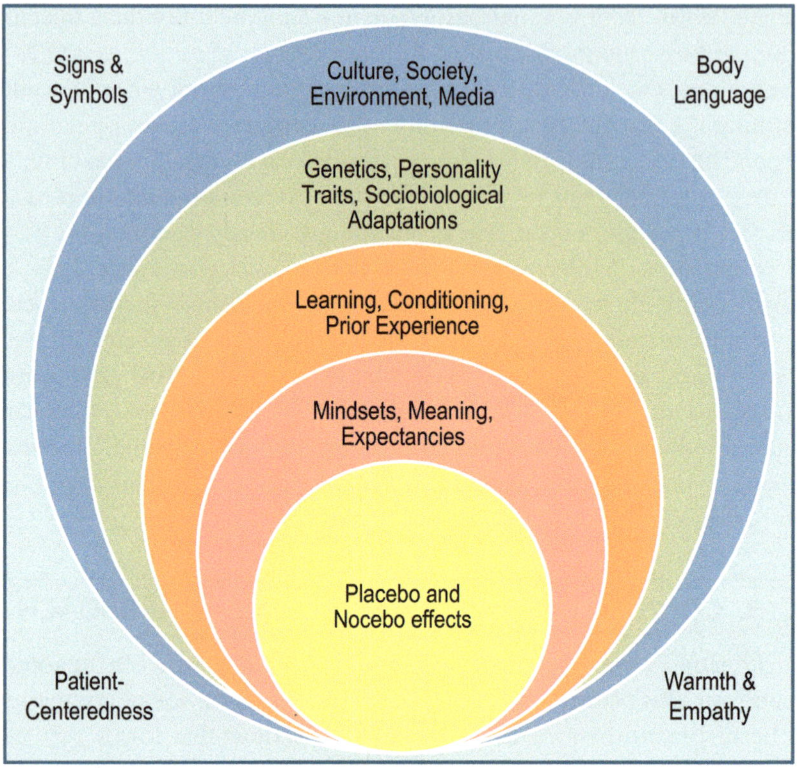

Fig. 1 Sociobiological and behavioral organization of placebo and nocebo effects. Like a socio-ecological model of human behavior and development, placebo and nocebo effects are preceded and modeled by many sociopsychobiological factors that influence patients' expectancies. On a macro scale, the environment, society, and culture play a very important role as they become one of the first filters of reality and day-to-day experience of human beings. Societal and cultural factors influence how an individual perceives disease, well-being, treatments, and even the figure of the physician. Since they can directly affect how a person is able to identify, cope, and function with illnesses, even those originating from genetic traits, psychobiological factors were located a level under the sociocultural tier. Placebo and nocebo effects are the product of a dynamic relational process with external forces that provide meaning and context to health, disease, the patient–practitioner relationship, and the therapeutic process. These meaningful contexts are influenced by an individual's priming and experience, and help shape the formation of expectancies, mindsets and the relevance of signs and symbols. At the center of all of these multi-layered factors lie the mechanisms of the placebo and nocebo effect. At the periphery, we find external vehicles and facilitators of messages that convey important information within the patient–practitioner interaction.

evidence that positive therapeutic relationships between patients and providers are correlated with improvements in quality of life, anxiety and depression levels, as well as patient satisfaction and treatment adherence (Kornhaber et al., 2016). A positive doctor–patient relationship can also help decrease the quantity of medication used, and therefore, experienced side effects. A commentary published in the New England Journal of Medicine suggested the application of placebos in the treatment for pediatric migraine in order to reduce side effects (Jackson, 2017). On the other hand, negative patient–practitioner relationships are characterized by psychological distress and the onset of negative feelings of invalidation and dehumanization (Kornhaber et al., 2016).

Interactions that hold a business-like character and that create doubt and uncertainty around the practitioner's skills and competence could lead to nocebo effects or to a reduction of placebo effects (Howe et al., 2017). HPA axis activation, anxiety, and the activity of the cholecystokinin system have been associated to the occurrence of nocebo effects (Benedetti, Amanzio, Casadio, Oliaro, & Maggi, 1997; Benedetti, Amanzio, Vighetti, & Asteggiano, 2006; Blasini et al., 2017). Nocebo effects are strong enough to decrease the efficacy of remifentanil, a short-acting and very potent opioid (Bingel et al., 2011). Therefore, preventing interactions that are uncaring, judgmental and insensitive, which send a message of invalidation and lack of warmth, while at the same time creating a safe and positive environment for patients to begin internalizing their own process of healing, can prevent the nocebo effect while promoting the activation of placebo mechanisms.

This "healing environment" can be concretely achieved through different approaches aiming to promote the formation and consolidation of positive expectancies. These positive expectancies do not refer to a use of illusory or deceptive information. Rather, positive expectancies can begin to be established through a clear identification of patient beliefs, concerns, and goals (also, see Darnall & Colloca, 2018). This patient-centered approach can help identify patient *mindsets*, which are views or constructs that orient the development of associations, convictions, and dispositions about situations and experiences. Mindsets can be formed and shaped by different factors, including culture, socialization processes, spirituality and religion, and media exposure (Crum & Zuckerman, 2017). These factors help direct decision-making, behaviors, and attitudes, and can function as a shortcut for mental processes and behaviors. Changes toward the establishment and modification of positive mindsets have been shown to provoke physiological changes in immunity (Howe et al., 2017), as well as in multiple hormonal and cardiovascular factors including cortisol (Crum, Salovey, & Achor, 2013) and hemoglobin A_{1C} (Crum & Zuckerman, 2017). Researchers have identified two main subtypes of mindsets that are particularly influential in the context of the patient–practitioner relationship: mindsets regarding *treatment efficacy* and individual *capacity to change*

(Crum & Zuckerman, 2017). When strategically approached, perhaps through the application of questioning and motivational interviewing skills, identifying patient mindsets may serve as an efficient way for evaluating a patient's stance on different aspects of the clinical interaction and setting, including previous experiences and expectations regarding treatment plans and other outcomes.

In addition to the practitioner's expression of certain psychosocial characteristics within their role, other actions can be taken to enhance the placebo and prevent the nocebo effect. For instance, a practitioner could avoid over-emphasizing negative information regarding a treatment by balancing it with information about the positive effects in a truthful and ethical manner that preserves patient autonomy (Colloca, 2017; Klinger, Blasini, Schmitz, & Colloca, 2017). Some research has suggested that providing patients with a rationale within an honest and positive interaction (Ballou et al., 2017; Kaptchuk et al., 2010), and explaining the mechanisms of action of a therapy (Colloca, 2017), boost placebo effects and prevent the formation of the nocebo effect.

It is imperative to consider the importance that cultural differences may have in the clinical setting, particularly in the identification of signs and cues within the patient–practitioner interaction. As stated by Dr. Moerman, "Much of our knowledge of the world is not an elicitation of what 'is,' but rather it is a construction laid atop the world of experience" (Moerman, 2002). Differences in the meaning of cultural signs and socialization lead to different psychosocial adaptations and subsequent physiologic responses to events related to health and illness. This occurs because these social processes form perceptions tailored to the culture and environment in which a person may be developing. Although this adds a degree of complexity to the practitioner who is part of the bi-cultural patient–practitioner relationship, training which focus on building linguistic and cultural competency skills can help overcome some of these challenges.

In summary, these described characteristics of a therapeutic interpersonal relationship harness the underlying mechanisms of the placebo effect via sociobiological modulation of physiologic responses. In the era of patient-centered outcomes in healthcare, physicians can be seen as "therapeutic agents" that can exert an effect through mechanisms associated with the placebo effect as a tool to help facilitate health and healing (Wampold, 2018). As stated in Kaptchuk's article, "supportive and attentive health care legitimately creates a therapeutic bias" in patients toward hope and an experience of relief (Fig. 2; Kaptchuk & Miller, 2015).

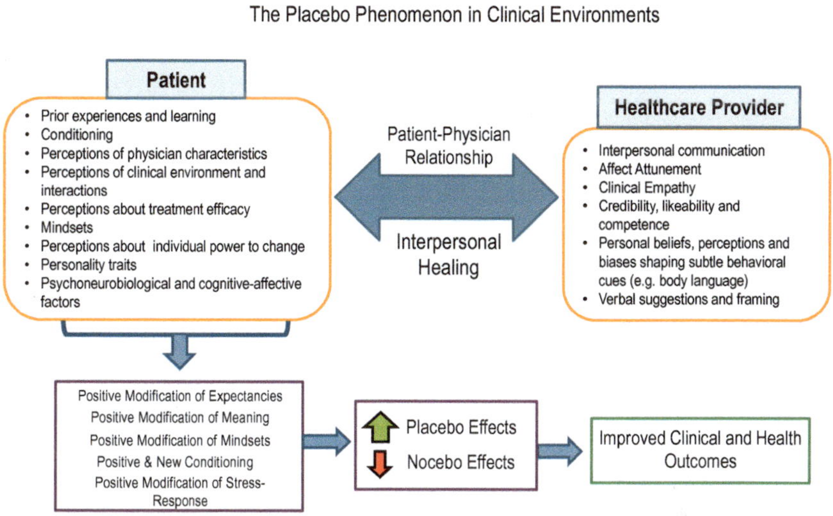

Fig. 2 The patient–practitioner interaction as a vehicle for the dynamic exchange of sociobiological information in the context of health and illness. A patient-centered process of relating sets the foundation for interpersonal healing to occur, and in doing so, it has the power to positively modify the expectancies that could influence a patient's health and clinical outcomes. This dyadic concept of the placebo phenomenon encompasses how a practitioner establishes a context of healing and well-being. The placebo and nocebo phenomena can be provoked and prevented, respectively, through a positive therapeutic alliance.

6. CONCLUSION

Medicine is a field that has been inherently built through rituals and symbols such as the act of going to a doctor or healer, writing prescriptions, and administering medication (Kaptchuk, 2011). As part of a larger contextual process of relating, the placebo effect can be implemented to help enhance treatments. Empathy, positive interpersonal relationships, cultural competence and sensitivity, and personalized interactions according to identified meaningful mindsets and conceptions, can become part of the therapeutic target as ways to improve clinical outcomes. Active engagement from physicians and health care providers can help patients positively reframe their mindsets, expectations, and expectancies in ways that activate both conscious and subconscious biological and coping mechanisms that can reduce symptom severity, promote patient satisfaction, and improve the natural physiological process of healing.

An attitude of caring, respect, and patient empowerment through meaningful patient-centered discussions within the patient–practitioner interaction can serve as a fundamental basis for engaging the placebo effect and preventing the nocebo effect in any treatment that is decided upon (Moerman, 2002). More research should be directed into how the characteristics of the patient–practitioner relationship help to improve health outcomes, for both objective and subjective measurements.

Although most of these concepts are not entirely new to the healthcare community, as health coaches and similar practitioners strategically utilize these during their interventions to promote and improve behavioral change, adherence, and lifestyle modifications (Wolever et al., 2013), they may lack practicality in clinical settings where there is limited time to spend with a patient or when there is insufficient communication between the various practitioners a patient may see. Many questions remain, as there are no specific guidelines that a practitioner can follow that will necessarily work exactly the same with each different patient. These questions include: Is there a specific order and method that can enhance the efficiency of the application of therapeutic interpersonal relationships in a system with limited time and resources? Could we specifically enhance the innate phenomenon of the placebo effect without the need of placebo interventions? How can these results differ across communities and cultures?

Despite these challenges, understanding the process of relating, as well as the bi-directional components that continually and dynamically feed into it, can help individual practitioners pragmatically and routinely engage these innate human mechanisms in an ethically compliant manner for the short- and long-term benefit of their patients. In the field of pain management, where clinically effective and lasting responses that protect a patient's quality of life and function can be so difficult at times, having an awareness of the potential benefits of positive patient–practitioner interactions can be a tool for producing substantial positive influences which could make a considerable difference in the lives of pain patients.

ACKNOWLEDGMENTS

This research was supported by the University of Maryland, Baltimore (L.C.) and the National Institute of Dental and Craniofacial Research (NIDCR, R01DE025946, L.C.).

REFERENCES

Ader, R., Mercurio, M. G., Walton, J., James, D., Davis, M., Ojha, V., et al. (2010). Conditioned pharmacotherapeutic effects: A preliminary study. *Psychosomatic Medicine*, *72*(2), 192–197. https://doi.org/10.1097/PSY.0b013e3181cbd38b.

Adler, H. M. (2002). The sociophysiology of caring in the doctor-patient relationship. *Journal of General Internal Medicine*, *17*(11), 874–881. Retrieved from http://www.ncbi.nlm.nih.gov/pubmed/12406360.

Adler, H. M. (2007). Toward a biopsychosocial understanding of the patient-physician relationship: An emerging dialogue. *Journal of General Internal Medicine, 22*(2), 280–285. https://doi.org/10.1007/s11606-006-0037-8.
Adler, H. M., & Hammett, V. B. (1973). The doctor-patient relationship revisited. An analysis of the placebo effect. *Annals of Internal Medicine, 78*(4), 595–598. Retrieved from http://www.ncbi.nlm.nih.gov/pubmed/4694043.
Ballou, S., Kaptchuk, T. J., Hirsch, W., Nee, J., Iturrino, J., Hall, K. T., et al. (2017). Open-label versus double-blind placebo treatment in irritable bowel syndrome: Study protocol for a randomized controlled trial. *Trials, 18*(1), 234. https://doi.org/10.1186/s13063-017-1964-x.
Beecher, H. K. (1955). The powerful placebo. *Journal of the American Medical Association, 159*(17), 1602. https://doi.org/10.1001/jama.1955.02960340022006.
Benedetti, F. (2009). *Placebo effects: Understanding the mechanisms in health and disease.* Oxford University Press.
Benedetti, F., Amanzio, M., Casadio, C., Oliaro, A., & Maggi, G. (1997). Blockade of nocebo hyperalgesia by the cholecystokinin antagonist proglumide. *Pain, 71*(2), 135–140. https://doi.org/10.1016/S0304-3959(97)03346-0.
Benedetti, F., Amanzio, M., Vighetti, S., & Asteggiano, G. (2006). The biochemical and neuroendocrine bases of the hyperalgesic nocebo effect. *Journal of Neuroscience, 26*(46), 12014–12022. https://doi.org/10.1523/JNEUROSCI.2947-06.2006.
Benedetti, F., Lanotte, M., Lopiano, L., & Colloca, L. (2007). When words are painful: Unraveling the mechanisms of the nocebo effect. *Neuroscience, 147*, 260–271. https://doi.org/10.1016/j.neuroscience.2007.02.020.
Benedetti, F., Pollo, A., & Colloca, L. (2007). Opioid-mediated placebo responses boost pain endurance and physical performance: Is it doping in sport competitions? *Journal of Neuroscience, 27*(44), 11934–11939. https://doi.org/10.1523/JNEUROSCI.3330-07.2007.
Bingel, U., Wanigasekera, V., Wiech, K., Ni Mhuircheartaigh, R., Lee, M. C., Ploner, M., et al. (2011). The effect of treatment expectation on drug efficacy: Imaging the analgesic benefit of the opioid remifentanil. *Science Translational Medicine, 3*(70), 70ra14. https://doi.org/10.1126/scitranslmed.3001244.
Blasini, M., Corsi, N., Klinger, R., & Colloca, L. (2017). Nocebo and pain: An overview of the psychoneurobiological mechanisms. *Pain Reports, 2*, 1–9. https://doi.org/10.1097/PR9.0000000000000585.
Branthwaite, A., & Cooper, P. (1981). Analgesic effects of branding in treatment of headaches. *British Medical Journal (Clinical Research Ed.), 282*(6276), 1576–1578. Retrieved from http://www.ncbi.nlm.nih.gov/pubmed/6786566.
Brody, H. (1997). Placebo response, sustained partnership, and emotional resilience in practice. *The Journal of the American Board of Family Practice, 10*(1), 72–74.
Carter, C. S. (1998). Neuroendocrine perspectives on social attachment and love. *Psychoneuroendocrinology, 23*(8), 779–818. Retrieved from http://www.ncbi.nlm.nih.gov/pubmed/9924738.
Carvalho, C., Caetano, J. M., Cunha, L., Rebouta, P., Kaptchuk, T. J., & Kirsch, I. (2016). Open-label placebo treatment in chronic low back pain: A randomized controlled trial. *Pain, 157*(12), 2766–2772. https://doi.org/10.1097/j.pain.0000000000000700.
Colagiuri, B., Schenk, L. A., Kessler, M. D., Dorsey, S. G., & Colloca, L. (2015). The placebo effect: From concepts to genes. *Neuroscience, 307*, 171–190. https://doi.org/10.1016/j.neuroscience.2015.08.017.
Colloca, L. (2014). *Placebo, nocebo, and learning mechanisms* pp. (17–35). Berlin, Heidelberg: Springer. https://doi.org/10.1007/978-3-662-44519-8_2.
Colloca, L. (2017). Tell me the truth and i will not be harmed: Informed consents and nocebo effects. *The American Journal of Bioethics: AJOB, 17*(6), 46–48. https://doi.org/10.1080/15265161.2017.1314057.

Colloca, L., & Benedetti, F. (2005). Placebos and painkillers: Is mind as real as matter? *Nature Reviews. Neuroscience*, *6*(7), 545–552. https://doi.org/10.1038/nrn1705.
Colloca, L., & Benedetti, F. (2007). Nocebo hyperalgesia: How anxiety is turned into pain. *Current Opinion in Anaesthesiology*, *20*(5), 435–439. https://doi.org/10.1097/ACO.0b013e3282b972fb.
Colloca, L., Benedetti, F., & Porro, C. A. (2008). *Experimental designs and brain mapping approaches for studying the placebo analgesic effect.* In *European Journal of Applied Physiology: Vol. 102.* Heidelberg: Springer-Verlag. Retrieved from http://www.academia.edu/23744290/Experimental_designs_and_brain_mapping_approaches_for_studying_the_placebo_analgesic_effect.
Colloca, L., Enck, P., & DeGrazia, D. (2016). Relieving pain using dose-extending placebos. *Pain*, *157*, 1. https://doi.org/10.1097/j.pain.0000000000000566.
Colloca, L., & Grillon, C. (2014). Understanding placebo and nocebo responses for pain management. *Current Pain and Headache Reports*, *18*(6), 419. https://doi.org/10.1007/s11916-014-0419-2.
Colloca, L., Lopiano, L., Lanotte, M., & Benedetti, F. (2004). Overt versus covert treatment for pain, anxiety, and Parkinson's disease. *The Lancet. Neurology*, *3*(11), 679–684. https://doi.org/10.1016/S1474-4422(04)00908-1.
Colloca, L., Miller, F. G., Harrington, A., Miller, F. G., Colloca, L., Peirce, C., et al. (2011). How placebo responses are formed: A learning perspective. *Philosophical Transactions of the Royal Society B: Biological Sciences*, *366*(1572), 1859–1869. https://doi.org/10.1098/rstb.2010.0398.
Colloca, L., Pine, D. S., Ernst, M., Miller, F. G., & Grillon, C. (2016). Vasopressin boosts placebo analgesic effects in women: A randomized trial. *Biological Psychiatry*, *79*(10), 794–802. https://doi.org/10.1016/j.biopsych.2015.07.019.
Corsi, N., & Colloca, L. (2017). Placebo and nocebo effects: The advantage of measuring expectations and psychological factors. *Frontiers in Psychology*, *8*, 308. https://doi.org/10.3389/fpsyg.2017.00308.
Coulehan, J. L. (1995). Tenderness and steadiness: Emotions in medical practice. *Literature and Medicine*, *14*(2), 222–236.
Crum, A. J., Salovey, P., & Achor, S. (2013). Rethinking stress: The role of mindsets in determining the stress response. *Journal of Personality and Social Psychology*, *104*(4), 716–733. https://doi.org/10.1037/a0031201.
Crum, A., & Zuckerman, B. (2017). Changing mindsets to enhance treatment effectiveness. *JAMA*, *317*(20), 2063. https://doi.org/10.1001/jama.2017.4545.
Darnall, B. D., & Colloca, L. (2018). Optimizing placebo and minimizing nocebo to reduce pain, catastrophizing, and opioid use: A review of the science and an evidence-informed clinical toolkit. In L. Colloca (Ed.), *Neurobiology of the placebo effect, part II, International review of neurobiology*: *Vol. 139* (pp. 129–157). New York: Academic Press.
Di Blasi, Z., Harkness, E., Ernst, E., Georgiou, A., & Kleijnen, J. (2001). Influence of context effects on health outcomes: A systematic review. *The Lancet (London, England)*, *357*(9258), 757–762. Retrieved from http://www.ncbi.nlm.nih.gov/pubmed/11253970.
Di Blasi, Z., & Reilly, D. (2005). Placebos in medicine: Medical paradoxes need disentangling. *BMJ (Clinical Research Ed.)*, *330*(7481), 45. https://doi.org/10.1136/bmj.330.7481.45-a.
Faria, V., Fredrikson, M., & Furmark, T. (2008). Imaging the placebo response: A neurofunctional review. *European Neuropsychopharmacology*, *18*(7), 473–485. https://doi.org/10.1016/j.euroneuro.2008.03.002.
Fife, B. L. (2005). The role of constructed meaning in adaptation to the onset of life-threatening illness. *Social Science & Medicine*, *61*(10), 2132–2143. https://doi.org/10.1016/j.socscimed.2005.04.026.

Frisaldi, E., Carlino, E., Zibetti, M., Barbiani, D., Dematteis, F., Lanotte, M., et al. (2017). The placebo effect on bradykinesia in Parkinson's disease with and without prior drug conditioning. *Movement Disorders*, *32*, 1474–1478. https://doi.org/10.1002/mds.27142.

Faude, J., Jones, C., & Robins, M. (1996). The affective life of infants: empirical and theoretical foundations. In D. Nathanson (Ed.), *Knowing feeling, affect, script, and psychotherapy* (pp. 219–256). New York: WW Norton & Co.

Gracely, R., Dubner, R., Deeter, W., & Wolskee, P. (1985). Clinicians' expectations influence placebo analgesia. *The Lancet*, *325*, 43. https://doi.org/10.1016/S0140-6736(85)90984-5.

Greenough, W., & Black, J. (1992). Induction of brain structure by experience: substrates for cognitive development. In M. Gunnar & C. Nelson (Eds.), *Minnesota symposia on child psychology: developmental neuroscience*, Vol. 24. Hillsdale, NJ: Lawrence Erlbaum.

Hofer, M. A. (1981). The roots of human behavior: An introduction to the psychobiology of early development. Myron A. Hofer. W. H. Freeman & Co., San Francisco, xiii + 331 pp. Cloth, $22.50. Paper, $11.50. *Developmental Psychobiology*, *15*(1), 89–91. https://doi.org/10.1002/dev.420150113.

Hofer, M. A. (1984). Relationships as regulators: A psychobiologic perspective on bereavement. *Psychosomatic Medicine*, *46*(3), 183–197. Retrieved from http://www.ncbi.nlm.nih.gov/pubmed/6739679.

Howe, L. C., Goyer, J. P., & Crum, A. J. (2017). Harnessing the placebo effect: Exploring the influence of physician characteristics on placebo response. *Health Psychology*, *36*, 1074–1082. https://doi.org/10.1037/hea0000499.

Insel, T. (1997). A neurological basis of social attachment. *American Journal of Psychiatry*, *154*, 726–735.

Jackson, J. L. (2017). Pediatric migraine headache—Still searching for effective treatments. *New England Journal of Medicine*, *376*(2), 169–170. https://doi.org/10.1056/NEJMe1614628.

Kandel, E. R. (1979). Psychotherapy and the single synapse: The impact of psychiatric thought on neurobiologic research. *The New England Journal of Medicine*, *301*(19), 1028–1037.

Kandel, E. R., & Squire, L. R. (2000). Neuroscience: Breaking down scientific barriers to the study of brain and mind. *Science*, *290*(5494), 1113–1120.

Kaptchuk, T. J. (2011). Placebo studies and ritual theory: A comparative analysis of Navajo, acupuncture and biomedical healing. *Philosophical Transactions of the Royal Society of London. Series B, Biological Sciences*, *366*(1572), 1849–1858. https://doi.org/10.1098/rstb.2010.0385.

Kaptchuk, T. J., Friedlander, E., Kelley, J. M., Sanchez, M. N., Kokkotou, E., Singer, J. P., et al. (2010). Placebos without deception: A randomized controlled trial in irritable bowel syndrome. *PLoS One*, *5*(12), e15591. https://doi.org/10.1371/journal.pone.0015591.

Kaptchuk, T. J., Kerr, C. E., & Zanger, A. (2009). Placebo controls, exorcisms, and the devil. *The Lancet (London, England)*, *374*(9697), 1234–1235. Retrieved from http://www.ncbi.nlm.nih.gov/pubmed/19839101.

Kaptchuk, T. J., & Miller, F. G. (2015). Placebo effects in medicine. *New England Journal of Medicine*, *373*(1), 8–9. https://doi.org/10.1056/NEJMp1504023.

Keller, A., Akintola, T., & Colloca, L. (2018). Placebo analgesia in rodents: Current and future research. *International Review of Neurobiology*, *138*, 1–15. https://doi.org/10.1016/bs.irn.2018.02.001.

Kirsch, I. (1997). Response expectancy theory and application: A decennial review. *Applied and Preventive Psychology*, *6*(2), 69–79. https://doi.org/10.1016/S0962-1849(05)80012-5.

Klinger, R., Blasini, M., Schmitz, J., & Colloca, L. (2017). Nocebo effects in clinical studies: Hints for pain therapy. *Pain Reports*, *2*(2), e586. https://doi.org/10.1097/PR9.0000000000000586.

Kornhaber, R., Walsh, K., Duff, J., & Walker, K. (2016). Enhancing adult therapeutic interpersonal relationships in the acute health care setting: An integrative review. *Journal of Multidisciplinary Healthcare*, *9*, 537–546. https://doi.org/10.2147/JMDH.S116957.

Larson, E. B., & Yao, X. (2005). Clinical empathy as emotional labor in the patient-physician relationship. *JAMA*, *293*(9), 1100. https://doi.org/10.1001/jama.293.9.1100.

Locher, C., Nascimento, A. F., Kirsch, I., Kossowsky, J., Meyer, A., & Gaab, J. (2017). Is the rationale more important than deception? A randomized controlled trial of openlabel placebo analgesia. *Pain*, *158*, 2320–2328. https://doi.org/10.1097/j.pain.0000000000001012.

McEwen, B. S. (1998). Stress, adaptation, and disease. Allostasis and allostatic load. *Annals of the New York Academy of Sciences*, *840*, 33–44. Retrieved from http://www.ncbi.nlm.nih.gov/pubmed/9629234.

Mehrabian, A. (1971). *Silent messages*. Belmont, CA: Wadsworth.

Miller, F. G., & Colloca, L. (2010). Semiotics and the placebo effect. *Perspectives in Biology and Medicine*, *53*(4), 509–516. https://doi.org/10.1353/pbm.2010.0004.

Miller, F. G., Colloca, L., & Kaptchuk, T. J. (2009). The placebo effect: Illness and interpersonal healing. *Perspectives in Biology and Medicine*, *52*(4), 518–539. https://doi.org/10.1353/pbm.0.0115.

Moerman, D. (2002). *Meaning, medicine, and the "placebo effect"*. New York: Cambridge University Press.

Moerman, D. E., & Jonas, W. B. (2002). Deconstructing the placebo effect and finding the meaning response. *Annals of Internal Medicine*, *136*(6), 471–476. Retrieved from http://www.ncbi.nlm.nih.gov/pubmed/11900500.

Murray, A. D. (1979). Infant crying as an elicitor of parental behavior: An examination of two models. *Psychological Bulletin*, *86*(1), 191–215. Retrieved from http://www.ncbi.nlm.nih.gov/pubmed/377352.

Nelso, E., & Panksepp, J. (1998). Brain substrates of infant–mother attachment: Contributions of opioids, oxytocin, and norepinephrine. *Neuroscience & Biobehavioral Reviews*, *22*(3), 437–452.

Petersen, G. L., Finnerup, N. B., Colloca, L., Amanzio, M., Price, D. D., Jensen, T. S., et al. (2014). The magnitude of nocebo effects in pain: A meta-analysis. *Pain*, *155*, 1426–1434. https://doi.org/10.1016/j.pain.2014.04.016.

Peirce, C. (1904). New elements. In T. P. E. Project (Ed.), *The essential Peirce. (Vol. 2*, pp. 300–324). Bloomington: Indiana University Press.

Peirce, C. (1940). Logic as semiotic: The theory of signs. In J. Bucheler (Ed.), *Philosophical writings of Peirce* (pp. 98–119). New York: Dover.

Porges, S. W. (1998). Love: An emergent property of the mammalian autonomic nervous system. *Psychoneuroendocrinology*, *23*(8), 837–861. Retrieved from http://www.ncbi.nlm.nih.gov/pubmed/9924740.

Price, D. D., Craggs, J., Nicholas Verne, G., Perlstein, W. M., & Robinson, M. E. (2007). Placebo analgesia is accompanied by large reductions in pain-related brain activity in irritable bowel syndrome patients. *Pain*, *127*(1), 63–72. https://doi.org/10.1016/j.pain.2006.08.001.

Shaibani, A., Frisaldi, E., & Benedetti, F. (2017). Placebo response in pain, fatigue, and performance: Possible implications for neuromuscular disorders. *Muscle & Nerve*, *56*(3), 358–367. https://doi.org/10.1002/mus.25635.

Shamay-Tsoory, S. G., Lester, H., Chisin, R., Israel, O., Bar-Shalom, R., Peretz, A., et al. (2005). The neural correlates of understanding the other's distress: A positron emission tomography investigation of accurate empathy. *NeuroImage*, *27*(2), 468–472. https://doi.org/10.1016/j.neuroimage.2005.05.012.

Shanafelt, T. D., Dyrbye, L. N., & West, C. P. (2017). Addressing physician burnout. *The Journal of the American Medical Association*, *317*(9), 901–902. https://doi.org/10.1001/jama.2017.0076.

Shanafelt, T. D., Hasan, O., Dyrbye, L. N., et al. (2015). Changes in burnout and satisfaction with work-life balance in physicians and the general US working population between 2011 and 2014. *Mayo Clinic Proceedings*, *90*(12), 1600–1613. https://doi.org/10.1016/j.mayocp.2015.08.023.

Sussex, R. (2018). Describing placebo phenomena in medicine: A linguistic approach. In L. Colloca (Ed.), *Neurobiology of the placebo effect, part II, International review of neurobiology*: *Vol. 139* (pp. 49–83). New York: Academic Press.

Tekampe, J., Van Middendorp, H., Meeuwis, S. H., Van Leusden, J. W. R., Pacheco-López, G., Hermus, A. R. M. M., et al. (2017). Conditioning immune and endocrine parameters in humans: A systematic review. *Psychotherapy and Psychosomatics*, *86*, 99–107. https://doi.org/10.1159/000449470.

Udupa, K., & Fox, S. H. (2015). Placebo effect in Parkinson's disease: Harnessing the mind in the treatment of PD. *Movement Disorders*, *30*(6), 786. https://doi.org/10.1002/mds.26222.

Young, L. J., Wang, Z., & Insel, T. R. (1998). Neuroendocrine bases of monogamy. *Trends in Neurosciences*, *21*(2), 71–75.

Wampold, B. E. (2018). The therapeutic value of the relationship for placebo effects and other healing practices. In L. Colloca (Ed.), *Neurobiology of the placebo effect, part II, International review of neurobiology*: *Vol. 139* (pp. 191–210). New York: Academic Press.

Wendt, L., Albring, A., & Schedlowski, M. (2014). *Learned placebo responses in neuroendocrine and immune functions* pp. (159–181). Berlin, Heidelberg: Springer. https://doi.org/10.1007/978-3-662-44519-8_10.

Wolever, R. Q., Simmons, L. A., Sforzo, G. A., Dill, D., Kaye, M., Bechard, E. M., et al. (2013). A systematic review of the literature on health and wellness coaching: Defining a key behavioral intervention in healthcare. *Global Advances in Health and Medicine*, *2*(4), 38–57. https://doi.org/10.7453/gahmj.2013.042.

FURTHER READING

Jensen, K. B., Petrovic, P., Kerr, C. E., Kirsch, I., Raicek, J., Cheetham, A., et al. (2014). Sharing pain and relief: Neural correlates of physicians during treatment of patients. *Molecular Psychiatry*, *19*(3), 392–398. https://doi.org/10.1038/mp.2012.195.

CHAPTER TEN

Placebo Effect: Theory, Mechanisms and Teleological Roots

Javeria Ali Hashmi[1]
Department of Anesthesia, Pain Management and Perioperative Medicine, Dalhousie University, Nova Scotia Health Authority, Halifax, NS, Canada
[1]Corresponding author: e-mail address: javeria.hashmi@dal.ca

Contents

Abstract

Why pain can be relieved with placebos is heavily debated. The term "placebo effect," implies that the placebo treatment induces pain relief which is imprecise because it is the mental cueing to the context of treatment and not the placebo itself that can reduce pain. This essay reverts to fundamentals of perception that have been used to explain how context generates predictions that can in turn effect the process of processing, organizing and interpreting of sensory inputs received from the periphery. We reinterpret placebo effect as a neurobiological phenomenon that occurs through the process of reward and aversive learning. The brain uses learnt information to generate predictions. The perceptual processes adjust the experience of pain to fit with the predictions generated from prior information. Placebo effect is thus understandably a result of the expectations and mental states that result from engaging in the process of treatment. These processes have teleological roots in ancient medicine and are the context that produces these responses is transforming with the evolution of modern medicine. Thus, when placebo effect is observed, the potent agent that induces pain reduction is not the placebo itself, but the mental cueing to the context of taking treatment.

International Review of Neurobiology, Volume 139
ISSN 0074-7742
https://doi.org/10.1016/bs.irn.2018.07.017

1. INTRODUCTION

Placebo effect is an intriguing but frequently misunderstood phenomenon. Despite a large number of convergent studies, the debate continues on validity of placebo effect as clinically useful (Enck, Bingel, Schedlowski, & Rief, 2013). That pain is reduced in response to placebos is a phenomenon widely witnessed and systematically recorded in clinical trials. Experimental and clinical studies have delved into this phenomenon to demonstrate that the body generates physiological responses that aid the process of treatment (Colloca, Enck, & DeGrazia, 2016). A major barrier for applying this knowledge into clinical strategies is the ambiguity around the term placebo response. In this article, first the issues with the term placebo effect will be discussed to make the assertion that veridical improvements that occur independent of drug action are formed by the contextual factors through a mental cueing process. Acknowledging this fact allows for a better conceptual and utilitarian model for understanding the physiological capacity that aids the therapeutic process. In addition, teleological role of mental cueing will be explained to postulate how we may have developed this response through persistence of cultures from antiquity. The significance of underlying brain mechanisms will be discussed with an emphasis on predictive processes and reward based learning that shape the mental cueing response.

2. DOES THE BRAIN RESPOND TO PLACEBOS OR TO MENTAL CUES?

Through the numerous studies on placebo mechanisms, it has been established that placebos can in fact trigger physiological changes in brain processes to alter pain (Colloca, Benedetti, & Porro, 2008). We associate the term placebo with the mental image of a dummy pill, and have been made to think that objects or procedures that is framed as an effective pain treatments can produce affects comparable to actual pain relievers. The word "placebo" means "to please" and has roots in earlier era of medicine when fake medicines were used as a tool to patronize and deceive the patient with the intent to make them think that they were being cared for. The term placebo originates from *placebo response* where the latter pertains to responses seen in the placebo arm of a randomized controlled trial (RCT). The canonical assumption for using RCTs is that unlike real medicines, placebos lack potency and hence serves as a control for non-drug related factors. For practical purposes, the factors to be controlled are non-specific and represent a list of irrelevant effects such as spontaneous fluctuations in pain, regression to mean, statistical artifacts, or anomalous psychosomatic responses (Enck et al., 2013; Enck & Klosterhalfen, 2013). Unlike real medicines, it should be clearly acknowledged that a placebo by itself lacks any potency or ability to alter mental or physiological processes. Beecher's assertion that placebo generate non-trivial responses that mimic pain relief (Beecher, 1955) led to an interest in the possibility that placebos have an impact on medical symptoms and the response was referred to as placebo response (Benedetti, Mayberg, Wager, Stohler, & Zubieta, 2005; Colloca, Klinger, Flor, & Bingel, 2013).

Placebos have been most thoroughly scrutinized in the context of randomized double-blind placebo controlled clinical trials have been repeatedly shown to reduce pain by 30–50%; this represents pain relief quantitatively similar to opioids and NSAIDS (Beecher, 1955; Colloca, Jonas, Killen, Miller, & Shurtleff, 2014; Enck et al., 2013; Enck & Klosterhalfen, 2013). Nevertheless, despite of the fact that placebos lack potency of an actual drug or treatment, the indubitable evidence for strong improvements in pain in the placebo arm suggests that there is a measurable and significant effect. A large body of psychophysical and neuroimaging studies have modeled the mechanisms that induce placebo response. Extensive scientific understanding of the validity of placebo effect has been established and an essential link has been drawn between placebo effect and learning systems of the brain (Buchel, Geuter, Sprenger, & Eippert, 2014; Colloca, 2014b; Egorova et al., 2015; Grahl, Onat, & Buchel, 2018; Hashmi et al., 2014; Jensen et al., 2015). However, the value of administering placebos, the longevity of responses and their clinical significance are useful questions that have still not been fully resolved.

The general perception of placebo effect is still clouded by obscurity. In the clinical realm, placebo effect can be misunderstood as pertaining to the use of sham medicines and as a result, the potential clinical value of placebo effect is often unclear and doubtful (Feys, Bekkering, Singh, & Devroey, 2014; Hrobjartsson, Kaptchuk, & Miller, 2011). The social and ethical ambiguities associated with the process of deception used in placebo is also problematic (Seminowicz, 2006). Here we contend that the responses observed with placebos are induced through physiological mechanisms and do not represent a response to placebo per se. The imprecise nomenclature is an important barrier for clinical conceptualization. As explained below, the clinical improvements are not owed to the placebo but to the mental cueing process that occurs through engaging with the context of the treatment process. Hence the response is ubiquitous and embedded in all treatment processes and hence should not be referred to commonly as placebo response. Despite its complexity, the motivation to understand how context and mental cueing contributes to treatment response is an important line of investigation with potential reward and benefits for patients and healthcare.

3. MENTAL CUEING RESPONSE

We now know that mental state such as expectations, attention and arousal are significantly altered based on the context of undergoing treatment (Corsi & Colloca, 2017; Enck, Benedetti, & Schedlowski, 2008; Vase et al., 2015). Aligning expectations and prior states with treatment goals plays a significant role in triggering changes in the neurobiological apparatus leading to changes in neurotransmitters and altering brain responses and connectivity (Atlas et al., 2012; Jensen et al., 2015; Klinger, Colloca, Bingel, & Flor, 2014; Schafer, Geuter, & Wager, 2018; Scott et al., 2008). Thus, the context of taking a treatment triggers top-down mental processes to become explicitly or implicitly aligned with the treatment goal and this process can be referred to as the mental cueing effect and the responses generated to as mental cueing response.

Mental cueing is useful as a term since it implies that the agency for inducing changes lies with the brain's ability to adjust perception based on changes in context. This is necessary because dummy medicines or sham devices are by definition inactive and hence ascribing a response to them is imprecise. The term placebo response is conflicted since the large observed in the placebo arm can be confused to mean that the placebo can cause a response. On the contrary, responses generated on starting a new treatment are not specific to the tool used for administering a sham or actual drug (Benedetti, 2014a). The device or tool itself is a small contributor relative to other factors such as prior knowledge, expectation and social effects (Benedetti, Carlino, & Piedimonte, 2016; Kong & Benedetti, 2014). The term mental cueing response thus more accurately suggests that the response occurs to a mental cueing to the context of initiating a treatment.

This distinction is important because if we consider that mental cueing is a significant phenomenon then we can begin to perceive its role in day to day clinical practice. Viewed from this vantage point, it becomes possible to acknowledge that drug activity is not the only necessary element for generating therapeutic responses. In other words, the responses normally attributed to analgesics are in reality a combination of pharmacological action of the drug and the mental cueing effect to varying degrees (Kirsch, 2014; Rutherford et al., 2017). Moreover, the mental cueing occurs not just during treatments with pills, but during any type of treatment, be it behavioral therapy such as CBT (cognitive behavioral therapy), surgery or alternative therapy (Jonas et al., 2015; Kaptchuk et al., 2006).

Another important dimension to mental cueing is the nocebo effect (Colloca, 2017). The tendency of some individuals to show worsening of symptoms after taking a drug is another layer of important evidence that mental cueing determines treatment success. The role of negative expectations is well demonstrated and nontrivial (Babel et al., 2017; Colloca, 2017; Corsi & Colloca, 2017). From a clinical perspective, it has been extensively noted that a negative context during the process of treatment can have a deleterious effect on the treatment outcomes by not only blocking the effect of the treatment, but may also worsen the symptoms (Klinger, Blasini, Schmitz, & Colloca, 2017). From a scientific and knowledge perspective, the possibility of negative mental cueing is in fact a demonstration of our physiological capacity for intrinsic adjustments in the disease process.

The factors that initiate mental cueing can be external or internal. The rituals that occur daily in an outpatient clinic are demonstrative evidence of *external* mental cueing. The doctor's bed side manner, the formalisms such as lab coats and scrubs, implicit or explicit verbal or facial suggestions made by clinicians about the effectiveness of the prescribed drug and encountering people who offer positive feedback verbally or implicitly are some examples of factors that are inherently positive external cues (Colloca et al., 2014; Jensen et al., 2014). The concept of "therapeutic encounter" points at the implicit therapeutic relationship is established between the clinicians and patients which in a positive situation asserts a level of confidence and reduces anxiety to trigger physiological healing processes.

Another set of factors are *internal* and related to the individual. Seeking or receiving medical attention, participating in the therapeutic encounter and undergoing treatment is inherently a substrate for triggering helpful physiological processes by reconfiguring the mind to prerequisite mental states. Thus, the agency lies not only with the treatments but with the patient as well. Therapy benefits from an internal locus of control and self efficacy (Elsenbruch & Enck, 2015). Entering the treatment process requires a shift in motivational state and expectancy toward getting better. Thus, seeking or eliciting help is a type of reward seeking behavior which needs a baseline intent for improvement. This assumption also implies that therapeutic process is not passive and that it requires mental and neurobiological engagement on the patient's behalf (Belcher, Ferre, Martinez, & Colloca, 2017). Inferences are generated and continually checked and updated based on new information that can have a major effect on expectations (Buchel et al., 2014; Schenk, Sprenger, Onat, Colloca, & Buchel, 2017). Inferences and conditioned responses are also updated based on positive and negative outcomes thus also contributing to an update in expectations.

Psychological factors such as anxiety and cognitive control and personality are factors that can interfere with and preclude the ability to generate the required intent and positive motivational state (Colloca, 2014a). Another important determining factor is prior experience with a particular medical system which can be affected due to medical history replete with treatment failures (Thomson & Jacques, 2009). For instance, sufferers of chronic condition may be disenfranchised from modern medicine and may turn toward alternative treatments as a recourse. An extreme recourse resulting from disenchantment from medical system coupled with a desire for relief can lead to eliciting and abusing prescribed or drugs such as opioids. The worst-case outcomes from these situations are addictions and overdose and a maladaptive mental cueing is potentially a major cause behind treatment failures and soliciting higher doses, multiple treatments and non-standard drugs. Chronic pain sufferers that are referred to tertiary pain clinics after failure of standard pain treatments benefit from undergoing comprehensive treatment approaches such as behavioral therapy and movement based treatments perhaps because they have not built negative expectations toward alternative approaches.

A complimentary aspect to actual experience is prior knowledge (Buchel et al., 2014; Colloca & Benedetti, 2006). Clinical information on drugs and diagnosis is now widely accessible on the internet and can play a major role in generating expectations. Patients have easy access to knowledge on disease and treatments which they can access on their own through media such as electronic information, news and books (Koban & Wager, 2016; Schmitz et al., 2017). Another resource for setting prior expectations is through social exchange like conversation, and also through social observation of other people engaging and showing overt signs of improvement (Egorova et al., 2015; Schenk, Krimmel, & Colloca, 2017). This aspect is in particular bound to vary between individuals since it is derived from our sociocultural world view and is subject to be different between ethnic groups and geographic distances.

Although there are many sources for positive mental cueing, and regardless of if a single type of cue is of specific importance or has the strongest role in triggering treatment response, how these factors will affect an individual will vary between individuals and will be contingent on patient history (Thomson & Jacques, 2009), prior experiences (Colloca & Benedetti, 2006), emotional states (Corsi & Colloca, 2017), personality traits (Corsi & Colloca, 2017), and brain connectivity (Hashmi, Baria, et al., 2012; Hashmi et al., 2014).

4. TELEOLOGICAL REASONS FOR MENTAL CUEING

Mental cueing that occurs during treatment and the responses generated by this phenomenon has deep teleological origins and are shaped by evolution of cultures. The cultural traditions of different ethnic and geographic groups may be distinct but the common aspect is the adaptation of the mind to derive meaning from symbols and icons. Therapy and treatment has taken on different types of symbolism that signify a force to ward off evil and initiate healing. Hence, mental cueing has an inherent adaptive value.

The use of iconic symbols and our neurobiological capacity to induce relief from such symbolism has deep roots in ancient culture (Benedetti, 2014b; Kaptchuk, 2011). Regardless of the type of treatment, the element of symbol-rich rituals was a key part of healing diseases. Ailments were considered as associated with bad spirits and the medicines and rituals were considered to be laden with magical properties. In addition to the symbolism in artifacts, the healer or shaman was positioned as the learned man with secret knowledge that he used for curing diseases. Symbolism, story-telling, and mythological references are used in many types of old healing traditions and such elements can have a stronger impact at many levels of information processing in the brain, including those levels that are not directly accessed during conscious processing.

It has been posited that we encounter symbolic external cues during the treatment process and the symbolism varies between cultures and ages, but these symbols are unified in that they signify relief or improvement in the medical condition (Colloca & Miller, 2011). The image of a nurse wearing a uniform or the doctor in a white coat are iconic symbols that are loaded with implicit meaning of relief from disease and resource for overcoming physical ailments. As such, the symbolic meaning is much greater than mere expectancy cues, and in many situations, the mental cueing is stronger than a mere expectation and the intensity of this effect is based on stronger priors represented by beliefs and faith. Expectations, especially those that are inferred on the short term are subject to re-evaluation, but some expectations, inferences and beliefs are based on long term learning. These beliefs normally go unquestioned because for practical purposes, our reality is immersed in the milieu of conceptual knowledge of how things occur. This knowledge is reinforced into common belies through our shared societal constructs and interpersonal subjectivity.

The topic of how treatment outcomes are influenced by unconscious mechanisms is explored in psychology, psychoanalysis and neuroscience but very little is known about how these mechanisms are connected to the deeper aspects of symbolism and faith. The belief in alternative medicine and knowledge ascribed to these traditions, as long as it is accepted as veridical by the subject can hold significant sway on inferential activity at the mental level. Thus, the persistence of faith based healing methods and some forms of alternative treatments is perhaps an indicator of their ability for mental cueing, albeit in some but not all people. This could explain why alternative treatments such as acupuncture, homeopathy and reiki hold a surprising significance with a large number of people and are widely popular (Bushnell, Ceko, & Low, 2013). Regardless of whether the explanation of mechanisms behind them do not match with modern science, their efficacy can be meaningful in a significant number of people and the outcomes may not be a direct result of the purported mechanisms but due to their potency in mentally cueing the suggestible individual. This suggestibility is rooted in culture. Old traditions and historical roots are entrenched in the local social milieu and can be effective at least in part because of the strong psychological impact they can have on a person who is interested more in relief and less in critiquing the source of treatment. These treatments serve as potent mental cues for initiating endogenous relief but the outcome albeit will be heavily contingent upon the credulity of the cues, the cultural and social belief systems and impressionability and suggestibility of the subject.

Our scientific knowledge of *disease* ontology and medicinal cures is relatively recent compared to the older culture of shamans and ritualistic healing. Instead of being sourced from superstition and faith, the cures are in general derived from physical causes and evidence collected from the scientific method. The dialectical suppositions that are assumed during the scientific process are constantly revised and renewed through the fast pace of technological and scientific progress. Our assumptions and inferences about the medical system ought not to be based on faith, and should instead be based on belief in the scientific process. This assumption derives from the fact that modern medicine should not be seen as a faith-based and more as a direct derivative of the scientific process. This especially holds credence for how we view pharmacological treatments that have succeeded the gold standard RCT challenge. The medical experts are not faith based healers, but are individuals trained in specialized scientific knowledge of practical and fact-based cures and are less rich in the social, interactive and symbolic aspects. This cut and dry method of administering cures is relatively new in relation to ancient healing cultures and hence, how we mentally cue patients in modern medicine is shaping up in new unknown dimensions across the world.

Interestingly, this dichotomy can be cosmetic since many so-called evidence-based treatments were introduced into medicine through a heuristic process and have not undergone gold standard clinical trials. Like classical alternative therapies, modern medicine is also laden with covert symbolism. That being said, icons in modern medicine follow certain common archetypes that are controlled to fall within that expected from a science-oriented and practical medical environment. Even the items used to furnish and decorate an office are standardized to be reasonably consistent and can serve as mental cues useful for generating conditioned responses. However, the effects of the relatively sparse symbolism inherent in modern medicine contrasts sharply to the older mythically rich, stimulating and interactive rituals of our past. The acceptance and belief systems in older systems used strong psychologically potent symbolism. Current symbols are relatively vacuous and responses to these types of cues are more associative and conditioned rather than hermeneutic. Taken together, there is a common historical theme that is delimited from the type of therapy and involves expectations and inferences that are derived from belief and faith in the medical tradition or system.

In contrast to modern medicine, many alternative therapies such acupuncture and yoga have a strong symbolism, ritualistic and interactive dimension (Nguyen, Davis, Kaptchuk, & Phillips, 2011). This can create a real dichotomy in the types of mental cues transmitted and how these cues are interpreted and will depend on the patient's biases or beliefs regarding science-based vs. tradition based therapies. Whether alternative treatments are effective independent of mental cueing effect, or if the mental cueing effect of alternative therapies is stronger than allopathic methods, is a debate beyond the scope of this discussion. But, the marked placebo responses observed in RCTs are a strong indicator that even a psychologically sparse and less interactive methods of therapy such as those used during clinical trials do have strong mental cueing impact. It can also be assumed that any differences in the mental cueing outcomes of these two types of medical approaches to medicine may be linked with the type of therapy to a lesser extent and more closely to the individual's characteristics such as past inferences on alternative therapy and openness to alternatives.

But regardless of the type of treatment, the above discussion underscores the fact that the more invasive, social, ritualistic, theatrical but convincing the treatment process, the stronger the mental cueing effect and stronger will be the treatment response. The veridical aspects of an effective treatment will also play a role in establishing the treatment outcome and the repeated success of a treatment will enhance the mental cueing effect. In this way, the active treatment and the mental cueing are integrally linked.

5. MENTAL CUEING RESPONSES RESULT FROM PREDICTIVE PROCESSES GENERATED BY REWARD AND AVERSIVE LEARNING

Placebo effects demonstrably shifts mental states, altering the patterns of brain activity and changes expectations that are conducive for generating strong therapeutic responses to both active pain analgesics and placebos (Colloca et al., 2008, 2013). Conversely, an inability to build positive expectations or appropriate mental states (i.e., the right "mind set") before starting treatment negatively affects treatment outcomes (Hashmi, Baliki, et al., 2012; Hashmi, Baria, et al., 2012; Hashmi et al., 2014). This is particularly true for analgesics because pain is a highly subjective percept and particularly malleable to context (Schafer et al., 2018). In general, placebo effect demonstrates our neurobiological capacity to translate mental cues, prior experience, and our conceptual beliefs in treatments into pain relief. These elements also direct treatment efficacy. The ability to generate prerequisite mental states from mental cues combined with conceptual inferences not only facilitate the treatment process, but a lack of these elements can also reduce and sometimes even completely block the efficacy of a drug. The impact of this phenomenon on pain responses is non-trivial and the level to which this mental-cueing effect can alter pain responses is putatively a demonstration of embedded evolutionary psychology and the neurobiological wiring of our expectation, inference and learning systems in the brain.

6. ROLE OF PREDICTIVE PROCESSING IN MENTAL CUEING EFFECTS

Complex and interlinked factors determine whether mental cues will be effective in triggering the prerequisite activity in neurobiological systems to induce positive responses. The factors can be external and internal and the juncture where these two classes of factors juxtapose is the nervous system. External cues received by the brain are processed within the context of prior mental state (Buchel et al., 2014). The concept of prior states and predictive processing is a useful model for understanding the placebo phenomenon. Perceptual processes are guided by prior information that creates thoughts and mental states that contain predictions, inferences and beliefs about future events (Atlas, Wielgosz, Whittington, & Wager, 2014; Blasini, Corsi, Klinger, & Colloca, 2017; Grahl et al., 2018; Koban, Kusko, & Wager, 2018; Kong et al., 2013; Koyama, McHaffie, Laurienti, & Coghill, 2005). Our brain uses predictions because they confer the ability to expect contingencies, anticipate what is ahead, and prepare for events. The concept of priors can be illustrated from relevant examples from semiotics and language studies, which describe that the brain uses cues and context as symbols to interpret new information based on prior knowledge in a goal directed manner (Pennartz, 2018; Tamir & Thornton, 2018). Hence, expectations reflect the capacity to which learning and past experience have prepared the brain to interpret and understand how a stimulus should be perceived based on context and adaptive needs. A similar phenomenon underlies placebo effect where the process of receiving treatment is inherently linked with positive or negative expectations. The strong inferential link between seeing a doctor and relief from ailment is implied by social observation and confirmed through experience. Hence the fact that we see systematic changes in pain to placebos in experiments and in clinical trials is not entirely surprising.

Expectations, prior knowledge, conditioning, motivation, and a healthy emotional state are the main ingredients mobilized by mental cues during the treatment process (Schafer et al., 2018; Weimer, Colloca, & Enck, 2015). From a neuroscientific perspective, the external cues induce expectations, orienting responses and increase arousal (alerting response) that can act as top-down factors. These prior states act as top-down signals for shaping and modulating pain perception. All of these contexts require mental cueing to elicit prior mental states that can then alter the perceived amount of pain.

Prior information that alters pain perception, referred to as top-down information, can take on many forms. It can be an explicit expectation that is conscious thought (inner talk), or it can be unconscious reflexive conditioned response. The intensity of pain is demonstrably a malleable and compliant perceptual system (Bushnell et al., 2013; Davis, Kucyi, & Moayedi, 2015; Hashmi & Davis, 2014) which is controlled by top-down expectations, i.e., we feel less pain when we expect less pain. Since the discovery of opioid systems in the brain, it has been known that the brain can adaptively adjust the intensity of pain based on the context and survival needs of an animal. This adjustment, or "modulation," of pain intensity by the brain has been observed systematically by testing effects of attention, expectation, conditioned pain modulation, and placebo responses (Atlas & Wager, 2012; Atlas et al., 2014; Kirsch et al., 2014; Kong & Benedetti, 2014). Distraction from a consciously attended pain stimulus, such as diverting attention away from the injection site during vaccination, has been quantified to reduce perceived pain intensity. The intensity of pain has been shown to be malleable and compliant to top-down expectations, i.e., we feel less pain when we expect less pain. Taken together, there is indubitable evidence that context can alter the perceptual reality of pain.

7. NEUROBIOLOGICAL CIRCUITRY FOR ADJUSTING PAIN BASED ON PRIOR PREDICTIONS

The ability to adjust pain based on context is linked directly with how pain pathways are organized in the nervous system. At a physiological level, the fact that "top-down" factors modulate pain was demonstrated by Sherrington who showed that reflex withdrawal from painful stimuli is more pronounced when the animal's brain is transected from the body (Woodworth & Sherrington, 1904). It was later shown in a series of studies, that analgesia could be produced by electrically stimulating areas such as the periaqueductal gray area (PAG) in the mid brain and in the reticular formation (RVM) in the pons (Fields & Basbaum, 1978). Subsequent studies identified circuits that bidirectionally connect the PAG, RVM, and sensory relays to processing units of the spinal dorsal horn including lamina 1 and lamina 5 (Eippert et al., 2009; Koutsikou, Parry, MacMillan, & Lumb, 2007; Waters & Lumb, 2008; Yelle, Oshiro, Kraft, & Coghill, 2009). These pathways were named the descending pain modulation systems. Moreover, connections of the PAG and RVM were also mapped to connect with subcortical structures in the brain, such as the amygdala and hypothalamus

(Paredes, Winters, Schneiderman, & McCabe, 2000; Pittman, Blume, & Renaud, 1981), and to cortical structures such as the anterior cingulate, and to orbital and lateral parts of the prefrontal cortex (An, Bandler, Ongur, & Price, 1998; Floyd, Price, Ferry, Keay, & Bandler, 2000; Neafsey, Hurley-Gius, & Arvanitis, 1986). The sub-cortical structures, especially those in the brain stem that are known for their role in pain processing and modulation, are bi-directionally connected to the spinal dorsal horn. In this way, the cortical and sub-cortical systems can exert their influence on nociceptive signals at early or at later stages of neural processing. These hierarchically organized interconnected structures can thus process and adjust nociceptive signals to generate adaptive responses to pain at the motor, autonomic and behavioral level. The activity in these pathways were shown to be fine-tuned by multiple neurotransmitters such as opioids, norepinephrine, dopamine and serotonin. Although the exact processes behind pain modulation are unclear, there is sufficient data demonstrating bidirectional circuits and pathways that connect higher order regions in the cerebral cortex that process contextual sensory information received from different modalities, to sub-cortical brain structures that process memory, associations, salience, reward and value. Animal electrophysiology and neuroimaging with fMRI have mapped activity in several structures such as the anterior cingulate, orbitofrontal cortex, insula, PAG/RVM lateral prefrontal cortex as important regions for pain modulation (Atlas & Wager, 2012; Colloca et al., 2013; Schafer et al., 2018).

Several studies have shown that positive expectations for pain relief result in greater benefit from analgesic medication (Colloca et al., 2013). Expectation is relatively well studied in the field of perception and as a concept, it permits deductive inquiries for understanding mechanisms of mental cueing at the psychophysical and neural level.

In experimental studies, we can quantify the extent to which expectations modify pain perception, and the change in pain reported is referred to as pain modulation. Expectations are generated in these studies by presenting cues as a context. The cues can be symbols, words or images of other people, or verbal suggestions, and are presented paired with a painful stimulus. Cues contain the prior information on the nature or intensity of the pain stimulus and are useful for building expectations toward incoming stimuli. Another technique for testing expectation induced pain modulation is to use cues for building expectations toward a treatment and quantifying pain reports after treatment with and without the expectation inducing cues. When expectations are built toward treatment, the models are called placebo analgesia models (Colagiuri, Schenk, Kessler, Dorsey, & Colloca, 2015).

Other models build expectations directly toward pain and are called expectancy models (Atlas, Bolger, Lindquist, & Wager, 2010; Kong et al., 2013; Koyama et al., 2005).

These types of models demonstrate show that expectations can significantly modulate pain, and the expectations can be induced to be positive or negative to either reduce or enhance pain thus mimicking placebo or nocebo responses respectively. But the size of the effect is often small and quickly extinguished. A relatively stronger and longer lasting effect of expectation on pain is observed when the expectations induced by the cues are validated with conditioning procedures (Babel et al., 2017; Grahl et al., 2018). For instance, in one type of model, after giving expectancy cues and the placebo treatment, expectations are validated by applying conditioning stimuli where the pain evoking stimulus is surreptitiously reduced in intensity and the weaker stimuli are inferred as a treatment effect (Hashmi et al., 2014). The conditioning stimulus is repeated to validate and reinforce the expectation. As a result of this process, when the expectation is tested, the stimulus is perceived as less painful when the stimulus intensity is increased.

Note that in most studies, expectations are produced with explicit cues, and the physical reinforcement validates the explicit verbal suggestion (Hashmi et al., 2014; Koban et al., 2018). On the other hand, some types of top-down effects on pain are encoded implicitly and the associations are not explicitly declared hence the expectations are formed but may or may are not be fully consciously known by the subject (Babel et al., 2017; Egorova et al., 2015; Jepma & Wager, 2015; Schafer et al., 2018). As is observed with procedural motor memories that are encoded at the sub-cortical level and occur outside of conscious perception, prior learning of associations of pain with other objects also contributes to top-down effects on pain. For instance, rats conditioned with morphine analgesia within a specific environmental context, show reduced pain response to saline injections administered within the same but not in a different environmental context (Everitt, Morris, O'Brien, & Robbins, 1991; Navratilova, Atcherley, & Porreca, 2015; Porreca & Navratilova, 2017). These learning effects may be subliminally triggered through unconsciously perceived external cues as was observed recently in a few studies done in humans (Egorova et al., 2015; Jensen et al., 2012). Illustrating this effect, a recent study presented cues that created expectations of less heat evoked pain which effectively reduce perceived pain; however, these positive cues were not presented alone but instead they were juxtaposed on an object that was subliminally perceivable in the background of the explicit cue (Jensen et al., 2015). These subliminal cues were effective for reducing perceived pain independently of the explicit cue albeit the effect was less pronounced relative to the explicit cues. Whether the participants created conscious and explicit expectations toward the background cues is unclear, but this underscores a putative top-down effect on pain that is offline or unconscious.

Based on vast evidence of procedural learning we know that prior experience and explicit/implicit learning results in adoption of procedures and responses in the form of habits. This offline processing of environmental and contextual information is useful for conserving brain resources and allows for complex mental functions such as learning and multitasking. As observed with habitual behaviors, offline associations control our behavior in the background: the expectations are not explicit and are instead more similar to unconscious motivations (Navratilova et al., 2015; Porreca & Navratilova, 2017). Although speculative, it is plausible that top-down expectation and beliefs guide our motivations for mental cueing and will adjust our pain responses to external and internal stimuli either with or without explicit expectations. Thus, pain modulation is another demonstration that the brain extracts situationally relevant meaning, engenders conscious or implicit expectations to predict subsequent events from cues in the surrounding and also adapts the perceived sensation to fit with the prediction and the current motivational state of the organism.

8. UTILITY OF STUDYING MENTAL CUEING AND POTENTIAL FOR ADAPTING THIS KNOWLEDGE FOR IMPROVING CLINICAL OUTCOMES

It is recognized that endogenous pain relief systems such as opioid circuitry contribute to placebo effect and several PET and fMRI have been consistently highlighting these brain circuits through consistent and reproducible findings (Colagiuri et al., 2015). But the role of this circuitry in treatment outcomes needs wider acknowledgement to be clinically useful. The new conceptual models for understanding these phenomenon and new imaging and analysis techniques are quickly revolutionizing our ability to understand the underlying brain mechanisms (Grahl et al., 2018; Roy, Shohamy, & Wager, 2012; Schenk, Sprenger, et al., 2017). In coming years, these efforts we will be able to use brain and behavioral data to predict the individual's capacity for mental cueing. Blue-sky research goals to deploy and enhance these intrinsic responses are quickly surging

(Belcher et al., 2017; Enck et al., 2013; Vase et al., 2015). The exact mechanisms that contribute to placebo effect remain to be properly mapped out, but with these advances, the mechanisms are being elucidated rapidly. An important implication of these new approaches is the possibility that the mental cueing phenomenon will no longer be seen as non-specific cryptic response relevant to clinical trials and devoid of clinical value. Instead, the brain circuits and neurotransmitter systems behind these responses are compelling us to think of the significance of physiological systems as a powerful resource for generating therapeutic responses.

9. CONCLUSION

The context of receiving a pain treatment alters pain perception and this effect is widely known because it has been systematically witnessed and the mechanisms have been linked to learning, motivational circuitry and descending pain inhibition. The placebo effect is an important indicator that analgesic responses to treatment are mediated by the active ingredient in part, and endogenous systems that reduce pain participate in shaping the extent to which a treatment will be effective in reducing pain.

Taken together, I have discussed that pain being a subjective and intransitive percept is particularly subject to reconfigurations of mental states during the process of eliciting and receiving treatment. As such, prior information in the form of inferences, beliefs, memories and conditioned responses are triggered by mental cues received during treatment and have an internal or external locus. These processes can occur on receiving placebos, but more importantly, they can also effect real treatments that have active effects. Thus by adopting a less restrictive terminology and thinking of this phenomenon as a mental cueing response that effects all types of treatments, we can shift direction toward thinking about the bodies intrinsic mechanisms that contribute to treatment outcome.

REFERENCES

An, X., Bandler, R., Ongur, D., & Price, J. L. (1998). Prefrontal cortical projections to longitudinal columns in the midbrain periaqueductal gray in macaque monkeys. *The Journal of Comparative Neurology*, *401*(4), 455–479.

Atlas, L. Y., Bolger, N., Lindquist, M. A., & Wager, T. D. (2010). Brain mediators of predictive cue effects on perceived pain. *The Journal of Neuroscience*, *30*(39), 12964–12977. https://doi.org/10.1523/JNEUROSCI.0057-10.2010.

Atlas, L. Y., & Wager, T. D. (2012). How expectations shape pain. *Neuroscience Letters*, *520*(2), 140–148. https://doi.org/10.1016/j.neulet.2012.03.039.

Atlas, L. Y., Whittington, R. A., Lindquist, M. A., Wielgosz, J., Sonty, N., & Wager, T. D. (2012). Dissociable influences of opiates and expectations on pain. *The Journal of Neuroscience, 32*(23), 8053–8064. https://doi.org/10.1523/JNEUROSCI.0383-12.2012.

Atlas, L. Y., Wielgosz, J., Whittington, R. A., & Wager, T. D. (2014). Specifying the non-specific factors underlying opioid analgesia: Expectancy, attention, and affect. *Psychopharmacology, 231*, 813–823. https://doi.org/10.1007/s00213-013-3296-1.

Babel, P., Bajcar, E. A., Adamczyk, W., Kicman, P., Lisinska, N., Swider, K., et al. (2017). Classical conditioning without verbal suggestions elicits placebo analgesia and nocebo hyperalgesia. *PLoS One, 12*(7), e0181856. https://doi.org/10.1371/journal.pone.0181856.

Beecher, H. K. (1955). The powerful placebo. *Journal of the American Medical Association, 159*(17), 1602–1606.

Belcher, A. M., Ferre, S., Martinez, P. E., & Colloca, L. (2017). Role of placebo effects in pain and neuropsychiatric disorders. *Progress in Neuro-Psychopharmacology & Biological Psychiatry*. https://doi.org/10.1016/j.pnpbp.2017.06.003.

Benedetti, F. (2014a). Drugs and placebos: What's the difference?: Understanding the molecular basis of the placebo effect could help clinicians to better use it in clinical practice. *EMBO Reports, 15*(4), 329–332. https://doi.org/10.1002/embr.201338399.

Benedetti, F. (2014b). Placebo effects: From the neurobiological paradigm to translational implications. *Neuron, 84*(3), 623–637. https://doi.org/10.1016/j.neuron.2014.10.023.

Benedetti, F., Carlino, E., & Piedimonte, A. (2016). Increasing uncertainty in CNS clinical trials: The role of placebo, nocebo, and Hawthorne effects. *Lancet Neurology, 15*(7), 736–747. https://doi.org/10.1016/S1474-4422(16)00066-1.

Benedetti, F., Mayberg, H. S., Wager, T. D., Stohler, C. S., & Zubieta, J. K. (2005). Neurobiological mechanisms of the placebo effect. *The Journal of Neuroscience, 25*(45), 10390–10402. https://doi.org/10.1523/JNEUROSCI.3458-05.2005.

Blasini, M., Corsi, N., Klinger, R., & Colloca, L. (2017). Nocebo and pain: An overview of the psychoneurobiological mechanisms. *Pain Reports, 2*(2). https://doi.org/10.1097/PR9.0000000000000585.

Buchel, C., Geuter, S., Sprenger, C., & Eippert, F. (2014). Placebo analgesia: A predictive coding perspective. *Neuron, 81*(6), 1223–1239. https://doi.org/10.1016/j.neuron.2014.02.042.

Bushnell, M. C., Ceko, M., & Low, L. A. (2013). Cognitive and emotional control of pain and its disruption in chronic pain. *Nature Reviews. Neuroscience, 14*(7), 502–511. https://doi.org/10.1038/nrn3516.

Colagiuri, B., Schenk, L. A., Kessler, M. D., Dorsey, S. G., & Colloca, L. (2015). The placebo effect: From concepts to genes. *Neuroscience, 307*, 171–190. https://doi.org/10.1016/j.neuroscience.2015.08.017.

Colloca, L. (2014a). Emotional modulation of placebo analgesia. *Pain, 155*(4), 651. https://doi.org/10.1016/j.pain.2014.01.009.

Colloca, L. (2014b). Placebo, nocebo, and learning mechanisms. *Handbook of Experimental Pharmacology, 225*, 17–35. https://doi.org/10.1007/978-3-662-44519-8_2.

Colloca, L. (2017). Nocebo effects can make you feel pain. *Science, 358*(6359), 44. https://doi.org/10.1126/science.aap8488.

Colloca, L., & Benedetti, F. (2006). How prior experience shapes placebo analgesia. *Pain, 124*(1–2), 126–133.

Colloca, L., Benedetti, F., & Porro, C. A. (2008). Experimental designs and brain mapping approaches for studying the placebo analgesic effect. *European Journal of Applied Physiology, 102*(4), 371–380.

Colloca, L., Enck, P., & DeGrazia, D. (2016). Relieving pain using dose-extending placebos: A scoping review. *Pain, 157*(8), 1590–1598. https://doi.org/10.1097/j.pain.0000000000000566.

Colloca, L., Jonas, W. B., Killen, J., Miller, F. G., & Shurtleff, D. (2014). Reevaluating the placebo effect in medical practice. *Zeitschrift für Psychologie, 222*(3), 124–127. https://doi.org/10.1027/2151-2604/a000177.

Colloca, L., Klinger, R., Flor, H., & Bingel, U. (2013). Placebo analgesia: Psychological and neurobiological mechanisms. *Pain, 154*(4), 511–514. https://doi.org/10.1016/j.pain.2013.02.002.

Colloca, L., & Miller, F. G. (2011). How placebo responses are formed: A learning perspective. *Philosophical Transactions of the Royal Society of London. Series B, Biological Sciences, 366*(1572), 1859–1869. https://doi.org/10.1098/rstb.2010.0398.

Corsi, N., & Colloca, L. (2017). Placebo and nocebo effects: The advantage of measuring expectations and psychological factors. *Frontiers in Psychology, 8*, 308. https://doi.org/10.3389/fpsyg.2017.00308.

Davis, K. D., Kucyi, A., & Moayedi, M. (2015). The pain switch: An "ouch" detector. *Pain, 156*(11), 2164–2166. https://doi.org/10.1097/j.pain.0000000000000303.

Egorova, N., Park, J., Orr, S. P., Kirsch, I., Gollub, R. L., & Kong, J. (2015). Not seeing or feeling is still believing: Conscious and non-conscious pain modulation after direct and observational learning. *Scientific Reports, 5*, 16809. https://doi.org/10.1038/srep16809.

Eippert, F., Bingel, U., Schoell, E. D., Yacubian, J., Klinger, R., Lorenz, J., et al. (2009). Activation of the opioidergic descending pain control system underlies placebo analgesia. *Neuron, 63*(4), 533–543.

Elsenbruch, S., & Enck, P. (2015). Placebo effects and their determinants in gastrointestinal disorders. *Nature Reviews. Gastroenterology & Hepatology, 12*(8), 472–485. https://doi.org/10.1038/nrgastro.2015.117.

Enck, P., Benedetti, F., & Schedlowski, M. (2008). New insights into the placebo and nocebo responses. *Neuron, 59*(2), 195–206.

Enck, P., Bingel, U., Schedlowski, M., & Rief, W. (2013). The placebo response in medicine: Minimize, maximize or personalize? *Nature Reviews. Drug Discovery, 12*(3), 191–204. https://doi.org/10.1038/nrd3923.

Enck, P., & Klosterhalfen, S. (2013). The placebo response in clinical trials-the current state of play. *Complementary Therapies in Medicine, 21*(2), 98–101. https://doi.org/10.1016/j.ctim.2012.12.010.

Everitt, B. J., Morris, K. A., O'Brien, A., & Robbins, T. W. (1991). The basolateral amygdala-ventral striatal system and conditioned place preference: Further evidence of limbic-striatal interactions underlying reward-related processes. *Neuroscience, 42*(1), 1–18.

Feys, F., Bekkering, G. E., Singh, K., & Devroey, D. (2014). Do randomized clinical trials with inadequate blinding report enhanced placebo effects for intervention groups and nocebo effects for placebo groups? *Systematic Reviews, 3*, 14. https://doi.org/10.1186/2046-4053-3-14.

Fields, H. L., & Basbaum, A. I. (1978). Brainstem control of spinal pain-transmission neurons. *Annual Review of Physiology, 40*, 217–248. https://doi.org/10.1146/annurev.ph.40.030178.001245.

Floyd, N. S., Price, J. L., Ferry, A. T., Keay, K. A., & Bandler, R. (2000). Orbitomedial prefrontal cortical projections to distinct longitudinal columns of the periaqueductal gray in the rat. *The Journal of Comparative Neurology, 422*(4), 556–578.

Grahl, A., Onat, S., & Buchel, C. (2018). The periaqueductal gray and Bayesian integration in placebo analgesia. *eLife, 7*. https://doi.org/10.7554/eLife.32930.

Hashmi, J., Baliki, M., Chanda, M., Huang, L., Parks, E., Schnitzer, T., et al. (2012). Lidocaine patch (5%) is no more potent than placebo in treating chronic back pain when tested in a randomised double blind placebo controlled brain imaging study. *Molecular Pain, 8*(1), 29.

Hashmi, J. A., Baria, A. T., Baliki, M. N., Huang, L., Schnitzer, T. J., & Apkarian, A. V. (2012). Brain networks predicting placebo analgesia in a clinical trial for chronic back pain. *Pain, 153*(12), 2393–2402. https://doi.org/10.1016/j.pain.2012.08.008.

Hashmi, J. A., & Davis, K. D. (2014). Deconstructing sex differences in pain sensitivity. *Pain, 155*(1), 10–13. https://doi.org/10.1016/j.pain.2013.07.039.

Hashmi, J. A., Kong, J., Spaeth, R., Khan, S., Kaptchuk, T. J., & Gollub, R. L. (2014). Functional network architecture predicts psychologically mediated analgesia related to treatment in chronic knee pain patients. *The Journal of Neuroscience, 34*(11), 3924–3936. https://doi.org/10.1523/JNEUROSCI.3155-13.2014.

Hrobjartsson, A., Kaptchuk, T. J., & Miller, F. G. (2011). Placebo effect studies are susceptible to response bias and to other types of biases. *Journal of Clinical Epidemiology, 64*(11), 1223–1229. https://doi.org/10.1016/j.jclinepi.2011.01.008.

Jensen, K. B., Kaptchuk, T. J., Chen, X., Kirsch, I., Ingvar, M., Gollub, R. L., et al. (2015). A neural mechanism for nonconscious activation of conditioned placebo and nocebo responses. *Cerebral Cortex, 25*(10), 3903–3910. https://doi.org/10.1093/cercor/bhu275.

Jensen, K. B., Kaptchuk, T. J., Kirsch, I., Raicek, J., Lindstrom, K. M., Berna, C., et al. (2012). Nonconscious activation of placebo and nocebo pain responses. *Proceedings of the National Academy of Sciences of the United States of America, 109*(39), 15959–15964. https://doi.org/10.1073/pnas.1202056109.

Jensen, K. B., Petrovic, P., Kerr, C. E., Kirsch, I., Raicek, J., Cheetham, A., et al. (2014). Sharing pain and relief: Neural correlates of physicians during treatment of patients. *Molecular Psychiatry, 19*, 392–398. https://doi.org/10.1038/mp.2012.195.

Jepma, M., & Wager, T. D. (2015). Conceptual conditioning: Mechanisms mediating conditioning effects on pain. *Psychological Science, 26*(11), 1728–1739. https://doi.org/10.1177/0956797615597658.

Jonas, W. B., Crawford, C., Colloca, L., Kaptchuk, T. J., Moseley, B., Miller, F. G., et al. (2015). To what extent are surgery and invasive procedures effective beyond a placebo response? A systematic review with meta-analysis of randomised, sham controlled trials. *BMJ Open, 5*(12), e009655. https://doi.org/10.1136/bmjopen-2015-009655.

Kaptchuk, T. J. (2011). Placebo studies and ritual theory: A comparative analysis of Navajo, acupuncture and biomedical healing. *Philosophical Transactions of the Royal Society of London. Series B, Biological Sciences, 366*(1572), 1849–1858. https://doi.org/10.1098/rstb.2010.0385.

Kaptchuk, T. J., Stason, W. B., Davis, R. B., Legedza, A. R., Schnyer, R. N., Kerr, C. E., et al. (2006). Sham device v inert pill: Randomised controlled trial of two placebo treatments. *BMJ, 332*(7538), 391–397.

Kirsch, I. (2014). The emperor's new drugs: Medication and placebo in the treatment of depression. *Handbook of Experimental Pharmacology, 225*, 291–303. https://doi.org/10.1007/978-3-662-44519-8_16.

Kirsch, I., Kong, J., Sadler, P., Spaeth, R., Cook, A., Kaptchuk, T., et al. (2014). Expectancy and conditioning in placebo analgesia: Separate or connected processes. *Psychology of Consciousness (Washington, D.C.), 1*(1), 51–59. https://doi.org/10.1037/cns0000007.

Klinger, R., Blasini, M., Schmitz, J., & Colloca, L. (2017). Nocebo effects in clinical studies: Hints for pain therapy. *Pain Reports, 2*(2). https://doi.org/10.1097/PR9.0000000000000586.

Klinger, R., Colloca, L., Bingel, U., & Flor, H. (2014). Placebo analgesia: Clinical applications. *Pain, 155*(6), 1055–1058. https://doi.org/10.1016/j.pain.2013.12.007.

Koban, L., Kusko, D., & Wager, T. D. (2018). Generalization of learned pain modulation depends on explicit learning. *Acta Psychologica, 184*, 75–84. https://doi.org/10.1016/j.actpsy.2017.09.009.

Koban, L., & Wager, T. D. (2016). Beyond conformity: Social influences on pain reports and physiology. *Emotion, 16*(1), 24–32. https://doi.org/10.1037/emo0000087.

Kong, J., & Benedetti, F. (2014). Placebo and nocebo effects: An introduction to psychological and biological mechanisms. *Handbook of Experimental Pharmacology, 225*, 3–15. https://doi.org/10.1007/978-3-662-44519-8_1.

Kong, J., Jensen, K., Loiotile, R., Cheetham, A., Wey, H. Y., Tan, Y., et al. (2013). Functional connectivity of the frontoparietal network predicts cognitive modulation of pain. *Pain, 154*(3), 459–467. https://doi.org/10.1016/j.pain.2012.12.004.

Koutsikou, S., Parry, D. M., MacMillan, F. M., & Lumb, B. M. (2007). Laminar organization of spinal dorsal horn neurones activated by C- vs. A-heat nociceptors and their descending control from the periaqueductal grey in the rat. *The European Journal of Neuroscience, 26*(4), 943–952. https://doi.org/10.1111/j.1460-9568.2007.05716.x.

Koyama, T., McHaffie, J. G., Laurienti, P. J., & Coghill, R. C. (2005). The subjective experience of pain: Where expectations become reality. *Proceedings of the National Academy of Sciences of the United States of America, 102*(36), 12950–12955.

Navratilova, E., Atcherley, C. W., & Porreca, F. (2015). Brain circuits encoding reward from pain relief. *Trends in Neurosciences, 38*(11), 741–750. https://doi.org/10.1016/j.tins.2015.09.003.

Neafsey, E. J., Hurley-Gius, K. M., & Arvanitis, D. (1986). The topographical organization of neurons in the rat medial frontal, insular and olfactory cortex projecting to the solitary nucleus, olfactory bulb, periaqueductal gray and superior colliculus. *Brain Research, 377*(2), 261–270.

Nguyen, L. T., Davis, R. B., Kaptchuk, T. J., & Phillips, R. S. (2011). Use of complementary and alternative medicine and self-rated health status: Results from a national survey. *Journal of General Internal Medicine, 26*(4), 399–404. https://doi.org/10.1007/s11606-010-1542-3.

Paredes, J., Winters, R. W., Schneiderman, N., & McCabe, P. M. (2000). Afferents to the central nucleus of the amygdala and functional subdivisions of the periaqueductal gray: Neuroanatomical substrates for affective behavior. *Brain Research, 887*(1), 157–173.

Pennartz, C. M. A. (2018). Consciousness, representation, action: The importance of being goal-directed. *Trends in Cognitive Sciences, 22*(2), 137–153. https://doi.org/10.1016/j.tics.2017.10.006.

Pittman, Q. J., Blume, H. W., & Renaud, L. P. (1981). Connections of the hypothalamic paraventricular nucleus with the neurohypophysis, median eminence, amygdala, lateral septum and midbrain periaqueductal gray: An electrophysiological study in the rat. *Brain Research, 215*(1–2), 15–28.

Porreca, F., & Navratilova, E. (2017). Reward, motivation, and emotion of pain and its relief. *Pain, 158*(Suppl. 1), S43–S49. https://doi.org/10.1097/j.pain.0000000000000798.

Roy, M., Shohamy, D., & Wager, T. D. (2012). Ventromedial prefrontal-subcortical systems and the generation of affective meaning. *Trends in Cognitive Sciences, 16*(3), 147–156. https://doi.org/10.1016/j.tics.2012.01.005.

Rutherford, B. R., Wall, M. M., Brown, P. J., Choo, T. H., Wager, T. D., Peterson, B. S., et al. (2017). Patient expectancy as a mediator of placebo effects in antidepressant clinical trials. *The American Journal of Psychiatry, 174*(2), 135–142. https://doi.org/10.1176/appi.ajp.2016.16020225.

Schafer, S. M., Geuter, S., & Wager, T. D. (2018). Mechanisms of placebo analgesia: A dual-process model informed by insights from cross-species comparisons. *Progress in Neurobiology, 160*, 101–122. https://doi.org/10.1016/j.pneurobio.2017.10.008.

Schenk, L. A., Krimmel, S. R., & Colloca, L. (2017). Observe to get pain relief: Current evidence and potential mechanisms of socially learned pain modulation. *Pain, 158*(11), 2077–2081. https://doi.org/10.1097/j.pain.0000000000000943.

Schenk, L. A., Sprenger, C., Onat, S., Colloca, L., & Buchel, C. (2017). Suppression of striatal prediction errors by the prefrontal cortex in placebo Hypoalgesia. *The Journal of Neuroscience, 37*(40), 9715–9723. https://doi.org/10.1523/JNEUROSCI.1101-17.2017.

Schmitz, J., Kamping, S., Wiegratz, J., Muller, M., Stork, J., Colloca, L., et al. (2017). Impact of patient information leaflets on pain medication intake behavior: A pilot study. *Pain Reports, 2*(6), e620. https://doi.org/10.1097/PR9.0000000000000620.

Scott, D. J., Stohler, C. S., Egnatuk, C. M., Wang, H., Koeppe, R. A., & Zubieta, J. K. (2008). Placebo and nocebo effects are defined by opposite opioid and dopaminergic responses. *Archives of General Psychiatry*, *65*(2), 220–231.

Seminowicz, D. A. (2006). Believe in your placebo. *The Journal of Neuroscience*, *26*(17), 4453–4454. https://doi.org/10.1523/JNEUROSCI.0789-06.2006.

Tamir, D. I., & Thornton, M. A. (2018). Modeling the predictive social mind. *Trends in Cognitive Sciences*, *22*(3), 201–212. https://doi.org/10.1016/j.tics.2017.12.005.

Thomson, S., & Jacques, L. (2009). Demographic characteristics of patients with severe neuropathic pain secondary to failed back surgery syndrome. *Pain Practice*, *9*(3), 206–215. https://doi.org/10.1111/j.1533-2500.2009.00276.x.

Vase, L., Vollert, J., Finnerup, N. B., Miao, X., Atkinson, G., Marshall, S., et al. (2015). Predictors of the placebo analgesia response in randomized controlled trials of chronic pain: A meta-analysis of the individual data from nine industrially sponsored trials. *Pain*, *156*(9), 1795–1802. https://doi.org/10.1097/j.pain.0000000000000217.

Waters, A. J., & Lumb, B. M. (2008). Descending control of spinal nociception from the periaqueductal grey distinguishes between neurons with and without C-fibre inputs. *Pain*, *134*(1–2), 32–40. https://doi.org/10.1016/j.pain.2007.03.025.

Weimer, K., Colloca, L., & Enck, P. (2015). Placebo effects in psychiatry: Mediators and moderators. *Lancet Psychiatry*, *2*(3), 246–257. https://doi.org/10.1016/S2215-0366(14)00092-3.

Woodworth, R. S., & Sherrington, C. S. (1904). A pseudaffective reflex and its spinal path. *The Journal of Physiology*, *31*(3–4), 234–243.

Yelle, M. D., Oshiro, Y., Kraft, R. A., & Coghill, R. C. (2009). Temporal filtering of nociceptive information by dynamic activation of endogenous pain modulatory systems. *The Journal of Neuroscience*, *29*(33), 10264–10271.

CHAPTER ELEVEN

Pain Modulation: From Conditioned Pain Modulation to Placebo and Nocebo Effects in Experimental and Clinical Pain

Janie Damien*, Luana Colloca[†,‡], Carmen-Édith Bellei-Rodriguez*, Serge Marchand*[,§,1]

*Research Center of the Centre hospitalier universitaire de Sherbrooke (CHUS), Department of Surgery, Faculty of Medicine and Health Sciences, Université de Sherbrooke, Sherbrooke, QC, Canada
[†]Department of Pain Translational Symptom Science, School of Nursing, University of Maryland, Baltimore, MD, United States
[‡]Departments of Psychiatry and Anesthesiology, School of Medicine, University of Maryland, Baltimore, MD, United States
[§]Fonds de Recherche du Québec—Santé (FRQS), Montréal, QC, Canada
[1]Corresponding author: e-mail address: serge.marchand@usherbrooke.ca

Contents

International Review of Neurobiology, Volume 139
ISSN 0074-7742
https://doi.org/10.1016/bs.irn.2018.07.024

Abstract

Accumulating evidence reveal important applications of endogenous pain modulation assessment in healthy controls and in patients in clinical settings, as dysregulations in the balance of pain modulatory circuits may facilitate pain and promote chronification of pain. This article reviews data on pain modulation, focusing on the mechanisms and translational aspects of pain modulation from conditioned pain modulation (CPM) to placebo and nocebo effects in experimental and clinical pain. The specific roles of expectations, learning, neural and neurophysiological mechanisms of the central nervous system are briefly reviewed herein. The interaction between CPM and placebo systems in pain inhibitory pathways is highly relevant in the clinic and in randomized controlled trials yet remains to be clarified. Examples of clinical implications of CPM and its relationship to placebo and nocebo effects are provided. A greater understanding of the role of pain modulation in various pain states can help characterize the manifestation and development of chronic pain and assist in predicting the response to pain-relieving treatments. Placebo and nocebo effects, intrinsic to every treatment, can be used to develop personalized therapeutic approaches that improve clinical outcomes while limiting unwanted effects.

1. INTRODUCTION

Recently, there have been many new developments in our understanding of the area of the endogenous mechanisms of pain modulation. From nociceptive stimulation to perception, multiple endogenous pain systems modulate our experience of pain. The perception of pain is the result of several endogenous pain inhibitory and facilitatory mechanisms that trigger pain at all levels of the central nervous system. Conditioned pain modulation (CPM) is one unique form of endogenous descending inhibitory pathway (Bannister & Dickenson, 2017; Willer, Le Bars, & De Broucker, 1990). The use of the CPM paradigm in scientific research for evaluating the inherent pain inhibition ability of individuals is highly relevant for understanding normal functioning and various disease states, including the development, persistency and treatment of chronic pain. The cognitive modulation of endogenous pain control mechanisms, illustrated by placebo analgesia, have greatly helped uncover its components and its neurophysiological mechanisms, which have generated enormous implications for personalizing pain management. Every treatment is inherently influenced by placebo and nocebo effects; the unpleasant component of pain as well as the expectation of pain or analgesia following an intervention contribute significantly to clinical outcomes. Aside from being relevant clinically, the study of CPM and placebo effects improved our understanding of the multiple factors and systems that interact and mediate the pain-modulatory effects of expectancies. While CPM and placebo effects are both influenced by expectations to a certain extent, evidence suggests they are not correlated (Skyt et al., 2018) and they implicate different mechanisms. This article aims to provide current and updated work on both mechanisms and translational aspects of endogenous mechanisms of pain modulation, from CPM to placebo and nocebo effects in experimental and clinical pain (also see Marchand, 2012).

2. THEORY OF PAIN MODULATION

Our current multidimensional and circular model of pain recognizes that information about pain, from sensory input to perception, does not travel in a linear way in the nervous system (Marchand, 2012). Nociceptive stimuli, first transformed into chemoelectric energy, travel as a nerve impulse along nerve fibers from the periphery to the spinal cord, from the spinal cord to the brainstem and the thalamus, and finally, from the thalamus to the cortex where the information is interpreted as pain. Along this path, the afferent noxious stimuli are modulated by neurons in the endogenous pain controls. Inhibitory cells, which modulate nociceptive inputs, are recruited by stimulation of non-nociceptive afferents (Aβ-fibers) and inhibited by stimulation of nociceptive afferents (Aδ- and C-fibers). The Gate Control Theory (Melzack & Wall, 1965) stipulates that a peripheral mechanism in the dorsal horns of the spinal cord as well as descending systems from the reticular formation modulate nociceptive information.

As described in Loeser's model, pain is the result of four components: nociception (nociceptive component), pain (sensory-discriminatory component), suffering (motivo-affective component) and pain behaviors (cognitive-behavioral component) (Loeser, 1980). Marchand (2012) highlighted how each component are best represented as a circular rather than linear model, existing independently, but also interacting with each other. Mere activation of nociceptive fibers does not necessarily cause pain. Inversely, phantom limb pain is an example of suffering that does not require nociceptive stimuli. The neuromatrix theory of pain is useful to outline the mechanisms of phantom or chronic pain (Melzack, 1990, 2001) that is produced, not solely by sensory input evoked by injury, inflammation, or pathology and the peripheral nervous system that surrounds it, but rather by the output of a neural network in the brain that is genetically determined and then modified by sensory experience, called the neuromatrix (Melzack, 2001). Placebo analgesia, for instance, could be understood as resulting from a complex set of sociocultural beliefs, expectations and conditioning behaviors that influence the neuromatrix of the brain.

3. ENDOGENOUS MECHANISMS OF PAIN MODULATION

Modulation of nociceptive information in the central nervous system may be excitatory or inhibitory of the nociceptive signal before it reaches the higher centers of the brain, resulting in pain perceived as more or less intense. Inhibition of nociceptive afferents can be best understood under three types of endogenous mechanisms: (1) spinal mechanisms, (2) descending inhibitory controls, and (3) mechanisms of the higher centers (Marchand, 2012). These mechanisms of endogenous pain modulation controls have greatly improved our understanding of normal and dysregulated modulation of pain in healthy individuals and in patients with chronic pain conditions. The conditioned pain modulation and placebo effects, which are the focus of this chapter, result from the activity of descending pain control systems and pain modulatory mechanisms from the higher centers, respectively.

CPM, previously known as "diffuse noxious inhibitory control (DNIC)" or "DNIC-like" effects (Yarnitsky, 2010), was first studied in animals by Le Bars, Dickenson, & Besson (1979a, 1979b). In 2015, experts recommended the term CPM for human application of DNIC protocols (Yarnitsky et al., 2015). A noxious stimulus applied in one part of the body can inhibit pain in another part of the body, by activating the descending inhibitory system (DNIC or CPM). In this sense, CPM refers to any approach based on the application of a localized painful stimulation that produce a diffuse analgesic effect beyond the stimulation period and body location (muscles, joints and viscera) independently of the site of application (diffuse), i.e., "pain inhibits pain" phenomenon or "counter-irritation."

This principle underlies numerous types of stimulation, ranging from acupuncture and deep massages to analgesic hyperstimulation and high intensity Transcutaneous Electrical Nerve Stimulators (TENS) or

TENS-acupuncture. These approaches apply intense painful peripheral stimuli, which recruit nociceptive afferences (Aδ- and C-fibers) and produce diffuse analgesic effects (Marchand, 2012).

4. ASSESSMENT OF ENDOGENOUS CPM

Pain modulation, in particular the inhibitory modulatory mechanisms (generally termed endogenous analgesia), can be investigated in experimental settings using quantitative sensory tests (QSTs) based on the psychophysical paradigm of CPM. Using "pain inhibits pain" models, CPM is tested using a noxious conditioning stimulus (CS) to influence another painful test stimulus (TS), upon which the analgesic effect is tested (Nir & Yarnitsky, 2015; Yarnitsky, 2010). The TS is typically of short duration, milliseconds to seconds, during which the pain response is typically recorded by means of subjective pain ratings or nociceptive withdrawal reflexes (RIII reflex) elicited by painful stimuli to the sural nerve (Jurth, Rehberg, & von Dincklage, 2014) and more recently by changes of cortical evoked brain potentials (Höffken, Özgül, Enax-Krumova, Tegenthoff, & Maier, 2017). The CPM effect correspond to the CS-induced changes in these pain responses. It has been suggested that reflex responses reflect spinal nociception while subjective pain ratings integrate more cognitive influences even though they are both reliable measures (Biurrun Manresa et al., 2014; Jurth et al., 2014).

In addition to being valuable for quantifying individuals' endogenous inhibition of pain, the evaluation of CPM effect in individuals has provided us important insight on the development of pain disorders and the response to analgesic treatment (Yarnitsky, 2015). Thus, CPM testing could help develop personalized pain management strategies based on individuals' endogenous analgesia capacity.

Although our knowledge on pain modulation has improved significantly over the years, the various CPM protocols employed and the lack of normative values limit the comparison of findings (Klyne, Schmid, Moseley, Sterling, & Hodges, 2015) and the utilization of CPM as a prognostic factor in experimental and clinical pain studies (Granovsky, Miller-Barmak, Goldstein, Sprecher, & Yarnitsky, 2016; Kennedy, Kemp, Ridout, Yarnitsky, & Rice, 2016). Further research on CPM methodology and its reliability is required to find a standard measurement of CPM and its main parameters such as the timing, modality, intensity, duration, and location of the stimuli.

5. TEST STIMULUS

The most studied test stimuli are pressure pain threshold and contact heat pain (Kennedy et al., 2016), but thermal, mechanical, electrical or chemical stimuli are also very common TS (Yarnitsky, 2015). The modality and intensity of test stimulus might influence CPM or DNIC reliability (Marcuzzi, Wrigley, Dean, Adams, & Hush, 2017) or magnitude (Okada, 2005), though it has been generally considered that CPM effect does not depend on characteristics of the TS as long as it is considered painful and not intolerable.

6. CONDITIONING STIMULUS

Researchers have also used various modalities as conditioning stimuli, such as cuff inflation, ischemia, capsaicin injection (Yarnitsky, 2015), with painful cold water immersion of a limb being one of the most potent (Nilsen, Olsen, Solem, & Matre, 2014) and reliable stimuli currently used (Kennedy et al., 2016; Lewis, Heales, Rice, Rome, & McNair, 2012). It has been found that the longer CS duration with the highest intensity (7/10 on a 0–10 pain rating scale) result in a more pronounced CPM in women only (Razavi, Hansson, Johansson, & Leffler, 2014), although these results need to be replicated. The intensity of the CS has been shown to be unrelated or positively correlated to the magnitude of CPM (Yarnitsky, 2015). It is believed that once CPM is induced by a minimal noxious load, even mild pain elicited by tonic heat pain with intensities around the pain threshold (Kunz et al., 2014), CPM might not be further activated by increased conditioning pain levels, hence CPM is considered a saturable phenomenon (Nir, Granovsky, Yarnitsky, Sprecher, & Granot, 2011).

7. CPM PROTOCOLS

CPM response is unrelated to individual sensitivity and adaptability to pain (Zheng, Wang, Yao, Xue, & Arendt-Nielsen, 2014). Nonetheless, modality-specific CPM responses are influenced by different psychological factors. Specifically, CPM responses evoked by pressure, heat and electrical stimuli were associated with anxiety, depression and catastrophizing levels, respectively (Nahman-Averbuch, Nir, Sprecher, & Yarnitsky, 2016).

Overall younger adult age, male gender, ovulatory phase of the menstrual cycle in women, positive expectations, attention to the CS, and carrier of the serotonin-transporter-linked polymorphic region (5-HTTLPR) long allele are other characteristics associated with larger CPM responses, according to a systematic review of 36 studies (Hermans et al., 2016). Data from a large population-based study of the Danish adult general population also found better CPM in males and in individuals with higher level of education and higher pain ratings on the visual analog scale (Skovbjerg et al., 2017). Therefore, it has been suggested that different underlying mechanisms might account for CPM responses (Nahman-Averbuch et al., 2016).

CPM is not influenced by stimuli location, except for ipsilateral homotopic sites (Klyne et al., 2015). To reflect an ascending–descending long tract activity, CPM is preferably triggered by a heterotopic nociceptive conditioning stimulation (HNCS) in which the TS and CS are applied on two remote and separate anatomic regions of the body. However, some studies using spatial summation ipsilaterally also activate endogenous inhibitions (Marchand & Arsenault, 2002). CPM measurement is reliable, but varies depending on stimulation parameters and study methodology (Kennedy et al., 2016). The combination of heat pain test or handheld or cuff pressure pain and the cold pressor test (CPT) are reliable methods to induce CPM, whereas electrical and heat pain combined with cuff do not yield significant CPM effects (Gehling et al., 2016; Imai, Petersen, Mørch, & Arendt Nielsen, 2016; Petersen, Vaegter, & Arendt-Nielsen, 2017).

The different test protocols limit generalization of findings and thus a group of experts in the field presented recommendations regarding a standard measurement of CPM in healthy individuals and pain patients (Yarnitsky et al., 2010, 2015). Preferably, mechanical and heat TS should be delivered either at a fixed pain intensity of 40 on a 0–100 pain rating scale or at an ascending intensity and discontinued when pain intensity of 40/100 is reached. Subsequently, to induce a conditioning analgesic effect, the CS needs to be at least mildly to moderately painful (>20/100 pain score). The CS should be given as rapidly as possible after the second TS as the duration of CPM effects is short (Yarnitsky et al., 2015) and varies across testing protocols. For instance, 6–9 min after the end of the conditioning period, the inhibition of the pain response (e.g., nociceptive reflex) recovered its baseline value (Willer et al., 1990). A sequential protocol in which the second TS is applied immediately at the end of the CS is preferred over a parallel method, wherein both stimuli are administered simultaneously. The latter method yields larger CPM effects, most likely from biases (distraction effects) (Yarnitsky, 2015).

8. HOW DOES IT WORK? THE ORIGINS

The information from nociceptive stimulation of a mechanical, chemical or thermal nature is sent from the periphery, through the primary neuron, and to the posterior horns of the spinal cord where there is a first synaptic contact with the secondary neuron. The second neuron will also establish synaptic contacts in different regions of the brainstem such as the periaqueductal gray (PAG) and the nucleus raphe magnus (NRM), a part of the rostral ventromedial medulla (RVM). Then the information passes through the medulla/spinal cord to the thalamus, where it establishes synaptic contact with the tertiary neuron before being transported to the higher centers of the brain. At each synaptic contact, the information is integrated and undergoes inhibitory or excitatory influences (Marchand, 2012).

In the thalamus, the nuclei of the ventro-basal complex receive their afferences from the spinothalamic pathway and project toward the primary and secondary somatosensory cortices (SI and SII), with precise receptor fields. The SI and SII regions are mainly responsible for the localization and the perception of the sensory-discriminative aspect of pain, a stable component related to the evaluation of the intensity of pain as well as to the spatial and temporal characteristics of pain. The other important group of nuclei in the thalamus, those of the centromedian or intralaminar complex of the somatosensory thalamus, receive their afferences from the spinoreticular pathway and project, with large receptor fields, to structures of the brainstem and the limbic system, including the thalamus, frontal cortex, anterior cingulate cortex (ACC) and insular cortex (IC). The latter connects with the SI, SII, cingulate cortex and limbic structures like the amygdala and the perirhinal complex. The spinoreticular pathway is mainly involved in the motivo-affective component of pain modulation that is related to the evaluation of the unpleasantness of pain. The PAG is partly responsible for the unpleasantness of pain; its stimulation causes strong uncomfortable feelings of distress to be perceived as pain (Marchand, 2012).

CPM is based on the DNIC spinobulbospinal mechanisms originally investigated in rats by Le Bars and colleagues (Le Bars et al., 1979a, 1979b; Villanueva & Le Bars, 1986) who argued that a painful stimulation conducts the nociceptive information to the higher centers of the brain via the spinothalamic pathway and the transmission of afferences to brainstem structures including the PAG and NRM, which send inhibitory efferences to spinal segments and finally produces diffuse inhibition (with the help of inhibitory interneurons). CPM triggered by HNCS affect all convergent neurons recorded in the dorsal horn of the spinal cord or the nucleus caudalis of the trigeminal system (Villanueva & Le Bars, 1986). Reynolds (1969) helped unravel the neurophysiological mechanisms of CPM when he realized CNS stimulation of the midbrain central gray (PAG) produced sufficient analgesia to perform abdominal surgery in the rat without any other form of anesthesia.

Numerous studies have completed this model. PAG and NRM have been targeted for serotonergic and noradrenergic descending pathways, respectively. These pathways recruit enkephalinergic interneurons in the spinal cord and produce an analgesic response by reducing the activity of nociceptive afferents. However, studies of lesions performed at different levels of the brainstem in rodents suggest that CIDN use caudal spinal bulb structures, and therefore would not require midbrain PAG input. A recent study using functional magnetic resonance imaging (fMRI) to determine the precise brainstem sites responsible for CPM in healthy individuals observed an association between expression of analgesia and reduction of signal in brainstem regions following counter-irritation: the caudalis subdivision of the spinal trigeminal nucleus, i.e., the primary synapse, the region of the subnucleus reticularis dorsalis (SRD) and the dorsolateral pons in the region of the parabrachial nucleus (Youssef, Macefield, & Henderson, 2016a). In comparison to subjects exhibiting CPM analgesia, those with impaired CPM showed greater signal intensity increases in the mid-cingulate and dorsolateral prefrontal cortices and increased functional connectivity with the SRD following counter-irritation (Youssef, Macefield, & Henderson, 2016b). The SRD conveys widespread descending inhibition to spinal secondary neurons via the dorsolateral funiculi (Nir & Yarnitsky, 2015). Descending pain modulation induces analgesia by activation of opioid receptors in the medullary reticularis nucleus dorsalis (MRD) (Villanueva, Bouhassira, & Le Bars, 1996). In animal studies, the opioid receptor antagonist naloxone injected into the MRD prevented DNIC, while naloxone injection into the RVM did not affect DNIC analgesia (de Resende, Silva, Sato, Arendt-Nielsen, & Sluka, 2011). Using a spatial summation paradigm, systemic naloxone blocked endogenous inhibitions (Julien & Marchand, 2006). Administration of lidocaine, a local analgesic, within the RVM, reversed allodynia induced by spinal nerve litigation in rats, suggesting that activation of descending inhibition from the RVM protects against chronic neuropathic pain in animals (De Felice et al., 2011). In patients with neuropathic pain and dynamic mechano-allodynia, the clinical CS (brushing or pressure within the allodynic area) reduced the pain sensation but did not inhibit the electrophysiological responses, while the experimental CS (cold pressor or tourniquet tests) inhibited both painful sensations and the RIII reflex, suggesting supraspinal mechanisms involved in the CPM effect of this pain condition (Bouhassira, Danziger, Attal, Guirimand, & Atta, 2003).

Pain inhibition by CPM rely partly on the orbitofrontal cortex (OFC) and the amygdala (Moont, Crispel, Lev, Pud, & Yarnitsky, 2011). Piché, Arsenault, and Rainville (2009) argue for at least two neural mechanisms underlying the effects of CPM on pain and spinal nociception in humans. Conditioning stimulus inhibits shock pain perception and RIII reflex amplitude (spinal nociception) only in a subset of individuals. Moreover, sustained activation of the OFC induced by the CS was predictive of pain decrease while sustained activity in SI and the PAG predicted nociceptive reflex modulation (Piché et al., 2009).

In another study, the conditioning stimulus induced reductions of blood oxygen level dependent (BOLD) responses in classical pain-responsive regions but increases in BOLD responses in sub-regions of the ACC. During the CS, decreased pain was positively correlated with the increase in strength of functional coupling between the subgenual ACC and structures of the descending pain control system. These results demonstrate the contribution of higher-order brain regions (supraspinal mechanisms) in the activation of CPM (Sprenger, Bingel, & Büchel, 2011). The strength of the PAG resting functional connectivity can explain some of the normal variability in CPM; higher resting connectivity between the PAG and cortical pain processing regions correlates with greater CPM efficacy (Harper et al., 2018). CPM scores are also correlated with the modulation of the laser-induced BOLD response in left posterior insula/SII (Bogdanov et al., 2015).

During CPM, the nociceptive stimulation activates the nociceptive neurons corresponding to the spinal segment they innervate but also other nociceptive neurons in the spinal cord serving the rest of the body. Descending modulatory effects act on wide dynamic range neurons in the spinal dorsal horn where nociceptive C-, Aδ- and Aβ-fibers converge. It has been demonstrated that during HNCS, the perceived intensity of nociceptive Aδ- and C-fiber nociceptive inputs, but also Aβ-non-nociceptive fiber inputs is reduced, as paralleled by inhibited pain perception and event-related potentials (ERPs) (Rustamov, Tessier, Provencher, Lehmann, & Piché, 2016). Since Aβ-fibers are related to primary afferents that send the noxious information upward directly through the dorsal columns to reach second-order neurons in the dorsal column nuclei without being relayed at spinal level. The pain modulatory responses may not be attributed only to descending spinal mechanisms (Torta, Churyukanov, Plaghki, & Mouraux, 2015).

CPM inhibits pain through cerebral/supraspinal and cerebrospinal mechanisms, reflecting activation of the endogenous analgesia system, where ascending pain-induced activity evokes descending pathways, which subsequently elicit inhibitory effects on spinal nociceptive inputs. However, the underlying mechanisms of CPM in human and DNIC in animals are still incompletely understood.

9. CLINICAL RELEVANCE OF PAIN MODULATION

9.1 Post-surgical Acute and Chronic Pain

An individual's endogenous pain inhibitory capacity can be reliably assessed by CPM protocols. However, the stability of CPM might be greater for women than for men (Martel, Wasan, & Edwards, 2013), requiring further investigation. Although DNIC seem to be more effective in animals after tissue injury, CPM is often impaired in humans with chronic pain (de Resende et al., 2011; Staud, 2012). Nevertheless, it is important to note that the criteria characterizing normal and deficient CPM using QST vary largely among the different studies (Gierthmühlen et al., 2015).

Chronic pain may develop following surgical and medical procedures, particularly those involving nerve injuries. The incidence and intensity of chronic neuropathic pain developed after a surgery may be predicted by pre-operative CPM efficacy (Granovsky, 2013). CPM efficacy may predict lower risk of developing chronic pain after thoracotomy (Yarnitsky et al., 2008), while less efficient CPM is sometimes associated with acute or chronic post-operative pain (Yarnitsky, 2010), emphasizing the clinical relevance of using CPM to help identify individuals at risk of developing chronic post-surgical pain. In contrast, others have found no consistent associations between most CPM parameters and prediction of pain after injuries (e.g., stroke) or surgeries (e.g., funnel chest repair) (Grosen, Vase, Pilegaard, Pfeiffer-Jensen, & Drewes, 2014; Roosink et al., 2011; Sangesland, Støren, & Vaegter, 2017).

9.2 Chronic Pain

Even though assessing endogenous analgesia in chronic pain conditions is complicated by the presence of spontaneous ongoing pain, Edwards, Ness, Weigent, and Fillingim (2003) have found CPM to be the only consistent predictor of health-related quality of life among several laboratory pain variables. Greater CPM responses were related to less clinical pain, better physical functioning, and better self-rated health (Edwards et al., 2003).

A growing body of knowledge has revealed consistent low pain inhibitory capacity underlying various chronic pain states such as fibromyalgia, irritable bowel syndrome, migraine, tension-type headache, temporomandibular joint disorders, osteoarthritis and muscle pain, whiplash-associated disorders, interstitial cystitis, cancer pain (Nir & Yarnitsky, 2015; Yarnitsky, 2010). However, the altered function of endogenous pain modulation mechanisms might depend on the pathophysiological mechanisms of clinical pain (Bouhassira et al., 2003).

9.3 Idiopathic Pain Syndromes

Substantial evidence supports a decreased CPM efficacy in idiopathic pain syndromes. In a recent meta-analysis and numerous studies, a significant alteration of pain modulation mechanisms (CPM or temporal summation) is present in patients with fibromyalgia (FM) or chronic widespread pain (O'Brien, Deitos, Triñanes Pego, Fregni, & Carrillo-de-la-Peña, 2018). While major depressive disorder is not associated with CPM deficits (Normand et al., 2011), patients with FM and comorbid depression have more pronounced deficits in pain inhibition (de Souza, Potvin, Goffaux, Charest, & Marchand, 2009). Moreover, pain facilitation during CPM procedure is reported by twice as many FM patients than healthy subjects (42% versus 21%) (Potvin & Marchand, 2016). Less gray matter volume in a cluster encompassing the PAG as well as greater PAG resting functional connectivity to the caudal pons/rostral medulla could explain the pain-facilitative CPM in FM (Harper et al., 2018). Endogenous pain inhibition mechanisms are significantly impaired in fibromyalgia patients, but only in a subgroup of them (Potvin & Marchand, 2016). In fact, the study showed that a group of patients with FM (e.g., 41.7%) experience pain facilitation during the CPM procedure that induces pain inhibition mechanisms in physiological conditions. Those patients experiencing pain facilitation during the CPM did not differ from FM patients who experienced pain inhibition. The two subgroups were comparable with respect to sociodemographic characteristics, pain intensity, depression, and psychophysical measures (except the CPM-induced change in pain perception).

In keeping with this line of findings, irritable bowel syndrome (IBS) and provoked vestibulodynia are other chronic pain conditions with a woman predominance in which facilitatory CPM are reported in only a subgroup of patients (Gougeon, Gaumond, Goffaux, Potvin, & Marchand, 2016). More women than men suffer from chronic pain conditions and experience less endogenous analgesia (Skovbjerg et al., 2017). In patients suffering from IBS, CPM is linked to the clinical severity of symptoms (Bouhassira et al., 2013). In addition, compared to healthy subjects, IBS patients have an increased risk of suffering from temporomandibular disorder (TMD), another chronic pain syndrome unknown origin (Gallotta et al., 2017). The function of CPM mechanisms is sometimes preserved in rheumatoid arthritis (up to 5 years after diagnosis) (Leffler, Kosek, Lerndal, Nordmark, & Hansson, 2002) and in osteoarthritis (OA), but impaired in temporomandibular joint (TMJ) arthralgia patients (Kothari et al., 2016) at sites with chronic pain but not at pain-free sites (Oono et al., 2014). Knee OA has sometimes been associated with reduced CPM and enhanced temporal summation of pain (Edwards et al., 2016), which predicts less pain relief after total knee replacement (Petersen, Graven-Nielsen, Simonsen, Laursen, & Arendt-Nielsen, 2016). Even though OA patients might not report a pain reduction following painful CS (provoked OA pain), magnetoencephalography and electroencephalography techniques indicate a decreased activation of the cingulate gyrus (Quante, Hille, Schofer, Lorenz, & Hauck, 2008).

9.4 Nociceptive

A pro-nociceptive pattern of amplified pain facilitation (central sensitization) and reduced endogenous pain inhibition is also common in carpal tunnel syndrome, chronic pancreatitis (Soon, Vicenzino, Schmid, & Coppieters, 2017), whiplash-associated disorders (Daenen et al., 2013; Ng, Pedler, Vicenzino, & Sterling, 2014; Yarnitsky, 2015) and in primary headaches such as migraine, chronic tension-type headache and chronic post-traumatic headache following mild traumatic brain injury (Sandrini et al., 2006). In patients with acute and chronic low back pain, CPM effect is triggered following counter-irritation, but its duration is significantly reduced compared to healthy individuals (Mlekusch et al., 2016). Moreover, CPM efficacy may predict the analgesic effect of pregabalin treatment for painful chronic pancreatitis (Olesen et al., 2013).

9.5 Neuropathic Pain

Evaluation of pain modulation capabilities by CPM protocols is particularly relevant in cases of chronic neuropathic pain (Granovsky, 2013) caused by a nerve lesion or disease of the somatosensory system, including peripheral fibers (Aβ-, Aδ- and C-fibers) and central neurons. In addition to being increasingly prevalent (currently affecting 7–10% of the general population), neuropathic pain is caused by a multitude of factors, including alterations in the modulation of pain by the CNS and in the excitatory and inhibitory somatosensory signaling, complexifying the treatment of its symptoms (Colloca et al., 2017). Dysregulated spinal endogenous inhibitory systems and ongoing primary afferent activity are related to the intensity and duration of central sensitization (dorsal horn neuron hypersensitivity) present in neuropathic pain (Taylor, 2009). Psychophysical and electrophysiological approaches reveal impaired potency and duration of CPM in patients with postherpetic neuralgia (Pickering, Pereira, Dufour, Soule, & Dubray, 2014), in individuals with acquired neuropathic pain following spinal cord injury (Albu, Gómez-Soriano, Avila-Martin, & Taylor, 2015), in patients with painful diabetic polyneuropathy (DPN) of short duration (less than 2 years) (Granovsky, Nahman-Averbuch, Khamaisi, & Granot, 2017) and in cancer survivors with painful neuropathy acquired after chemotherapy (Nahman-Averbuch et al., 2011). The duration of DPN pain correlated positively with CPM efficacy, suggesting improvement of endogenous analgesia with chronicity of the pain syndrome (Granovsky et al., 2017). Moreover, in patients with DPN, deficient descending pain inhibition measured by CPM responses improved following four weeks of daily treatment with tapentadol sustained-release (Niesters et al., 2014).

In sum, CPM deficits could reflect a predisposing factor primary to acquisition of chronic pain (Pickering et al., 2014) and help identify patients at risk for developing chronic pain (van Wijk & Veldhuijzen, 2010), but mechanisms and neurochemical correlates of this deficit are still only partially understood (Yarnitsky, 2010). It may also help optimize pain management, as chronic pain patients may benefit more from pharmacological agents that target and enhance the specific dysfunctional pain modulation mechanism. The descending pain inhibitory system is impaired in chronic pain and it is important to know how analgesics interact with this system (Nir & Yarnitsky, 2015). A systematic literature study including 12 articles of good to moderate quality provide limited evidence suggesting that decreased CPM efficacy is improved after chronic pain (and not acute pain) is surgically removed (Goubert et al., 2015). An improvement in clinical pain levels by analgesic treatments is accompanied by an improvement in CPM specific to the treated mechanism and not to a more generalized response to pain relief (Nir & Yarnitsky, 2015).

10. A FEW OTHER PHARMACOLOGICAL AND THERAPEUTIC APPLICATIONS

According to a recent systematic literature review, some pain inhibitory medication (e.g., ketamine) and oral contraceptives inhibit the CPM mechanism (Goubert et al., 2015) and facilitate pain responses (Niesters et al., 2011). In a sample of chronic neuropathic pain patients with inefficient CPM, treatment with ketamine, morphine, and placebo were effective in improving significantly CPM responses (Niesters, Aarts, Sarton, & Dahan, 2013). The opioid treatments buprenorphine and fentanyl also potentiate CPM in healthy male volunteers (Arendt-Nielsen et al., 2012). Pregabalin, a GABAergic medication used to treat neuropathic pain and fibromyalgia among other conditions, may also produce its analgesic effect by enhancing CPM. Importantly, pregabalin efficacy is influenced by initial CPM efficacy, such that individuals with less efficient CPM benefit more from the drug (Sugimine, Saito, Araki, Yamamoto, & Obata, 2017).

In summary, dysfunctions in pain modulation systems exist in individuals with and without chronic pain, and importantly, these deficits may be improved with drugs modulating the dysfunctional pain modulatory system, especially in individuals with less efficient CPM initially, highlighting the implication of CPM enhancement in pain management.

10.1 Neurotransmitters of Pain Modulation

Multiple neurotransmitters, including monoamines and opioids, are involved in the modulation of pain by endogenous descending inhibitory pathway.

10.1.1 Monoamines: Serotonin, Noradrenaline and Dopamine

Endogenous descending pain inhibitory systems originating in midbrain and brainstem regions (e.g., PAG and NRM) rich in amino biogenic transmitters, including serotonin and norepinephrine, are activated by bulbospinal monoaminergic pathways that project onto the spinal cord. In rats, DNIC is affected by clinical doses of adrenergic agonists (e.g., dexmedetomidine and phenylephrine), suggesting an involvement of adrenergic neurons in DNIC (Sanada, Kohase, Makino, & Umino, 2009). Catecholamines like norepinephrine and epinephrine signal through the α2-adrenergic receptors (Vo & Drummond, 2016) concentrated in the spinal cord to produce an inhibitory effect. Neurons of the dorsal horn in the spinal cord are directly inhibited by serotonin (from serotoninergic neurons) which, through its action on facilitatory spinal 5-HT3 receptors, influences the expression of CPM (Bannister & Dickenson, 2017). Measures of monoamines in men before a prostate surgery, the levels of monoamines, particularly adrenergics, was correlated to the efficacy of CPM (Parent et al., 2015).

10.1.2 Dopamine

A growing body of literature support the role of dopaminergic neurotransmission in the modulation of pain, however, the underlying mechanisms need further investigation. In healthy subjects, CPM effect was significantly more important following apomorphine, a non-specific dopamine agonist, than placebo administration (27% versus 4%, respectively) (Treister, Pud, & Eisenberg, 2013). Inversely, the inhibition of dopaminergic neurotransmission in healthy human subjects using acute phenylalanine and tyrosine depletion (APTD) induced decreases of cerebral dopaminergic activity accompanied by increases of pain unpleasantness but not pain intensity (Tiemann, Heitmann, Schulz, Baumkötter, & Ploner, 2014). Even though parts of the descending pain inhibitory system involve dopaminergic pathways, no dysregulations in CPM contribute to altered pain processing in Parkinson's disease (PD), a pathology mainly characterized by lower secretion of dopamine. However, the akinetic-rigid subtype of PD is associated, at a trend level, to impairment of pain inhibition (Grashorn et al., 2015).

10.1.3 Catecholamines

Some patients with chronic pain conditions such as fibromyalgia display reduced CPM efficacy as well as lower concentration of serotonin and noradrenaline in the cerebrospinal fluid (Marchand, 2012). A recent study found a relationship between CPM efficacy and basal monoamine levels in blood, specifically blood-bound norepinephrine and metanephrine concentrations. Patients with chronic pain and deficits in CPM presented lower peripherally (plasma) acting norepinephrine and metanephrine concentrations (Parent et al., 2015). Lower CPM-mediated pain inhibition has also been linked to expression of a polymorphism in the serotonin transporter

(5-HTT) gene (SLC6A4) (Lindstedt et al., 2011), but this association is not observed in all samples (Locke et al., 2014).

Pharmacological treatments engaging descending noradrenergic and serotonergic control pathways have shown some efficacy in the treatment of pain, demonstrating the importance of serotonin and norepinephrine in pain modulation (Bannister & Dickenson, 2017; Kirkpatrick et al., 2015). Antidepressants such as tricyclic antidepressants, serotonin–norepinephrine reuptake inhibitors and selective serotonin reuptake inhibitors are used for pain management in chronic pain conditions such as neuropathic pain, migraines, and fibromyalgia (Dharmshaktu, Tayal, & Kalra, 2012). Duloxetine, a serotonin–norepinephrine reuptake inhibitor (SNRI), benefited more patients with painful diabetic polyneuropathy and deficient CPM than subjects with efficient CPM. Furthermore, drug efficacy correlated with improvement in CPM efficacy (Yarnitsky, Granot, Nahman-Averbuch, Khamaisi, & Granovsky, 2012).

10.1.4 Opioids

The functioning of descending controls induced by counter-irritation include the analgesic action of endogenous opioid release, including enkephalins, B-endorphin, and dynorphins. Opioid receptors are contained on several primary afferents in the dorsal horn of the spinal cord, in particular the PAG and probably also the rostro-ventral gray matter of the bulb. Dorsal horn neurons mediated by serotonergic neurons act on interneurons that contain enkephalins which limit the discharge transmitted from the primary afferents to the cells of the dorsal horn (Marchand, 2012). In rats, plasticity of spinal serotonergic neurotransmission can modulate spinal mu-opioid receptors mechanisms (Aira et al., 2012).

In humans, pain inhibition and decrease in the nociceptive reflex induced by CPM can be completely blocked by naloxone hydrochloride, a non-selective and competitive opioid receptor antagonist after 5 min from the administration. These findings led Willer et al. (1990) to believe that the supraspinal CPM phenomenon involves an opioidergic system. Similarly, the endogenous pain inhibitory systems activated by spatial summation is mediated by an opioid pathway (Julien & Marchand, 2006). Basal mu-opioid receptor availability in the amygdala predicts pain-related brain activity (on the P260 component of somatosensory evoked potentials reflecting activity in the ACC) during CPM but does not affect the nociceptive reflex. The authors argue that the inhibition of pain-related brain activity induced by activation of mu-opioid receptors in the amygdala may depend on a cerebral mechanism different from descending modulation (Piché et al., 2014). King et al. (2013) further reported only a partial blockage of the CPM effect by naltrexone, a long-term opioid antagonist, highlighting the importance of endogenous opioid release in the expression of CPM with the effects of naltrexone being moderated by levels of pain catastrophizing.

In healthy individuals, the mu-opioid receptor *agonist* morphine impairs CPM expression, while tapentadol, a combined mu-opioid receptor agonist and noradrenaline reuptake inhibitor, does not affect CPM (Martini et al., 2015). Conversely, tapentadol enhances pain inhibition in chronic pain patients with diabetic polyneuropathy (Niesters et al., 2014). In patients with chronic fatigue syndrome/fibromyalgia and rheumatoid arthritis, morphine produced anti-hyperalgesia effects (comparable to placebo), but neither morphine nor naloxone affected CPM. Findings suggest that the opioid system might not be responsible for impaired CPM in these patients (Hermans et al., 2018). Repeated use of opioids in the course of chronic pain treatment may lead to decreased capacity for endogenous analgesia. Patients with chronic pain using oral opioids express smaller CPM responses than patients with chronic pain using other analgesics (Ram, Eisenberg, Haddad, & Pud, 2008). Moreover, the dosage and duration of opioid treatment correlate negatively with CPM efficacy in men. The finding that CPM is affected by opioid use might hint at mechanisms of pain modulation and opioid-induced hyperalgesia phenomenon (Ram et al., 2008).

10.1.5 Cannabinoids

In producing antinociception, the neural circuitry activated by endogenous opioids and cannabinoids overlap (Ibrahim et al., 2005; Manning, Martin, & Meng, 2003; Sáez-Cassanelli, Fontanella, Delgado-García, & Carrión, 2007). Evidence of an implication of endogenous cannabinoids (endocannabinoids) in inhibitory pain modulation (Grotenhermen, 2004) arises primarily from animal studies. In the rat, the endogenous descending pain-modulatory pathways that includes PAG, its projection to downstream RVM neurons, and their inhibitory projections to the dorsal horn of the spinal cord (Palazzo, Luongo, Novellis, De Rossi, & Maione, 2010), play an important role in antinociception modulated by endocannabinoids (Calejesan, Kim, & Zhuo, 2000; Escobar et al., 2012; Hohmann, Tsou, & Walker, 1999; Manning et al., 2003) by actions primarily at the cannabinoid receptors type 1 (CB1) (Hohmann et al., 1999; Palazzo et al., 2010).

Cannabinoid receptor activation by endocannabinoids regulate the mode and the probability of g-aminobutyric acid (GABA) release within the midbrain PAG (Aubrey, Drew, Jeong, Lau, & Vaughan, 2017) and modulate GABA release in the RVM neurons (Li, Suchland, & Ingram, 2017). Cannabinoids produce antinociception in the superficial dorsal horn also by modulation of the descending noradrenergic pathways (Gutierrez et al., 2003). CB1 receptors on GABAergic neurons play a role in the electroacupuncture effect of pain alleviation and improvement of DNIC function in a mouse model of knee osteoarthritis (Yuan et al., 2018). Neuropathic pain in rats may be reduced by action of the spinal endocannabinoid anandamide (AEA) on CB1 or transient receptor potential vanilloid 1 (TRPV1) receptors, depending on its local concentration (Maione et al., 2006; Starowicz et al., 2012). Several ventrolateral PAG neurons co-express TRPV1 and CB1 receptors (Maione et al., 2006; Palazzo, Rossi, & Maione, 2008) to produce spinal antinociception (Horvath, Kekesi, Nagy, & Benedek, 2008). Some of the therapeutic action of the pain treatment metamizole (Escobar et al., 2012) or the triptans for migraines may act via endocannabinoid containing neurons and CB1 receptors in the lateral–ventrolateral PAG (Akerman, Holland, Lasalandra, & Goadsby, 2013).

11. ROLE AND MANIPULATION OF EXPECTANCIES

Physiological pathways interact with psychological-cognitive ones to produce subjective pain experience. Neuroimaging studies support the modulation of CPM by certain regions of the limbic system (e.g., anterior cingulate cortex (ACC)) and the insula involved in the motivational-affective component of pain (Moont, Pud, Sprecher, Sharvit, & Yarnitsky, 2010). Reports link pain modulation with cognitive factors, including expectation, distraction, and attention, even though CPM effects are not due to attentional or distraction processes (Lautenbacher, Prager, & Rollman, 2007; Moont et al., 2010). Several studies have shown that conditioned modulation of pain is mediated by changes in expectancies (Jepma & Wager, 2015; Koban & Wager, 2016).

Expectancy refers to beliefs that something is going to happen or is likely to happen. Expectancies can be implicit or explicit. Implicit expectancies refer to beliefs based on what usually occurs in the individual's environment (Schwarz, Pfister, & Büchel, 2016). These beliefs can also be built by social observation (Colloca & Benedetti, 2009; Schwarz et al., 2016) where people adapt their behavior partially based on these sort of statistical probabilities. Explicit expectancies refer to personal beliefs about oneself, the others and the environment. There are two categories of explicit expectancies: the first one refers to the beliefs consciously used to make decisions and the second one refers to the beliefs that will "incidentally" influence the decision making (Schwarz et al., 2016). Oneself expectancies can also be influenced by someone else's expectancies (Pygmalion effect), like a therapist's expectation concerning a patient or a teacher's expectation concerning a student (Schwarz et al., 2016).

Expectancies can modulate conditioned pain modulation. When individuals with and without chronic back pain are asked, prior to testing if they expect ischemic pain (CS) would increase, decrease or not change their heat pain (TS), those who anticipate lower pain during ischemia report significantly more pain inhibition in heat pain than those with no expectancies or with pain augmentation expectancies. Administration of placebo, naloxone, or morphine did not affect these results. These findings reveal an important contribution of expectancy in CPM that is not mediated by opioid mechanisms (France et al., 2016), supporting evidence from previous studies. Expectancies of pain augmentation can counteract an analgesic treatment (Cormier, Piché, & Rainville, 2013; Goffaux, Redmond, Rainville, & Marchand, 2007). *A priori* expectancies strongly correlate with CPM-induced analgesia, such that expectancies of hyperalgesia verbally induced through suggestions can enhance the nociceptive reflex amplitude and shock pain (thus blocking the analgesia normally activated by CPM), while suggestions of analgesia enhance the decrease in nociceptive reflex amplitude (Cormier et al., 2013). Changes in expected pain predict changes in pain outcomes mediated by descending pain modulation affecting spinal processes (Taylor, Chang, Rainville, & Roy, 2017). Findings suggest that while positive expectancies (analgesia) and lower stress are related to a greater pain reduction induced by CPM, negative expectancies (e.g., expectation of enhanced pain, hyperalgesia) enhance pain intensity. Interestingly, expectation of decreased or enhanced pain following a conditioning stimulation might produce the corresponding effect only in women (Bjørkedal & Flaten, 2012). However, further studies are needed to confirm a potential dysmorphic effect. The influence of expectancies on CPM seems to decrease in elderly individuals, while the deterioration of descending pain inhibitory mechanisms occurs (Grashorn, Sprenger, Forkmann, Wrobel, & Bingel, 2013).

Evidence of the effect of expectancies on pain modulation was elegantly provided by Goffaux et al. (2007). Expectancies of hyperalgesia completely blocked the typical CPM response on subjective pain ratings as well as spinal nociceptive reflexes and somatosensory evoked brain potentials; whereas expectancies of analgesia did not affect spinal nociception. The blockade of the analgesic effect by expectancies both at the spinal level (nociceptive reflex) and at the cortical level (somatosensory evoked potentials) suggests a physiological effect of expectancies related to spinal nociceptive signal processing rather than to the reinterpretation of nociceptive messages by the cortex (Goffaux et al., 2007).

Analgesia induced by CPM activation might depend mainly on the subjective perception of the painfulness of the CS rather than its physical intensity. Participants randomized to one of four cognitive interventions aimed at manipulating only the perception of pain intensity of the CPM conditioning component (and not its physical component); placebo (CS less painful after application of a local anesthetic cream); nocebo (CS more painful); and the informed control groups. Placebo and nocebo interventions respectively decreased and increased the perceived CS painfulness, however, only the placebo suggestions were paralleled by changes in CPM magnitude (Nir, Yarnitsky, Honigman, & Granot, 2012). These opposite effects of placebo analgesia and nocebo hyperalgesia are supported by brain mapping studies which generate a divergent pattern of cerebral activation after the administration of placebo and nocebo (Nir et al., 2012).

Expanding on these findings, Rattanavong (2013) manipulated the motivational-affective component of pain, while rendering untouched the intensity of pain, the sensory-discriminative component of pain. Suggestions of increase, decrease or no change in CS unpleasantness following inactive cream application were made. Consequently, a decrease in perceived unpleasantness significantly reduces the effectiveness of CPM mechanisms. The manipulation of the affective component of pain, through specific suggestions, modulates the analgesic efficacy of CPM, however, an initially effective analgesic response is not enhanced by increased unpleasantness of the stimulus, indicating a ceiling effect (Rattanavong, 2013).

It is hypothesized that a change in pain intensity affects pain unpleasantness. The sensory and affective components of pain can be dissociated by hypnotic suggestions. When it is suggested, via hypnosis, that pain induced by a CS will be as intense as the previous one, but not unpleasant, the corresponding changes are observed in pain ratings and brain activity. The reduction of pain unpleasantness is associated with reduced activity in the cortical regions involved in the emotional component of pain (e.g., ACC), while the unaltered sensory component of pain is consistent with the absence of change in the activity of the SI. In contrast, hypnotic suggestions used to modulate the sensory component of pain reduce perception and brain activity associated with both the sensory and affective components of pain (Rainville, Carrier, Hofbauer, Bushnell, & Duncan, 1999). The analgesic effect of hypnosis is thought to activate descending inhibitory mechanisms, among other mechanisms (Kiernan, Dane, Phillips, & Price, 1995). Moreover, CPM-induced analgesia is more pronounced in subjects highly susceptible to hypnosis (Fidanza, Varanini, Ciaramella, Carli, & Santarcangelo, 2017).

According to the literature, expectancies and CPM effects are generally correlated (Bjørkedal & Flaten, 2012; Hermans et al., 2016). It has been suggested that "reduced inhibitory CPM can be due to contextually induced cognitive and emotional factors and not necessarily to a dysfunction of descending inhibitory pathways" (Bjørkedal & Flaten, 2012).

A recent study has examined the relationship between CPM responses and the magnitude of placebo effects induced by open and hidden applications of lidocaine and a control analgesic in individuals with neuropathic pain. Even though patients expressed efficient CPM and placebo effects, the CPM response did *not* predict the occurrence and magnitude of individual placebo effects, suggesting similar behavioral effects but most likely different mechanisms underlying CPM and placebo effects (Skyt et al., 2018). A very recent study in our laboratory also suggest that placebo-induced analgesia does not vary as a function of CPM in healthy adults, suggesting different pain inhibitory mechanisms (Damien, Léonard, Chalaye, Colloca, & Marchand, 2018). Evidence on the capacity of placebo and nocebo effects to act on descending pain modulation mechanisms is important for the prevention and treatment of pain considering that the capacity to trigger endogenous pain modulation mechanisms may help predict chronic pain. The extensive research on the placebo effect has contributed to the recognition of its strength and its ability to produce significant therapeutic benefits in managing experimental and clinical pain. In fact, as there are different CPM, there exists many placebo effects, with different mechanisms (Benedetti, Carlino, & Pollo, 2011). Recent experimental and clinical studies begin to unravel the psychoneurobiological mechanisms underlying placebo analgesic effects (Elsenbruch & Labrenz, 2018; Medoff & Colloca, 2015).

While placebo effect varies between people with different characteristics (Corsi & Colloca, 2017; Geers, Kosbab, Helfer, Weiland, & Wellman, 2007) and within the same person when receiving more than one treatment (Whalley, Hyland, & Kirsch, 2008), the evaluation of the effect of expectancies is a challenge in research (Schwarz et al., 2016). Nevertheless, as mentioned elsewhere, numerous studies reveal that expectancies also modulate placebo and nocebo effects (Bartels et al., 2014; Bartels, van Laarhoven, Heijmans, et al., 2017; Bartels, van Laarhoven, Stroo, et al., 2017; Bingel et al., 2011; Bräscher, Witthöft, & Becker, 2018; Fiorio et al., 2012). Expectancies are induced by verbal instructions, social observation (for reviews see, Colloca & Grillon, 2014; Medoff & Colloca, 2015) and distinct forms of learning (for example, Benedetti et al., 2003; Egorova et al., 2015; Morton, El-Deredy, & Jones, 2014; Reicherts, Gerdes, Pauli, & Wieser, 2016). Expectancies and placebo-related mechanisms and specific effects of therapies or drugs may partially overlap to affect therapeutic outcome (Benedetti et al., 2011). Drug actions and placebo effects may interact additively or synergistically in some conditions, such that a treatment is less effective if the placebo component is absent. Therefore, patients' beliefs and expectancies should be exploited to optimize treatment outcomes (Bingel et al., 2011) in adult as well as pediatric populations (Simmons et al., 2014).

The extensive research on the placebo effect has contributed to the recognition of its strength and its ability to produce significant therapeutic benefits in managing experimental and clinical pain. There exists different CPM and placebo effects (Benedetti et al., 2011). Not only is placebo research helpful for personalizing and optimizing pain management, it also contributes to the understanding of the endogenous mechanisms of analgesia involved in normal functioning and pain states (Benedetti, 2007; Colloca & Grillon, 2014) and in the context of clinical trials (Gourion & Mouchabac, 2016).

12. EXPERIMENTAL AND CLINICAL PLACEBO STUDIES

Placebo effects have been described both in different experimental settings and multiple clinical conditions (Murray & Stoessl, 2013). A positive placebo effect is observed in almost half of patients with Parkinson's disease (PD), pain syndromes, and depression, particularly in states of advanced disease or during invasive procedures (Diederich & Goetz, 2008).

The placebo and nocebo effects in clinical trials for migraines are also substantial (Antonaci, Chimento, Diener, Sances, & Bono, 2007). Placebo analgesic effects are usually greater in studies investigating placebo analgesic mechanisms compared with clinical trials in which placebos served as a control (Vase, Riley, & Price, 2002).

Placebo effects have been suggested to affect more strongly postoperative or clinical pain than experimental pain. Several studies demonstrate higher and more enduring placebo effects in patients as compared to healthy participants (Sauro & Greenberg, 2005). Expectancy-induced placebo effects elicited a large placebo analgesic effect to both acute experimental and chronic pain, however, the placebo effects were unrelated to each other (Müller et al., 2016). Charron, Rainville, and Marchand (2006) examined changes in the intensity and unpleasantness of pain in patients with low back pain undergoing a placebo treatment consisting of a saline injection presented as a potent painkiller in one session or as an inactive substance in the control session. Following the injection, comparable pain relief was expected for clinical and experimental pain. Nonetheless, reported ratings of pain intensity, pain unpleasantness, and perceived relief demonstrated a more potent placebo effect in clinical pain versus experimental pain (Charron et al., 2006).

12.1 Clinical Pain

Longer exposure to fibromyalgia pain is associated with lower placebo-induced analgesia in addition to low pain inhibitory capacity, which is depicted by less efficient CPM (Kosek et al., 2017; Yarnitsky, 2010), highlighting the need for early interventions (Kosek et al., 2017). Similarly, patients with Alzheimer's disease or dementia with impaired prefrontal lobe functions show reduced expectancy-induced placebo responses, underlying the need to compensate for the loss of the endogenous placebo-related mechanisms in this condition in order to optimize therapeutic benefits (Benedetti et al., 2006, 2011). Placebo effects obtained after a conditioning procedure on pain tolerance can be transferred to other modalities such as motor endurance, emphasizing the potential implications of placebo analgesic procedures in pathological conditions such as chronic fatigue or Parkinson's disease (Carlino, Guerra, & Piedimonte, 2016).

In contrast, major depressive disorder is a condition that is highly responsive to placebos in clinical trials; placebo response rates can attain up to 30–40% (Gourion & Mouchabac, 2016). In this condition, placebo effects can partially mimic selective serotonin reuptake inhibitor-mediated brain activation (Diederich & Goetz, 2008). Depressive symptoms as well as trait anxiety, pain catastrophizing, and pain disability appear to be associated to deficits in endogenous opioid function, which may be improved by opioid analgesic treatment (e.g., morphine) (Burns et al., 2017). Placebo effects and particularly nocebo effects are important in clinical trials for sexual dysfunctions and migraine or acute headache treatments (Colloca & Miller, 2011; Diener, Schorn, Bingel, & Dodick, 2008; Mitsikostas, 2012).

12.1.1 Open-Label Placebos Clinical Trials

According to a recent systematic reviews and meta-analysis, open-label placebos (without deception) positively influence clinical effects (e.g., mean global improvement scores, severity of symptoms and pain relief) in comparison to the absence of treatment across many clinical conditions: irritable bowel syndrome, depression, allergic rhinitis, back pain, and attention deficit hyperactivity disorder (Charlesworth et al., 2017; Colloca & Howick, 2018; Kaptchuk et al., 2010). Open-label and dose-extending placebos do not raise the typical ethical dilemmas usually associated with deceptive placebos because patients are explicitly informed of the nature the placebo treatment and the rationale of receiving the placebo (Carvalho et al., 2016; Colloca & Howick, 2018). Thus, the use of clinically meaningful placebo effects with potential benefits and limited adverse events (Colloca & Miller, 2011) may be a helpful strategy for pain management (Carvalho et al., 2016). Laboratory and clinical studies on placebo analgesia might benefit from using open-label and dose-extending placebos to potentiate the placebo effect in therapeutic regimes (Colloca & Howick, 2018).

In clinical practice, patients always expect to receive an active drug, whereas they are informed about the possibility of receiving an active drug or a placebo in clinical trials, resulting in reduced treatment efficacy. Nonetheless, active placebos inducing minor sensations versus inert/passive placebo substances can make participants believe they received an active drug and enhance pain relief induced by placebo effects (Rief & Glombiewski, 2012).

In clinical trials, placebo responses have increased thereby making it difficult to attain superior analgesic effects with new drugs (Tuttle et al., 2015; Vase et al., 2015). Various individual factors influence the placebo response in clinical trials, such as patients' expectation, baseline pain intensity, age, gender, geographical distribution. General characteristics and methodology of randomized controlled trials (RCTs) impact the placebo response, including design (studies with a parallel design yield higher placebo response than cross-over studies), blinding, randomization ratio, treatment allocation, route of application of drugs (Diener et al., 2008; Vase et al., 2015), adverse effects and discontinuation, washout length and number of planned face-to-face visits (Vase et al., 2015) and countries where the trials are ran (Tuttle et al., 2015).

A pioneering Cochrane review including randomized placebo trials on 60 clinical conditions (including pain, nausea, phobia, asthma, depression, smoking, dementia, obesity, hypertension, insomnia and anxiety) found a very variable effect of placebo interventions on pain, from negligible to clinically important. Larger effects of placebo interventions were found in studies with physical placebo interventions (e.g., sham acupuncture), outcomes reported by patients rather than by observers, small trials, and trials that did not inform patients about the possibility of placebo treatment (Hróbjartsson & Gøtzsche, 2010).

In contrast, a meta-analysis of RCTs in the treatment of depression with selective serotonin reuptake inhibitors (SSRIs) found no significant difference in effect size between clinical trials that excluded placebo responders after the placebo run-in phase and those trials that did not (Lee, Walker, Jakul, & Sexton, 2004). A recent study has shown that it is possible to estimate patients' expectancies toward treatment efficacy and hence predict the magnitude of the placebo response in RCTs (Vase et al., 2015).

Weimer, Colloca, and Enck (2015) evaluated 75 systematic reviews and meta-analyses on the predictors of the placebo response in major medical areas (neurology, psychiatry, internal medicine) known for high placebo response rates and concluded that a younger age may contribute to placebo response rate in some conditions (psychiatric conditions but not in depression, and internal medicine but not in gastroenterology), but found no supporting evidence of an effect of sex or older age in placebo response. Importantly, lower symptom severity at baseline, a randomization ratio that selected more patients to drugs than to placebo, and more frequent study visits were common predictors of placebo response (Weimer et al., 2015).

12.1.2 Psychological Variables

The large variability in placebo effects and responses is partially mediated by individual variables, although the factors identified are unlikely to be exhaustive. For instance, aspects of pain catastrophizing including feelings of helplessness and magnification of pain are associated with the efficacy of pharmaceutical and natural placebos in reducing pain intensity (Watkinson, Chapman, & Horne, 2017). Higher anxiety also contributes to greater placebo and nocebo effects, in part, via activation of the CCK pathway (Colloca & Miller, 2011). Heightened anxiety and autonomic arousal are involved in the persistence of nocebo hyperalgesia (Colagiuri & Quinn, 2017). Personality traits such as resiliency, altruism, straightforwardness, and hostility explain some variance in placebo analgesic responses that is related to activations in endogenous opioid systems (Peciña et al., 2013). Biological markers contributing to interindividual variations in placebo-induced pain modulation and the pain-related neurotransmission of endogenous opioid and dopamine have been identified. For instance, the A118G mu-opioid receptor gene (*OPRM1*) polymorphism has been linked to psychological factors associated with placebo analgesia and brain activity changes (Peciña, Love, Stohler, Goldman, & Zubieta, 2015). Similar endogenous MOR mechanisms are activated by the analgesic effects of placebo and real transcranial direct current stimulation (tDCS) of the primary motor cortex M1 (DosSantos et al., 2014).

The history of treatment is another important factor to consider in experimental and clinical modulation of placebo and nocebo effects. In healthy individuals, prior positive experiences with effective treatment enhance the placebo effect. In patients with chronic pain, a more negative pain-related treatment history was associated with larger placebo effects for the relief of clinical pain (Müller et al., 2016). Clinically, the efficacy of sham rTMS for neuropathic pain relief could be improved by being administered following a successful active rTMS (André-Obadia, Magnin, & Garcia-Larrea, 2011). Furthermore, the persistence of placebo and nocebo effects is associated to the length of conditioning (i.e., number of exposures) to prior effective (and ineffective) interventions (Colloca, Petrovic, Wager, Ingvar, & Benedetti, 2010). In any treatment, learned placebo responses following exposure to an active therapy or drug should be maximized by using a conditioned strategy of therapeutic effects, so that placebos can acquire properties of active treatments (like in dose-extending placebos) and extend their clinical benefits, especially in conditions of chronic diseases (Colloca, Enck, & DeGrazia, 2016).

Translational placebo research should also consider the common nocebo effects that can arise from information about side effects. Nocebo effects can negatively influence clinical outcomes, by attenuating the perceived pain relief, consequently causing a decreased effectiveness of pain-management interventions and increase nonadherence or discontinuation of trials

(Colloca & Miller, 2011). To improve patient care, Klinger, Blasini, Schmitz, and Colloca (2017) suggest multiple strategies to minimize nocebo effects in the treatment of patients with chronic pain, focusing on communication and interactions between patients and clinicians during treatment and minimizing negative information, cues, context and lack of positive information.

13. CONCLUSION

In recent years, there have been many new exciting discoveries to advance our understanding of the area of the endogenous mechanisms of pain modulation. Cumulative evidence in research on pain modulation substantiate similar and divergent mechanisms in the various conditioned pain modulation and placebo effects. CPM and placebo effects are both influenced by expectancies of analgesia built up through verbal suggestions, past personal experience, contextual cues, socio-affective and learning processes (e.g., conditioning and social observation). Expectancies of pain relief in CPM and placebo effects activate subcortical nuclei in the descending pain inhibitory system, including brainstem regions such as the RVM and PAG, which send inhibitory projections to the spine and produce diffuse analgesic responses important for endogenous pain modulation (Fazeli & Büchel, 2018). Multiple neurotransmitters, including endogenous monoamines, opioids and cannabinoids, are involved in the modulation of pain by endogenous inhibitory pathway in animals and humans. In contrast, nocebo hyperalgesic effects activate release of the cholecystokinin system (CCK), among other neurotransmitters. Although CPM and placebo or nocebo effects induce similar behavioral pain reduction or enhancements, the related analgesic effects are not correlated and they activate distinct endogenous pain inhibitory systems (e.g., activation of descending pain control systems from spinobulbospinal and higher centers, respectively).

Placebo effects inherently contribute to every treatment and clinical outcomes. Pre-treatment assessment of CPM efficacy and the functioning of descending pain inhibitory pathways may provide the ability to predict the development of clinical pain in patients and the magnitude of analgesic effects in response to treatments. Therefore, the knowledge of the interaction between placebo and CPM systems in the descending pain inhibitory pathways is of high clinical relevance yet remains to be fully elucidated. Exploring the action of placebo and nocebo effects along with CPM on different parts of descending pain inhibition systems can reveal whether the engagement of potential common pathways limits the overall analgesic effect (e.g., a ceiling effect), or alternatively, results in a potentiation of the inhibitory response (e.g., additive effects). The ultimate goal of this research is to minimize the experience of pain in patients and help optimize pain management strategies based on inhibitory pain modulation mechanisms.

REFERENCES

Aira, Z., Buesa, I., García del Caño, G., Salgueiro, M., Mendiable, N., Mingo, J., et al. (2012). Selective impairment of spinal mu-opioid receptor mechanism by plasticity of serotonergic facilitation mediated by 5-HT2A and 5-HT2B receptors. *Pain, 153*(7), 1418–1425. https://doi.org/10.1016/j.pain.2012.03.017.

Akerman, S., Holland, P. R., Lasalandra, M. P., & Goadsby, P. J. (2013). Endocannabinoids in the brainstem modulate dural trigeminovascular nociceptive traffic via CB1 and "triptan" receptors: Implications in migraine. *The Journal of Neuroscience: The Official Journal of the Society for Neuroscience, 33*(37), 14869–14877. https://doi.org/10.1523/JNEUROSCI.0943-13.2013.

Albu, S., Gómez-Soriano, J., Avila-Martin, G., & Taylor, J. (2015). Deficient conditioned pain modulation after spinal cord injury correlates with clinical spontaneous pain measures. *Pain, 156*(2), 260–272. https://doi.org/10.1097/01.j.pain.0000460306.48701.f9.

André-Obadia, N., Magnin, M., & Garcia-Larrea, L. (2011). On the importance of placebo timing in rTMS studies for pain relief. *Pain, 152*(6), 1233–1237. https://doi.org/10.1016/j.pain.2010.12.027.

Antonaci, F., Chimento, P., Diener, H.-C., Sances, G., & Bono, G. (2007). Lessons from placebo effects in migraine treatment. *The Journal of Headache and Pain, 8*(1), 63–66. https://doi.org/10.1007/s10194-007-0360-4.

Arendt-Nielsen, L., Andresen, T., Malver, L. P., Oksche, A., Mansikka, H., & Drewes, A. M. (2012). A double-blind, placebo-controlled study on the effect of buprenorphine and fentanyl on descending pain modulation: A human experimental study. *The Clinical Journal of Pain, 28*(7), 623–627. https://doi.org/10.1097/AJP.0b013e31823e15cb.

Aubrey, K. R., Drew, G. M., Jeong, H.-J., Lau, B. K., & Vaughan, C. W. (2017). Endocannabinoids control vesicle release mode at midbrain periaqueductal grey inhibitory synapses. *The Journal of Physiology, 595*(1), 165–178. https://doi.org/10.1113/JP272292.

Bannister, K., & Dickenson, A. H. (2017). The plasticity of descending controls in pain: Translational probing. *The Journal of Physiology, 595*(13), 4159–4166. https://doi.org/10.1113/JP274165.

Bartels, D. J. P., van Laarhoven, A. I. M., Haverkamp, E. A., Wilder-Smith, O. H., Donders, A. R. T., van Middendorp, H., et al. (2014). Role of conditioning and verbal suggestion in placebo and nocebo effects on itch. *PLoS One, 9*(3), e91727. https://doi.org/10.1371/journal.pone.0091727.

Bartels, D. J. P., van Laarhoven, A. I. M., Heijmans, N., Hermans, D., Debeer, E., van de Kerkhof, P. C. M., et al. (2017). Cognitive schemas in placebo and nocebo responding: Role of autobiographical memories and expectations. *Clinical Therapeutics, 39*(3) 502–512.e1. https://doi.org/10.1016/j.clinthera.2017.02.004.

Bartels, D. J. P., van Laarhoven, A. I. M., Stroo, M., Hijne, K., Peerdeman, K. J., Donders, A. R. T., et al. (2017). Minimizing nocebo effects by conditioning with verbal suggestion: A randomized clinical trial in healthy humans. *PLoS One, 12*(9), e0182959. https://doi.org/10.1371/journal.pone.0182959.

Benedetti, F. (2007). Placebo and endogenous mechanisms of analgesia. *Handbook of Experimental Pharmacology*, *177*, 393–413. https://doi.org/10.1007/978-3-540-33823-9_14.

Benedetti, F., Arduino, C., Costa, S., Vighetti, S., Tarenzi, L., Rainero, I., et al. (2006). Loss of expectation-related mechanisms in Alzheimer's disease makes analgesic therapies less effective. *Pain*, *121*(1–2), 133–144. https://doi.org/10.1016/j.pain.2005.12.016.

Benedetti, F., Carlino, E., & Pollo, A. (2011). How placebos change the patient's brain. *Neuropsychopharmacology*, *36*(1), 339–354. https://doi.org/10.1038/npp.2010.81.

Benedetti, F., Pollo, A., Lopiano, L., Lanotte, M., Vighetti, S., & Rainero, I. (2003). Conscious expectation and unconscious conditioning in analgesic, motor, and hormonal placebo/nocebo responses. *The Journal of Neuroscience: The Official Journal of the Society for Neuroscience*, *23*(10), 4315–4323. https://doi.org/10.1523/JNEUROSCI.23-10-04315.2003.

Bingel, U., Wanigasekera, V., Wiech, K., Ni Mhuircheartaigh, R., Lee, M. C., Ploner, M., et al. (2011). The effect of treatment expectation on drug efficacy: Imaging the analgesic benefit of the opioid remifentanil. *Science Translational Medicine*, *3*(70), 70ra14. https://doi.org/10.1126/scitranslmed.3001244.

Biurrun Manresa, J. A., Fritsche, R., Vuilleumier, P. H., Oehler, C., Mørch, C. D., Arendt-Nielsen, L., et al. (2014). Is the conditioned pain modulation paradigm reliable? A test-retest assessment using the nociceptive withdrawal reflex. *PLoS One*, *9*(6), e100241. https://doi.org/10.1371/journal.pone.0100241.

Bjørkedal, E., & Flaten, M. A. (2012). Expectations of increased and decreased pain explain the effect of conditioned pain modulation in females. *Journal of Pain Research*, *5*, 289–300. https://doi.org/10.2147/JPR.S33559.

Bogdanov, V. B., Viganò, A., Noirhomme, Q., Bogdanova, O. V., Guy, N., Laureys, S., et al. (2015). Cerebral responses and role of the prefrontal cortex in conditioned pain modulation: An fMRI study in healthy subjects. *Behavioural Brain Research*, *281*, 187–198. https://doi.org/10.1016/j.bbr.2014.11.028.

Bouhassira, D., Danziger, N., Attal, N., Guirimand, F., & Atta, N. (2003). Comparison of the pain suppressive effects of clinical and experimental painful conditioning stimuli. *Brain: A Journal of Neurology*, *126*(Pt. 5), 1068–1078.

Bouhassira, D., Moisset, X., Jouet, P., Duboc, H., Coffin, B., & Sabate, J.-M. (2013). Changes in the modulation of spinal pain processing are related to severity in irritable bowel syndrome. *Neurogastroenterology and Motility: The Official Journal of the European Gastrointestinal Motility Society*, *25*(7), 623. https://doi.org/10.1111/nmo.12123.

Bräscher, A.-K., Witthöft, M., & Becker, S. (2018). The underestimated significance of conditioning in placebo hypoalgesia and nocebo hyperalgesia. *Pain Research & Management*, *2018*, 6841985. https://doi.org/10.1155/2018/6841985.

Burns, J. W., Bruehl, S., France, C. R., Schuster, E., Orlowska, D., Buvanendran, A., et al. (2017). Psychosocial factors predict opioid analgesia through endogenous opioid function. *Pain*, *158*(3), 391–399. https://doi.org/10.1097/j.pain.0000000000000768.

Calejesan, A. A., Kim, S. J., & Zhuo, M. (2000). Descending facilitatory modulation of a behavioral nociceptive response by stimulation in the adult rat anterior cingulate cortex. *European Journal of Pain (London, England)*, *4*(1), 83–96. https://doi.org/10.1053/eujp.1999.0158.

Carlino, E., Guerra, G., & Piedimonte, A. (2016). Placebo effects: From pain to motor performance. *Neuroscience Letters*, *632*, 224–230. https://doi.org/10.1016/j.neulet.2016.08.046.

Carvalho, C., Caetano, J. M., Cunha, L., Rebouta, P., Kaptchuk, T. J., & Kirsch, I. (2016). Open-label placebo treatment in chronic low back pain: A randomized controlled trial. *Pain*, *157*(12), 2766–2772. https://doi.org/10.1097/j.pain.0000000000000700.

Charlesworth, J. E. G., Petkovic, G., Kelley, J. M., Hunter, M., Onakpoya, I., Roberts, N., et al. (2017). Effects of placebos without deception compared with no treatment: A systematic review and meta-analysis. *Journal of Evidence-Based Medicine, 10*(2), 97–107. https://doi.org/10.1111/jebm.12251.

Charron, J., Rainville, P., & Marchand, S. (2006). Direct comparison of placebo effects on clinical and experimental pain. *The Clinical Journal of Pain, 22*(2), 204–211.

Colagiuri, B., & Quinn, V. (2017). Autonomic arousal as a mechanism of the persistence of nocebo hyperalgesia. *The Journal of Pain: Official Journal of the American Pain Society, 19*(5), 476–486. https://doi.org/10.1016/j.jpain.2017.12.006.

Colloca, L., & Benedetti, F. (2009). Placebo analgesia induced by social observational learning. *Pain, 144*(1–2), 28–34. https://doi.org/10.1016/j.pain.2009.01.033.

Colloca, L., Enck, P., & DeGrazia, D. (2016). Relieving pain using dose-extending placebos: A scoping review. *Pain, 157*(8), 1590–1598. https://doi.org/10.1097/j.pain.0000000000000566.

Colloca, L., & Grillon, C. (2014). Understanding placebo and nocebo responses for pain management. *Current Pain and Headache Reports, 18*(6), 419. https://doi.org/10.1007/s11916-014-0419-2.

Colloca, L., & Howick, J. (2018). Placebos without deception: Outcomes, mechanisms, and ethics. *International Review of Neurobiology, 138*, 219–240. https://doi.org/10.1016/bs.irn.2018.01.005.

Colloca, L., Ludman, T., Bouhassira, D., Baron, R., Dickenson, A. H., Yarnitsky, D., et al. (2017). Neuropathic pain. *Nature Reviews Disease Primers, 3*, 17002. https://doi.org/10.1038/nrdp.2017.2.

Colloca, L., & Miller, F. G. (2011). Harnessing the placebo effect: The need for translational research. *Philosophical Transactions of the Royal Society of London. Series B, Biological Sciences, 366*(1572), 1922–1930. https://doi.org/10.1098/rstb.2010.0399.

Colloca, L., Petrovic, P., Wager, T. D., Ingvar, M., & Benedetti, F. (2010). How the number of learning trials affects placebo and nocebo responses. *Pain, 151*(2), 430–439. https://doi.org/10.1016/j.pain.2010.08.007.

Cormier, S., Piché, M., & Rainville, P. (2013). Expectations modulate heterotopic noxious counter-stimulation analgesia. *The Journal of Pain: Official Journal of the American Pain Society, 14*(2), 114–125. https://doi.org/10.1016/j.jpain.2012.10.006.

Corsi, N., & Colloca, L. (2017). Placebo and nocebo effects: The advantage of measuring expectations and psychological factors. *Frontiers in Psychology, 8*, 308. https://doi.org/10.3389/fpsyg.2017.00308.

Daenen, L., Nijs, J., Roussel, N., Wouters, K., Van Loo, M., & Cras, P. (2013). Dysfunctional pain inhibition in patients with chronic whiplash-associated disorders: An experimental study. *Clinical Rheumatology, 32*(1), 23–31. https://doi.org/10.1007/s10067-012-2085-2.

Damien, J., Léonard, G., Chalaye, P., Colloca, L., & Marchand, S. (2018). In *Sex differences in analgesia induced by conditioned pain modulation, hypnosis and placebo treatments. Presented at the Canadian Pain Society's (CPS) 39th Annual Scientific Meeting, Montreal: Canadian Journal of Pain.* https://doi.org/10.1080/24740527.2018.1476313.

De Felice, M., Sanoja, R., Wang, R., Vera-Portocarrero, L., Oyarzo, J., King, T., et al. (2011). Engagement of descending inhibition from the rostral ventromedial medulla protects against chronic neuropathic pain. *Pain, 152*(12), 2701–2709. https://doi.org/10.1016/j.pain.2011.06.008.

de Resende, M. A., Silva, L. F. S., Sato, K., Arendt-Nielsen, L., & Sluka, K. A. (2011). Blockade of opioid receptors in the medullary reticularis nucleus dorsalis, but not the rostral ventromedial medulla, prevents analgesia produced by diffuse noxious inhibitory control in rats with muscle inflammation. *The Journal of Pain: Official Journal of the American Pain Society, 12*(6), 687–697. https://doi.org/10.1016/j.jpain.2010.12.009.

de Souza, J. B., Potvin, S., Goffaux, P., Charest, J., & Marchand, S. (2009). The deficit of pain inhibition in fibromyalgia is more pronounced in patients with comorbid depressive symptoms. *The Clinical Journal of Pain*, *25*(2), 123–127. https://doi.org/10.1097/AJP.0b013e318183cfa4.

Dharmshaktu, P., Tayal, V., & Kalra, B. S. (2012). Efficacy of antidepressants as analgesics: A review. *Journal of Clinical Pharmacology*, *52*(1), 6–17. https://doi.org/10.1177/0091270010394852.

Diederich, N. J., & Goetz, C. G. (2008). The placebo treatments in neurosciences: New insights from clinical and neuroimaging studies. *Neurology*, *71*(9), 677–684. https://doi.org/10.1212/01.wnl.0000324635.49971.3d.

Diener, H.-C., Schorn, C. F., Bingel, U., & Dodick, D. W. (2008). The importance of placebo in headache research. *Cephalalgia: An International Journal of Headache*, *28*(10), 1003–1011. https://doi.org/10.1111/j.1468-2982.2008.01660.x.

DosSantos, M. F., Martikainen, I. K., Nascimento, T. D., Love, T. M., DeBoer, M. D., Schambra, H. M., et al. (2014). Building up analgesia in humans via the endogenous μ-opioid system by combining placebo and active tDCS: A preliminary report. *PLoS One*, *9*(7), e102350. https://doi.org/10.1371/journal.pone.0102350.

Edwards, R. R., Dolman, A. J., Martel, M. O., Finan, P. H., Lazaridou, A., Cornelius, M., et al. (2016). Variability in conditioned pain modulation predicts response to NSAID treatment in patients with knee osteoarthritis. *BMC Musculoskeletal Disorders*, *17*, 284. https://doi.org/10.1186/s12891-016-1124-6.

Edwards, R. R., Ness, T. J., Weigent, D. A., & Fillingim, R. B. (2003). Individual differences in diffuse noxious inhibitory controls (DNIC): Association with clinical variables. *Pain*, *106*(3), 427–437. https://doi.org/10.1016/j.pain.2003.09.005.

Egorova, N., Park, J., Orr, S. P., Kirsch, I., Gollub, R. L., & Kong, J. (2015). Not seeing or feeling is still believing: Conscious and non-conscious pain modulation after direct and observational learning. *Scientific Reports*, *5*, 16809. https://doi.org/10.1038/srep16809.

Elsenbruch, S., & Labrenz, F. (2018). Nocebo effects and experimental models in visceral pain. *International Review of Neurobiology*, *138*, 285–306. https://doi.org/10.1016/bs.irn.2018.01.010.

Escobar, W., Ramirez, K., Avila, C., Limongi, R., Vanegas, H., & Vazquez, E. (2012). Metamizol, a non-opioid analgesic, acts via endocannabinoids in the PAG-RVM axis during inflammation in rats. *European Journal of Pain (London, England)*, *16*(5), 676–689. https://doi.org/10.1002/j.1532-2149.2011.00057.x.

Fazeli, S., & Büchel, C. (2018). Pain related expectation and prediction error signals in the anterior insula are not related to aversiveness. *The Journal of Neuroscience: The Official Journal of the Society for Neuroscience*, *38*(29), 6461–6474. https://doi.org/10.1523/JNEUROSCI.0671-18.2018.

Fidanza, F., Varanini, M., Ciaramella, A., Carli, G., & Santarcangelo, E. L. (2017). Pain modulation as a function of hypnotizability: Diffuse noxious inhibitory control induced by cold pressor test vs explicit suggestions of analgesia. *Physiology & Behavior*, *171*, 135–141. https://doi.org/10.1016/j.physbeh.2017.01.013.

Fiorio, M., Recchia, S., Corrà, F., Simonetto, S., Garcia-Larrea, L., & Tinazzi, M. (2012). Enhancing non-noxious perception: Behavioural and neurophysiological correlates of a placebo-like manipulation. *Neuroscience*, *217*, 96–104. https://doi.org/10.1016/j.neuroscience.2012.04.066.

France, C. R., Burns, J. W., Gupta, R. K., Buvanendran, A., Chont, M., Schuster, E., et al. (2016). Expectancy effects on conditioned pain modulation are not influenced by naloxone or morphine. *Annals of Behavioral Medicine: A Publication of the Society of Behavioral Medicine*, *50*(4), 497–505. https://doi.org/10.1007/s12160-016-9775-y.

Gallotta, S., Bruno, V., Catapano, S., Mobilio, N., Ciacci, C., & Iovino, P. (2017). High risk of temporomandibular disorder in irritable bowel syndrome: Is there a correlation with greater illness severity? *World Journal of Gastroenterology*, *23*(1), 103–109. https://doi.org/10.3748/wjg.v23.i1.103.

Geers, A. L., Kosbab, K., Helfer, S. G., Weiland, P. E., & Wellman, J. A. (2007). Further evidence for individual differences in placebo responding: An interactionist perspective. *Journal of Psychosomatic Research*, *62*(5), 563–570. https://doi.org/10.1016/j.jpsychores.2006.12.005.

Gehling, J., Mainka, T., Vollert, J., Pogatzki-Zahn, E. M., Maier, C., & Enax-Krumova, E. K. (2016). Short-term test-retest-reliability of conditioned pain modulation using the cold-heat-pain method in healthy subjects and its correlation to parameters of standardized quantitative sensory testing. *BMC Neurology*, *16*, 125. https://doi.org/10.1186/s12883-016-0650-z.

Gierthmühlen, J., et al. (2015). Who is healthy? Aspects to consider when including healthy volunteers in QST-based studies—A consensus statement by the EUROPAIN and NEUROPAIN consortia. *Pain*, *156*(11), 2203–2211. https://doi.org/10.1097/j.pain.0000000000000227.

Goffaux, P., Redmond, W. J., Rainville, P., & Marchand, S. (2007). Descending analgesia—When the spine echoes what the brain expects. *Pain*, *130*(1), 137–143. https://doi.org/10.1016/j.pain.2006.11.011.

Goubert, D., Danneels, L., Cagnie, B., Van Oosterwijck, J., Kolba, K., Noyez, H., et al. (2015). Effect of pain induction or pain reduction on conditioned pain modulation in adults: A systematic review. *Pain Practice: The Official Journal of World Institute of Pain*, *15*(8), 765–777. https://doi.org/10.1111/papr.12241.

Gougeon, V., Gaumond, I., Goffaux, P., Potvin, S., & Marchand, S. (2016). Triggering descending pain inhibition by observing ourselves or a loved-one in pain. *The Clinical Journal of Pain*, *32*(3), 238–245. https://doi.org/10.1097/AJP.0000000000000244.

Gourion, D., & Mouchabac, S. (2016). Placebo effect: Clinical, biological and therapeutical involvements in depression. *L'Encephale*, *42*(1 Suppl. 1), 1S24-30. https://doi.org/10.1016/S0013-7006(16)30016-1. 1S24–1S30.

Granovsky, Y. (2013). Conditioned pain modulation: A predictor for development and treatment of neuropathic pain. *Current Pain and Headache Reports*, *17*(9), 361. https://doi.org/10.1007/s11916-013-0361-8.

Granovsky, Y., Miller-Barmak, A., Goldstein, O., Sprecher, E., & Yarnitsky, D. (2016). CPM test–retest reliability: "Standard" vs "single test-stimulus" protocols. *Pain Medicine*, *17*(3), 521–529. https://doi.org/10.1111/pme.12868.

Granovsky, Y., Nahman-Averbuch, H., Khamaisi, M., & Granot, M. (2017). Efficient conditioned pain modulation despite pain persistence in painful diabetic neuropathy. *Pain Reports*, *2*(3), e592. https://doi.org/10.1097/PR9.0000000000000592.

Grashorn, W., Schunke, O., Buhmann, C., Forkmann, K., Diedrich, S., Wesemann, K., et al. (2015). Influence of dopaminergic medication on conditioned pain modulation in Parkinson's disease patients. *PLoS One*, *10*(8), e0135287. https://doi.org/10.1371/journal.pone.0135287.

Grashorn, W., Sprenger, C., Forkmann, K., Wrobel, N., & Bingel, U. (2013). Age-dependent decline of endogenous pain control: Exploring the effect of expectation and depression. *PLoS One*, *8*(9), e75629. https://doi.org/10.1371/journal.pone.0075629.

Grosen, K., Vase, L., Pilegaard, H. K., Pfeiffer-Jensen, M., & Drewes, A. M. (2014). Conditioned pain modulation and situational pain catastrophizing as preoperative predictors of pain following chest wall surgery: A prospective observational cohort study. *PLoS One*, *9*(2), e90185. https://doi.org/10.1371/journal.pone.0090185.

Grotenhermen, F. (2004). Pharmacology of cannabinoids. *Neuro Endocrinology Letters, 25*(1–2), 14–23.

Gutierrez, T., Nackley, A. G., Neely, M. H., Freeman, K. G., Edwards, G. L., & Hohmann, A. G. (2003). Effects of neurotoxic destruction of descending noradrenergic pathways on cannabinoid antinociception in models of acute and tonic nociception. *Brain Research, 987*(2), 176–185. https://doi.org/10.1016/S0006-8993(03)03324-9.

Harper, D. E., Ichesco, E., Schrepf, A., Hampson, J. P., Clauw, D. J., Schmidt-Wilcke, T., et al. (2018). Resting functional connectivity of the periaqueductal gray is associated with normal inhibition and pathological facilitation in conditioned pain modulation. *The Journal of Pain: Official Journal of the American Pain Society, 19*(6), 635.e1–635.e15. https://doi.org/10.1016/j.jpain.2018.01.001.

Hermans, L., Nijs, J., Calders, P., De Clerck, L., Moorkens, G., Hans, G., et al. (2018). Influence of morphine and naloxone on pain modulation in rheumatoid arthritis, chronic fatigue syndrome/fibromyalgia, and controls: A double-blind, randomized, placebo-controlled, cross-over study. *Pain Practice: The Official Journal of World Institute of Pain, 18*(4), 418–430. https://doi.org/10.1111/papr.12613.

Hermans, L., Oosterwijck, J. V., Goubert, D., Goudman, L., Crombez, G., Calders, P., et al. (2016). Inventory of personal factors influencing conditioned pain modulation in healthy people: A systematic literature review. *Pain Practice: The Official Journal of World Institute of Pain, 16*(6), 758–769. https://doi.org/10.1111/papr.12305.

Höffken, O., Özgül, Ö. S., Enax-Krumova, E. K., Tegenthoff, M., & Maier, C. (2017). Evoked potentials after painful cutaneous electrical stimulation depict pain relief during a conditioned pain modulation. *BMC Neurology, 17*(1), 167. https://doi.org/10.1186/s12883-017-0946-7.

Hohmann, A. G., Tsou, K., & Walker, J. M. (1999). Cannabinoid suppression of noxious heat-evoked activity in wide dynamic range neurons in the lumbar dorsal horn of the rat. *Journal of Neurophysiology, 81*(2), 575–583. https://doi.org/10.1152/jn.1999.81.2.575.

Horvath, G., Kekesi, G., Nagy, E., & Benedek, G. (2008). The role of TRPV1 receptors in the antinociceptive effect of anandamide at spinal level. *Pain, 134*(3), 277–284. https://doi.org/10.1016/j.pain.2007.04.032.

Hróbjartsson, A., & Gøtzsche, P. C. (2010). Placebo interventions for all clinical conditions. *The Cochrane Database of Systematic Reviews, 3*, CD003974. https://doi.org/10.1002/14651858.CD003974.pub3.

Ibrahim, M. M., Porreca, F., Lai, J., Albrecht, P. J., Rice, F. L., Khodorova, A., et al. (2005). CB2 cannabinoid receptor activation produces antinociception by stimulating peripheral release of endogenous opioids. *Proceedings of the National Academy of Sciences of the United States of America, 102*(8), 3093–3098. https://doi.org/10.1073/pnas.0409888102.

Imai, Y., Petersen, K. K., Mørch, C. D., & Arendt Nielsen, L. (2016). Comparing test-retest reliability and magnitude of conditioned pain modulation using different combinations of test and conditioning stimuli. *Somatosensory & Motor Research, 33*(3–4), 169–177. https://doi.org/10.1080/08990220.2016.1229178.

Jepma, M., & Wager, T. D. (2015). Conceptual conditioning: Mechanisms mediating conditioning effects on pain. *Psychological Science, 26*(11), 1728–1739. https://doi.org/10.1177/0956797615597658.

Julien, N., & Marchand, S. (2006). Endogenous pain inhibitory systems activated by spatial summation are opioid-mediated. *Neuroscience Letters, 401*(3), 256–260. https://doi.org/10.1016/j.neulet.2006.03.032.

Jurth, C., Rehberg, B., & von Dincklage, F. (2014). Reliability of subjective pain ratings and nociceptive flexion reflex responses as measures of conditioned pain modulation. *Pain Research & Management, 19*(2), 93–96.

Kaptchuk, T. J., Friedlander, E., Kelley, J. M., Sanchez, M. N., Kokkotou, E., Singer, J. P., et al. (2010). Placebos without deception: A randomized controlled trial in irritable bowel syndrome. *PLOS ONE*, *5*(12), e15591. https://doi.org/10.1371/journal.pone.0015591.

Kennedy, D. L., Kemp, H. I., Ridout, D., Yarnitsky, D., & Rice, A. S. C. (2016). Reliability of conditioned pain modulation: A systematic review. *Pain*, *157*(11), 2410–2419. https://doi.org/10.1097/j.pain.0000000000000689.

Kiernan, B. D., Dane, J. R., Phillips, L. H., & Price, D. D. (1995). Hypnotic analgesia reduces R-III nociceptive reflex: Further evidence concerning the multifactorial nature of hypnotic analgesia. *Pain*, *60*(1), 39–47.

King, C. D., Goodin, B., Kindler, L. L., Caudle, R. M., Edwards, R. R., Gravenstein, N., et al. (2013). Reduction of conditioned pain modulation in humans by naltrexone: An exploratory study of the effects of pain catastrophizing. *Journal of Behavioral Medicine*, *36*(3), 315–327. https://doi.org/10.1007/s10865-012-9424-2.

Kirkpatrick, D. R., McEntire, D. M., Hambsch, Z. J., Kerfeld, M. J., Smith, T. A., Reisbig, M. D., et al. (2015). Therapeutic basis of clinical pain modulation. *Clinical and Translational Science*, *8*(6), 848–856. https://doi.org/10.1111/cts.12282.

Klinger, R., Blasini, M., Schmitz, J., & Colloca, L. (2017). Nocebo effects in clinical studies: Hints for pain therapy. *Pain Reports*, *2*(2), e586. https://doi.org/10.1097/PR9.0000000000000586.

Klyne, D. M., Schmid, A. B., Moseley, G. L., Sterling, M., & Hodges, P. W. (2015). Effect of types and anatomic arrangement of painful stimuli on conditioned pain modulation. *The Journal of Pain: Official Journal of the American Pain Society*, *16*(2), 176–185. https://doi.org/10.1016/j.jpain.2014.11.005.

Koban, L., & Wager, T. D. (2016). Beyond conformity: Social influences on pain reports and physiology. *Emotion (Washington, DC)*, *16*(1), 24–32. https://doi.org/10.1037/emo0000087.

Kosek, E., Rosen, A., Carville, S., Choy, E., Gracely, R. H., Marcus, H., et al. (2017). Lower placebo responses after long-term exposure to fibromyalgia pain. *The Journal of Pain: Official Journal of the American Pain Society*, *18*(7), 835–843. https://doi.org/10.1016/j.jpain.2017.02.434.

Kothari, S. F., Baad-Hansen, L., Hansen, L. B., Bang, N., Sørensen, L. H., Eskildsen, H. W., et al. (2016). Pain profiling of patients with temporomandibular joint arthralgia and osteoarthritis diagnosed with different imaging techniques. *The Journal of Headache and Pain*, *17*(1), 61. https://doi.org/10.1186/s10194-016-0653-6.

Kunz, M., Mohammadian, P., Renner, B., Roscher, S., Kobal, G., & Lautenbacher, S. (2014). Chemo-somatosensory evoked potentials: A sensitive tool to assess conditioned pain modulation? *Somatosensory & Motor Research*, *31*(2), 100–110. https://doi.org/10.3109/08990220.2014.887562.

Lautenbacher, S., Prager, M., & Rollman, G. B. (2007). Pain additivity, diffuse noxious inhibitory controls, and attention: A functional measurement analysis. *Somatosensory & Motor Research*, *24*(4), 189–201. https://doi.org/10.1080/08990220701637638.

Le Bars, D., Dickenson, A. H., & Besson, J. M. (1979a). Diffuse noxious inhibitory controls (DNIC). I. Effects on dorsal horn convergent neurones in the rat. *Pain*, *6*(3), 283–304.

Le Bars, D., Dickenson, A. H., & Besson, J. M. (1979b). Diffuse noxious inhibitory controls (DNIC). II. Lack of effect on non-convergent neurones, supraspinal involvement and theoretical implications. *Pain*, *6*(3), 305–327.

Lee, S., Walker, J. R., Jakul, L., & Sexton, K. (2004). Does elimination of placebo responders in a placebo run-in increase the treatment effect in randomized clinical trials? A meta-analytic evaluation. *Depression and Anxiety*, *19*(1), 10–19. https://doi.org/10.1002/da.10134.

Leffler, A.-S., Kosek, E., Lerndal, T., Nordmark, B., & Hansson, P. (2002). Somatosensory perception and function of diffuse noxious inhibitory controls (DNIC) in patients suffering from rheumatoid arthritis. *European Journal of Pain (London, England)*, *6*(2), 161–176. https://doi.org/10.1053/eujp.2001.0313.

Lewis, G. N., Heales, L., Rice, D. A., Rome, K., & McNair, P. J. (2012). Reliability of the conditioned pain modulation paradigm to assess endogenous inhibitory pain pathways. *Pain Research & Management*, *17*(2), 98–102. https://doi.org/10.1155/2012/610561.

Li, M.-H., Suchland, K. L., & Ingram, S. L. (2017). Compensatory activation of cannabinoid CB2 receptor inhibition of GABA release in the rostral ventromedial medulla in inflammatory pain. *The Journal of Neuroscience: The Official Journal of the Society for Neuroscience*, *37*(3), 626–636. https://doi.org/10.1523/JNEUROSCI.1310-16.2016.

Lindstedt, F., Berrebi, J., Greayer, E., Lonsdorf, T. B., Schalling, M., Ingvar, M., et al. (2011). Conditioned pain modulation Is associated with common polymorphisms in the serotonin transporter gene. *PLoS One*, *6*(3), e18252. https://doi.org/10.1371/journal.pone.0018252.

Locke, D., Gibson, W., Moss, P., Munyard, K., Mamotte, C., & Wright, A. (2014). Analysis of meaningful conditioned pain modulation effect in a pain-free adult population. *The Journal of Pain: Official Journal of the American Pain Society*, *15*(11), 1190–1198. https://doi.org/10.1016/j.jpain.2014.09.001.

Loeser, J. D. (1980). Perspectives on pain. In P. Turner, C. Padgham, & A. Hedges (Eds.), *Clinical pharmacology & therapeutics* (pp. 313–316). London: Palgrave Macmillan.

Maione, S., Bisogno, T., de Novellis, V., Palazzo, E., Cristino, L., Valenti, M., et al. (2006). Elevation of endocannabinoid levels in the ventrolateral periaqueductal grey through inhibition of fatty acid amide hydrolase affects descending nociceptive pathways via both cannabinoid receptor type 1 and transient receptor potential vanilloid type-1 receptors. *The Journal of Pharmacology and Experimental Therapeutics*, *316*(3), 969–982. https://doi.org/10.1124/jpet.105.093286.

Manning, B. H., Martin, W. J., & Meng, I. D. (2003). The rodent amygdala contributes to the production of cannabinoid-induced antinociception. *Neuroscience*, *120*(4), 1157–1170. https://doi.org/10.1016/S0306-4522(03)00356-7.

Marchand, S. (2012). *The phenomenon of pain*. IASP Press.

Marchand, S., & Arsenault, P. (2002). Spatial summation for pain perception: Interaction of inhibitory and excitatory mechanisms. *Pain*, *95*(3), 201–206.

Marcuzzi, A., Wrigley, P. J., Dean, C. M., Adams, R., & Hush, J. M. (2017). The long-term reliability of static and dynamic quantitative sensory testing in healthy individuals. *Pain*, *158*(7), 1217–1223. https://doi.org/10.1097/j.pain.0000000000000901.

Martel, M., Wasan, A., & Edwards, R. (2013). Sex differences in the stability of conditioned pain modulation (CPM) among patients with chronic pain. *Pain Medicine (Malden, Mass.)*, *14*(11), 1757–1768. https://doi.org/10.1111/pme.12220.

Martini, C., van Velzen, M., Drewes, A., Aarts, L., Dahan, A., & Niesters, M. (2015). A randomized controlled trial on the effect of tapentadol and morphine on conditioned pain modulation in healthy volunteers. *PLoS One*, *10*(6), e0128997. https://doi.org/10.1371/journal.pone.0128997.

Medoff, Z. M., & Colloca, L. (2015). Placebo analgesia: Understanding the mechanisms. *Pain Management*, *5*(2), 89–96. https://doi.org/10.2217/pmt.15.3.

Melzack, R. (1990). Phantom limbs and the concept of a neuromatrix. *Trends in Neurosciences*, *13*(3), 88–92.

Melzack, R. (2001). Pain and the neuromatrix in the brain. *Journal of Dental Education*, *65*(12), 1378–1382.

Melzack, R., & Wall, P. D. (1965). Pain mechanisms: A new theory. *Science (New York, N.Y.)*, *150*(3699), 971–979.

Mitsikostas, D. D. (2012). Nocebo in headaches: Implications for clinical practice and trial design. *Current Neurology and Neuroscience Reports, 12*(2), 132–137. https://doi.org/10.1007/s11910-011-0245-4.

Mlekusch, S., Neziri, A. Y., Limacher, A., Jüni, P., Arendt-Nielsen, L., & Curatolo, M. (2016). Conditioned pain modulation in patients with acute and chronic low back pain. *The Clinical Journal of Pain, 32*(2), 116–121. https://doi.org/10.1097/AJP.0000000000000238.

Moont, R., Crispel, Y., Lev, R., Pud, D., & Yarnitsky, D. (2011). Temporal changes in cortical activation during conditioned pain modulation (CPM), a LORETA study. *Pain, 152*(7), 1469–1477. https://doi.org/10.1016/j.pain.2011.01.036.

Moont, R., Pud, D., Sprecher, E., Sharvit, G., & Yarnitsky, D. (2010). "Pain inhibits pain" mechanisms: Is pain modulation simply due to distraction? *Pain, 150*(1), 113–120. https://doi.org/10.1016/j.pain.2010.04.009.

Morton, D. L., El-Deredy, W., & Jones, A. K. P. (2014). Placebo analgesia: Cognition or perception. *Handbook of Experimental Pharmacology, 225*, 71–80. https://doi.org/10.1007/978-3-662-44519-8_4.

Müller, M., Kamping, S., Benrath, J., Skowronek, H., Schmitz, J., Klinger, R., et al. (2016). Treatment history and placebo responses to experimental and clinical pain in chronic pain patients. *European Journal of Pain (London, England), 20*(9), 1530–1541. https://doi.org/10.1002/ejp.877.

Murray, D., & Stoessl, A. J. (2013). Mechanisms and therapeutic implications of the placebo effect in neurological and psychiatric conditions. *Pharmacology & Therapeutics, 140*(3), 306–318. https://doi.org/10.1016/j.pharmthera.2013.07.009.

Nahman-Averbuch, H., Nir, R.-R., Sprecher, E., & Yarnitsky, D. (2016). Psychological factors and conditioned pain modulation: A meta-analysis. *The Clinical Journal of Pain, 32*(6), 541–554. https://doi.org/10.1097/AJP.0000000000000296.

Nahman-Averbuch, H., Yarnitsky, D., Granovsky, Y., Sprecher, E., Steiner, M., Tzuk-Shina, T., et al. (2011). Pronociceptive pain modulation in patients with painful chemotherapy-induced polyneuropathy. *Journal of Pain and Symptom Management, 42*(2), 229–238. https://doi.org/10.1016/j.jpainsymman.2010.10.268.

Ng, T. S., Pedler, A., Vicenzino, B., & Sterling, M. (2014). Less efficacious conditioned pain modulation and sensory hypersensitivity in chronic whiplash-associated disorders in Singapore. *The Clinical Journal of Pain, 30*(5), 436–442. https://doi.org/10.1097/AJP.0b013e3182a03940.

Niesters, M., Aarts, L., Sarton, E., & Dahan, A. (2013). Influence of ketamine and morphine on descending pain modulation in chronic pain patients: A randomized placebo-controlled cross-over proof-of-concept study. *British Journal of Anaesthesia, 110*(6), 1010–1016. https://doi.org/10.1093/bja/aes578.

Niesters, M., Dahan, A., Swartjes, M., Noppers, I., Fillingim, R. B., Aarts, L., et al. (2011). Effect of ketamine on endogenous pain modulation in healthy volunteers. *Pain, 152*(3), 656–663. https://doi.org/10.1016/j.pain.2010.12.015.

Niesters, M., Proto, P. L., Aarts, L., Sarton, E. Y., Drewes, A. M., & Dahan, A. (2014). Tapentadol potentiates descending pain inhibition in chronic pain patients with diabetic polyneuropathy. *British Journal of Anaesthesia, 113*(1), 148–156. https://doi.org/10.1093/bja/aeu056.

Nilsen, K. B., Olsen, I. C., Solem, A. N., & Matre, D. (2014). A large conditioned pain modulation response is not related to a large blood pressure response: A study in healthy men. *European Journal of Pain (London, England), 18*(9), 1271–1279. https://doi.org/10.1002/j.1532-2149.2014.486.x.

Nir, R.-R., Granovsky, Y., Yarnitsky, D., Sprecher, E., & Granot, M. (2011). A psychophysical study of endogenous analgesia: the role of the conditioning pain in the induction and magnitude of conditioned pain modulation. *European Journal of Pain (London, England), 15*(5), 491–497. https://doi.org/10.1016/j.ejpain.2010.10.001.

Nir, R.-R., & Yarnitsky, D. (2015). Conditioned pain modulation. *Current Opinion in Supportive and Palliative Care*, *9*(2), 131–137. https://doi.org/10.1097/SPC.0000000000000126.

Nir, R.-R., Yarnitsky, D., Honigman, L., & Granot, M. (2012). Cognitive manipulation targeted at decreasing the conditioning pain perception reduces the efficacy of conditioned pain modulation. *Pain*, *153*(1), 170–176. https://doi.org/10.1016/j.pain.2011.10.010.

Normand, E., Potvin, S., Gaumond, I., Cloutier, G., Corbin, J.-F., & Marchand, S. (2011). Pain inhibition is deficient in chronic widespread pain but normal in major depressive disorder. *The Journal of Clinical Psychiatry*, *72*(2), 219–224. https://doi.org/10.4088/JCP.08m04969blu.

O'Brien, A. T., Deitos, A., Triñanes Pego, Y., Fregni, F., & Carrillo-de-la-Peña, M. T. (2018). Defective endogenous pain modulation in fibromyalgia: A meta-analysis of temporal summation and conditioned pain modulation paradigms. *The Journal of Pain: Official Journal of the American Pain Society*, *19*(8), 819–836. https://doi.org/10.1016/j.jpain.2018.01.010.

Okada, C. (2005). The relationship between diffuse noxious inhibitory control effects and the magnitude of noxious stimuli. *Kokubyo Gakkai Zasshi. The Journal of the Stomatological Society, Japan*, *72*(1), 49–55.

Olesen, S. S., Graversen, C., Bouwense, S. A. W., van Goor, H., Wilder-Smith, O. H. G., & Drewes, A. M. (2013). Quantitative sensory testing predicts pregabalin efficacy in painful chronic pancreatitis. *PLoS One*, *8*(3), e57963 https://doi.org/10.1371/journal.pone.0057963.

Oono, Y., Wang, K., Baad-Hansen, L., Futarmal, S., Kohase, H., Svensson, P., et al. (2014). Conditioned pain modulation in temporomandibular disorders (TMD) pain patients. *Experimental Brain Research*, *232*(10), 3111–3119. https://doi.org/10.1007/s00221-014-3997-7.

Palazzo, E., Luongo, L., Novellis, V., De Rossi, F., & Maione, S. (2010). The role of cannabinoid receptors in the descending modulation of pain. *Pharmaceuticals (Basel, Switzerland)*, *3*(8), 2661–2673. https://doi.org/10.3390/ph3082661.

Palazzo, E., Rossi, F., & Maione, S. (2008). Role of TRPV1 receptors in descending modulation of pain. *Molecular and Cellular Endocrinology*, *286*(1–2 Suppl. 1), S79–S83. https://doi.org/10.1016/j.mce.2008.01.013.

Parent, A. J., Beaudet, N., Daigle, K., Sabbagh, R., Sansoucy, Y., Marchand, S., et al. (2015). Relationship between blood- and cerebrospinal fluid-bound neurotransmitter concentrations and conditioned pain modulation in pain-free and chronic pain subjects. *The Journal of Pain: Official Journal of the American Pain Society*, *16*(5), 436–444. https://doi.org/10.1016/j.jpain.2015.01.007.

Peciña, M., Azhar, H., Love, T. M., Lu, T., Fredrickson, B. L., Stohler, C. S., et al. (2013). Personality trait predictors of placebo analgesia and neurobiological correlates. *Neuropsychopharmacology*, *38*(4), 639–646. https://doi.org/10.1038/npp.2012.227.

Peciña, M., Love, T., Stohler, C. S., Goldman, D., & Zubieta, J.-K. (2015). Effects of the mu opioid receptor polymorphism (OPRM1 A118G) on pain regulation, placebo effects and associated personality trait measures. *Neuropsychopharmacology*, *40*(4), 957–965. https://doi.org/10.1038/npp.2014.272.

Petersen, K. K., Graven-Nielsen, T., Simonsen, O., Laursen, M. B., & Arendt-Nielsen, L. (2016). Preoperative pain mechanisms assessed by cuff algometry are associated with chronic postoperative pain relief after total knee replacement. *Pain*, *157*(7), 1400–1406. https://doi.org/10.1097/j.pain.0000000000000531.

Petersen, K. K., Vaegter, H. B., & Arendt-Nielsen, L. (2017). An updated view on the reliability of different protocols for the assessment of conditioned pain modulation. *Pain*, *158*(5), 988. https://doi.org/10.1097/j.pain.0000000000000833.

Piché, M., Arsenault, M., & Rainville, P. (2009). Cerebral and cerebrospinal processes underlying counterirritation analgesia. *The Journal of Neuroscience: The Official Journal of the Society for Neuroscience*, *29*(45), 14236–14246. https://doi.org/10.1523/JNEUROSCI.2341-09.2009.

Piché, M., Watanabe, N., Sakata, M., Oda, K., Toyohara, J., Ishii, K., et al. (2014). Basal μ-opioid receptor availability in the amygdala predicts the inhibition of pain-related brain activity during heterotopic noxious counter-stimulation. *Neuroscience Research*, *81–82*, 78–84. https://doi.org/10.1016/j.neures.2014.02.006.

Pickering, G., Pereira, B., Dufour, E., Soule, S., & Dubray, C. (2014). Impaired modulation of pain in patients with postherpetic neuralgia. *Pain Research & Management*, *19*(1), e19–e23.

Potvin, S., & Marchand, S. (2016). Pain facilitation and pain inhibition during conditioned pain modulation in fibromyalgia and in healthy controls. *Pain*, *157*(8), 1704–1710. https://doi.org/10.1097/j.pain.0000000000000573.

Quante, M., Hille, S., Schofer, M. D., Lorenz, J., & Hauck, M. (2008). Noxious counter-irritation in patients with advanced osteoarthritis of the knee reduces MCC but not SII pain generators: A combined use of MEG and EEG. *Journal of Pain Research*, *1*, 1–8.

Rainville, P., Carrier, B., Hofbauer, R. K., Bushnell, M. C., & Duncan, G. H. (1999). Dissociation of sensory and affective dimensions of pain using hypnotic modulation. *Pain*, *82*(2), 159–171. https://doi.org/10.1016/S0304-3959(99)00048-2.

Ram, K. C., Eisenberg, E., Haddad, M., & Pud, D. (2008). Oral opioid use alters DNIC but not cold pain perception in patients with chronic pain—New perspective of opioid-induced hyperalgesia. *Pain*, *139*(2), 431–438. https://doi.org/10.1016/j.pain.2008.05.015.

Rattanavong, V. (2013). *Manipulation de la perception de la composante affective de la douleur* (Master's thesis). Quebec, Canada: Université de Sherbrooke, Sherbrooke.

Razavi, M., Hansson, P. T., Johansson, B., & Leffler, A.-S. (2014). The influence of intensity and duration of a painful conditioning stimulation on conditioned pain modulation in volunteers. *European Journal of Pain (London, England)*, *18*(6), 853–861. https://doi.org/10.1002/j.1532-2149.2013.00435.x.

Reicherts, P., Gerdes, A. B. M., Pauli, P., & Wieser, M. J. (2016). Psychological placebo and nocebo effects on pain rely on expectation and previous experience. *The Journal of Pain: Official Journal of the American Pain Society*, *17*(2), 203–214. https://doi.org/10.1016/j.jpain.2015.10.010.

Reynolds, D. V. (1969). Surgery in the rat during electrical analgesia induced by focal brain stimulation. *Science (New York, NY)*, *164*(3878), 444–445.

Rief, W., & Glombiewski, J. A. (2012). The hidden effects of blinded, placebo-controlled randomized trials: An experimental investigation. *Pain*, *153*(12), 2473–2477. https://doi.org/10.1016/j.pain.2012.09.007.

Roosink, M., Renzenbrink, G. J., Buitenweg, J. R., van Dongen, R. T. M., Geurts, A. C. H., & Ijzerman, M. J. (2011). Somatosensory symptoms and signs and conditioned pain modulation in chronic post-stroke shoulder pain. *The Journal of Pain: Official Journal of the American Pain Society*, *12*(4), 476–485. https://doi.org/10.1016/j.jpain.2010.10.009.

Rustamov, N., Tessier, J., Provencher, B., Lehmann, A., & Piché, M. (2016). Inhibitory effects of heterotopic noxious counter-stimulation on perception and brain activity related to Aβ-fibre activation. *The European Journal of Neuroscience*, *44*(1), 1771–1778. https://doi.org/10.1111/ejn.13258.

Sáez-Cassanelli, J. L., Fontanella, G. H., Delgado-García, J. M., & Carrión, A. M. (2007). Functional blockage of the cannabinoid receptor type 1 evokes a kappa-opiate-dependent analgesia. *Journal of Neurochemistry*, *103*(6), 2629–2639. https://doi.org/10.1111/j.1471-4159.2007.05000.x.

Sanada, T., Kohase, H., Makino, K., & Umino, M. (2009). Effects of alpha-adrenergic agonists on pain modulation in diffuse noxious inhibitory control. *Journal of Medical and Dental Sciences*, *56*(1), 17–24.

Sandrini, G., Rossi, P., Milanov, I., Serrao, M., Cecchini, A. P., & Nappi, G. (2006). Abnormal modulatory influence of diffuse noxious inhibitory controls in migraine and chronic tension-type headache patients. *Cephalalgia: An International Journal of Headache*, *26*(7), 782–789. https://doi.org/10.1111/j.1468-2982.2006.01130.x.

Sangesland, A., Støren, C., & Vaegter, H. B. (2017). Are preoperative experimental pain assessments correlated with clinical pain outcomes after surgery? A systematic review. *Scandinavian Journal of Pain*, *15*, 44–52. https://doi.org/10.1016/j.sjpain.2016.12.002.

Sauro, M. D., & Greenberg, R. P. (2005). Endogenous opiates and the placebo effect: A meta-analytic review. *Journal of Psychosomatic Research*, *58*(2), 115–120. https://doi.org/10.1016/j.jpsychores.2004.07.001.

Schwarz, K. A., Pfister, R., & Büchel, C. (2016). Rethinking explicit expectations: Connecting placebos, social cognition, and contextual perception. *Trends in Cognitive Sciences*, *20*(6), 469–480. https://doi.org/10.1016/j.tics.2016.04.001.

Simmons, K., Ortiz, R., Kossowsky, J., Krummenacher, P., Grillon, C., Pine, D., et al. (2014). Pain and placebo in pediatrics: A comprehensive review of laboratory and clinical findings. *Pain*, *155*(11), 2229–2235. https://doi.org/10.1016/j.pain.2014.08.036.

Skovbjerg, S., Jørgensen, T., Arendt-Nielsen, L., Ebstrup, J. F., Carstensen, T., & Graven-Nielsen, T. (2017). Conditioned pain modulation and pressure pain sensitivity in the adult Danish general population: The DanFund study. *The Journal of Pain: Official Journal of the American Pain Society*, *18*(3), 274–284. https://doi.org/10.1016/j.jpain.2016.10.022.

Skyt, I., Moslemi, K., Baastrup, C., Grosen, K., Svensson, P., Jensen, T. S., et al. (2018). Does conditioned pain modulation predict the magnitude of placebo effects in patients with neuropathic pain? *European Journal of Pain (London, England)*, *22*(4), 784–792. https://doi.org/10.1002/ejp.1164.

Soon, B., Vicenzino, B., Schmid, A. B., & Coppieters, M. W. (2017). Facilitatory and inhibitory pain mechanisms are altered in patients with carpal tunnel syndrome. *PLoS One*, *12*(8), e0183252. https://doi.org/10.1371/journal.pone.0183252.

Sprenger, C., Bingel, U., & Büchel, C. (2011). Treating pain with pain: Supraspinal mechanisms of endogenous analgesia elicited by heterotopic noxious conditioning stimulation. *Pain*, *152*(2), 428–439. https://doi.org/10.1016/j.pain.2010.11.018.

Starowicz, K., Makuch, W., Osikowicz, M., Piscitelli, F., Petrosino, S., Di Marzo, V., et al. (2012). Spinal anandamide produces analgesia in neuropathic rats: Possible CB(1)- and TRPV1-mediated mechanisms. *Neuropharmacology*, *62*(4), 1746–1755. https://doi.org/10.1016/j.neuropharm.2011.11.021.

Staud, R. (2012). Abnormal endogenous pain modulation is a shared characteristic of many chronic pain conditions. *Expert Review of Neurotherapeutics*, *12*(5), 577–585. https://doi.org/10.1586/ern.12.41.

Sugimine, S., Saito, S., Araki, T., Yamamoto, K., & Obata, H. (2017). Endogenous analgesic effect of pregabalin: A double-blind and randomized controlled trial. *European Journal of Pain (London, England)*, *21*(6), 997–1006. https://doi.org/10.1002/ejp.1007.

Taylor, B. K. (2009). Spinal inhibitory neurotransmission in neuropathic pain. *Current Pain and Headache Reports*, *13*(3), 208–214.

Taylor, V. A., Chang, L., Rainville, P., & Roy, M. (2017). Learned expectations and uncertainty facilitate pain during classical conditioning. *Pain*, *158*(8), 1528–1537. https://doi.org/10.1097/j.pain.0000000000000948.

Tiemann, L., Heitmann, H., Schulz, E., Baumkötter, J., & Ploner, M. (2014). Dopamine precursor depletion influences pain affect rather than pain sensation. *PLoS One*, *9*(4), e96167. https://doi.org/10.1371/journal.pone.0096167.

Torta, D. M., Churyukanov, M. V., Plaghki, L., & Mouraux, A. (2015). The effect of heterotopic noxious conditioning stimulation on Aδ-, C- and Aβ-fibre brain responses in humans. *The European Journal of Neuroscience*, *42*(9), 2707–2715. https://doi.org/10.1111/ejn.13071.

Treister, R., Pud, D., & Eisenberg, E. (2013). The dopamine agonist apomorphine enhances conditioned pain modulation in healthy humans. *Neuroscience Letters*, *548*, 115–119. https://doi.org/10.1016/j.neulet.2013.05.041.

Tuttle, A. H., Tohyama, S., Ramsay, T., Kimmelman, J., Schweinhardt, P., Bennett, G. J., et al. (2015). Increasing placebo responses over time in U.S. clinical trials of neuropathic pain. *Pain*, *156*(12), 2616–2626. https://doi.org/10.1097/j.pain.0000000000000333.

van Wijk, G., & Veldhuijzen, D. S. (2010). Perspective on diffuse noxious inhibitory controls as a model of endogenous pain modulation in clinical pain syndromes. *The Journal of Pain: Official Journal of the American Pain Society*, *11*(5), 408–419. https://doi.org/10.1016/j.jpain.2009.10.009.

Vase, L., Riley, J. L., & Price, D. D. (2002). A comparison of placebo effects in clinical analgesic trials versus studies of placebo analgesia. *Pain*, *99*(3), 443–452. https://doi.org/10.1016/S0304-3959(02)00205-1.

Vase, L., Vollert, J., Finnerup, N. B., Miao, X., Atkinson, G., Marshall, S., et al. (2015). Predictors of the placebo analgesia response in randomized controlled trials of chronic pain: A meta-analysis of the individual data from nine industrially sponsored trials. *Pain*, *156*(9), 1795. https://doi.org/10.1097/j.pain.0000000000000217.

Villanueva, L., Bouhassira, D., & Le Bars, D. (1996). The medullary subnucleus reticularis dorsalis (SRD) as a key link in both the transmission and modulation of pain signals. *Pain*, *67*(2–3), 231–240.

Villanueva, L., & Le Bars, D. (1986). Indirect effects of intrathecal morphine upon diffuse noxious inhibitory controls (DNICs) in the rat. *Pain*, *26*(2), 233–243.

Vo, L., & Drummond, P. D. (2016). Involvement of α2-adrenoceptors in inhibitory and facilitatory pain modulation processes. *European Journal of Pain (London, England)*, *20*(3), 386–398. https://doi.org/10.1002/ejp.736.

Watkinson, A., Chapman, S. C. E., & Horne, R. (2017). Beliefs about pharmaceutical medicines and natural remedies explain individual variation in placebo analgesia. *The Journal of Pain: Official Journal of the American Pain Society*, *18*(8), 908–922. https://doi.org/10.1016/j.jpain.2017.02.435.

Weimer, K., Colloca, L., & Enck, P. (2015). Age and sex as moderators of the placebo response—An evaluation of systematic reviews and meta-analyses across medicine. *Gerontology*, *61*(2), 97–108. https://doi.org/10.1159/000365248.

Whalley, B., Hyland, M. E., & Kirsch, I. (2008). Consistency of the placebo effect. *Journal of Psychosomatic Research*, *64*(5), 537–541. https://doi.org/10.1016/j.jpsychores.2007.11.007.

Willer, J. C., Le Bars, D., & De Broucker, T. (1990). Diffuse noxious inhibitory controls in man: Involvement of an opioidergic link. *European Journal of Pharmacology*, *182*(2), 347–355.

Yarnitsky, D. (2010). Conditioned pain modulation (the diffuse noxious inhibitory control-like effect): Its relevance for acute and chronic pain states. *Current Opinion in Anesthesiology*, *23*(5), 611. https://doi.org/10.1097/ACO.0b013e32833c348b.

Yarnitsky, D. (2015). Role of endogenous pain modulation in chronic pain mechanisms and treatment. *Pain*, *156*(Suppl. 1), S24–S31. https://doi.org/10.1097/01.j.pain.0000460343.46847.58.

Yarnitsky, D., Arendt-Nielsen, L., Bouhassira, D., Edwards, R. R., Fillingim, R. B., Granot, M., et al. (2010). Recommendations on terminology and practice of psychophysical DNIC testing. *European Journal of Pain (London, England)*, *14*(4), 339. https://doi.org/10.1016/j.ejpain.2010.02.004.

Yarnitsky, D., Bouhassira, D., Drewes, A. m., Fillingim, R. B., Granot, M., Hansson, P., et al. (2015). Recommendations on practice of conditioned pain modulation (CPM) testing. *European Journal of Pain, 19*(6), 805–806. https://doi.org/10.1002/ejp.605.

Yarnitsky, D., Crispel, Y., Eisenberg, E., Granovsky, Y., Ben-Nun, A., Sprecher, E., et al. (2008). Prediction of chronic post-operative pain: Pre-operative DNIC testing identifies patients at risk. *Pain, 138*(1), 22–28. https://doi.org/10.1016/j.pain.2007.10.033.

Yarnitsky, D., Granot, M., Nahman-Averbuch, H., Khamaisi, M., & Granovsky, Y. (2012). Conditioned pain modulation predicts duloxetine efficacy in painful diabetic neuropathy. *Pain, 153*(6), 1193–1198. https://doi.org/10.1016/j.pain.2012.02.021.

Youssef, A. M., Macefield, V. G., & Henderson, L. A. (2016a). Pain inhibits pain; human brainstem mechanisms. *NeuroImage, 124*(Pt. A), 54–62. https://doi.org/10.1016/j.neuroimage.2015.08.060.

Youssef, A. M., Macefield, V. G., & Henderson, L. A. (2016b). Cortical influences on brainstem circuitry responsible for conditioned pain modulation in humans. *Human Brain Mapping, 37*(7), 2630–2644. https://doi.org/10.1002/hbm.23199.

Yuan, X.-C., Zhu, B., Jing, X.-H., Xiong, L.-Z., Wu, C.-H., Gao, F., et al. (2018). Electroacupuncture potentiates cannabinoid receptor-mediated descending inhibitory control in a mouse model of knee osteoarthritis. *Frontiers in Molecular Neuroscience, 11*, 112. https://doi.org/10.3389/fnmol.2018.00112.

Zheng, Z., Wang, K., Yao, D., Xue, C. C. L., & Arendt-Nielsen, L. (2014). Adaptability to pain is associated with potency of local pain inhibition, but not conditioned pain modulation: A healthy human study. *Pain, 155*(5), 968–976. https://doi.org/10.1016/j.pain.2014.01.024.

FURTHER READING

Amanzio, M., & Benedetti, F. (1999). Neuropharmacological dissection of placebo analgesia: Expectation-activated opioid systems versus conditioning-activated specific subsystems. *The Journal of Neuroscience: The Official Journal of the Society for Neuroscience, 19*(1), 484–494.

Benedetti, F., Amanzio, M., Baldi, S., Casadio, C., & Maggi, G. (1999). Inducing placebo respiratory depressant responses in humans via opioid receptors. *The European Journal of Neuroscience, 11*(2), 625–631.

CHAPTER TWELVE

Modulation of the Motor System by Placebo and Nocebo Effects

Mirta Fiorio[1]
Department of Neuroscience, Biomedicine and Movement Sciences, University of Verona, Verona, Italy
[1]Corresponding author: e-mail address: mirta.fiorio@univr.it

Contents

Abstract

There is strong behavioral evidence that placebo and nocebo effects can influence aspects of motor performance like speed, force, and resistance to fatigue in athletes and non-athletes alike. These behavioral studies were essential for extending experimental investigation of the placebo and nocebo effects from the pain to the motor domain and to reveal how verbal suggestions and experiential learning are involved in shaping modulatory systems and related behavioral responses. However, the neural underpinnings of these effects in the motor domain are still largely unknown. Studies in healthy subjects demonstrated that the placebo-induced enhancement of force is associated with increased activity in the corticospinal system and that the placebo-induced reduction of fatigue can be disclosed by recording the readiness potential, an electrophysiological sign of movement preparation. Further evidence derives from studies in patients with Parkinson's disease that have directly demonstrated that placebo-induced improvements in motor symptoms are related to changes in subcortical neural firing activity and dopamine release. Future investigations are needed to better clarify the complex neural architecture underpinning the placebo and nocebo effects in the motor domain.

International Review of Neurobiology, Volume 139
ISSN 0074-7742
https://doi.org/10.1016/bs.irn.2018.07.012

LIST OF ABBREVIATIONS

CSP cortical silent period
DBS deep brain stimulation
MEP motor evoked potential
M1 primary motor cortex
SNr substantia nigra pars reticulate
STN subthalamic nucleus
TENS transcutaneous electrical nerve stimulation
TMS transcranial magnetic stimulation
VA ventral anterior nucleus of the thalamus
VLa anterior ventrolateral nucleus of the thalamus

1. INTRODUCTION

Motor performance is a multifaceted definition that refers to several different dimensions, including force production, precision control, movement speed, resistance to fatigue, motor adaptation, and motor skill learning. To date, many behavioral studies have demonstrated the efficacy of placebo and nocebo procedures in influencing some aspects of motor performance (e.g., endurance, strength, speed), while others remain unexplored (e.g., precision control or motor skill learning). The beneficial effects of placebos on physical performance extend even to critical conditions, such as high altitudes in which oxygen pressure is low (Benedetti, Barbiani, & Camerone, 2018).

Descriptions of the beneficial effects of placebos on physical performance date back to the 1970s. Nonetheless, controlled experimental studies in this field are more recent and are not limited to well-trained athletes but extend to untrained, naïve subjects tested in controlled laboratory settings. Moreover, the investigation of placebo and nocebo effects in the motor domain concerns not only movements performed with the whole body, as in sports, but also more controlled and localized movements performed with a single muscle or a restricted muscle group, such as abduction or flexion of the finger and the arm or extension of the leg. Restricting the investigation to specific muscles allows to conduct precise neurophysiological investigations.

1.1 Verbal Suggestions and Motor Performance

Cognitive functions are of fundamental importance for motor performance. Athletes' expectations and beliefs about their abilities and the effectiveness of ergogenic aids (e.g., substances or devices) exert a positive effect on motor outcome. Expectations in sports can be variously induced. In a study by McKay, Lewthwaite, and Wulf (2012), healthy participants were asked to complete questionnaires that they thought served to measure their ability to perform well under pressure (while they actually measured other aspects of personality). After completing the questionnaires, a group of participants was told they had scored high on the questionnaires, meaning that they were able to handle difficulties and perform well under pressure. Differently, a control group was not debriefed about the questionnaire scores. Before and after this experimental manipulation, all participants were evaluated for their accuracy to perform baseball throws in two conditions: under low and under high pressure. Both groups showed equal performance in the initial phase and in the low pressure condition, whereas after the manipulation and in the high pressure condition the enhanced-expectancy group displayed a significant increase in accuracy as compared to the control group (McKay et al., 2012). These positive effects were obtained even when the induced expectation was unrelated to movement performance per se. This demonstrates that the positive effects of general, non-specific expectations of being able to perform well under pressure affect motor performance on a specific task.

The role of expectation as a psychological mechanism that can influence sport performance has been extensively demonstrated with placebo manipulations. In 1972, Ariel and Saville first described the positive effects of placebos in sports by showing that after the administration of an inert substance, masked by anabolic steroids, weightlifters improved their performance by 9.5% compared to the pre-placebo phase (Ariel & Saville, 1972). Similar findings were replicated almost 30 years later by Maganaris and colleagues (Maganaris, Collins, & Sharp, 2000), who observed improvement in the performance of weightlifters who were told they had ingested an anabolic steroid (actually a placebo). This improvement returned to baseline values after the subjects were told the truth about the real inert nature of the substance. As demonstrated in a later study in non-athletes, however, the placebo effect on strength in weightlifting appears to be beneficial even after overt debriefing as compared with the performance of a control group that received no treatment (Kalasountas, Reed, & Fitzpatrick, 2007). A more recent study involving untrained participants further demonstrated the effects of positive and negative expectations for the effect of a drink in modulating the power and perception of discomfort during incremental arm crank ergometry (Bottoms, Buscombe, & Nicholettos, 2014).

The beneficial effects of placebos on motor performance have also been demonstrated in cycling and running (Beedie, Coleman, & Foad, 2007; Beedie, Stuart, Coleman, & Foad, 2006; Clark, Hopkins, Hawley, & Burke, 2000; Foster, Felker, Porcari, et al., 2004; McClung & Collins, 2007). Clark et al. (2000) evaluated the effects of real and placebo supplement of carbohydrates on the resistance capacity of athletes in a 40-km cycling race. The cyclists were assigned to three groups: one was told it received the carbohydrate supplement; one was told it received a non-caloric substance (placebo); and one was told it had a 50% chance of receiving the supplement or not. Unbeknown to the subjects, half of the participants in each group were given the supplement and the other half received the placebo substance. The results showed a clear increase in power in the first group and very small changes in the third group, similar to those of the second group. The placebo effect was equal to 3.8% of improvement in power, whereas the actual effect of the carbohydrate supplement was equal to 0.3%. Uncertainty in the second group had the effect of reducing the subjects' expectations, with the perception of a greater sense of effort or even worse performance (Clark et al., 2000). This is consistent with the observation that the deceptive administration of placebo (e.g., caffeine) induces greater expectancies, resulting in stronger placebo effects, than double-blind administration (Kirsch & Weixel, 1988).

In runners, the placebo-induced improvement of performance appears to be more evident in less-trained athletes, suggesting a role of level of physical activity in modulating such effects (Porcari, Otto, Felker, Mikat, & Foster, 2006). Moreover, there is some evidence that the placebo effect resembles the effects obtained with a real substance (McClung & Collins, 2007). For example, McClung and Collins (2007) used sodium bicarbonate to improve the performance of expert runners running repetitions over a distance of 1000 m. The study confirmed the effectiveness of the active substance in inducing significant improvements in performance, but it also highlighted that the mere expectation of receiving the substance can have an effect similar to that produced by the substance itself: what athletes were told about the effects of a substance influences their performance more than what they were actually given (McClung & Collins, 2007). These findings were not fully confirmed in cyclists, however, in whom the effects of the real substance (caffeine in this case) were stronger than the placebo effect (Foad, Beedie, & Coleman, 2008).

Moreover, the amount of active ingredient that participants think is present in the placebo substance can differentially modulate the placebo effect in motor performance. This phenomenon was demonstrated by Beedie et al. (2006) in well-trained cyclists who were tested in a simulation of a 10-km race. The cyclists were told about the efficacy of caffeine in improving motor performance and were then tested in three experimental conditions in which they expected they'd receive either an inert substance or different doses of caffeine (4.5 or 9.0 mg/kg). But only placebo was given in all three conditions. When they believed they had ingested the inert substance, their power decreased by 1.4% over baseline; when they thought they had ingested caffeine their performance improved depending on the dose: the power increased by 1.3% when they thought they had received the low dose caffeine (4.5 mg/kg) and was 3.1% when they thought they had received the high dose (9.0 mg/kg). This suggests that, as what happens with real drugs, there is a dose-dependent response even for the placebo effect.

Words can also have a negative effect on motor performance. Runners influenced by negative beliefs about the ergogenic properties of a pill worsened their speed by 1.57% (Beedie et al., 2007). These findings suggest that positive and negative verbally-induced expectations can significantly affect physical performance in opposite directions and that the nocebo effect should not be considered limited to illnesses but it can even extend to physiological situations, like motor activity.

Taken together, these findings suggest a role for expectation in inducing placebo (and likely nocebo) effects that can modify motor performance similarly to active substances. But expectation can also result inversely related to the uncertainty of the effects and to the level of an athlete's experience. A meta-analysis of 14 studies in sports revealed that placebos have small to moderate effects on performance (Bérdi, Koteles, Szabo, & Bardos, 2011). Although not so large, these effects can make a significant difference for the athlete. Further studies are needed to estimate the size of these effects in placebo responders, who would most likely benefit from placebo treatments (Beedie, Foad, & Coleman, 2008; Bérdi et al., 2011).

1.2 The Effect of Conditioning on Motor Performance

Other psychological mechanisms besides verbally-induced expectation that can induce placebo effects are prior experience and learning (e.g. classical, operant, and partially reinforced conditioning). The bulk of the studies in the placebo literature has focused on classical conditioning as a method to induce conditioned placebo effects (Colloca, Petrovic, Wager, Ingvar, & Benedetti, 2010; Montgomery & Kirsch, 1997; Voudouris, Peck, & Coleman, 1989,

1990), but recent evidence shows that also social observation can be effective in eliciting placebo effects (Colloca & Benedetti, 2009; Swider & Babel, 2013; Vogtle, Barke, & Kroner-Herwig, 2013; for a review, see Schenk, Krimmel, & Colloca, 2017). In the field of pain research, the placebo effects obtained with a conditioning procedure seem to be more robust than those obtained with verbal suggestions alone (Colloca et al., 2008). According to a cognitive interpretation of classical conditioning (Reiss, 1980; Rescorla, 1988), after previous exposure to the effects of a treatment, individuals learn what to expect so that the conditioning procedure allows the formation of reinforced expectations (Kirsch, 2018; Kirsch, Lynn, Vigorito, & Miller, 2004).

In the context of motor performance, conditioning has positive effects on muscle work, especially in instances in which verbal suggestion is so low (i.e., a 25% probability to receive caffeine) that alone it does not induce a beneficial effect (Carlino, Benedetti, & Pollo, 2014). Conversely, high verbal suggestions of motor improvement can work even without conditioning (Carlino et al., 2014). Moreover, through a conditioning procedure the placebo effect induced in the pain modality (pain tolerance) can be transferred to the motor domain (i.e., motor endurance) (Carlino, Guerra, & Piedimonte, 2016). In the study by Carlino et al. (2016), participants were informed about the analgesic and ergogenic effects of an electrical stimulation treatment that would have in turn increased pain tolerance and motor endurance and were then tested in both domains. In the one condition, the participants were also conditioned about the effects of the treatment with a surreptitious reduction of the pain stimulation. Results showed that the conditioning in the pain modality induced transferability of the placebo effect also to the motor endurance, by inducing an increase of the number of movement repetitions before exhaustion (Carlino et al., 2016).

Conditioning procedures in motor performance can be induced with pharmacological or non-pharmacological approaches. Benedetti, Pollo, and Colloca (2007) applied pharmacological conditioning with morphine to investigate the placebo effect in resistance to pain during the simulation of a sports competition. In this study, healthy participants were divided into four groups, each of which underwent a different training program in the 2 weeks prior to the competition: teams A and B did not receive any kind of treatment, while teams C and D underwent a conditioning procedure with morphine. On the day of the competition, group A was not treated with any substance, groups B and C were given placebo and told that they were receiving morphine, while group D surreptitiously received naloxone, an opioid antagonist, together with the placebo administration.

Pain tolerance on the day of competition was higher in group C (conditioned with morphine) than in group B (not conditioned), suggesting that a conditioning procedure can increase the placebo effect (Benedetti et al., 2007). This study also demonstrated that these effects were mediated by endogenous opioids: the administration of opioid antagonist naloxone counteracted the beneficial effects of the conditioning phase in group D (Benedetti et al., 2007). Hence, the administration of morphine before the competition and its substitution with an inert substance on the day of competition was able to produce an analgesic response mediated by endogenous opioids (Benedetti et al., 2007).

The effect of conditioning in inducing placebo effects in the motor domain was demonstrated also with the use of a non-pharmacological substance (Pollo, Carlino, & Benedetti, 2008). In this study, the authors investigated the effects of caffeine as an ergogenic placebo on the work of the quadriceps muscle and on the perception of fatigue during repeated leg extensions executed to lift a weight (Pollo et al., 2008). Healthy volunteers were involved in two different experiments. In the first, they were given a placebo and told that it was a high dose of caffeine. In the second, the intake of placebo caffeine was accompanied by a conditioning procedure in which the weight to be lifted was reduced unknown to the participants. This was done so that the participants believed in the efficacy of the ergogenic agent. In the final test, the load was surreptitiously set again to the baseline value. The results showed a significant increase in muscle work but no variation in the perception of fatigue in the first experiment (verbal suggestions alone). In the second experiment (verbal suggestions and conditioning manipulation), there was not only a further improvement in quadriceps muscle work but also a substantial decrease in perceived fatigue. This suggests that the combination of verbal suggestions and conditioning manipulation may induce a stronger placebo effect on motor outcome than verbal suggestions alone, supporting the important role of learning (and reinforced expectations) in modulating these effects (Pollo et al., 2008).

Also the execution of a simple motor task appears to be boosted with a conditioning procedure better than with verbal suggestions alone. In a study in non-athletes, the placebo procedure consisted of the application of an inert treatment (low-frequency transcutaneous electrical nerve stimulation, TENS) together with verbal information about its powerful effect in increasing force (Fiorio, Andani, Marotta, Classen, & Tinazzi, 2014). Force was produced by pressing the right index finger against a piston with abduction movements and TENS was applied over the hand muscle involved in the task (i.e., the first dorsal interosseous). Participants were randomly assigned to four different groups: two placebo and two control groups. One placebo group received only verbal suggestion about the positive effects of TENS, whereas another placebo group was additionally conditioned with a surreptitious amplification of visual feedback that represented the level of force. In this way, the subjects' expectation about the effect of TENS was reinforced with the exposure to its effects. One control group received TENS and was explicitly told that it was not active and another control group performed the motor task without any treatment. The results showed that force increased after the placebo procedure in both placebo groups, with a more prominent enhancement in the group that had also been conditioned (Fiorio et al., 2014). This was not found in the two control groups (Fig. 1).

Motor performance can also be diminished through nocebo procedures (Beedie et al., 2007; Pollo, Carlino, Vase, & Benedetti, 2012). A study by Pollo et al. (2012) demonstrated that negative verbal suggestion and conditioning can have a negative effect on quadriceps muscle work and on subjective perception of exertion in healthy participants (Pollo et al., 2012). Moreover, negative verbal information about the effects of a treatment in reducing force, together with a conditioning procedure, induces a decrease in force and an increase in the sense of effort (Emadi Andani, Tinazzi, Corsi, & Fiorio, 2015). Nocebo effects in motor performance are associated with the subjective perception of treatment efficacy and personality traits (Corsi, Emadi Andani, Tinazzi, & Fiorio, 2016). For example, participants who more consistently perceive a negative effect of treatment in reducing force not only display a more robust nocebo effect (reduction of force) but are also characterized by such personality traits as higher levels of trait anxiety and harm avoidance and less optimism (Corsi et al., 2016).

These studies highlight the role of learning mechanisms in modulating the placebo and nocebo responses and underline the importance of different conditioning procedures, pharmacological or non-pharmacological, through which an inert substance or device can promote or impair the motor outcome. As found for pain, it appears that a combination of verbal suggestions and conditioning induces stronger motor-related placebo effects than verbal suggestions alone (Fiorio et al., 2014; Pollo et al., 2008).

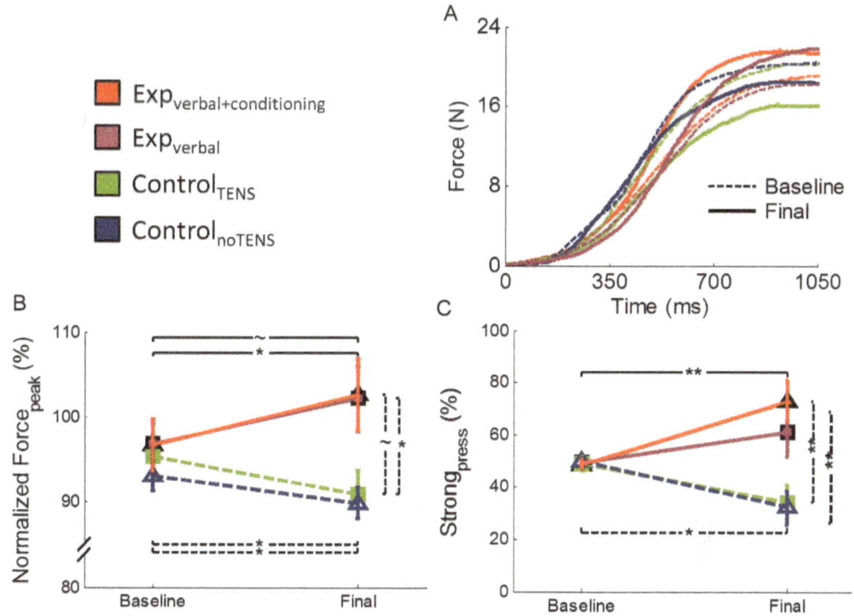

Fig. 1 (A) Force profile before (*dashed line*, baseline) and after (*solid line*, final) the placebo procedure in representative participants of the placebo group with conditioning (*red line*, $Exp_{verbal+conditioning}$), the placebo group with only verbal suggestion (*pink line*, Exp_{verbal}), the control group with TENS (*green*, $Control_{TENS}$), and the control group without TENS (*blue*, $Control_{noTENS}$). (B) Force peak normalized to maximum voluntary force was higher in the $Exp_{verbal+conditioning}$ group in the final as compared to the baseline session. A trend toward significance was observed in the Exp_{verbal} group. The two control groups showed a decrease in force. (C) The percentage of strong pressure was higher in the $Exp_{verbal+conditioning}$ group in the final than in the baseline session. The $Control_{noTENS}$ showed a decrease in force. (For interpretation of the references to color in this figure legend, the reader is referred to the web version of this article.) *Reproduced with permission from the Society for Neuroscience.*

2. NEURAL CORRELATES OF PLACEBO AND NOCEBO EFFECTS IN THE MOTOR DOMAIN

Despite mounting behavioral evidence for the power of placebo and nocebo effects in influencing physical performance, our knowledge of the neural correlates of these effects lags behind. To gain a deeper understanding of how placebos and nocebos influence motor behavior, we need to investigate whether changes can be detected in the complex neural circuitry subtending the cognitive control of movements. A challenge for placebo research in the motor domain is choosing the best approach to uncover the neurobiological bases. To date very few studies in healthy participants have demonstrated the involvement of cortical brain areas in motor-related placebo and nocebo effects (Emadi Andani et al., 2015; Fiorio et al., 2014; Piedimonte, Benedetti, & Carlino, 2015). More have been conducted in patients with motor disorders such as Parkinson's disease (Benedetti et al., 2004, 2016, 2009; de la Fuente-Fernández et al., 2002, 2001; Frisaldi, Carlino, Lanotte, Lopiano, & Benedetti, 2014; Keitel, Ferrea, Südmeyer, Schnitzler, & Wojtecki, 2013; Lidstone et al., 2010; Lou et al., 2013, 2005; Pollo et al., 2002; Strafella, Ko, & Monchi, 2006). Parkinson's disease has proved to be an interesting model for furthering our understanding of the complex neurobiological mechanisms that characterize the placebo effect (Frisaldi et al., 2014). Since the pathophysiology of the disease is localized mainly in the basal ganglia, which are important nodes in a brain network involved in motor control, it may clarify the role of these subcortical structures in the motor placebo effect.

2.1 The Role of Cortical Brain Regions

One key brain area involved in movement execution is the primary motor cortex (M1). Transcranial magnetic stimulation (TMS) provides a good non-invasive and safe method to study the excitability of the human M1. Single-pulse TMS entails delivering a short magnetic pulse through a coil positioned over the M1 area that controls for a specific muscle. Through an electromyographic system, the amplitude of motor evoked potentials (MEP) and the duration of the cortical silent period (CSP) can be recorded with surface electrodes positioned over the muscle of interest. The MEP amplitude represents both cortical and spinal motor circuits (Rösler & Magistris, 2008), whereas the CSP duration is mainly determined at the cortical level in M1 (Wolters, Ziemann, & Benecke, 2008). To investigate how M1 is involved in the placebo effect in the motor domain, single-pulse TMS was applied over M1 in healthy participants while they performed a simple motor task (index finger force pressure applied against a piston) and underwent a placebo procedure (application of the TENS device) (Fiorio et al., 2014). After the placebo procedure, the MEP amplitude increased in both the group treated with only verbal suggestion and the group treated also with conditioning, whereas a reduction in CSP duration was observed only in the conditioned placebo group (Fig. 2). The higher MEP amplitude in both groups may represent an overall increased excitability of the corticospinal

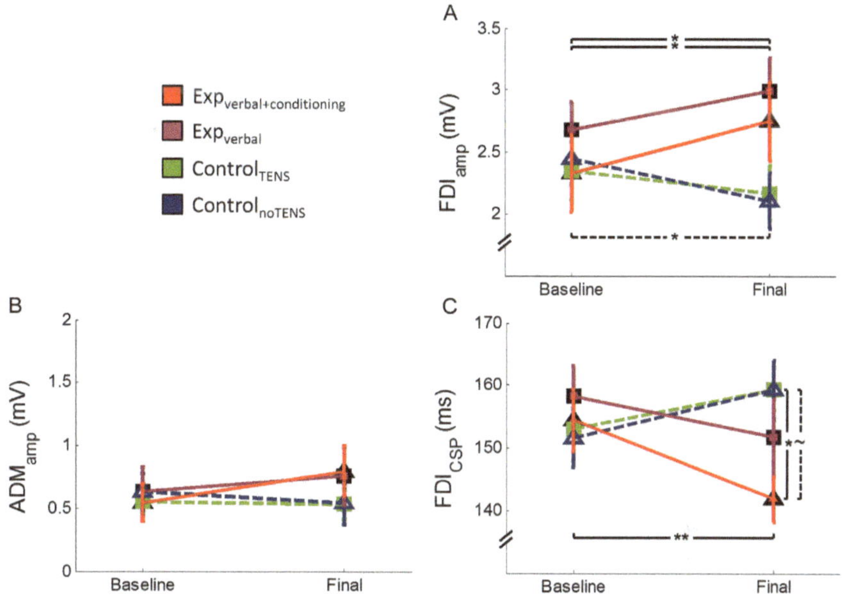

Fig. 2 (A) MEP amplitude recorded from the muscle involved in the task (i.e., the first interosseous, FDI) was greater after the placebo procedure in the two placebo groups (Exp$_{verbal+conditioning}$ and Exp$_{verbal}$). Conversely, MEP amplitude was decreased in the Control$_{noTENS}$. (B) MEP amplitude recorded from the control muscle abductor digiti minimi (ADM) was stable across sessions. (C) Duration of the cortical silent period (CSP) recorded from the FDI was shorter after the placebo procedure only in the Exp$_{verbal+conditioning}$ group. *Reproduced with permission from the Society for Neuroscience.*

system, while the shorter CSP duration only in the conditioned group may suggest that reinforced expectation is necessary to modulate the cortical inhibitory circuits (Fiorio et al., 2014). The TMS pulse was delivered when all participants exerted the same amount of force, corresponding to 30% of the maximal voluntary force calculated at the beginning of the procedure, thus ruling out peripheral factors like the amount of force impressed at the time of stimulation. On the basis of these findings, it was suggested that the placebo procedure induced an expectation of better motor performance and increased the voluntary drive to the motor cortex, resulting in enhanced corticospinal excitability. These effects hint at the existence for a top-down, central mechanism associated with the placebo effect in motor performance (Fiorio et al., 2014).

Changes in MEP amplitude were also found in patients with Parkinson's disease who were told they had a 0%, 50%, or 100% probability of receiving an antiparkinsonian drug (Lou et al., 2013). In this study, the MEP amplitude was reduced in the patients who thought they had a 50% and a 100% probability of receiving the drug and it was increased in the group that did not expect receiving the drug. These findings are consistent with evidence that levodopa reduces MEP amplitude in patients with Parkinson's disease (Lou, Benice, Kearns, Sexton, & Nutt, 2003) and suggest that the placebo effect resembled the effect of the real treatment (Lou et al., 2005).

The hypothesis for a central mechanism underlying the placebo effect in motor performance is further sustained by another study in which the readiness potential, a wave that emerges from the electroencephalographic signal before a voluntary movement is executed (Shibasaki & Hallett, 2006), was recorded as a sign of movement preparation (Piedimonte et al., 2015). Associated with movement preparation, the readiness potential seems to be generated in the supplementary motor area and in M1 (Deecke, 1996; Shibasaki & Hallett, 2006). The amplitude of the readiness potential is modulated by fatigue, as demonstrated by the fact that an increase in fatigue is associated with an increase in amplitude of this potential (Schillings et al., 2006; Slobounov, Hallett, & Newell, 2004). Exploring this notion, Piedimonte et al. (2015) designed an experiment in which participants performed a fatigue-inducing task like repeated flexion movements with the right index finger to lift a weight. The readiness potential was recorded along with perception of fatigue before and after a placebo procedure consisting of placebo caffeine intake associated with verbal information about its positive effects in reducing fatigue. The assumption was that if the placebo effect modulates the preparatory phase of movements, then the expectation of a reduction in fatigue induced through the placebo procedure would affect the amplitude of the readiness potential. Accordingly, a decrease in the perception of fatigue with no increase in the amplitude of the readiness potential was noted for the placebo group. Conversely, the control group displayed an increase in fatigue and in the amplitude of the readiness potential (Fig. 3). This is an elegant demonstration that the placebo-induced decrease in fatigue is regulated centrally before movement execution (Piedimonte et al., 2015).

Our knowledge on the neural correlates of the placebo effect in motor performance is still limited, and even less is known about the neural correlates of the nocebo effect. A TMS study showed that the nocebo-induced decrease in force was associated with a shortening of the CSP and a stable MEP amplitude (Emadi Andani et al., 2015). The shorter CSP duration was similar to that found following a placebo procedure described above (Fiorio et al., 2014). This suggests that cortical inhibitory circuits in the

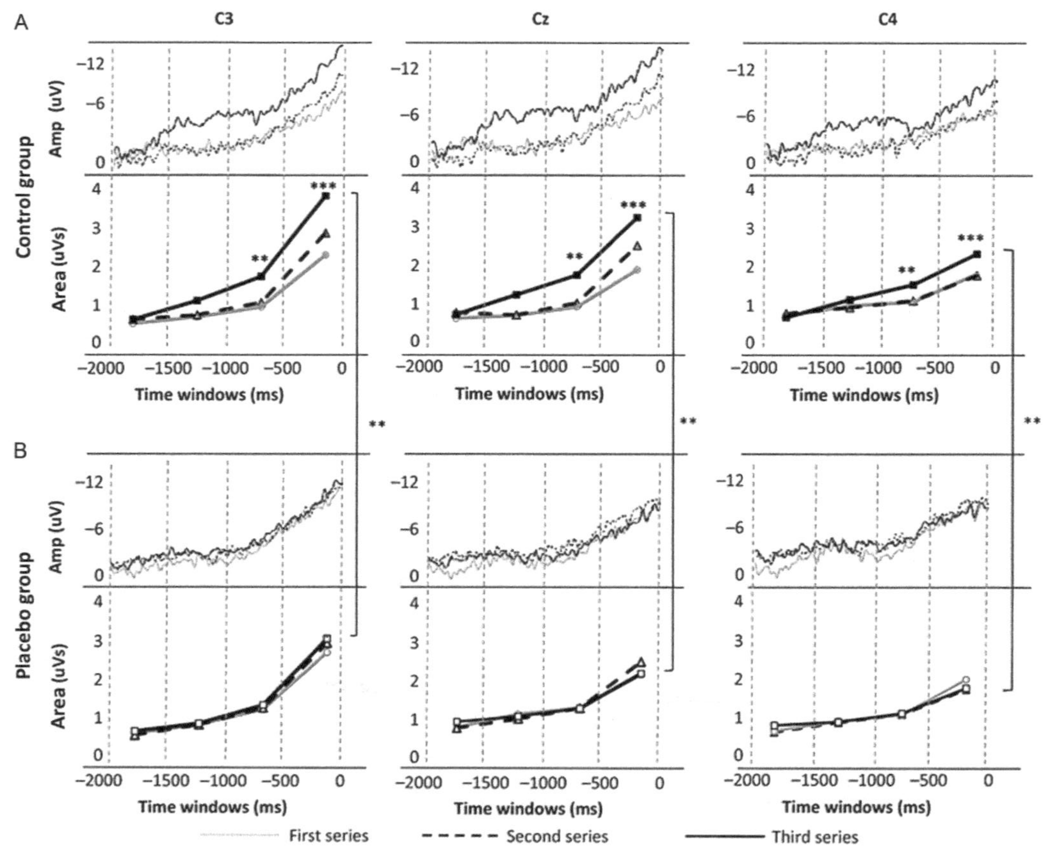

Fig. 3 (A) Amplitude of the readiness potential (*upper panels*) and the area under the curve (*lower panels*) in the control group increased with increasing fatigue (*third series*) in recording sites C3, Cz, C4. (B) The readiness potential did not increase in the placebo group and was significantly lower than in the control group. *Reproduced with permission from John Wiley and Sons.*

primary motor cortex may be engaged by nocebo and placebo procedures similarly but independent of the direction of the behavioral outcome (decrease or increase in force).

These studies hint at involvement of cortical brain areas in the placebo and nocebo effects in motor performance. Placebo-induced enhancement of force involves the inhibitory and excitatory circuits in M1 (Fiorio et al., 2014), whereas a placebo-induced decrease in fatigue involves areas associated with movement preparation like the supplementary motor area (Piedimonte et al., 2015). Since motor control requires a complex brain network of cortical and subcortical regions, it is reasonable that other brain areas will also be involved in these effects. For instance, control of force requires M1 and other cortical regions (i.e., the supplementary motor area, the premotor cortex), as well as subcortical structures like the cerebellum (Dettmers et al., 1995). In a similar vein, fatigue regulation is a complex function influenced by emotions, expectations, and prior experience. As such, it requires cortical and subcortical areas like M1, the orbitofrontal, prefrontal and anterior cingulate cortex, the limbic system, the basal ganglia, and the thalamus (Noakes, 2012). Hence, it is reasonable to hypothesize that placebos and nocebos in the motor domain act by modulating the activity of a complex brain network.

2.2 The Role of Subcortical Brain Structures

Not only cortical but also subcortical brain structures are involved in the placebo effect in the motor domain. Our knowledge derives from studies in patients with a movement disorder like Parkinson's disease. A neurodegenerative disorder of the basal ganglia, Parkinson's disease is characterized by the loss of dopaminergic neurons in the substantia nigra pars compacta. It represents a good model to investigate the involvement of subcortical structures in different motor and cognitive functions (Mallet et al., 2007; Mink, 1996). With regard to placebo, Parkinson's disease allows to investigate the effects from different perspectives: objective evaluation of motor symptom improvements following a placebo procedure (Goetz, Laska, et al., 2008; Goetz, Leurgans, & Raman, 2002; Goetz, Leurgans, Raman, & Stebbins, 2000; Goetz, Wuu, et al., 2008) and direct recording of the neurophysiological properties of subcortical neurons during the placebo response and involvement of the dopaminergic system (Benedetti et al., 2004, 2016, 2009; de la Fuente-Fernández et al., 2002, 2001).

A study by Pollo et al. (2002) has shown that the verbally-induced expectation of good or bad motor performance modulates the therapeutic effect of deep brain stimulation (DBS) of the subthalamic nucleus (STN) in patients with Parkinson's disease. Movements of the right hand were faster when patients believed that they had received DBS at normal or high intensity, whereas movement speed worsened when they knew that the DBS intensity was actually reduced (Pollo et al., 2002). These effects occurred within a few minutes, suggesting that the neuronal changes induced by expectations act very rapidly (Pollo et al., 2002). Benedetti et al. (2003) conditioned patients with Parkinson's disease to experience worsening of motor symptoms by switching off the DBS stimulator. On the day of the test, negative expectations were induced by informing the patients that the stimulator was switched off. In this condition, the patients showed worsening of movement speed even though the stimulator was actually on (Benedetti et al., 2003). Conversely, when the patients were told that the stimulator intensity was increased, even though they have been conditioned to experience negative effects, movement velocity did not decrease, suggesting that the placebo procedure prevented a nocebo effect on the test day (Benedetti et al., 2003). Mercado et al. (2006) obtained similar findings and demonstrated in addition that, as compared to other symptoms like tremor and rigidity, the most relevant motor symptom sensitive to the placebo and nocebo manipulation in Parkinson's disease is bradykinesia (Mercado et al., 2006). A more recent study confirmed the positive effect of placebo on bradykinesia and revealed a prominent role of conditioning (Frisaldi et al., 2017). Since bradykinesia is usually correlated with a lack of dopamine in the striatum (Vingerhoets, Schulzer, Calne, & Snow, 1997), the positive effect of the placebo procedure on this symptom was interpreted as a dopamine-mediated response (Mercado et al., 2006). It was demonstrated that in parkinsonian patients with tremor, a placebo procedure can reduce tremor amplitude objectively measured with an accelerometer placed on the patient's hand (Barbagallo et al., 2018). This reduction was more pronounced in placebo responders and resembled the effect of the exogenous dopaminergic stimulation obtained with the injection of apomorphine (Barbagallo et al., 2018).

The role of dopamine release in the placebo effect in parkinsonian patients was directly demonstrated by positron emission tomography studies with raclopride, an antagonist of D2–D3 dopaminergic receptors (de la Fuente-Fernández et al., 2002, 2001). After the placebo procedure (saline injection), the raclopride binding in the dorsal (de la Fuente-Fernández et al., 2001) and ventral (de la Fuente-Fernández et al., 2002) striatum was decreased, reflecting the presence of endogenous dopamine. The release of dopamine in the dorsal striatum, involved in movement planning, was associated with the patients' perception of motor improvement. This relation was not present in the ventral striatum, which is more involved in the reward circuit and less in motor performance (de la Fuente-Fernández, Schulzer, & Stoessl, 2004). The authors concluded that the dopamine release in the ventral striatum might reflect the patient's expectation of a reward (clinical benefit), before its occurrence (de la Fuente-Fernández, 2009; de la Fuente-Fernández et al., 2004). These findings were confirmed by a subsequent study, in which sham repetitive transcranial magnetic stimulation was used as placebo therapy for patients with Parkinson's disease (Strafella et al., 2006). Even this type of placebo treatment was able to induce variations in raclopride binding in both the dorsal and the ventral striatum (Strafella et al., 2006). The level of expectation of clinical benefit following placebo administration is strictly correlated with the amount of dopamine release in the striatum. In a study by Lidstone et al. (2010), the strength of patients' expectation was manipulated by telling them they had a certain probability to receive an active treatment (25%, 50%, 75%, or 100%) but actually received a placebo. The difference in dopamine release was significant only when the patients thought they had at least a 75% probability of receiving the treatment (Lidstone et al., 2010). This suggests that the efficacy of a hidden therapy (unexpected) may differ from that of an open therapy (expected).

A deeper understanding of the involvement of subcortical structures in the placebo effect has been gained by investigating the firing properties of single neurons in awake patients. This was possible with the use of recording techniques that can precisely localize the electrodes implanted for DBS interventions in parkinsonian patients. In this way, it was possible to demonstrate that placebo procedures can induce changes in the pattern of subcortical neuron activity. Benedetti et al. (2004) applied a pre-conditioning procedure by injecting the dopamine agonist apomorphine, an antiparkinsonian drug, in patients with Parkinson's disease selected for neurosurgical intervention with DBS (Benedetti et al., 2004). On the day of surgery, neuron activity in the STN was directly recorded while the patients underwent a placebo procedure consisting of injection of saline solution together with verbal suggestion of motor improvement. The results showed a clear reduction in the discharge frequency and a shift from bursting to nonbursting activity in the subthalamic neurons in the patients who also displayed a reduction in muscular rigidity and improvement in perceived well-being (Benedetti et al., 2004).

Because the subthalamic nucleus is connected with other nuclei of the basal ganglia and of the thalamus involved in motor control, it was hypothesized that placebo-induced STN changes could be associated also with changes in other subcortical structures (Benedetti et al., 2009). Consistent with this hypothesis, Benedetti et al. (2009) extended their investigation to include direct recording from neurons of the STN, the substantia nigra pars reticulata (SNr), the ventral anterior (VA), and the anterior ventrolateral (VLa) nuclei of the thalamus in Parkinson patients selected for DBS and treated with a placebo procedure (Benedetti et al., 2009). The results showed a decrease in the firing rate in STN and SNr neurons and an increase in the VA and VLa neurons in the patients in which the placebo procedure induced a reduction in muscle rigidity (Benedetti et al., 2009). These findings were explained by the modulatory activity of the STN on the other structures by means of excitatory and inhibitory connections. With regard to involvement of the thalamus in the placebo effect, it was demonstrated that a conditioning procedure with real drug administration (apomorphine) is necessary to induce clinical improvement and changes in VA and VLa neural discharge (Benedetti et al., 2016). Moreover, the number of drug administrations in the conditioning phase was associated with the strength of the clinical and neural changes induced by placebo, hinting at the role of learning mechanisms (likely conditioning) in these effects (Benedetti et al., 2016).

The basal ganglia make up part of different brain circuits: they are part of a network that, together with the orbitofrontal cortex (Petrovic, Kalso, Petersson, & Ingvar, 2002) and the dorsolateral prefrontal cortex (Mayberg et al., 2002), processes reward expectation (Schultz, 1998). This relates the placebo effect to reward mechanisms (de la Fuente-Fernández et al., 2004) and extends the neural underpinnings to other brain areas. Moreover, brain regions consistently active in placebo responses and receiving dopaminergic projections, like the cingulate cortex (Mayberg et al., 2002; Petrovic et al., 2002), may also play an important role in the placebo modulation of motor control.

Whether the findings obtained from patients can be applied also to healthy individuals is not straightforward. Because the electrophysiological functioning of the basal ganglia is altered in parkinsonian patients (Leblois et al., 2007; Mallet et al., 2008), it is difficult to translate the findings obtained in patients to physiological conditions. Nonetheless, direct recording of subcortical activity in patients undergoing DBS electrode implantation can uncover the involvement of these structures in the motor placebo effect.

3. CONCLUDING REMARKS

The behavioral effects of placebos and nocebos on motor performance have been described in sport disciplines, including running, cycling, and weightlifting (Beedie & Foad, 2009; Pollo, Carlino, & Benedetti, 2011), as well as in more controlled laboratory settings. The cognitive mechanisms widely described for placebo analgesia and nocebo hyperalgesia, such as verbal suggestion and a variety of learning paradigms (e.g. classical conditioning, modeling, partial reinforcement, operant conditioning), also apply to the motor domain.

Our knowledge of the neurobiological bases of these effects is still limited. Studies in patients with Parkinson's disease suggest a role of subcortical structures in the clinical response to placebo (Benedetti et al., 2004, 2009). The main neurobiological mechanisms of placebo-induced improvement of motor symptoms are related to dopamine release in the striatum (de la Fuente-Fernández et al., 2001) and changes in the firing rate of a subcortical network connecting different structures, the subthalamic nucleus (Benedetti et al., 2004) and the thalamus (Benedetti et al., 2016).

Thorough investigation of the neurobiological mechanisms underlying the influence of placebos and nocebos in a healthy motor system is still lacking. The few experimental studies conducted to date in healthy participants have revealed changes in the excitability of the corticospinal system (Emadi Andani et al., 2015; Fiorio et al., 2014) and in the readiness potential (Piedimonte et al., 2015), hinting at a top-down modulation of movement execution and preparation. There may be a role for the primary motor cortex and the supplementary motor area, although it is conceivable that, the motor control being a complex function relying on different brain networks, other brain regions will also be involved.

Several ethical issues need to be addressed when considering the use of placebo procedures to improve motor performance in sports. First, adopting a conditioning procedure (even non-pharmacological) may instill in young athletes the need to search for an external aid to feel confident during a competition. Second, a pharmacological conditioning procedure may raise additional concerns about the use of illegal substances before a competition. The World Anti-Doping Agency (WADA) regulations prohibit the intake of doping substances that can artificially increase efficiency during sport competitions and even before competition in some cases. Possible ways to overcome these caveats could be to develop the use of words as placebo and to increase the use of open-label placebo procedures (Carvalho et al., 2016; Kaptchuk et al., 2010; Schaefer, Harke, & Denke, 2016). In both cases, positive expectations can be induced with a beneficial effect on motor outcome. A survey among elite athletes revealed that they believe in the power of placebo in improving performance and they have a positive attitude toward receiving a placebo substance (Bérdi, Köteles, Hevesi, Bárdos, & Szabo, 2015). Research along this line is needed to deepen our understanding of the underlying mechanisms and to develop suitable paradigms for healthy individuals and patients with motor deficits.

REFERENCES

Ariel, G., & Saville, W. (1972). Anabolic steroids: The physiological effects of placebos. *Medicine and Science in Sports and Exercise*, *4*, 124–126.

Barbagallo, G., Nisticò, R., Vescio, B., Cerasa, A., Olivadese, G., Nigro, S., et al. (2018). The Placebo effect on resting tremor in Parkinson's disease: An electrophysiological study. *Parkinsonism & Related Disorders*. Mar 14. pii: S1353-8020(18)30127-5. https://doi.org/10.1016/j.parkreldis.2018.03.012 [Epub ahead of print].

Beedie, C. J., Coleman, D. A., & Foad, A. J. (2007). Positive and negative placebo effects resulting from the deceptive administration of an ergogenic aid. *International Journal of Sport Nutrition and Exercise Metabolism*, *17*, 259–269.

Beedie, C. J., & Foad, A. J. (2009). The placebo effect in sports performance. *Sports Medicine*, *39*(4), 313–329.

Beedie, C. J., Foad, A. J., & Coleman, D. A. (2008). Identification of placebo responsive participants in 40 km laboratory cycling performance. *Journal of Sports Science & Medicine*, 7, 166.

Beedie, C. J., Stuart, E. M., Coleman, D. A., & Foad, A. J. (2006). Placebo effect of caffeine in cycling performance. *Medicine and Science in Sports and Exercise*, *38*, 2159–2164.

Benedetti, F., Barbiani, D., & Camerone, E. (2018). Critical life functions: Can placebo replace oxygen? *International Review of Neurobiology*, *138*, 201–218.

Benedetti, F., Colloca, L., Torre, E., Lanotte, M., Melcarne, A., Pesare, M., et al. (2004). Placebo-responsive Parkinson patients show decreased activity in single neurons of subthalamic nucleus. *Nature Neuroscience*, 7, 587–588.

Benedetti, F., Frisaldi, E., Carlino, E., Giudetti, L., Pampallona, A., Zibetti, M., et al. (2016). Teaching neurons to respond to placebos. *Journal of Physiology*, *594*, 5647–5660.

Benedetti, F., Lanotte, M., Colloca, L., Ducati, A., Zibetti, M., & Lopiano, L. (2009). Electrophysiological properties of thalamic, subthalamic and nigral neurons during the anti-parkinsonian placebo response. *Journal of Physiology*, *587*, 3869–3883.

Benedetti, F., Pollo, A., & Colloca, L. (2007). Opioid-mediated placebo responses boost pain endurance and physical performance: Is it doping in sport competitions? *Journal of Neuroscience*, *27*, 11934–11939.

Benedetti, F., Pollo, A., Lopiano, L., Lanotte, M., Vighetti, S., & Rainero, I. (2003). Conscious expectation and unconscious conditioning in analgesic, motor, and hormonal placebo/nocebo responses. *Journal of Neuroscience*, *23*(10), 4315–4323.

Bérdi, M., Köteles, F., Hevesi, K., Bárdos, G., & Szabo, A. (2015). Elite athletes' attitudes towards the use of placebo-induced performance enhancement in sports. *European Journal of Sport Science*, *15*, 315–321.

Bérdi, M., Koteles, F., Szabo, A., & Bardos, G. (2011). Placebo effects in sport and exercise: A meta-analysis. *European Journal of Mental Health*, *6*, 196–212.

Bottoms, L., Buscombe, R., & Nicholettos, A. (2014). The placebo and nocebo effects on peak minute power during incremental arm crank ergometry. *European Journal of Sport Science, 14*, 362–367.

Carlino, E., Benedetti, F., & Pollo, A. (2014). The effects of manipulating verbal suggestions on physical performance. *Zeitschrift füer Psychologie, 222*, 154–164.

Carlino, E., Guerra, G., & Piedimonte, A. (2016). Placebo effects: From pain to motor performance. *Neuroscience Letters, 632*, 224–230.

Carvalho, C., Caetano, J. M., Cunha, L., Rebouta, P., Kaptchuk, T. J., & Kirsch, I. (2016). Open-label placebo treatment in chronic low back pain: A randomized controlled trial. *Pain, 157*, 2766–2772.

Clark, V. R., Hopkins, W. G., Hawley, J. A., & Burke, L. M. (2000). Placebo effect of carbohydrate feedings during a 40-km cycling time trial. *Medicine and Science in Sports and Exercise, 32*, 1642–1647.

Colloca, L., & Benedetti, F. (2009). Placebo analgesia induced by social observational learning. *Pain, 144*, 28–34.

Colloca, L., Petrovic, P., Wager, T. D., Ingvar, M., & Benedetti, F. (2010). How the number of learning trials affects placebo and nocebo responses. *Pain, 151*(2), 430–439.

Colloca, L., Tinazzi, M., Recchia, S., Le Pera, D., Fiaschi, A., Benedetti, F., et al. (2008). Learning potentiates neurophysiological and behavioral placebo analgesic responses. *Pain, 139*, 306–314.

Corsi, N., Emadi Andani, M., Tinazzi, M., & Fiorio, M. (2016). Changes in perception of treatment efficacy are associated to the magnitude of the nocebo effect and to personality traits. *Scientific Reports, 6*, 30671.

de la Fuente-Fernández, R. (2009). The placebo-reward hypothesis: Dopamine and the placebo effect. *Parkinsonism & Related Disorders, 15*(Suppl. 3), S72–S74.

de la Fuente-Fernández, R., Phillips, A. G., Zamburlini, M., Sossi, V., Calne, D. B., Ruth, T. J., et al. (2002). Dopamine release in human ventral striatum and expectation of reward. *Behavioural Brain Research, 136*, 359–363.

de la Fuente-Fernández, R., Ruth, T. J., Sossi, V., Schulzer, M., Calne, D. B., & Stoessl, A. J. (2001). Expectation and dopamine release: Mechanism of the placebo effect in Parkinson's disease. *Science, 293*, 1164–1166.

de la Fuente-Fernández, R., Schulzer, M., & Stoessl, A. J. (2004). Placebo mechanisms and reward circuitry: Clues from Parkinson's disease. *Biological Psychiatry, 56*, 67–71 [Review].

Deecke, L. (1996). Planning, preparation, execution, and imagery of volitional action. *Brain Research. Cognitive Brain Research, 3*, 59–64.

Dettmers, C., Fink, G. R., Lemon, R. N., Stephan, K. M., Passingham, R. E., Silbersweig, D., et al. (1995). Relation between cerebral activity and force in the motor areas of the human brain. *Journal of Neurophysiology, 74*, 802–815.

Emadi Andani, M., Tinazzi, M., Corsi, N., & Fiorio, M. (2015). Modulation of inhibitory corticospinal circuits induced by a nocebo procedure in motor performance. *PLoS One, 10*, e0125223.

Fiorio, M., Andani, M. E., Marotta, A., Classen, J., & Tinazzi, M. (2014). Placebo-induced changes in excitatory and inhibitory corticospinal circuits during motor performance. *Journal of Neuroscience, 34*, 3993–4005.

Foad, A. J., Beedie, C. J., & Coleman, D. A. (2008). Pharmacological and psychological effects of caffeine ingestion in 40-km cycling performance. *Medicine and Science in Sports and Exercise, 40*, 158–165.

Foster, C., Felker, H., Porcari, J. P., et al. (2004). The placebo effect on exercise performance [abstract]. *Medicine and Science in Sports and Exercise, 36*(Suppl. 5), S171.

Frisaldi, E., Carlino, E., Lanotte, M., Lopiano, L., & Benedetti, F. (2014). Characterization of the thalamic–subthalamic circuit involved in the placebo response through single-neuron recording in Parkinson patients. *Cortex, 60*, 3–9.

Frisaldi, E., Carlino, E., Zibetti, M., Barbiani, D., Dematteis, F., Lanotte, M., et al. (2017). The placebo effect on bradykinesia in Parkinson's disease with and without prior drug conditioning. *Movement Disorders*, *32*, 1474–1478.

Goetz, C. G., Laska, E., Hicking, C., Damier, P., Müller, T., Nutt, J., et al. (2008). Placebo influences on dyskinesia in Parkinson's disease. *Movement Disorders*, *23*, 700–707.

Goetz, C. G., Leurgans, S., & Raman, R. (2002). Placebo-associated improvements in motor function: Comparison of subjective and objective sections of the UPDRS in early Parkinson's disease. *Movement Disorders*, *17*, 283–288.

Goetz, C. G., Leurgans, S., Raman, R., & Stebbins, G. T. (2000). Objective changes in motor function during placebo treatment in PD. *Neurology*, *54*, 710–714.

Goetz, C. G., Wuu, J., McDermott, M. P., Adler, C. H., Fahn, S., Freed, C. R., et al. (2008). Placebo response in Parkinson's disease: Comparisons among 11 trials covering medical and surgical interventions. *Movement Disorders*, *23*, 690–699.

Kalasountas, V., Reed, J., & Fitzpatrick, J. (2007). The effect of placebo-induced changes in expectancies on maximal force production in college students. *Journal of Applied Sport Psychology*, *19*, 116–124.

Kaptchuk, T. J., Friedlander, E., Kelley, J. M., Sanchez, M. N., Kokkotou, E., Singer, J. P., et al. (2010). Placebos without deception: A randomized controlled trial in irritable bowel syndrome. *PLoS One*, *5*, e15591.

Keitel, A., Ferrea, S., Südmeyer, M., Schnitzler, A., & Wojtecki, L. (2013). Expectation modulates the effect of deep brain stimulation on motor and cognitive function in tremor-dominant Parkinson's disease. *PLoS One*, *8*, e81878.

Kirsch, I. (2018). Response expectancy and the placebo effect. *International Review of Neurobiology*, *138*, 81–93.

Kirsch, I., Lynn, S. J., Vigorito, M., & Miller, R. R. (2004). The role of cognition in classical and operant conditioning. *Journal of Clinical Psychology*, *60*, 369–392.

Kirsch, I., & Weixel, L. J. (1988). Double-blind versus deceptive administration of a placebo. *Behavioral Neuroscience*, *102*, 319–323.

Leblois, A., Meissner, W., Bioulac, B., Gross, C. E., Hansel, D., & Boraud, T. (2007). Late emergence of synchronized oscillatory activity in the pallidum during progressive parkinsonism. *European Journal of Neuroscience*, *26*, 1701–1713.

Lidstone, S. C., Schulzer, M., Dinelle, K., Mak, E., Sossi, V., Ruth, T. J., et al. (2010). Effects of expectation on placebo induced dopamine release in Parkinson disease. *Archives of General Psychiatry*, *67*, 857–865.

Lou, J. S., Benice, T., Kearns, G., Sexton, G., & Nutt, J. (2003). Levodopa normalizes exercise related corticomotoneuron excitability abnormalities in Parkinson's disease. *Clinical Neurophysiology*, *114*, 930–937.

Lou, J. S., Dimitrova, D. M., Hammerschlag, R., Nutt, J., Hunt, E. A., Eaton, R. W., et al. (2013). Effect of expectancy and personality on cortical excitability in Parkinson's disease. *Movement Disorders*, *28*, 1257–1262.

Lou, J. S., Eaton, R., Blaine, K., Coakley, B., Sexton, G., & Nutt, J. (2005). Motor-evoked potential as a physiological marker for the placebo effect in PD. *Annals of Neurology*, *58*, S53.

Maganaris, C. N., Collins, D., & Sharp, M. (2000). Expectancy effects and strength training: Do steroids make a difference? *Sport Psychology*, *14*, 272–278.

Mallet, N., Pogosyan, A., Sharott, A., Csicsvari, J., Bolam, J. P., Brown, P., et al. (2008). Disrupted dopamine transmission and the emergence of exaggerated beta oscillations in subthalamic nucleus and cerebral cortex. *Journal of Neuroscience*, *28*, 4795–4806.

Mallet, L., Schupbach, M., N'Diaye, K., Remy, P., Bardinet, E., Czernecki, V., et al. (2007). Stimulation of subterritories of the subthalamic nucleus reveals its role in the integration of the emotional and motor aspects of behavior. *Proceedings of the National Academy of Sciences of the United States of America*, *104*, 10661–10666.

Mayberg, H. S., Silva, J. A., Brannan, S. K., Tekell, J. L., Mahurin, R. K., McGinnis, S., et al. (2002). The functional neuroanatomy of the placebo effect. *American Journal of Psychiatry, 159*, 728–737.

McClung, M., & Collins, D. (2007). "Because I know it will!": Placebo effects of an ergogenic aid on athletic performance. *Journal of Sport and Exercise Psychology, 29*, 382–394.

McKay, B., Lewthwaite, R., & Wulf, G. (2012). Enhanced expectancies improve performance under pressure. *Frontiers in Psychology, 3*, 8. eCollection.

Mercado, R., Constantoyannis, C., Mandat, T., Kumar, A., Schulzer, M., Stoessl, A. J., et al. (2006). Expectation and the placebo effect in Parkinson's disease patients with subthalamic nucleus deep brain stimulation. *Movement Disorders, 21*, 1457–1461.

Mink, J. W. (1996). The basal ganglia: Focused selection and inhibition of competing motor programs. *Progress in Neurobiology, 50*, 381–425.

Montgomery, G. H., & Kirsch, I. (1997). Classical conditioning and the placebo effect. *Pain, 72*, 107–113.

Noakes, T. D. (2012). Fatigue is a brain-derived emotion that regulates the exercise behavior to ensure the protection of whole body homeostasis. *Frontiers in Physiology, 3*, 82.

Petrovic, P., Kalso, E., Petersson, K. M., & Ingvar, M. (2002). Placebo and opioid analgesia—Imaging a shared neuronal network. *Science, 295*, 1737–1740.

Piedimonte, A., Benedetti, F., & Carlino, E. (2015). Placebo-induced decrease in fatigue: Evidence for a central action on the preparatory phase of movement. *European Journal of Neuroscience, 41*, 492–497.

Pollo, A., Carlino, E., & Benedetti, F. (2008). The top-down influence of ergogenic placebos on muscle work and fatigue. *European Journal of Neuroscience, 28*, 379–388.

Pollo, A., Carlino, E., & Benedetti, F. (2011). Placebo mechanisms across different conditions: From the clinical setting to physical performance. *Philosophical Transactions of the Royal Society, B: Biological Sciences, 366*, 1790–1798.

Pollo, A., Carlino, E., Vase, L., & Benedetti, F. (2012). Preventing motor training through nocebo suggestions. *European Journal of Applied Physiology, 112*, 3893–3903.

Pollo, A., Torre, E., Lopiano, L., Rizzone, M., Lanotte, M., Cavanna, A., et al. (2002). Expectation modulates the response to subthalamic nucleus stimulation in parkinsonian patients. *Neuroreport, 13*, 1383–1386.

Porcari, J. P., Otto, J., Felker, H., Mikat, R. P., & Foster, C. (2006). The placebo effect on exercise performance. *Journal of Cardiopulmonary Rehabilitation and Prevention, 26*, 269.

Reiss, S. (1980). Pavlovian conditioning and human fear: An expectancy model. *Behavior Therapy, 11*, 380–396.

Rescorla, R. A. (1988). Pavlovian conditioning: Its not what you think it is. *American Psychologist, 43*, 151–160.

Rösler, K. M., & Magistris, M. R. (2008). The size of motor evoked potentials: Influencing parameters and quantification. In E. M. Wassermann, C. M. Epstein, U. Ziemann, V. Walsh, T. Paus, & S. H. Lisanby (Eds.), *The Oxford handbook of transcranial stimulation* (pp. 77–89). Oxford: University Press.

Schaefer, M., Harke, R., & Denke, C. (2016). Open-label placebos improve symptoms in allergic rhinitis: A randomized controlled trial. *Psychotherapy and Psychosomatics, 85*, 373–374.

Schenk, L. A., Krimmel, S. R., & Colloca, L. (2017). Observe to get pain relief: Current evidence and potential mechanisms of socially learned pain modulation. *Pain, 158*, 2077–2081.

Schillings, M. L., Kalkman, J. S., van der Werf, S. P., Bleijenberg, G., van Engelen, B. G. M., & Zwarts, M. J. (2006). Central adaptations during repetitive contractions assessed by the readiness potential. *European Journal of Applied Physiology, 97*, 521–526.

Schultz, W. (1998). Predictive reward signal of dopamine neurons. *Journal of Neurophysiology, 80*, 1–27.

Shibasaki, H., & Hallett, M. (2006). What is the Bereitschaftspotential? *Clinical Neurophysiology, 117*, 2341–2356.
Slobounov, S., Hallett, M., & Newell, K. M. (2004). Perceived effort in force production as reflected in motor-related cortical potentials. *Clinical Neurophysiology, 115*, 2391–2402.
Strafella, A. P., Ko, J. H., & Monchi, O. (2006). Therapeutic application of transcranial magnetic stimulation in Parkinson's disease: The contribution of expectation. *NeuroImage, 31*, 1666–1672.
Swider, K., & Babel, P. (2013). The effect of the sex of a model on nocebo hyperalgesia induced by social observational learning. *Pain, 154*, 1312–1317.
Vingerhoets, F. J., Schulzer, M., Calne, D. B., & Snow, B. J. (1997). Which clinical sign of Parkinson's disease best reflects the nigrostriatal lesion? *Annals of Neurology, 41*, 58–64.
Vogtle, E., Barke, A., & Kroner-Herwig, B. (2013). Nocebo hyperalgesia induced by social observational learning. *Pain, 154*, 1427–1433.
Voudouris, N. J., Peck, C. L., & Coleman, G. (1989). Conditioned response models of placebo phenomena: Further support. *Pain, 38*, 109–116.
Voudouris, N. J., Peck, C. L., & Coleman, G. (1990). The role of conditioning and verbal expectancy in the placebo response. *Pain, 43*, 121–128.
Wolters, A., Ziemann, U., & Benecke, R. (2008). The cortical silent period. In E. M. Wassermann, C. M. Epstein, U. Ziemann, V. Walsh, T. Paus, & S. H. Lisanby (Eds.), *The oxford handbook of transcranial stimulation* (pp. 91–102). Oxford: University Press.

FURTHER READING

de la Fuente-Fernández, R., & Stoessl, A. J. (2002). The placebo effect in Parkinson's disease. *Trends in Neurosciences, 25*, 302–306.

CHAPTER THIRTEEN

The Role of Sleep in Learning Placebo Effects

Florian Chouchou*,†, Thien Thanh Dang-Vu*,†, Pierre Rainville†,‡, Gilles Lavigne‡,§,1

*PERFORM Center & Department of Exercise Science, Concordia University, Montreal, QC, Canada
†Centre de Recherche de l'Institut Universitaire de Gériatrie de Montréal, Université de Montréal, Montreal, QC, Canada
‡Faculté de Médecine Dentaire, Université de Montréal, Montreal, QC, Canada
§Centre d'Etudes Avancées en Médecine du Sommeil, Hôpital du Sacré-Coeur de Montréal, Montreal, QC, Canada
[1]Corresponding author: e-mail address: gilles.lavigne@umontreal.ca

Contents

Abstract

The placebo effect is a psychobiological phenomenon producing clinical benefits attributed to a wide range of neurobiological mechanisms. Independently from placebo effects, these mechanisms may also be under the influence of processes that can take place during sleep. The relationship between sleep and placebo effects has received very little attention. Three experimental studies, conducted on healthy subjects, have examined sleep changes following placebo conditioning associated with analgesic suggestions and the effects of sleep deprivation on placebo effects. A relation between rapid eye movement (REM) sleep, expectations of relief and placebo analgesia was observed in which REM sleep deprivation seems to improve placebo-induced expectations and analgesia. Moreover, analgesic expectations developed before sleep produced a reduction in cortical arousals evoked by noxious stimuli during REM sleep. In this article, we describe sleep and pain/analgesia interactions, the relationship between sleep and placebo analgesia, and finally the potential mechanisms underlying this relationship.

International Review of Neurobiology, Volume 139
ISSN 0074-7742
https://doi.org/10.1016/bs.irn.2018.07.013

1. INTRODUCTION

The placebo effect is a well-recognized and clinically-important effect in medicine (Colloca, 2018). Over the past years, there has been considerable research conducted to elucidate the mechanisms underlying the placebo effect (Amanzio, Benedetti, Porro, Palermo, & Cauda, 2013; Benedetti, 2014; Finniss, Kaptchuk, Miller, & Benedetti, 2010). In particular, these studies have shown the relevance of placebo effect in the treatment of acute and chronic pain (Enck, Bingel, Schedlowski, & Rief, 2013; Petersen, Finnerup, Nørskov, et al., 2012) and some evidence indicated that placebo effect is potentially involved in the therapeutic outcome (Colloca & Benedetti, 2005).

In fact, the placebo effect is based on a psychobiological phenomenon typically observed following the administration of an inert substance or of a sham treatment, accompanied by verbal suggestions of clinical benefits (Carlino, Frisaldi, & Benedetti, 2014; Colloca, Enck, & DeGrazia, 2016). While placebo effects may also contribute to the total therapeutic effect of active treatment, this component of the clinical response can be attributable to the psychological context surrounding the psychobiological response (Benedetti, 2014). A more complete understanding of this placebo effect may help clinicians optimize patients' response to therapeutic treatment, thereby leading to even greater efficacy of these treatments.

These studies have contributed in understanding how the patient–clinician relationship, patients' expectations and learning in general can influence the efficacy of an analgesic intervention (Colloca, Klinger, Flor, & Bingel, 2013). Albeit a high variance in placebo effects, meta-analyses have demonstrated the strength of placebo analgesia in clinical settings (Vase, Petersen, Riley, & Price, 2009; Vase, Riley, & Price, 2002). This led to proposals of placebo use in some clinical situation to optimize treatment outcomes (Finniss, Nicholas, & Benedetti, 2009; Klinger, Colloca, Bingel, & Flor, 2014). In this view, Kong and colleagues have recently found that enhancing expectation of acupuncture analgesia improves pain relief in comparison to the standard acupuncture for osteoarthritis knee pain, an effect associated with the strengthening of the connectivity between the brain reward/motivation system, descending pain modulation system, and sensory and affective pain processes (Kong, Wang, Leiser, et al., 2018).

Interestingly, almost all mechanisms thought to underlie the placebo effect are also related to sleep (Fig. 1). For example, learning, which is one of the main potential mechanisms of the placebo effect is well known

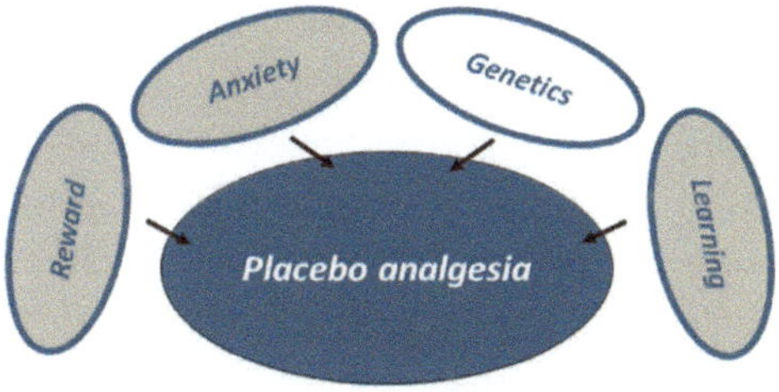

Fig. 1 Model of psychosocial and psychobiological factors that modulate placebo response. The majority of these factors, such as learning, reward or anxiety, are themselves modulated by sleep.

to be consolidated after a period of sleep (Diekelmann & Born, 2010; Dudai, Karni, & Born, 2015). From these observations, it seems pertinent to investigate the involvement of sleep in the development and consolidation of the placebo effect. However, a relationship between sleep and placebo effect has once again received very little attention. Two studies have examined changes in pain perception following conditioning manipulation and suggestions of analgesia with and without sleep deprivation in healthy subjects (Chouchou, Chauny, Rainville, & Lavigne, 2015; Laverdure-Dupont, Rainville, Montplaisir, & Lavigne, 2009). More recently, a third study reported that analgesic expectations developed before sleep reduced nocturnal pain perception and subjective sleep perturbations (Laverdure-Dupont, Rainville, Renancio, Montplaisir, & Lavigne, 2018). In this article, we will review the state of knowledge on the potential role of sleep in consolidation of expectation-mediated placebo effect, and its potential mechanisms. To facilitate this comprehension, we will start by summarizing recent findings on sleep and its interaction with pain and analgesia treatment.

1.1 Normal Sleep in Humans

Sleep is a natural and unavoidable state characterized by a decrease in consciousness and behavioral responsiveness (Bayne, Hohwy, & Owen, 2016; Boly, Phillips, Tshibanda, et al., 2008; Sanders, Tononi, Laureys, & Sleigh, 2012) occurring in all vertebrates, including fishes, birds, and reptiles (Chiu & Prober, 2013; Libourel & Herrel, 2016; Vorster & Born, 2015). Perceptions of external or internal stimulations, regardless of their nature, are decreased but not abolished during sleep (Chouchou, Pichot, Barthélémy, Bastuji, & Roche, 2014; Chouchou, Pichot, Perchet, et al., 2011; Dang-Vu, Schabus, Desseilles, et al., 2010). These seemingly simple behaviors hide complex processes. Sleep, periodically, takes place in a nycthemeral rhythm alternating with wakefulness, and is organized in two main sleep stages, as illustrated in Fig. 2: non-REM sleep and REM sleep ("stage R" according to the American Association of Sleep Medicine (Iber, Chesson, & Quan, for the AASM, 2007)). In humans, non-REM sleep is further divided into three stages: from the lightest stages 1 "N1" and 2 "N2" to the deepest slow wave sleep (SWS or "N3"). These sleep stages are defined by their patterns of electroencephalographic (EEG), electromyographic (EMG), and electroculographic (EOG) activities.

Sleep stage 1 is considered as a transitional stage between wakefulness and sleep, usually of short duration (a few minutes). The alpha-EEG rhythm (8–12 Hz), predominant during quiet wakefulness (eyes closed), slows down and disappears gradually in favor of theta rhythm (4–8 Hz) during stage 1, associated with some vertex sharp waves and slow eye movements (Porte, 2004). This stage is quickly followed by sleep stage 2, with a distinct EEG pattern consisting of waxing-and-waning oscillations called sleep spindles (12–14 Hz), and K-complexes (St Louis Erik, Frey, Britton, et al., 2016). Some slow waves or delta rhythm, defined by the 0.5–4.0-Hz frequency band, may appear but are still infrequent (less than 20%). As sleep deepens, delta waves (0.5–4 Hz) become more frequent (in SWS, more than 20%), and include the 1-Hz slow oscillations (Mölle, Marshall, Gais, & Born, 2002; Weiner & Dang-Vu, 2016).

In addition, this slowing in electrophysiological activities during non-REM sleep is accompanied by metabolic changes in the brain. Imaging studies have shown an overall drop of brain activity during non-REM sleep (Dang-Vu, Schabus, et al., 2010; Nofzinger, Buysse, Miewald, et al., 2002; Zoccoli, Walker, Lenzi, & Franzini, 2002); this drop is estimated at around 40% during SWS compared to wakefulness (Dang-Vu, Schabus, et al., 2010). The reductions of brain activity are mainly located in regions involved in arousal, including subcortical regions, such as brainstem, basal forebrain, thalamus and basal ganglia, and cortical regions including anterior cingulate cortex, prefrontal cortex, and precuneus (Dang-Vu, Schabus, et al., 2010; Zoccoli et al., 2002).

At last, the EEG during REM sleep resembles the waking EEG more closely than non-REM sleep. REM sleep is characterized by fast- and low-amplitude oscillatory brain activity, resembling waking. But these fast- and low-amplitude EEG oscillations are combined with muscle atonia and rapid-eye movements, and sometimes with fronto-central, sharply contoured theta frequencies called sawtooth waves (St Louis Erik et al., 2016). As for the electrophysiological activities, brain metabolic activities during

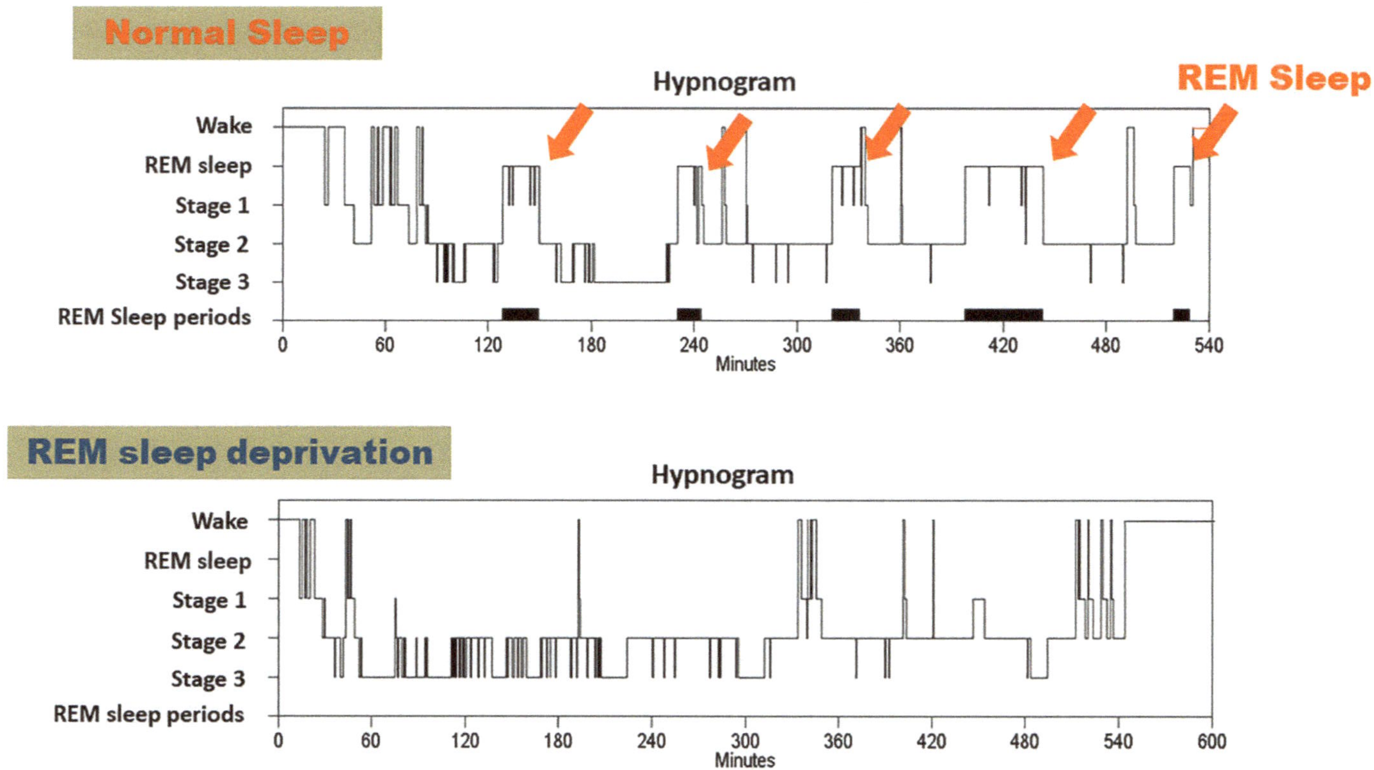

Fig. 2 Hypnograms of normal sleep (up) and rapid eye-movement sleep deprivation (below). Normal sleep is divided into several sleep stages non-REM sleep, composed by three sleep stages: from the lightest stages 1 and 2 to the deepest stages 3 (slow wave sleep). The REM sleep and non-REM sleep stages alternate to form ultradian cycles, which repeat throughout the sleep period. Slow wave sleep dominates in the first part, and REM sleep in the last part of the sleep period. A pharmacological REM sleep deprivation design, using clonidine was applied because clonidine, an alpha-2 adrenergic receptor agonist, was known to suppress most of REM sleep in a majority of subjects, with no changes in sleep duration or sleep fragmentation (Autret, Minz, Beillevaire, Cathala, & Castaigne, 1976; Gentili, Godschalk, Gheorghiu, et al., 1996; N. Huynh, Kato, Rompré, et al., 2006). REM sleep: rapid eye-movement sleep. Arrows indicate REM sleep periods.

REM sleep are drastically different from those of non-REM sleep. Quantification of whole brain glucose metabolism in REM sleep showed an overall activity that is not different from wakefulness (Maquet, Dive, Salmon, et al., 1990; Maquet, Péters, Aerts, et al., 1996). However regional increases and decreases were observed during REM sleep as compared to wakefulness. Among the structures that increased their activity during REM sleep in comparison to wakefulness were the amygdala, thalamus, basal forebrain, hippocampus, anterior cingulate cortex, temporo-occipital regions and pontine tegmentum. By contrast, a decrease in activity was observed in the following structures: precuneus, dorsolateral prefrontal cortex, posterior cingulate gyrus, and inferior parietal cortex. These differences in brain metabolic activity during non-REM and REM sleep could be interpreted in light of the mental content during these stages. As a clarification it is now known that dreams occur during both REM sleep and SWS. However, when subjects are awakened from REM sleep, they are more prone to report a perception of being in a dreaming state (80–85% of the time) compared to when they are awakened from SWS, a state with low perception of being in a dreaming state (10–15%) (Hobson, 1990). REM sleep dreams are also reported to be longer, more vivid, bizarre, emotionally intense, and illogical than SWS dreams (Desseilles, Dang-Vu, Sterpenich, & Schwartz, 2011).

The majority of sleep time in adult humans is spent during the lighter stages of non-REM sleep, i.e., stage N2. N2 sleep stage is usually sequentially followed by SWS, which is predominant during the early part of sleep period and decreases in intensity and duration across the sleep period. The REM sleep stage, although present for a short duration in the first non-REM cycle (after about 90 min from sleep onset) becomes more intense and extensive toward the end of the sleep period. Non-REM and REM sleep alternate in cycles of 90–120 min repeated 3 to 6 times during the night, this is called the ultradian cycle. Time spent in stage N1 represents less than 5% of the total sleep duration, stage N2 represents 45–55%, and both SWS and REM sleep represent 20–25% each (St Louis Erik et al., 2016). Sleep is homeostatically regulated: a lack of sleep results in subsequently prolonged sleep (Borbély & Achermann, 1999; Sforza, Chapotot, Pigeau, Paul, & Buguet, 2004). This homeostatic process is regulated by the circadian pacemaker (Borbély, Daan, Wirz-Justice, & Deboer, 2016).

Sleep deprivation and sleep disruptions cause severe cardiovascular, cognitive and emotional problems (Brown, 2012; Chouchou, Pichot, Pépin, et al., 2013; Killgore, 2010). Following deprivation of sleep for several weeks, animals ultimately died of tissue lesions or infections

(Rechtschaffen & Bergmann, 1995). However, from an evolutionary perspective, reduced responsiveness to potentially threatening stimuli during sleep is a danger. The fact that almost all animals sleep argues in favor of an adaptive role of sleep in increasing the overall fitness of an organism, although its exact functions are still a matter of debate (Siegel, 2005). Sleep has been proposed as serving an energy-saving function (Berger & Phillips, 1995; Webb, 1988), thermoregulation (Rechtschaffen & Bergmann, 1995), the repairing of cell tissue (Oswald, 1980), cardiovascular and metabolic regulation (Knutson, Spiegel, Penev, & Van Cauter, 2007; Van Cauter, Eve, Tasali, & Leproult, 2008), or immune functions (Lange, Dimitrov, & Born, 2010). However, the decreased responsiveness to the external environment during sleep strongly speaks for the notion that sleep is essential for brain functions (Hobson, 2005; Kavanau, 1997).

Sleep is important for memory consolidation. Many studies have demonstrated that memory formation benefits from the off-line reprocessing during sleep (for review, see Ackermann & Rasch, 2014; Boyce, Williams, & Adamantidis, 2017; Diekelmann & Born, 2010; Xia & Storm, 2017). While encoding of memories take place during waking, sleep provides optimal conditions for an effective consolidation and stabilization of memories and their integration into long-term memory (Rasch & Born, 2013). Evidence acquired supports the view of a differential role of sleep stages in distinct memory processes, relating SWS to simple declarative memories and REM sleep to complex procedural and emotional memories.

The role of SWS in declarative memory consolidation is supported by experimental findings. For example, Alger and colleagues showed that only a nap containing SWS improved memory consolidation from future interference in a paired association task (Alger, Lau, & Fishbein, 2012). The importance of SWS for declarative memory consolidation is also supported by studies using instrumental manipulation of sleep oscillations. For example, a study showed that the induction of slow oscillations by transcranial electrical stimulation increased spontaneous slow oscillatory activity during sleep and improved declarative memory consolidation (Marshall, Helgadóttir, Mölle, & Born, 2006). Rauchs, Bertran, Guillery-Girard, et al. (2004), using a protocol of partial sleep deprivation (deprivation of the first or second half of the night, dominated by SWS and REM sleep, respectively), have shown a beneficial effect of SWS for the consolidation of temporal aspects of memory and a beneficial effect of REM sleep for spatial aspects and details (perceptions, associations of ideas made during encoding) (Rauchs et al., 2004). Brain imaging studies have shown the role of SWS in the consolidation of declarative episodic memory. For example, Peigneux, Laureys, Fuchs, et al. (2004) showed that the hippocampus was activated by a spatial navigation task during wakefulness, and was then subsequently reactivated during SWS of the night following the learning process. Moreover, hippocampal activity during SWS was positively correlated with improved memory consolidation across the night (Peigneux et al., 2004). All these data support the idea that SWS consolidates declarative episodic memories. REM sleep could also participate in this process but perhaps by consolidating certain aspects of declarative memories, such as the emotional dimension.

REM sleep seems indeed more important for procedural memories and for the emotional modulation of declarative memories (Wagner, Gais, & Born, 2001). Studies consistently showed that selective REM sleep deprivation impaired the consolidation of cognitive procedural tasks like the tower of Hanoi task or word fragment completion, while the consolidation of simple declarative memories remained unchanged (Smith, 2001). Furthermore, selective increases in REM sleep after learning have been particularly observed after learning of novel complex motor activity or other tasks containing a procedural component (Ackermann & Rasch, 2014). Memory for emotionally loaded information could also benefit from REM sleep (Tempesta, Socci, De Gennaro, & Ferrara, 2017). Research on relations between sleep and emotional processing have accompanied the growing evidence supporting that sleep is critical for the formation of episodic memories including emotionally-valenced information throughout all the stages. However, many studies have pointed to REM sleep as the specific sleep stage responsible for optimal emotional processing. Indeed, the functional and neurochemical changes that characterize REM sleep show an association with the brain mechanisms and substrates of emotional memory processing (Deliens, Gilson, & Peigneux, 2014; Groch, Wilhelm, Diekelmann, & Born, 2013). In this respect, imaging data showed a dysregulation of the emotional brain imposed by REM sleep loss (Walker & van der Helm, 2009).

At last, some evidence in humans and animals suggests that sleep can promote consolidation of conditioning and REM sleep is proposed to be particularly important to fear conditioning (Pace-Schott, Germain, & Milad, 2015). The neurobiology of REM sleep appears to be well suited for this function since it involves selective activation of the same limbic structures implicated in the learning and memory of fear conditioning. However, the literature about the effects of REM on the various components of fear learning is ambiguous; some studies showed that REM sleep increased recall of previously encoded fear (Menz, Rihm, Salari, et al., 2013), while others showed that REM sleep decreased fear recall (Fu, Li, Ouyang, et al., 2007), and yet others suggested that REM sleep increased the rate of extinction learning and its later preservation (Spoormaker, Schröter, Andrade, et al., 2012). Given the multiple consequences of experimental sleep deprivation and possible biases of sleep study, the exact role and function of a single sleep stage or the benefit of sleep continuity on memory processing is probably an open area of research that has not yet been fully explored.

1.2 Sleep and Pain Interaction

Clinical and experimental research provide evidence that pain disrupts sleep and that poor sleep affects baseline pain perception and may impair pain regulatory processes. Chronic painful conditions are frequently associated with sleep disturbances, including changes in sleep continuity and sleep architecture as well as increased daytime fatigue and sleepiness (Bjurstrom & Irwin, 2016; Weingarten, Dubrovsky, Basner, et al., 2016). To understand this relationship between pain and sleep in chronic painful conditions, nociceptive stimulations were used to disturb sleep in experimental settings, and in another stream, sleep deprivation, restriction, and disruption were used to explore the impact of altered sleep on pain. It is well recognized that pain may disturb sleep (Bastuji, Perchet, Legrain, Montes, & Garcia-Larrea, 2008; Chouchou et al., 2011; Khoury, Chouchou, Amzica, et al., 2013), and we also know that disruption or deprivation of sleep may affect nociceptive processes in the brain (Schuh-Hofer, Baumgärtner, & Treede, 2015; Schuh-Hofer, Wodarski, Pfau, et al., 2013; Simpson, Scott-Sutherland, Gautam, Sethna, & Haack, 2018). In this part, we will review the state of the knowledge on the potential role of pain on sleep disruptions and the effect of sleep disruptions on pain perception.

1.2.1 Does Pain Disturb Sleep?

The decreased responsiveness during sleep is underpinned by sensory information processing precociously inhibited during sleep (Claude, Chouchou, Prados, et al., 2015), suggesting a sensorial isolation of the sleeper. During sleep, these mechanisms protect the sleeper from awakening by irrelevant sensory inputs. In the case of painful stimulation, pain stimulations induce more arousals than non-nociceptive ones (Lavigne, Zucconi, Castronovo, et al., 2000). Indeed, pain stimuli produce significantly more arousal reactions (27–48% of stimuli) (Bastuji et al., 2008; Lavigne, Brousseau, Kato, et al., 2004; Lavigne et al., 2000) than auditory stimuli

(Kato, Montplaisir, & Lavigne, 2004). Thus, intensity range allowing sleep continuity for nociceptive stimulation seems more limited than the one for acoustic stimulation. In healthy subjects, these nociceptive stimuli induce cortical arousals in all sleep stages (Bastuji et al., 2008; Lavigne et al., 2004, 2000). However, during SWS higher stimulation intensities are necessary to produce arousal as compared to stage 2 and PS the temperature intensity required to awaken participants in sleep stage 2 must be increased by 1 °C to produce a similar response during SWS or REM sleep (Bentley, Newton, & Zio, 2003).

If painful stimuli delivered during sleep are followed by arousals in only 27–48% of stimuli, they can trigger a variety of other reactions (Bastuji et al., 2008; Lavigne et al., 2000). Central nervous system reactivity to nociceptive stimuli delivered during sleep are accompanied by a concomitant sympathetic reactivity, well known during wakefulness (Burton, Birznieks, Bolton, Henderson, & Macefield, 2009) and which changes the physiological processes under autonomic control (Chouchou et al., 2011; Lavigne, Zucconi, Castronovo, et al., 2001). This sympathetic reactivity, which is of higher amplitude than with non-nociceptive stimuli, is also preserved in all sleep stages, and is closely related to electroencephalographic activation. However, when nociceptive stimulations do not produce cortical arousals, sympathetic reactivity remains at lower magnitude (Chouchou et al., 2011). This autonomic reactivity might be considered as the first mechanism that allows potential behavioral response and could take part in occurrence of cortical arousals. Also, Sandrini and colleagues have studied the effect of sleep on spinal nociceptive reflex (RIII reflex). The threshold of the RIII reflex is considered as the pain threshold and the size of the reflex as the level of pain perception. This reflex might constitute a useful tool in the investigation of pain processing. With respect to wakefulness, the RIII reflex threshold is increased from stage 2 to SWS sleep and exhibits highest values during REM sleep. The latency of the reflex is delayed during SWS and REM sleep (Sandrini, Milanov, Rossi, et al., 2001).

These studies showed that experimental pain during sleep in healthy participants induces sleep fragmentation, characterized by abrupt activation of the central nervous system, producing EEG activation (evoked potentials) and concomitant changes in physiologic parameters under autonomic control, such as cardiovascular and respiratory systems.

1.2.2 Does Sleep Disruption Influence Pain Perception?

The first experimental study of sleep deprivation (e.g., partial or total sleep time reduction) and the consequences on pain perception was conducted in

1935 by Cooperman, Mullin, and Kleitman (1935) who observed that total sleep deprivation induced a lower pain threshold in healthy participants. Since this pioneering work, several experimental studies have explored the relationship between sleep and pain.

The relationship between sleep disruptions and pain processes was also tested in healthy participants (Schuh-Hofer et al., 2015, 2013; Simpson et al., 2018) and patients with different diseases (Busch, Haas, Crönlein, et al., 2012; Haack, Scott-Sutherland, Santangelo, et al., 2012; Irwin, Olmstead, Carrillo, et al., 2012; Schey, Dickman, Parthasarathy, et al., 2007). Overall, the results indicated that sleep disruption, specific sleep stage deprivation or sleep continuity disturbances altered pain perception and caused spontaneous pain in healthy individuals (Schuh-Hofer et al., 2015, 2013; Simpson et al., 2018). These results demonstrated that total and partial sleep deprivation (Arima, Svensson, Rasmussen, et al., 2001; Azevedo, Manzano, Silva, et al., 2011; Haack, Lee, Cohen, & Mullington, 2009; Lentz, Landis, Rothermel, & Shaver, 1999; Moldofsky, Scarisbrick, England, & Smythe, 1975; Older, Battafarano, Danning, et al., 1998; Schuh-Hofer et al., 2015; Smith, Edwards, McCann, & Haythornthwaite, 2007) induced pain-related discomfort such as stiffness, joint pain, back pain, and headache in healthy participants, whereas recovery nights were associated with decreased intensity of discomfort (Irwin et al., 2012; Moldofsky et al., 1975; Smith et al., 2007). Two studies used heat laser stimulation to show that the same stimulation was perceived as more intense after partial or total sleep deprivation (Azevedo et al., 2011; Tiede, Magerl, Baumgärtner, et al., 2010). One more study showed that total sleep deprivation resulted in a generalized decrease in pain threshold (Schuh-Hofer et al., 2013) suggesting that sleep deprivation induces hyperalgesia. In contrast, Roehrs, Harris, Randall, and Roth (2012) found that lengthening sleep duration, in healthy subjects, increased the heat pain threshold (decreased pain sensitivity). Taken together, these studies showed that both sleep disruption and deprivation induce pain-related discomfort and hyperalgesia. Experimental studies in healthy participants suggest that sleep improvement in patients with sleep disorders may reduce their hyperalgesia and pain-related discomfort. The reductions in hyperalgesia and pain-related discomfort associated with sleep improvements observed in experimental studies in healthy participants warrants further studies in patients suffering from chronic pain in sleep disorders.

It was also proposed that SWS and REM sleep play specific roles in diurnal pain control (Roehrs, Hyde, Blaisdell, Greenwald, & Roth, 2006). Specific SWS deprivation caused a decrease in the mechanical pain threshold (Arima et al., 2001; Lentz et al., 1999; Older et al., 1998; Onen, Alloui, Gross, Eschallier, & Dubray, 2001; Roehrs et al., 2006; Zhang, Lam, Li, et al., 2012) and induced pain-related discomfort (Arima et al., 2001; Lentz et al., 1999; Moldofsky et al., 1975). Only a few studies have addressed REM sleep deprivation (Azevedo et al., 2011; Roehrs et al., 2006), and only one showed no change in heat pain threshold (Onen et al., 2001). Although only a few studies have focused on REM sleep, these divergent results do not support a role of a specific sleep stage in diurnal pain control. Moreover, Onen et al. (2001) deprived participants of both SWS and REM sleep and found no specific effect of either sleep stage deprivation on pain perception. Smith et al. (2007) found that the *disruption of sleep continuity* impaired the endogenous pain-inhibitory function leading to an increase in pain-related discomfort, an effect not replicated with simple sleep restriction. Thus, it remains unclear as to whether alterations in the pain process are due to specific sleep stage deprivation or to sleep continuity disturbance.

These results are consistent with our knowledge of the sleep in patients suffering from pain. Chronic painful conditions are frequently associated with sleep disturbances, including changes in sleep *continuity* and sleep *architecture* as well as increased daytime fatigue and sleepiness (Bjurstrom & Irwin, 2016; Weingarten et al., 2016). In turn, sleep quality in chronic pain conditions is considered a consistent predictor of pain the next day (Tang, Goodchild, Sanborn, Howard, & Salkovskis, 2012). This vicious cycle has also been reported by patients with chronic pain due to conditions such as fibromyalgia (Choy, 2015) or severe skin burns (Raymond, Ancoli-Israel, & Choinière, 2004). However, the relationship between pain and sleep involves multiple unidirectional or bidirectional interactions in which acute and chronic pain are associated with sleep disruptions and disturbed or shortened sleep can in turn alter pain perception (Afolalu, Ramlee, & Tang, 2017; Lavigne, Khoury, Chauny, & Desautels, 2015; Sivertsen, Lallukka, Petrie, et al., 2015). The sensitivity to pain and sleep deprivation can vary considerably between individuals, in relation with genetic factors (Maire, Reichert, Gabel, et al., 2014; Matic, van den Bosch, de Wildt, Tibboel, & van Schaik, 2016), psychological factors such as anxiety and catastrophizing (Karmann, Lauer, Ziegler, et al., 2018), age (Eslami, Zimmerman, Grewal, Katz, & Lipton, 2016) and chronic sleep disorders (Jank, Gallee, Boeckle, Fiegl, & Pieh, 2017) as well as loud environments (Efe, Dikmen, Altaş, & Boneval, 2013) could render patients vulnerable to pain exacerbation by sleep disruptions (see Fig. 3). These factors could contribute to the wide inter-individual differences in the intensity and duration of reported sleep disturbances and complaints.

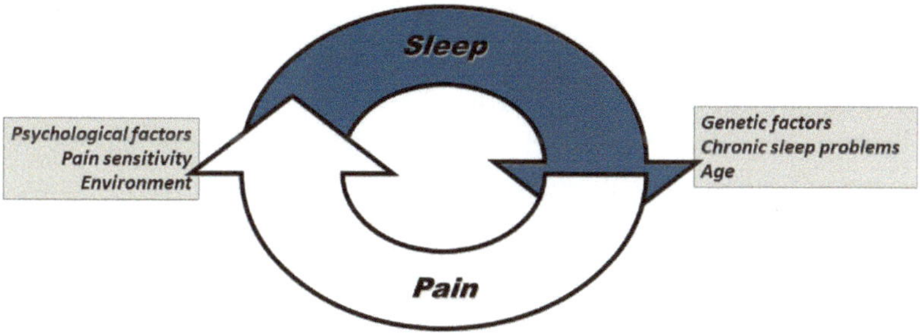

Fig. 3 Model of the relationship between sleep and pain, and respective contributing factors.

Recently, pain perception and daily reported pain were assessed in patients with primary insomnia (Haack et al., 2012; Tang et al., 2012). Under heat and mechanical pain, insomnia patients presented decreased pain threshold compared to healthy participants. They also experienced more frequent and more intense pain and presented a dysfunctional pain-inhibition system, measured by a diffuse noxious inhibitory controls paradigm, in which perception of a main experimental pain is inhibited with parallel pain stimulation by endogenous pain pathway activation. Consistently, in another study, forced awakening paradigm (i.e., subjects awakened 20 min every hour of sleep plus one full hour during sleep) showed greater capacity to cause spontaneous pain and induce the dysfunctional pain-inhibition system compared to simple sleep restriction in healthy women (Smith et al., 2007). This further reinforces the notion that sleep discontinuity perturbs pain regulatory processes.

A possible explanation for altered pain control caused by sleep disruptions could reflect impaired attention processes, which is an important modulator of the pain experience (Campbell, Bounds, Simango, et al., 2011; Quiton & Greenspan, 2007). Tiede et al. (2010) showed that sleep disruptions were associated with a significantly weaker ability to disengage oneself from painful stimulation. In accordance, Campbell et al. (2011) showed that shorter self-reported habitual sleep duration may reduce or delay the onset of distraction-induced analgesia. Inflammation was also proposed as a contributor to alter pain processes that may be related to sleep homeostasis. Haack et al. (2009) reported, in an experimental sleep protocol including healthy subjects, that pro-inflammatory mediators (IL-6) increased in prolonged sleep restriction, associated with pain-related discomfort. This interesting finding was replicated in chronic pain conditions (Heffner, France, Trost, Ng, & Pigeon, 2011; Park & Chung, 2016), and inflammatory processes have been proposed as playing a significant role in the interaction between pain and sleep disturbance. Finally, pain is influenced by anxiety and emotional state (Garcia-Larrea & Peyron, 2013). Sleep loss not only enhances pain but also alters emotional well-being, which is an integral part of the pain experience (Walker, 2010), and sleep disruption in particular (Gujar, Yoo, Hu, & Walker, 2011) is known to promote a negative emotional state such as anxiety, suggesting that it indirectly amplifies pain.

In summary, acute or chronic pain can induce poor sleep quality and in turn, poor sleep quality can alter how individuals perceive the pain, and may interfere with pain regulation. These data concur that sleep deprivation or disruption contributes to disrupt pain control, exacerbates pain perception and limits the effects of analgesic treatments (Campbell et al., 2011; Cho, Lee, Lee, Shin, & Hwang, 2015; Doufas, Tian, Padrez, et al., 2013). Several mechanisms may be involved, some of which include high-level cognitive functions. In accordance with these results and questions, studies are needed to examine the relationship between sleep and pain relief.

1.3 Sleep and Placebo-Induced Expectations and Analgesia

As described in other articles of this volume (Klinger, Stuhlreyer, Schwartz, Schmitz, & Colloca, 2018), the placebo neurobiological response may contribute to the therapeutic effects of some active treatments, particularly for the relief of acute and chronic pain (Vase et al., 2002). It depends on multiple complex brain processes (Fig. 1) (Benedetti, Carlino, & Pollo, 2011; Colloca et al., 2013; Finniss et al., 2009; Hall, Loscalzo, & Kaptchuk, 2015; Peerdeman, van Laarhoven, Keij, et al., 2016; Schafer, Geuter, & Wager, 2018), among which expectation appears to play a pivotal role. For example, the expectation of a positive outcome may affect an individual's physiological pain responses. Thus, the expectation of relief may help reduce a subject's anxiety about a future treatment and/or activate the neuronal networks of reward mechanisms (Harden, Saracoglu, Connolly, et al., 2016). By contrast, the expectation of a negative outcome involves the anticipation of a possible threat, thus increasing anxiety. Indeed, there is ample support for a role of both anxiety and reward mechanisms in placebo responsiveness (Benedetti et al., 2011). Moreover, pain relief may be conceived as a reward that we can learn to expect implicitly through conditioning or explicitly through suggestions and experience, as previously illustrated in imaging studies of Parkinson's disease (de la Fuente-Fernández, Ruth, Sossi, et al., 2001), in healthy participants (Schweinhardt, Seminowicz,

Jaeger, Duncan, & Bushnell, 2009) or in chronic pain conditions (Baliki, Geha, Fields, & Vania Apkarian, 2010). At last, learning is a way by which humans build representations of self-environment interactions which may be reinforced when outcomes are consistent with predictions or updated/revised whenever expectations are violated (Chouchou & Lavigne, 2014; Laverdure-Dupont, Rainville, Montplaisir, & Lavigne, 2010; Thorwart & Livesey, 2016).

Many of the factors involved with placebo effects are related to sleep quality including anxiety (Palmer, Oosterhoff, Bower, Kaplow, & Alfano, 2018), reward (Perogamvros, Dang-Vu, Desseilles, & Schwartz, 2013) and learning (Diekelmann & Born, 2010). Sleep deprivation can also lead to changes in reward, anxiety and motivation systems that induce a higher state of increased anxiety (Pires, Bezerra, Tufik, & Andersen, 2016), increasing the risk of substance abuse (Wong, Brower, & Zucker, 2009) and appetitive behavior (Benedict, Brooks, O'Daly, et al., 2012). Finally, as we have described above, deprivation or disruption of sleep alters memory consolidation, including procedural, episodic or declarative memory (Ackermann & Rasch, 2014).

From these observations and from the fact that pain and sleep maintain close reciprocal interrelationships, it seems highly plausible that sleep may be involved in the development and consolidation of placebo analgesia. Following this perspective, our research group completed three experimental studies exploring the interaction between sleep and placebo analgesia.

1.3.1 Expectation-Mediated Placebo Analgesia in Relation to Sleep

Laverdure-Dupont et al. (2009) examined the possibility that sleep might contribute to the consolidation of new expectations and consequently influence expectation mediated-placebo analgesia. Thirty-eight healthy participants were recruited and all underwent placebo conditioning and test following standardized procedures (Bingel, Lorenz, Schoell, Weiller, & Büchel, 2006; Price, Milling, Kirsch, et al., 1999; Wager, Rilling, Smith, et al., 2004).

The conditioning manipulation was intended to provide subjects with an unambiguous experience of analgesia, coming from the sensory experience during the painful stimulations and the information given to the subject. The test phase was used to test the placebo effect in the morning. For that, at the beginning of the conditioning and test, the same inert cream was applied to the control and treatment sites identified on each arm. The cream applied to the treated site (placebo site) was described as a topical analgesic, whereas the cream applied to the control site was described as an inert cream used to control for nonspecific effects of the vehicle compound of the analgesic cream. In the conditioning phase, successive phasic stimuli were delivered to the two sites, but the temperature of the stimuli applied to the placebo site was surreptitiously decreased by 2 °C compared with the control site. In the test phase, subjects received five thermal stimuli of the same predetermined moderate pain stimulation level on each arm.

Then, the thirty-eight healthy participants were divided into three groups: the night, the day and the incongruence night group. In the *night group*, the conditioning procedure was performed in the evening and the placebo analgesia was only evaluated the following morning (see Fig. 4). In the *day group*, the conditioning procedure was performed in the morning and the placebo analgesia was tested in the evening of the same day (10–12 h delay, no sleep). This group controlled for sleep-independent effects associated with the simple passage of time. In the *incongruence night group*, a first test was performed in the evening, after conditioning, and it was repeated in the next morning. During the test phases, the stimulus intensity was matched between the placebo site and the control site. The addition of an evening placebo test in the *incongruence night group* was used to provide subjects with a more ambiguous experience of analgesia in the evening compared to conditioning alone and may be conceived as an extinction phase according to classical models of conditioned learning. The ambiguity comes from the discrepancy between bottom-up sensory signals and top-down expectation induced by the placebo conditioning procedure (Legrain, Van Damme, Eccleston, et al., 2009).

Expectations and placebo analgesic effects were measured the following morning and compared with those of a control daytime group without sleep. Although placebo effects were observed in both groups, correlation analysis suggests that the mediating effect of expectations on placebo effects was strongest in the *night group*. Moreover, after exposure to a convincing analgesia experience, the relative duration of REM sleep was lower in subjects showing higher analgesic expectations and placebo effects the next morning. In the *incongruence night group*, expectations reported in the morning were comparable with other groups. However, expectations were *positively* correlated with the duration of REM sleep and did not emerge as a significant mediator of the analgesic effect. In other words, these results show that when congruence between sensory experience and expectations is high, the lower relative duration of REM sleep may contribute to the preservation or the reinforcement of expectation-induced placebo analgesia that may be more

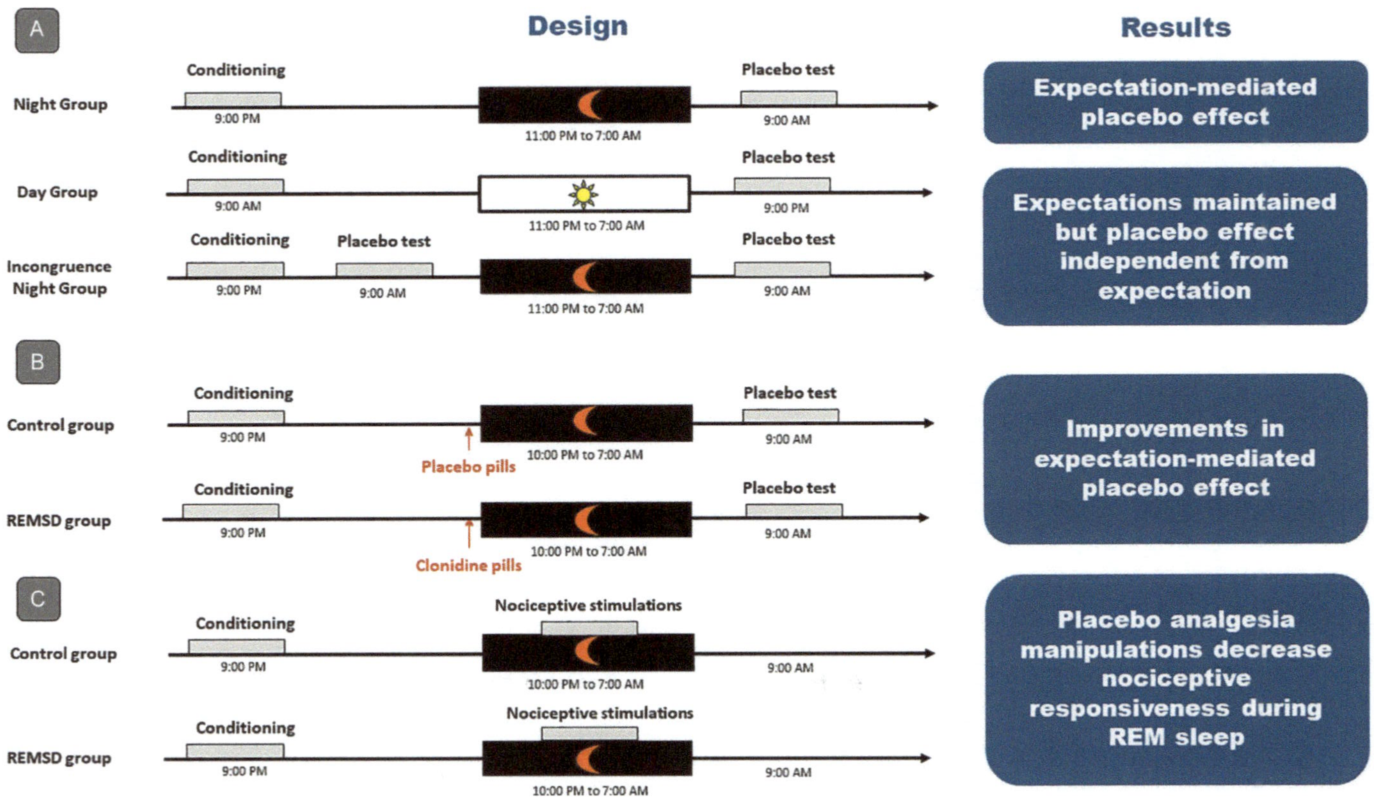

Fig. 4 See legend on next page.

resistant to extinction. On the contrary, when there is an incongruence between sensory experience and expectations, the need for REM sleep may be higher and the expectations of relief was found to have less influence on the placebo effect.

Together, these findings show that sleep-related processes may influence the association between expectations and placebo analgesia and that REM sleep may be involved in the off-line reprocessing of placebo-induced expectations of pain relief or more generally in the efficacy of pain-regulatory processes. As shown previously (Benedetti et al., 2011; Price et al., 1999), expectations were a significant mediator of the analgesic effects, but their relative contribution varied considerably when the post-conditioning delay included a sleep period. In this condition, expectations were tightly correlated with placebo analgesia and predicted as much as 77% of the effect. The acquisition of expectations leading to changes in perception can be viewed in the general framework of episodic memory formation in which new information is acquired, consolidated, and recalled. In the present study, expectations of relief induced experimentally were gradually integrated into an existing network of personal beliefs and past experience and were evoked in subsequent testing.

Among possible underlying mechanisms, growing evidence suggests that placebo effect could be explained by predictive coding models (e.g., Rief & Petrie, 2016). According to classical models, prediction error is a key factor in reinforcement learning: when the bottom-up sensory input is as expected from top-down signal driven by prior experience, beliefs or suggestions, there is low prediction error and high concordance (Rescorla & Wagner, 1972). When the bottom-up signal is inconsistent with top-down

Fig. 4 Time course of placebo and sleep studies. (A) Conditioning manipulations were performed in the evening while the test phase was performed in the morning, 12 h later, in the night group ($n = 12$). In the day group ($n = 13$), the conditioning and test phases were separated by 12 h of daytime. At last, in the night incongruence group ($n = 13$), both the conditioning manipulations and test phases were performed in the evening and the test phase to assess placebo effects was repeated in the morning (Laverdure-Dupont et al., 2009). (B) Rapid eye-movement (REM) sleep deprivation was applied to study its effect on the expectation-mediated placebo analgesia. Either clonidine pills were given to participants in REM sleep deprivation group (REMSD, $n = 13$) or placebo pills in control group ($n = 13$) immediately before going to bed (Chouchou et al., 2015). The graphs illustrate the impact of sleep, incongruence and REMSD on the expectation-mediated placebo effect. (C) After placebo analgesia manipulations, responsiveness to nociceptive stimulations during REM sleep was decreased, whereas during SWS the responsiveness was increased.

expectation, the brain generates a strong prediction error signal reflecting the high discordance, which may in turn affect pain regulation (e.g., Taylor, Chang, Rainville, & Roy, 2017). Illustrating the involvement of predictor error in placebo effect, Peciña, Stohler, and Zubieta (2014) tested an experimental conditioning protocol that showed larger placebo effects in individuals with low expectations and high subjective analgesic effectiveness (i.e., large positive prediction error signal) whereas nocebo hyperalgesic responses were observed in those reporting high expectations and low reported analgesic effectiveness (i.e., large negative prediction error signal). The magnitude of μ-opioid system activation in regions relevant to error detection was further associated with placebo-induced analgesia. Moreover, these data confirmed that the effect of expectation and outcome associations on behavioral placebo effects is mediated by endogenous opioid release and μ-opioid receptor system activation in the anterior cingulate cortex (Peciña et al., 2014). In this vein, we speculate that REM sleep is a key factor influencing the evaluation of expectation according to bottom-up signal, and leading to changes in the internal representation. In the present context, an experience of analgesia that is not fully consistent with expectations before sleep may generate a negative prediction error (i.e., more pain is experienced than expected) which may be reprocessed during REM sleep thereby reducing expectation-related placebo effects the next day. Reciprocally, the reduction of REM sleep may prevent a revision of the internal model and preserve or reinforce expectation-related effects.

1.4 REM Sleep Deprivation and Placebo-Induced Expectations and Analgesia

To better understand this relationship between expectation-mediated placebo analgesia and sleep, we tested the hypothesis that placebo analgesia would be improved by REM sleep deprivation (REMSD) in a follow-up experiment conducted on healthy participants (Chouchou et al., 2015).

1.4.1 Effect of Experimental REM Sleep Deprivation on Placebo Analgesia

Following an adaptive night in a sleep laboratory, twenty-six healthy participants underwent classical experimental placebo analgesic conditioning in the evening and were randomized to either clonidine, an alpha-2 adrenergic receptor agonist known to suppress most of REM sleep (Autret et al., 1976; Huynh, Lavigne, Lanfranchi, Montplaisir, & de Champlain, 2006) or inert control pills in a double-blind fashion. As used in the previous study

(Laverdure-Dupont et al., 2010) and mentioned above, the conditioning procedure was intended to provide subjects with an unambiguous experience of analgesia, and the test phase was used to test the placebo responsiveness in the morning.

Healthy subjects were invited for a second recording night 1 week later, and were administered either clonidine or an inert control pill in a double-blind randomized sequence to create two groups: one experimental (REMSD, $n = 13$) and one control (controls, $n = 13$). The procedure was identical for both groups and was administered by the same investigator, who was blind to the experimental condition. We used a pharmacological REM sleep deprivation (REMSD) design with the alpha-2 adrenergic receptor agonist clonidine (see Fig. 5). REMSD is particularly relevant for examining the relationship between sleep and placebo analgesia, because sleep disruptions (e.g., total and partial sleep restriction, awakenings during sleep) are known to alter pain sensitivity the next day (Chouchou, Khoury, Chauny, Denis, & Lavigne, 2014; Lavigne, Okura, Abe, et al., 2011). A pharmacological REMSD model is more suitable than the usual awakening from REM sleep method for investigating pain perception and placebo analgesia, without change in sleep duration or sleep fragmentation (Autret et al., 1976; Gentili et al., 1996; Huynh, Lavigne, et al., 2006). In this study, REM sleep duration was shorter in REMSD group (Controls: $16.9 \pm 0.7\%$; REMSD group: $0.9 \pm 0.4\%$). Conversely, REMSD group showed longer sleep stage 2 compared to the control group (Controls: $48.6 \pm 1.1\%$; REMSD group: $61.8 \pm 2.5\%$). However, total sleep and SWS durations did not differ between groups, nor did sleep fragmentation. Overall, all subjects in both groups reported comparably good sleep quality. As expected, these results indicate that clonidine medication strongly decreases or suppresses REM sleep, with no alterations in sleep continuity, architecture, or quality.

Placebo analgesia, tested in the morning increased following REMSD (see Fig. 5). Interestingly, expectations of pain relief did not differ between the REMSD and the control group but placebo analgesia correlated positively and strongly with pain relief expectations only in REMSD but not in the control group (Fig. 5). These findings strengthen our suggestion that the reprocessing of expectations during REM sleep may reduce rather than consolidate expectation-mediated placebo analgesia. Importantly, the REMSD group also displayed a compensatory increase in NREM sleep. This implies that NREM sleep may contribute to the consolidation of expectation-congruent effects while REM sleep may involve a reprocessing of expectation leading to a facilitation of extinction on the next day.

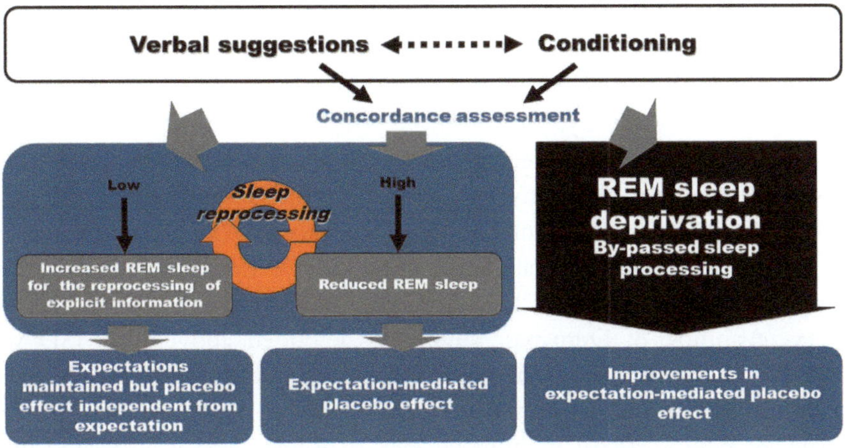

Fig. 5 Updated model of sleep-dependent processing of placebo-induced expectations. *Induction of relief expectation*: relief expectations associated with a placebo treatment are generated by verbal suggestion and reinforced by conditioning manipulations. *Off-line processing*: during sleep, concordance during verbal suggestion and sensory experiences was re-assess by participants. Depending on the level of concordance between the expectations initially created and the sensory experience, the need for offline reprocessing during sleep may vary: high level of concordance is associated with a reduction of REM sleep and a low level with an increase in REM sleep. *Generalization of expectation*: when level of concordance is high, and REM sleep reduced, placebo effects are mediated by expectations, but when level of concordance is low, and REM sleep lengthened, expectations and placebo effects were dissociated. At last, REM sleep deprivation bypassed the sleep processing of expectation, and whatever the concordance between sensory experience and expectation mediated placebo effect.

Consistent with our results, Spoormaker and colleagues demonstrated altered prediction error signaling during recall of fear extinction after REM sleep deprivation (Spoormaker et al., 2012). Prediction error processing is known to be related to learning and motivational behavior (Steinberg, Keiflin, Boivin, et al., 2013), which are proposed to be involved in placebo and pain perception (Tracey, 2010), suggesting that this complex interrelationship may be a relevant underlying mechanism of the placebo analgesic response following REMSD, as observed in our study. In a brain imaging study, Scott, Stohler, Egnatuk, et al. (2007) tested the correlation between placebo responsiveness and monetary reward and found that placebo responsiveness was related to dopamine activation in the nucleus accumbens. Subjects were then tested for monetary responses in the nucleus accumbens, revealing positive correlations between placebo and monetary responses. Similarly, a recent study showed correlations between ventromedial prefrontal cortex activity during fear conditioning and during REM sleep in healthy participants (Spoormaker, Gvozdanovic, Sämann, & Czisch, 2014), and that ventromedial prefrontal cortex activity during REM sleep mediates the relationship between fear conditioning and extinction the next day. Furthermore, the motivational/reward system is activated during sleep, especially REM sleep, whereby the reward-related mesolimbic regions are activated, including the ventral tegmental area, nucleus accumbens, and anterior cingulate cortex (Perogamvros et al., 2013). REMSD leads to reward and motivation system dysfunction, resulting in significant alternations in behavioral and emotional regulation (Sandyk, Tsagas, Anninos, & Derpapas, 1992). As described above, REMSD was suggested to contribute to increase the risk of substance abuse (Aalto & Kiianmaa, 1984) and appetitive behavior (Hanlon, Andrzejewski, Harder, Kelley, & Benca, 2005). In humans, lack of REM sleep is associated with enhanced emotional reactivity (Rosales-Lagarde, Armony, Del Río-Portilla, et al., 2012). Collectively, these data suggest that REM may be involved in the off-line reprocessing of expectations and that REMSD may prevent the extinction of expectation-driven changes in outcomes.

1.5 Placebo Analgesia During Sleep

It is established today that the brain processes external stimuli to some extent during sleep (Cote, Etienne, & Campbell, 2001). Despite the decrease in behavioral responsiveness regardless of the nature of the stimulation during sleep, autonomic (Chouchou et al., 2011; Lavigne et al., 2001) and cortical reactivities (Bastuji et al., 2008; Lavigne et al., 2000) show that the brain continues to process nociceptive stimulations. More recently, Jensen and colleagues demonstrated that placebo effect may be triggered without conscious awareness of the associated cues (Jensen, Kaptchuk, Chen, et al., 2015; Jensen, Kaptchuk, Kirsch, et al., 2012). Considering that the brain still reacts to nociceptive stimuli during sleep (Bastuji et al., 2008; Chouchou et al., 2011; Lavigne et al., 2001) and placebo effect can be demonstrated without conscious processing of the placebo inducer, Laverdure-Dupont and colleagues examined if placebo effect persists during sleep. The authors studied nociceptive reactivity during sleep following placebo conditioning (Laverdure-Dupont et al., 2018).

1.5.1 Persistence of Conditioned Placebo Analgesia During Sleep

In this study, nine healthy participants were recruited and after a habituation night, they underwent two overnight sessions, one experimental and one control in within-subject design with a counterbalanced order (Fig. 4).

Before sleeping in the experimental session, all participants underwent the conditioning manipulation described above. Expected, concurrent and retrospective pain ratings were collected for each site as well as expectation of nocturnal pain, sleep disruption and anxiety. During the following sleep, in both conditions, nociceptive stimulations were administered with the contact probe stably attached to the participant's forearm. Nociceptive stimulations of equal intensity, corresponding to moderate pain during wakefulness, were applied in the control and the placebo night thereby allowing us to assess placebo analgesic effects. Stimulations were delivered pseudorandomly over sleep stage 2, SWS and REM sleep, in an attempt to obtain a comparable distribution in the two nights. Although Laverdure-Dupont et al. (2018) did not observe any changes in sleep architecture between experimental and control nights, it was observed that the conditioning manipulation before sleep modified the proportion of arousals produced by the noxious stimulations during REM sleep and SWS. During REM sleep, placebo treatment was associated with a lower rate of cortical EEG arousals consistent with placebo analgesia (Placebo: 83.2% vs. control: 93.0%). Conversely, placebo treatment was associated with a *higher* rate of arousals during SWS (Placebo: 88.8% vs. control: 74.8%). This suggests that the placebo treatment did affect nociceptive processing during sleep but that the direction of the effect depends on the sleep stage. Additional analyses of the changes in cortical EEG reactivity to the noxious stimuli administered during sleep were further assessed following the conditioning procedure. Although these findings are derived from a small sample, results suggested an increase in theta activity during the noxious stimuli administered in REM sleep during the placebo night compared to the control night (Fig. 6). Interestingly, larger relative increases in theta reactivity during REM sleep in the placebo night were predicted by lower ratings of expected pain unpleasantness and pain-related sleep disturbances obtained in the evening immediately after conditioning and before sleep. This suggested that the reduction in arousals during REM sleep in the placebo night may involve expectation-related theta activity. In the morning, participants further reported a significant reduction in their subjective estimation of nocturnal pain intensity, pain unpleasantness, anxiety, and sleep disturbance following the placebo night. This effect mirrored the reduction in arousals observed in REM sleep.

These experimental findings could also be interpreted in light of neurobiological substrate of pain perception and brain activity according to sleep stages. From imaging findings, neurobiological substrate of pain perception,

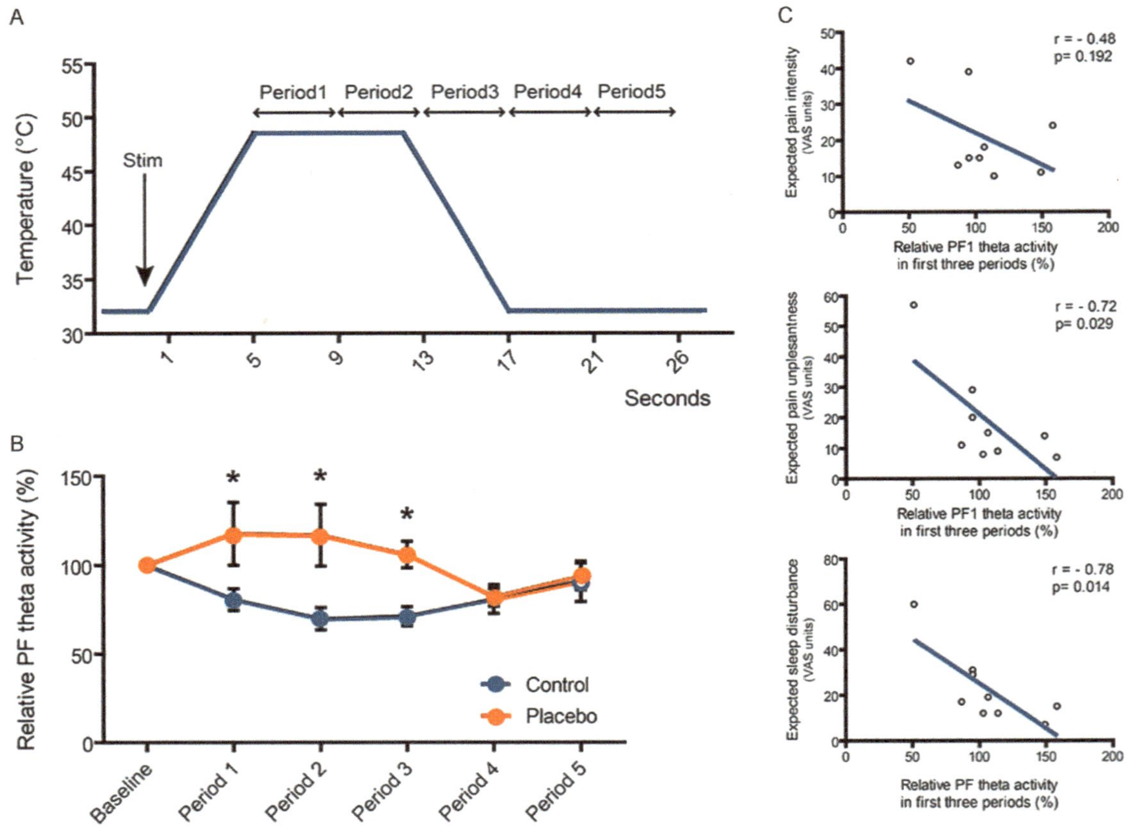

Fig. 6 Modulation of left prefrontal (PF1) theta activity associated with stimuli presentation during REM sleep. (A) Time course of nocturnal stimuli. Stimulation onset (stim) and five periods of 4 s following the beginning of the plateau are marked by arrows. (B) Increase of relative left prefrontal theta activity in the placebo night compared to the control night. (C) The relative amount of left prefrontal theta activity (mean of Periods 1–3) in the placebo night is correlated with the level of expected pain and pain-related sleep disturbances noted in that evening.

defined as the pain matrix (Melzack, 1990, 1999), is characterized by a nociceptive responsive network of brain structures receiving spinothalamic projections, mainly the posterior operculoinsular areas (Dum, Levinthal, & Strick, 2009), subserving the bodily specificity of pain. A second-order network, including posterior parietal, prefrontal and anterior insular areas, allows cortical nociception to become conscious pain and modulations of pain perception (Apkarian, Catherine Bushnell, Treede, & Zubieta, 2005; Garcia-Larrea & Peyron, 2013; Peyron, Laurent, & García-Larrea, 2000). During non-REM sleep, imaging and electrophysiological studies have shown a slowing in electrophysiological activities associated to an overall drop of brain metabolism (Dang-Vu, Schabus, et al., 2010; Nofzinger et al., 2002; Zoccoli et al., 2002), in particular cortical regions including anterior cingulate cortex, prefrontal cortex, and precuneus (Dang-Vu, McKinney, Buxton, Solet, & Ellenbogen, 2010; Zoccoli et al., 2002). In contrast, REM sleep is characterized by fast- and low-amplitude oscillatory brain activity such as theta oscillations, resembling waking (St Louis Erik et al., 2016), with an overall metabolic activity increase in particular in some brain regions such as amygdala, thalamus, basal forebrain, hippocampus, anterior cingulate cortex, temporo-occipital regions and pontine tegmentum, whereas others decrease such as precuneus, dorsolateral prefrontal cortex, posterior cingulate gyrus, and inferior parietal cortex (Maquet et al., 1996). Taken together, these works might suggest that during sleep, the reactivity of second-order structure of this pain matrix, particularly prefrontal and frontal structures is diminished (Velluti, 1997). However, the motivational/reward system is activated during sleep, especially during REM sleep, including the ventral tegmental area, nucleus accumbens, and anterior cingulate cortex (Perogamvros et al., 2013). Overall these studies supports a complex processing of nociceptive inputs by the brain during REM sleep and the overactivity of the reward system during REM sleep but not during non-REM sleep could explain the difference in cortical reactivity between these two sleep stages, associated with a reactivation of second-order structure of this pain matrix during REM sleep. Another explanation may come from the increase in theta oscillations following nociceptive stimulation during sleep. Indeed, the frontal theta response during wakefulness represents a complex set of cognitive processes whereby selective attention becomes focused on the stimuli and higher cognitive process. It may be speculated that the theta response to the stimuli following the conditioning procedures may be a result of the critical role that the stimuli play in sensory-perceptual and attentional processing (Karakaş, Arikan, Cakmak, et al., 2006; Karakaş,

Cakmak, Bekçi, & Aydin, 2007). This theta response could mark a high-level cerebral process evoking a re-evaluation of sensory imputation during sleep in light of the expectation of relief built during placebo analgesia.

According to predictive coding theories (Peciña et al., 2014; Rescorla & Wagner, 1972), this evidence suggests that the internal model generating analgesic expectations may be reactivated implicitly during REM sleep (associated with frontal theta) and reduce pain-related responses (i.e., reduced arousal and next day ratings). In contrast, the placebo-induced increase in arousal during SWS could reflect a stronger prediction error during the night in which participants are exposed to the conditioning manipulation (i.e., more bottom-up pain signal than expected) or a mechanism protecting sleep from pain in the control night (expecting more pain may trigger a protective response) consistent with the usual lower responsiveness observed in SWS in comparison with other sleep stages (Lavigne et al., 2000). Obviously, although three nights of sleep recording was used for each subjects to generate these findings, replication is warranted to confirm the placebo effect during sleep and the putative role of prefrontal theta activity in the production of this placebo analgesia.

2. CONCLUSION AND FUTURE PERSPECTIVES

To conclude, REM sleep appears to be a potential actor in regulating placebo analgesia. In particular, discordance between verbal suggestions and sensory experiences in the evening might have led to a re-assessment of expectations during sleep. Depending on the level of concordance between the expectations initially created and the sensory experience, the need for offline reprocessing during sleep may vary with high level of concordance being associated with a reduction of REM sleep and a low level with an increase in REM sleep. When the level of concordance is high, and REM sleep reduced, placebo effects are mediated by expectations, but when the level of concordance is low, and REM sleep lengthened, expectations and placebo effects were dissociated. REM sleep deprivation bypasses the sleep processing of expectation, reinforcing the expectations of relief and the subsequent placebo effect the next day. There is a link between placebo analgesia and sleep, more specifically REM sleep and the reactivation of a memory trace of relief expectations during REM sleep. These results suggest that the internal model generating analgesic expectations may be reactivated implicitly during REM sleep (associated with frontal theta) reducing pain-related responses.

Future studies are required to confirm the role of REM sleep and its clinical implications in larger samples and diverse populations, including patients suffering from chronic pain conditions (Klinger et al., 2018; Vase, Vollert, Finnerup, et al., 2015) or sleep disorders (Winkler & Rief, 2015). We anticipate that the increasing influence of predictive-coding theoretical models and the advance in the understanding of neurobiological mechanisms underlying the role of sleep in associative learning and particularly the offline reprocessing of prediction error will provide fertile grounds to understand further the role of sleep in placebo effects.

REFERENCES

Aalto, J., & Kiianmaa, K. (1984). Increased voluntary alcohol drinking concurrent with REM-sleep deprivation. *Alcohol (Fayetteville, NY)*, *1*(1), 77–79.

Ackermann, S., & Rasch, B. (2014). Differential effects of non-REM and REM sleep on memory consolidation? *Current Neurology and Neuroscience Reports*, *14*(2), 430.

Afolalu, E. F., Ramlee, F., & Tang, N. K. Y. (2017). Effects of sleep changes on pain-related health outcomes in the general population: A systematic review of longitudinal studies with exploratory meta-analysis. *Sleep Medicine Reviews*, *39*, 82–97.

Alger, S. E., Lau, H., & Fishbein, W. (2012). Slow wave sleep during a daytime nap is necessary for protection from subsequent interference and long-term retention. *Neurobiology of Learning and Memory*, *98*(2), 188–196.

Amanzio, M., Benedetti, F., Porro, C. A., Palermo, S., & Cauda, F. (2013). Activation likelihood estimation meta-analysis of brain correlates of placebo analgesia in human experimental pain. *Human Brain Mapping*, *34*(3), 738–752.

Apkarian, A. V., Catherine Bushnell, M., Treede, R.-D., & Zubieta, J.-K. (2005). Human brain mechanisms of pain perception and regulation in health and disease. *European Journal of Pain (London, England)*, *9*(4), 463–484.

Arima, T., Svensson, P., Rasmussen, C., et al. (2001). The relationship between selective sleep deprivation, nocturnal jaw-muscle activity and pain in healthy men. *Journal of Oral Rehabilitation*, *28*(2), 140–148.

Autret, A., Minz, M., Beillevaire, T., Cathala, H. P., & Castaigne, P. (1976). Suppression of paradoxal sleep by clonidine in man. *Comptes Rendus Hebdomadaires Des Seances De l'Academie Des Sciences. Série D: Sciences Naturelles*, *283*(8), 955–957.

Azevedo, E., Manzano, G. M., Silva, A., et al. (2011). The effects of total and REM sleep deprivation on laser-evoked potential threshold and pain perception. *Pain*, *152*(9), 2052–2058.

Baliki, M. N., Geha, P. Y., Fields, H. L., & Vania Apkarian, A. (2010). Predicting value of pain and analgesia: Nucleus accumbens response to noxious stimuli changes in the presence of chronic pain. *Neuron*, *66*(1), 149–160.

Bastuji, H., Perchet, C., Legrain, V., Montes, C., & Garcia-Larrea, L. (2008). Laser evoked responses to painful stimulation persist during sleep and predict subsequent arousals. *Pain*, *137*(3), 589–599.

Bayne, T., Hohwy, J., & Owen, A. M. (2016). Are there levels of consciousness? *Trends in Cognitive Sciences*, *20*(6), 405–413.

Benedetti, F. (2014). Placebo effects: From the neurobiological paradigm to translational implications. *Neuron*, *84*(3), 623–637.

Benedetti, F., Carlino, E., & Pollo, A. (2011). How placebos change the patient's brain. *Neuropsychopharmacology: Official Publication of the American College of Neuropsychopharmacology*, *36*(1), 339–354.

Benedict, C., Brooks, S. J., O'Daly, O. G., et al. (2012). Acute sleep deprivation enhances the brain's response to hedonic food stimuli: An FMRI study. *The Journal of Clinical Endocrinology & Metabolism*, *97*(3), E443–E447.

Bentley, A. J., Newton, S., & Zio, C. D. (2003). Sensitivity of sleep stages to painful thermal stimuli. *Journal of Sleep Research*, *12*(2), 143–147.

Berger, R. J., & Phillips, N. H. (1995). Energy conservation and sleep. *Behavioural Brain Research*, *69*(1–2), 65–73.

Bingel, U., Lorenz, J., Schoell, E., Weiller, C., & Büchel, C. (2006). Mechanisms of placebo analgesia: RACC recruitment of a subcortical antinociceptive network. *Pain*, *120*(1–2), 8–15.

Bjurstrom, M. F., & Irwin, M. R. (2016). Polysomnographic characteristics in nonmalignant chronic pain populations: A review of controlled studies. *Sleep Medicine Reviews*, *26*, 74–86.

Boly, M., Phillips, C., Tshibanda, L., et al. (2008). Intrinsic brain activity in altered states of consciousness: How conscious is the default mode of brain function? *Annals of the New York Academy of Sciences*, *1129*, 119–129.

Borbély, A. A., & Achermann, P. (1999). Sleep homeostasis and models of sleep regulation. *Journal of Biological Rhythms*, *14*(6), 557–568.

Borbély, A. A., Daan, S., Wirz-Justice, A., & Deboer, T. (2016). The two-process model of sleep regulation: A reappraisal. *Journal of Sleep Research*, *25*(2), 131–143.

Boyce, R., Williams, S., & Adamantidis, A. (2017). REM sleep and memory. *Current Opinion in Neurobiology*, *44*, 167–177.

Brown, L. K. (2012). Can sleep deprivation studies explain why human adults sleep? *Current Opinion in Pulmonary Medicine*, *18*(6), 541–545.

Burton, A. R., Birznieks, I., Bolton, P. S., Henderson, L. A., & Macefield, V. G. (2009). Effects of deep and superficial experimentally induced acute pain on muscle sympathetic nerve activity in human subjects. *The Journal of Physiology*, *587*(Pt. 1), 183–193.

Busch, V., Haas, J., Crönlein, T., et al. (2012). Sleep deprivation in chronic somatoform pain-effects on mood and pain regulation. *Psychiatry Research*, *195*(3), 134–143.

Campbell, C. M., Bounds, S. C., Simango, M. B., et al. (2011). Self-reported sleep duration associated with distraction analgesia, hyperemia, and secondary hyperalgesia in the heat-capsaicin nociceptive model. *European Journal of Pain (London, England)*, *15*(6), 561–567.

Carlino, E., Frisaldi, E., & Benedetti, F. (2014). Pain and the context. *Nature Reviews. Rheumatology*, *10*(6), 348–355.

Chiu, C. N., & Prober, D. A. (2013). Regulation of zebrafish sleep and arousal states: Current and prospective approaches. *Frontiers in Neural Circuits*, *7*, 58.

Cho, C.-H., Lee, S.-W., Lee, Y.-K., Shin, H.-K., & Hwang, I. (2015). Effect of a sleep aid in analgesia after arthroscopic rotator cuff repair. *Yonsei Medical Journal*, *56*(3), 772–777.

Chouchou, F., Chauny, J.-M., Rainville, P., & Lavigne, G. J. (2015). Selective REM sleep deprivation improves expectation-related placebo analgesia. Andrej A. Romanovsky, ed. *PLOS ONE*, *10*(12), e0144992.

Chouchou, F., Khoury, S., Chauny, J.-M., Denis, R., & Lavigne, G. J. (2014). Postoperative sleep disruptions: A potential catalyst of acute pain? *Sleep Medicine Reviews*, *18*(3), 273–282.

Chouchou, F., & Lavigne, G.-J. (2014). Analgésie placebo et sommeil. *Pathologie Biologie*, *62*(5), 270–275.

Chouchou, F., Pichot, V., Barthélémy, J.-C., Bastuji, H., & Roche, F. (2014). Cardiac sympathetic modulation in response to apneas/hypopneas through heart rate variability analysis. Andrea Barbuti, ed. *PLoS ONE*, *9*(1), e86434.

Chouchou, F., Pichot, V., Pépin, J. L., et al. (2013). Sympathetic overactivity due to sleep fragmentation is associated with elevated diurnal systolic blood pressure in healthy elderly subjects: The PROOF-SYNAPSE study. *European Heart Journal*, *34*(28), 2122–2131.

Chouchou, F., Pichot, V., Perchet, C., et al. (2011). Autonomic pain responses during sleep: A study of heart rate variability. *European Journal of Pain (London, England)*, *15*(6), 554–560.

Choy, E. H. S. (2015). The role of sleep in pain and fibromyalgia. *Nature Reviews. Rheumatology*, *11*(9), 513–520.

Claude, L., Chouchou, F., Prados, G., et al. (2015). Sleep spindles and human cortical nociception: A surface and intracerebral electrophysiological study: Sleep spindles do not inhibit pain responses in humans. *The Journal of Physiology*, *593*(22), 4995–5008.

Colloca, L. (2018). Neurobiology of the placebo effect, part I. In L. Colloca (Ed.), *Vol. 138. International review of neurobiology* (1st ed.): Academic Press (Elsevier). https://www.elsevier.com/books/neurobiology-of-the-placebo-effect-part-i/colloca/978-0-12-814325-4. accessed June 30, 2018.

Colloca, L., & Benedetti, F. (2005). Placebos and painkillers: Is mind as real as matter? *Nature Reviews. Neuroscience*, *6*(7), 545–552.

Colloca, L., Enck, P., & DeGrazia, D. (2016). Relieving pain using dose-extending placebos: A scoping review. *Pain*, *157*(8), 1590–1598.

Colloca, L., Klinger, R., Flor, H., & Bingel, U. (2013). Placebo analgesia: Psychological and neurobiological mechanisms. *Pain*, *154*(4), 511–514.

Cooperman, N. R., Mullin, F. J., & Kleitman, N. (1935). Further observations on the effects of prolonged sleeplessness. *The American Journal of Physiology*, *107*(589), e93.

Cote, K. A., Etienne, L., & Campbell, K. B. (2001). Neurophysiological evidence for the detection of external stimuli during sleep. *Sleep*, *24*(7), 791–803.

Dang-Vu, T. T., McKinney, S. M., Buxton, O. M., Solet, J. M., & Ellenbogen, J. M. (2010). Spontaneous brain rhythms predict sleep stability in the face of noise. *Current Biology: CB*, *20*(15), R626–R627.

Dang-Vu, T. T., Schabus, M., Desseilles, M., et al. (2010). Functional neuroimaging insights into the physiology of human sleep. *Sleep*, *33*(12), 1589–1603.

de la Fuente-Fernández, R., Ruth, T. J., Sossi, V., et al. (2001). Expectation and dopamine release: Mechanism of the placebo effect in Parkinson's disease. *Science (New York, NY)*, *293*(5532), 1164–1166.

Deliens, G., Gilson, M., & Peigneux, P. (2014). Sleep and the processing of emotions. *Experimental Brain Research*, *232*(5), 1403–1414.

Desseilles, M., Dang-Vu, T. T., Sterpenich, V., & Schwartz, S. (2011). Cognitive and emotional processes during dreaming: A neuroimaging view. *Consciousness and Cognition*, *20*(4), 998–1008.

Diekelmann, S., & Born, J. (2010). The memory function of sleep. *Nature Reviews. Neuroscience*, *11*(2), 114–126.

Doufas, A. G., Tian, L., Padrez, K. A., et al. (2013). Experimental pain and opioid analgesia in volunteers at high risk for obstructive sleep apnea. *PLoS One*, *8*(1), e54807. https://www.ncbi.nlm.nih.gov/pmc/articles/PMC3558510/. accessed March 6, 2018.

Dudai, Y., Karni, A., & Born, J. (2015). The consolidation and transformation of memory. *Neuron*, *88*(1), 20–32.

Dum, R. P., Levinthal, D. J., & Strick, P. L. (2009). The spinothalamic system targets motor and sensory areas in the cerebral cortex of monkeys. *The Journal of Neuroscience: The Official Journal of the Society for Neuroscience*, *29*(45), 14223–14235.

Efe, E., Dikmen, S., Altaş, N., & Boneval, C. (2013). Turkish pediatric surgical nurses' knowledge and attitudes regarding pain assessment and nonpharmacological and environmental methods in newborns' pain relief. *Pain Management Nursing: Official Journal of the American Society of Pain Management Nurses*, *14*(4), 343–350.

Enck, P., Bingel, U., Schedlowski, M., & Rief, W. (2013). The placebo response in medicine: Minimize, maximize or personalize? *Nature Reviews. Drug Discovery*, *12*(3), 191–204.

Eslami, V., Zimmerman, M. E., Grewal, T., Katz, M., & Lipton, R. B. (2016). Pain grade and sleep disturbance in older adults: Evaluation the role of pain, and stress for depressed and non-depressed individuals. *International Journal of Geriatric Psychiatry*, *31*(5), 450–457.
Finniss, D. G., Kaptchuk, T. J., Miller, F., & Benedetti, F. (2010). Biological, clinical, and ethical advances of placebo effects. *Lancet (London, England)*, *375*(9715), 686–695.
Finniss, D. G., Nicholas, M. K., & Benedetti, F. (2009). Placebo analgesia-understanding the mechanisms and implications for clinical practice. *Reviews in Pain*, *3*(2), 15–19.
Fu, J., Li, P., Ouyang, X., et al. (2007). Rapid eye movement sleep deprivation selectively impairs recall of fear extinction in hippocampus-independent tasks in rats. *Neuroscience*, *144*(4), 1186–1192.
Garcia-Larrea, L., & Peyron, R. (2013). Pain matrices and neuropathic pain matrices: A review. *Pain*, *154*(Suppl. 1), S29–S43.
Gentili, A., Godschalk, M. F., Gheorghiu, D., et al. (1996). Effect of clonidine and yohimbine on sleep in healthy men: A double-blind, randomized, controlled trial. *European Journal of Clinical Pharmacology*, *50*(6), 463–465.
Groch, S., Wilhelm, I., Diekelmann, S., & Born, J. (2013). The role of REM sleep in the processing of emotional memories: Evidence from behavior and event-related potentials. *Neurobiology of Learning and Memory*, *99*, 1–9.
Gujar, N., Yoo, S.-S., Hu, P., & Walker, M. P. (2011). Sleep deprivation amplifies reactivity of brain reward networks, biasing the appraisal of positive emotional experiences. *The Journal of Neuroscience: The Official Journal of the Society for Neuroscience*, *31*(12), 4466–4474.
Haack, M., Lee, E., Cohen, D. A., & Mullington, J. M. (2009). Activation of the prostaglandin system in response to sleep loss in healthy humans: Potential mediator of increased spontaneous pain. *Pain*, *145*(1–2), 136–141.
Haack, M., Scott-Sutherland, J., Santangelo, G., et al. (2012). Pain sensitivity and modulation in primary insomnia. *European Journal of Pain (London, England)*, *16*(4), 522–533.
Hall, K. T., Loscalzo, J., & Kaptchuk, T. J. (2015). Genetics and the placebo effect: The placebome. *Trends in Molecular Medicine*, *21*(5), 285–294.
Hanlon, E. C., Andrzejewski, M. E., Harder, B. K., Kelley, A. E., & Benca, R. M. (2005). The effect of REM sleep deprivation on motivation for food reward. *Behavioural Brain Research*, *163*(1), 58–69.
Harden, R. N., Saracoglu, M., Connolly, S., et al. (2016). "Managing" the placebo effect: The single-blind placebo lead-in response in two pain models. *Pain Medicine (Malden, MA)*, *17*(12), 2305–2310.
Heffner, K. L., France, C. R., Trost, Z., Ng, H. M., & Pigeon, W. R. (2011). Chronic low back pain, sleep disturbance, and interleukin-6. *The Clinical Journal of Pain*, *27*(1), 35–41.
Hobson, J. A. (1990). Sleep and dreaming. *The Journal of Neuroscience: The Official Journal of the Society for Neuroscience*, *10*(2), 371–382.
Hobson, J. A. (2005). Sleep is of the brain, by the brain and for the brain. *Nature*, *437*(7063), 1254–1256.
Huynh, N., Kato, T., Rompré, P. H., et al. (2006). Sleep bruxism is associated to micro-arousals and an increase in cardiac sympathetic activity. *Journal of Sleep Research*, *15*(3), 339–346.
Huynh, N., Lavigne, G. J., Lanfranchi, P. A., Montplaisir, J. Y., & de Champlain, J. (2006). The effect of 2 sympatholytic medications—Propranolol and clonidine—On sleep bruxism: Experimental randomized controlled studies. *Sleep*, *29*(3), 307–316.
Iber, C., Chesson, A., Quan, S. F., & For the AASM. (2007). *The AASM manual for the scoring of sleep and associated events*. Westchester, IL: American Academy of Sleep Medicine.
Irwin, M. R., Olmstead, R., Carrillo, C., et al. (2012). Sleep loss exacerbates fatigue, depression, and pain in rheumatoid arthritis. *Sleep*, *35*(4), 537–543.
Jank, R., Gallee, A., Boeckle, M., Fiegl, S., & Pieh, C. (2017). Chronic pain and sleep disorders in primary care. *Pain Research and Treatment*, *2017*, 9081802.

Jensen, K. B., Kaptchuk, T. J., Chen, X., et al. (2015). A neural mechanism for nonconscious activation of conditioned placebo and nocebo responses. *Cerebral Cortex (New York, NY: 1991)*, *25*(10), 3903–3910.

Jensen, K. B., Kaptchuk, T. J., Kirsch, I., et al. (2012). Nonconscious activation of placebo and nocebo pain responses. *Proceedings of the National Academy of Sciences of the United States of America*, *109*(39), 15959–15964.

Karakaş, S., Arikan, O., Cakmak, E. D., et al. (2006). Early gamma response of sleep is sensory/perceptual in origin. *International Journal of Psychophysiology: Official Journal of the International Organization of Psychophysiology*, *62*(1), 152–167.

Karakaş, S., Cakmak, E. D., Bekçi, B., & Aydin, H. (2007). Oscillatory responses representing differential auditory processing in sleep. *International Journal of Psychophysiology: Official Journal of the International Organization of Psychophysiology*, *65*(1), 40–50.

Karmann, A. J., Lauer, C., Ziegler, E., et al. (2018). Associations of nocturnal sleep with experimental pain and pain catastrophizing in healthy volunteers. *Biological Psychology*, *135*, 1–7.

Kato, T., Montplaisir, J. Y., & Lavigne, G. J. (2004). Experimentally induced arousals during sleep: A cross-modality matching paradigm. *Journal of Sleep Research*, *13*(3), 229–238.

Kavanau, J. L. (1997). Memory, sleep and the evolution of mechanisms of synaptic efficacy maintenance. *Neuroscience*, *79*(1), 7–44.

Khoury, S., Chouchou, F., Amzica, F., et al. (2013). Rapid EEG activity during sleep dominates in mild traumatic brain injury patients with acute pain. *Journal of Neurotrauma*, *30*(8), 633–641.

Killgore, W. D. S. (2010). Effects of sleep deprivation on cognition. *Progress in Brain Research*, *185*, 105–129.

Klinger, R., Colloca, L., Bingel, U., & Flor, H. (2014). Placebo analgesia: Clinical applications. *Pain*, *155*(6), 1055.

Klinger, R., Stuhlreyer, J., Schwartz, M., Schmitz, J., & Colloca, L. (2018). Clinical use of placebo effects in patients with pain disorders. In L. Colloca (Ed.), *Neurobiology of the placebo effect, part II, International review of neurobiology: Vol. 139* (pp. 107–128). New York: Academic Press.

Knutson, K. L., Spiegel, K., Penev, P., & Van Cauter, E. (2007). The metabolic consequences of sleep deprivation. *Sleep Medicine Reviews*, *11*(3), 163–178.

Kong, J., Wang, Z., Leiser, J., et al. (2018). Enhancing treatment of osteoarthritis knee pain by boosting expectancy: A functional neuroimaging study. *NeuroImage. Clinical*, *18*, 325–334.

Lange, T., Dimitrov, S., & Born, J. (2010). Effects of sleep and circadian rhythm on the human immune system. *Annals of the New York Academy of Sciences*, *1193*, 48–59.

Laverdure-Dupont, D., Rainville, P., Montplaisir, J., & Lavigne, G. (2009). Changes in rapid eye movement sleep associated with placebo-induced expectations and analgesia. *The Journal of Neuroscience: The Official Journal of the Society for Neuroscience*, *29*(38), 11745–11752.

Laverdure-Dupont, D., Rainville, P., Montplaisir, J., & Lavigne, G. (2010). Relief expectation and sleep. *Reviews in the Neurosciences*, *21*(5), 381–395.

Laverdure-Dupont, D., Rainville, P., Renancio, C., Montplaisir, J., & Lavigne, G. (2018). Placebo analgesia persists during sleep: An experimental study. *Progress in Neuro-Psychopharmacology and Biological Psychiatry*, *85*, 33–38.

Lavigne, G., Brousseau, M., Kato, T., et al. (2004). Experimental pain perception remains equally active over all sleep stages. *Pain*, *110*(3), 646–655.

Lavigne, G., Khoury, S., Chauny, J.-M., & Desautels, A. (2015). Pain and sleep in post-concussion/mild traumatic brain injury. *Pain*, *156*(Suppl. 1), S75–S85.

Lavigne, G. J., Okura, K., Abe, S., et al. (2011). Gender specificity of the slow wave sleep lost in chronic widespread musculoskeletal pain. *Sleep Medicine*, *12*(2), 179–185.

Lavigne, G., Zucconi, M., Castronovo, C., et al. (2000). Sleep arousal response to experimental thermal stimulation during sleep in human subjects free of pain and sleep problems. *Pain, 84*(2–3), 283–290.

Lavigne, G. J., Zucconi, M., Castronovo, V., et al. (2001). Heart rate changes during sleep in response to experimental thermal (nociceptive) stimulations in healthy subjects. *Clinical Neurophysiology: Official Journal of the International Federation of Clinical Neurophysiology, 112*(3), 532–535.

Legrain, V., Van Damme, S., Eccleston, C., et al. (2009). A neurocognitive model of attention to pain: Behavioral and neuroimaging evidence. *Pain, 144*(3), 230–232.

Lentz, M. J., Landis, C. A., Rothermel, J., & Shaver, J. L. (1999). Effects of selective slow wave sleep disruption on musculoskeletal pain and fatigue in middle aged women. *The Journal of Rheumatology, 26*(7), 1586–1592.

Libourel, P.-A., & Herrel, A. (2016). Sleep in amphibians and reptiles: A review and a preliminary analysis of evolutionary patterns. *Biological Reviews of the Cambridge Philosophical Society, 91*(3), 833–866.

Maire, M., Reichert, C. F., Gabel, V., et al. (2014). Sleep ability mediates individual differences in the vulnerability to sleep loss: Evidence from a PER3 polymorphism. *Cortex; a Journal Devoted to the Study of the Nervous System and Behavior, 52*, 47–59.

Maquet, P., Dive, D., Salmon, E., et al. (1990). Cerebral glucose utilization during sleep-wake cycle in man determined by positron emission tomography and [18F]2-fluoro-2-deoxy-D-glucose method. *Brain Research, 513*(1), 136–143.

Maquet, P., Péters, J., Aerts, J., et al. (1996). Functional neuroanatomy of human rapid-eye-movement sleep and dreaming. *Nature, 383*(6596), 163–166.

Marshall, L., Helgadóttir, H., Mölle, M., & Born, J. (2006). Boosting slow oscillations during sleep potentiates memory. *Nature, 444*(7119), 610–613.

Matic, M., van den Bosch, G. E., de Wildt, S. N., Tibboel, D., & van Schaik, R. H. N. (2016). Genetic variants associated with thermal pain sensitivity in a paediatric population. *Pain, 157*(11), 2476–2482.

Melzack, R. (1990). Phantom limbs and the concept of a neuromatrix. *Trends in Neurosciences, 13*(3), 88–92.

Melzack, R. (1999). From the gate to the neuromatrix. *Pain*, (Suppl. 6), S121–S126.

Menz, M. M., Rihm, J. S., Salari, N., et al. (2013). The role of sleep and sleep deprivation in consolidating fear memories. *NeuroImage, 75*, 87–96.

Moldofsky, H., Scarisbrick, P., England, R., & Smythe, H. (1975). Musculosketal symptoms and non-REM sleep disturbance in patients with "fibrositis syndrome" and healthy subjects. *Psychosomatic Medicine, 37*(4), 341–351.

Mölle, M., Marshall, L., Gais, S., & Born, J. (2002). Grouping of spindle activity during slow oscillations in human non-rapid eye movement sleep. *The Journal of Neuroscience: The Official Journal of the Society for Neuroscience, 22*(24), 10941–10947.

Nofzinger, E. A., Buysse, D. J., Miewald, J. M., et al. (2002). Human regional cerebral glucose metabolism during non-rapid eye movement sleep in relation to waking. *Brain: A Journal of Neurology, 125*(Pt. 5), 1105–1115.

Older, S. A., Battafarano, D. F., Danning, C. L., et al. (1998). The effects of delta wave sleep interruption on pain thresholds and fibromyalgia-like symptoms in healthy subjects; correlations with insulin-like growth factor I. *The Journal of Rheumatology, 25*(6), 1180–1186.

Onen, S. H., Alloui, A., Gross, A., Eschallier, A., & Dubray, C. (2001). The effects of total sleep deprivation, selective sleep interruption and sleep recovery on pain tolerance thresholds in healthy subjects. *Journal of Sleep Research, 10*(1), 35–42.

Oswald, I. (1980). Sleep as restorative process: Human clues. *Progress in Brain Research, 53*, 279–288.

Pace-Schott, E. F., Germain, A., & Milad, M. R. (2015). Effects of sleep on memory for conditioned fear and fear extinction. *Psychological Bulletin, 141*(4), 835–857.

Palmer, C. A., Oosterhoff, B., Bower, J. L., Kaplow, J. B., & Alfano, C. A. (2018). Associations among adolescent sleep problems, emotion regulation, and affective disorders: Findings from a nationally representative sample. *Journal of Psychiatric Research, 96*, 1–8.
Park, J. W., & Chung, J. W. (2016). Inflammatory cytokines and sleep disturbance in patients with temporomandibular disorders. *Journal of Oral & Facial Pain and Headache, 30*(1), 27–33.
Peciña, M., Stohler, C. S., & Zubieta, J.-K. (2014). Neurobiology of placebo effects: Expectations or learning? *Social Cognitive and Affective Neuroscience, 9*(7), 1013–1021.
Peerdeman, K. J., van Laarhoven, A. I. M., Keij, S. M., et al. (2016). Relieving patients' pain with expectation interventions: A meta-analysis. *Pain, 157*(6), 1179–1191.
Peigneux, P., Laureys, S., Fuchs, S., et al. (2004). Are spatial memories strengthened in the human hippocampus during slow wave sleep? *Neuron, 44*(3), 535–545.
Perogamvros, L., Dang-Vu, T. T., Desseilles, M., & Schwartz, S. (2013). Sleep and dreaming are for important matters. *Frontiers in Psychology, 4*, 474.
Petersen, G. L., Finnerup, N. B., Nørskov, K. N., et al. (2012). Placebo manipulations reduce hyperalgesia in neuropathic pain. *Pain, 153*(6), 1292–1300.
Peyron, R., Laurent, B., & García-Larrea, L. (2000). Functional imaging of brain responses to pain. A review and meta-analysis (2000). *Neurophysiologie Clinique = Clinical Neurophysiology, 30*(5), 263–288.
Pires, G. N., Bezerra, A. G., Tufik, S., & Andersen, M. L. (2016). Effects of acute sleep deprivation on state anxiety levels: A systematic review and meta-analysis. *Sleep Medicine, 24*, 109–118.
Porte, H. S. (2004). Slow horizontal eye movement at human sleep onset. *Journal of Sleep Research, 13*(3), 239–249.
Price, D. D., Milling, L. S., Kirsch, I., et al. (1999). An analysis of factors that contribute to the magnitude of placebo analgesia in an experimental paradigm. *Pain, 83*(2), 147–156.
Quiton, R. L., & Greenspan, J. D. (2007). Sex differences in endogenous pain modulation by distracting and painful conditioning stimulation. *Pain, 132*(Suppl. 1), S134–S149.
Rasch, B., & Born, J. (2013). About sleep's role in memory. *Physiological Reviews, 93*(2), 681–766.
Rauchs, G., Bertran, F., Guillery-Girard, B., et al. (2004). Consolidation of strictly episodic memories mainly requires rapid eye movement sleep. *Sleep, 27*(3), 395–401.
Raymond, I., Ancoli-Israel, S., & Choinière, M. (2004). Sleep disturbances, pain and analgesia in adults hospitalized for burn injuries. *Sleep Medicine, 5*(6), 551–559.
Rechtschaffen, A., & Bergmann, B. M. (1995). Sleep deprivation in the rat by the disk-over-water method. *Behavioural Brain Research, 69*(1–2), 55–63.
Rescorla, A., & Wagner, A. (1972). A theory of Pavlovian conditioning: Variation in the effectiveness of reinforcement and nonreinforcement | BibSonomy. In *Classical conditioning II: current research and theory* (Appelton-Century-Crofts, pp. 64–99). New York: Black AH & Prokasy WF.
Rief, W., & Petrie, K. J. (2016). Can psychological expectation models be adapted for placebo research? *Frontiers in Psychology, 7*, 1876.
Roehrs, T. A., Harris, E., Randall, S., & Roth, T. (2012). Pain sensitivity and recovery from mild chronic sleep loss. *Sleep, 35*(12), 1667–1672.
Roehrs, T., Hyde, M., Blaisdell, B., Greenwald, M., & Roth, T. (2006). Sleep loss and REM sleep loss are hyperalgesic. *Sleep, 29*(2), 145–151.
Rosales-Lagarde, A., Armony, J. L., Del Río-Portilla, Y., et al. (2012). Enhanced emotional reactivity after selective REM sleep deprivation in humans: An FMRI study. *Frontiers in Behavioral Neuroscience, 6*, 25.
Sanders, R. D., Tononi, G., Laureys, S., & Sleigh, J. W. (2012). Unresponsiveness ≠ Unconsciousness. *Anesthesiology, 116*(4), 946–959.
Sandrini, G., Milanov, I., Rossi, B., et al. (2001). Effects of sleep on spinal nociceptive reflexes in humans. *Sleep, 24*(1), 13–17.

Sandyk, R., Tsagas, N., Anninos, P. A., & Derpapas, K. (1992). Magnetic fields mimic the behavioral effects of REM sleep deprivation in humans. *The International Journal of Neuroscience*, *65*(1–4), 61–68.

Schafer, S. M., Geuter, S., & Wager, T. D. (2018). Mechanisms of placebo analgesia: A dual-process model informed by insights from cross-species comparisons. *Progress in Neurobiology*, *160*, 101–122.

Schey, R., Dickman, R., Parthasarathy, S., et al. (2007). Sleep deprivation is hyperalgesic in patients with gastroesophageal reflux disease. *Gastroenterology*, *133*(6), 1787–1795.

Schuh-Hofer, S., Baumgärtner, U., & Treede, R.-D. (2015). Effect of sleep deprivation on the electrophysiological signature of habituation to noxious laser stimuli. *European Journal of Pain (London, England)*, *19*(8), 1197–1209.

Schuh-Hofer, S., Wodarski, R., Pfau, D. B., et al. (2013). One night of Total sleep deprivation promotes a state of generalized hyperalgesia: A surrogate pain model to study the relationship of insomnia and pain. *Pain*, *154*(9), 1613–1621.

Schweinhardt, P., Seminowicz, D. A., Jaeger, E., Duncan, G. H., & Bushnell, M. C. (2009). The anatomy of the mesolimbic reward system: A link between personality and the placebo analgesic response. *The Journal of Neuroscience: The Official Journal of the Society for Neuroscience*, *29*(15), 4882–4887.

Scott, D. J., Stohler, C. S., Egnatuk, C. M., et al. (2007). Individual differences in reward responding explain placebo-induced expectations and effects. *Neuron*, *55*(2), 325–336.

Sforza, E., Chapotot, F., Pigeau, R., Paul, P. N., & Buguet, A. (2004). Effects of sleep deprivation on spontaneous arousals in humans. *Sleep*, *27*(6), 1068–1075.

Siegel, J. M. (2005). Clues to the functions of mammalian sleep. *Nature*, *437*(7063), 1264–1271.

Simpson, N. S., Scott-Sutherland, J., Gautam, S., Sethna, N., & Haack, M. (2018). Chronic exposure to insufficient sleep alters processes of pain habituation and sensitization. *Pain*, *159*(1), 33–40.

Sivertsen, B., Lallukka, T., Petrie, K. J., et al. (2015). Sleep and pain sensitivity in adults. *Pain*, *156*(8), 1433–1439.

Smith, C. (2001). Sleep states and memory processes in humans: Procedural versus declarative memory systems. *Sleep Medicine Reviews*, *5*(6), 491–506.

Smith, M. T., Edwards, R. R., McCann, U. D., & Haythornthwaite, J. A. (2007). The effects of sleep deprivation on pain inhibition and spontaneous pain in women. *Sleep*, *30*(4), 494–505.

Spoormaker, V. I., Gvozdanovic, G. A., Sämann, P. G., & Czisch, M. (2014). Ventromedial prefrontal cortex activity and rapid eye movement sleep are associated with subsequent fear expression in human subjects. *Experimental Brain Research*, *232*(5), 1547–1554.

Spoormaker, V. I., Schröter, M. S., Andrade, K. C., et al. (2012). Effects of rapid eye movement sleep deprivation on fear extinction recall and prediction error signaling. *Human Brain Mapping*, *33*(10), 2362–2376.

St Louis Erik, K., Frey, L. C., Britton, J. W., et al. (2016). *The normal EEG*. American Epilepsy Society. *https://www.ncbi.nlm.nih.gov/books/NBK390343/*. Accessed 15 September 2017.

Steinberg, E. E., Keiflin, R., Boivin, J. R., et al. (2013). A causal link between prediction errors, dopamine neurons and learning. *Nature Neuroscience*, *16*(7), 966–973.

Tang, N. K. Y., Goodchild, C. E., Sanborn, A. N., Howard, J., & Salkovskis, P. M. (2012). Deciphering the temporal link between pain and sleep in a heterogeneous chronic pain patient sample: A multilevel daily process study. *Sleep*, *35*(5), 675–687A.

Taylor, V. A., Chang, L., Rainville, P., & Roy, M. (2017). Learned expectations and uncertainty facilitate pain during classical conditioning. *Pain*, *158*(8), 1528–1537.

Tempesta, D., Socci, V., De Gennaro, L., & Ferrara, M. (2017). Sleep and emotional processing. *Sleep Medicine Reviews*, *40*, 183–195.

Thorwart, A., & Livesey, E. J. (2016). Three ways that non-associative knowledge may affect associative learning processes. *Frontiers in Psychology*, 7, 2024.

Tiede, W., Magerl, W., Baumgärtner, U., et al. (2010). Sleep restriction attenuates amplitudes and attentional modulation of pain-related evoked potentials, but augments pain ratings in healthy volunteers. *Pain*, *148*(1), 36–42.

Tracey, I. (2010). Getting the pain you expect: Mechanisms of placebo, nocebo and reappraisal effects in humans. *Nature Medicine*, *16*(11), 1277–1283.

Van Cauter, E., Spiegel, K., Tasali, E., & Leproult, R. (2008). Metabolic consequences of sleep and sleep loss. *Sleep Medicine*, *9*(Suppl. 1), S23–S28.

Vase, L., Petersen, G. L., Riley, J. L., & Price, D. D. (2009). Factors contributing to large analgesic effects in placebo mechanism studies conducted between 2002 and 2007. *Pain*, *145*(1–2), 36–44.

Vase, L., Riley, J. L., & Price, D. D. (2002). A comparison of placebo effects in clinical analgesic trials versus studies of placebo analgesia. *Pain*, *99*(3), 443–452.

Vase, L., Vollert, J., Finnerup, N. B., et al. (2015). Predictors of the placebo analgesia response in randomized controlled trials of chronic pain: A meta-analysis of the individual data from nine industrially sponsored trials. *Pain*, *156*(9), 1795–1802.

Velluti, R. A. (1997). Interactions between sleep and sensory physiology. *Journal of Sleep Research*, *6*(2), 61–77.

Vorster, A. P., & Born, J. (2015). Sleep and memory in mammals, birds and invertebrates. *Neuroscience and Biobehavioral Reviews*, *50*, 103–119.

Wager, T. D., Rilling, J. K., Smith, E. E., et al. (2004). Placebo-induced changes in FMRI in the anticipation and experience of pain. *Science (New York, NY)*, *303*(5661), 1162–1167.

Wagner, U., Gais, S., & Born, J. (2001). Emotional memory formation is enhanced across sleep intervals with high amounts of rapid eye movement sleep. *Learning & Memory (Cold Spring Harbor, NY)*, *8*(2), 112–119.

Walker, M. P. (2010). Sleep, Memory and emotion. *Progress in Brain Research*, *185*, 49–68.

Walker, M. P., & van der Helm, E. (2009). Overnight therapy? The role of sleep in emotional brain processing. *Psychological Bulletin*, *135*(5), 731–748.

Webb, W. B. (1988). An objective behavioral model of sleep. *Sleep*, *11*(5), 488–496.

Weiner, O. M., & Dang-Vu, T. T. (2016). Spindle oscillations in sleep disorders: A systematic review. *Neural Plasticity*, *2016*, 7328725.

Weingarten, J. A., Dubrovsky, B., Basner, R. C., et al. (2016). Polysomnographic measurement of sleep duration and bodily pain perception in the sleep heart health study. *Sleep*, *39*(8), 1583–1589.

Winkler, A., & Rief, W. (2015). Effect of placebo conditions on polysomnographic parameters in primary insomnia: A meta-analysis. *Sleep*, *38*(6), 925–931.

Wong, M. M., Brower, K. J., & Zucker, R. A. (2009). Childhood sleep problems, early onset of substance use and behavioral problems in adolescence. *Sleep Medicine*, *10*(7), 787–796.

Xia, Z., & Storm, D. (2017). Role of circadian rhythm and REM sleep for memory consolidation. *Neuroscience Research*, *118*, 13–20.

Zhang, J., Lam, S.-P., S. X., L., et al. (2012). Insomnia, sleep quality, pain, and somatic symptoms: Sex differences and shared genetic components. *Pain*, *153*(3), 666–673.

Zoccoli, G., Walker, A. M., Lenzi, P., & Franzini, C. (2002). The cerebral circulation during sleep: Regulation mechanisms and functional implications. *Sleep Medicine Reviews*, *6*(6), 443–455.

CHAPTER FOURTEEN

Are Blue Pills Better Than Green? How Treatment Features Modulate Placebo Effects

Karin Meissner*[,1], Klaus Linde[†]

*Division of Integrative Health Promotion, Coburg University of Applied Sciences, Coburg, Germany
[†]Institute of General Practice, Technical University of Munich, München, Germany
[1]Corresponding author: e-mail address: karin.meissner@hs-coburg.de

Contents

Abstract

Treatment-related expectations are important predictors for placebo effects in various medical conditions. They are formed by verbal and nonverbal cues during the administration of treatments, such as verbal suggestions, conscious and unconscious associations with previous treatments, characteristics of patients and health-care providers as well as perceptual characteristics of the treatment. This review provides an overview of studies that aimed to elucidate the impact of treatment characteristics on expectations and placebo effects. Results show that high price, brand label, the use of invasive placebos and high placebo dose increase the expectations toward the beneficial effect of a placebo treatment, which in turn boost the placebo effect. Neuroimaging studies provide first insights into the neurobiological mechanisms underlying these differential placebo effects.

International Review of Neurobiology, Volume 139
ISSN 0074-7742
https://doi.org/10.1016/bs.irn.2018.07.014

1. INTRODUCTION

Treatment-related expectations have been shown to be an important predictor for placebo effects in various medical conditions (Meissner et al., 2011; Price, Finniss, & Benedetti, 2008; Schedlowski, Enck, Rief, & Bingel, 2015). For example, a pill administered with verbal information that it is a potent analgesic can significantly reduce pain, even when the pill is pharmacologically inert. Expectations regarding the effectiveness of treatments can be formed via different psychological mechanisms, including verbal suggestions and associative learning (Colloca & Miller, 2011). Research in the recent decades has clearly shown that expectation effects are deeply rooted in our brains. For example, placebo analgesia is accompanied by activation of a specific pain modulating network that extends from cortical areas to the spinal cord (Geuter, Koban, & Wager, 2017). In addition to pain, placebo effects have been demonstrated for nausea (Müller, Remus, Hoffmann, Tschöp, & Meissner, 2016; Quinn & Colagiuri, 2014), depression and anxiety (Meyer et al., 2015; Peciña et al., 2015; Rutherford et al., 2016), gastrointestinal and cardiovascular functions (Kaptchuk et al., 2008; Meissner, 2009; Meissner & Ziep, 2011; Rief et al., 2017), Parkinson's disease (De la Fuente-Fernández et al., 2001), and immune responses (Hadamitzky, Sondermann, Benson, & Schedlowski, 2018; Pacheco-López, Engler, Niemi, & Schedlowski, 2006). In this regard, it is important to emphasize that placebo effects can be elicited and exploited not only by the administration of placebo interventions, but also by maximizing the placebo component of any active intervention. This has been impressively demonstrated by studies in which patients received pharmacological drugs either overtly or covertly (i.e., via an intravenous line from a separate room): The improvements were larger and occurred earlier when the patients knew when to expect a beneficial effect of a treatment (Colloca, Lopiano, Lanotte, & Benedetti, 2004).

However, not only verbal information accompanying treatments, but also nonverbal cues can affect the size of expectancies toward a medical treatment, and thus the size of the placebo effect. For example, there is increasing evidence from experimental and clinical studies that more invasive and more expensive placebo interventions are systematically associated with larger placebo effects (Bannuru et al., 2015; Faasse, Martin, Grey, Gamble, & Petrie, 2016; Fehse, Maikowski, Simmank, Gutyrchik, & Meissner, 2015; Geuter, Eippert, Attar, & Büchel, 2013; Kaptchuk et al., 2006; Meissner et al., 2013). A thorough review of these studies is important to get an idea of what an optimized placebo intervention might look like.

2. COLOR OF DRUGS

The influence of the color of drugs on their effectiveness is a popular example in the placebo literature and the media to illustrate the existence of the placebo effect. But how robust and reproducible are these results?

Early studies investigated the role of the drug's perceptual characteristics including color for the perceived potency of drugs. Many of these studies have been driven by the desire to increase patient adherence through the most appropriate color of drugs. An early experiment was performed by Sallis and Buckalew (1984). Twenty participants between the ages of 20 and 30 were asked to rate the perceived efficacy of seven capsules of different color. The task was not related to a particular medical condition. Numeric ratings scales ranged from 1 (strongest) to 7 (weakest). Results indicate that red and black capsules were associated with the highest perceived effect (mean, 2.3 and 2.9, respectively) and white capsules with the lowest (5.3), while yellow, green, blue, and orange capsules were rated as moderately effective (4.2–4.6). Statistical tests to substantiate these findings, however, were missing.

Almost 30 years later, Wan, Woods, Salgado-Montejo, Velasco, and Spence (2015) used an online survey approach to explore the influence of 7 different colors and 3 different shapes of drugs (resulting in 21 different drugs) on their expected efficacy. In this case, researchers limited the task to the treatment of headache. Participants had to arrange the 21 pills in a virtual box, in which the x-axis served as a visual analogue scale (i.e., positions on the extreme left were given a value of zero, while positions on the extreme right were given a value of 100). In a first experiment in 97 participants, white pills were perceived as significantly more effective in treating headache as any of the other colors, while light-green pills received the lowest scores. In a follow-up experiment, 337 participants from three different countries (112 USA, 104 China, 121 Colombia) performed the same tasks as in the first experiment, except that three drug colors were different (red, blue, green, white, pink, yellow, orange). Again, white pills were perceived as the most effective drug in combating headaches, and this was true in all three countries. This clear and "new" effect is probably due to the white color of the worldwide known headache drug "aspirin," which has been intensely marketed since Buckalew's study (Wan et al., 2015). This example shows how strongly the association between colors of drugs and their perceived potency is influenced by contextual factors, such as advertisement and recognition of a particular drug.

Several studies addressed the perceived association between colors of drugs and their pharmacological effects. Jacobs and Nordan (1979) handled 50 students (mean age 19 years) and 50 nonstudents (mean age 41 years) six bottles with capsules of different colors and asked them to classify these capsules as either a depressant/tranquilizer, or a stimulant/antidepressant, or a hallucinogen. Participants were informed that the study's aim was to investigate "how well someone unfamiliar with the medications can correctly classify them on the basis of color alone" (Jacobs & Nordan, 1979, p. 369). Results revealed a significant association between blue color of capsules and an expected depressant/tranquilizer effect, while reddish colors (yellow in students, red in nonstudents) were associated with a stimulant/antidepressant action. Furthermore, only the student group associated black pills with a depressant action, which the authors attributed to their specific knowledge on "black mollies," i.e., psychedelic drugs known for their depressant effect. No specific pharmacological effect could be attributed to the white and green pills. In a slightly adapted approach, Buckalew and Coffield (1982b) asked 68 college students to assign 12 capsules with different colors to one of four actions: stimulant/antidepressant, analgesic/narcotic, depressant/sedative, or psychedelic/hallucinogenic. Results revealed that reddish colors (orange and yellow) were associated with a stimulant/antidepressant effect, while white color was linked to an analgesic/narcotic effect. In addition, lavender capsules were expected to have a psychedelic/hallucinogenic action. In a replication study in 50 black undergraduate students, however, these associations were rather different: White capsules were associated with a stimulant/antidepressant effect, black color was linked to an analgesic/narcotic effect, and light-green color was attributed to a depressant/sedative effect. The authors explained this discrepancy by ethnic/cultural differences in meanings ascribed to colors. A decade later, Buckalew and Ross (1991) asked 102 volunteers between 20 and 69 years to assign 29 differently colored drugs to one of 10 different physiological systems, or to a "general" category. Specific associations were found between red and scarlet color and the cardiovascular system, scarlet color and the hemic/lymphatic system, and tan, beige and burnt-orange color and the skin. Furthermore, white colored drugs were typically assigned to the "general" category, which the authors interpreted as to have a more wide-spread effect on the body.

Wan et al. (2015; see above) also looked for associations between drug color and expected drug effects in a total of 434 participants from three different countries. Results further confirmed that red tablets are perceived as stimulating. Finally, a recent study examined the role of drug color for expected pharmacological effects in a Chinese cohort (Tao, Wang, & Wang, 2017). Thirteen pictures of identical but differently colored capsules—7 single colored and 6 two-colored—were provided to 80 Chinese volunteers (mean age, 30 years). Participants were asked to assign the colored drugs to one of four possible effects: stimulant, analgesic, hallucinogenic, or depressant. Findings confirmed the previously reported associations between red color and a stimulant action and blue and green colors and a depressant effect. In addition, white drugs were perceived as depressant and yellow drugs as hallucinogenic, while the effects of multicolored drugs varied more widely. In addition to possible cultural differences, the sample size of 80 participants appears rather low to provide consistent and reliable results for presumably "weaker" color–drug associations.

Since placebo research has shown that specific expectations can pave the way for clinically relevant changes in health, it is reasonable to assume that expectations generated by the drug's colors can likewise modulate drug and placebo responses. Schapira, McClelland, Griffiths, and Newell (1970) conducted a randomized crossover study in 48 patients with anxiety symptoms. All patients received benzodiazepines for three consecutive weeks, with the color of the tablets changing weekly. The improvements of the various symptoms differed slightly between the colors blue, green, or yellow; none of these differences, however, reached statistical significance. Blackwell, Bloomfield, and Buncher (1972) administered 56 medical students either blue or pink placebos in a single-blind design. They were informed that they could receive either a stimulant or a sedative drug and were asked to rate the severity of pre-defined symptoms on four-point Likert scales before and one hour after taking the pill. Sixty-six percent of participants reported that they felt less alert after taking the pink pills compared to 26% with the blue capsules. In addition, 72% of participants on blue capsules felt drowsier compared to 37% on pink capsules. The results of this early but well-controlled study suggest that blue pills not only suggest but also induce sedative effects, while reddish (pink) colors can actually be stimulating. Two clinical studies further suggest that appropriate colors of pills may induce a placebo effect. Lucchelli, Cattaneo, and Zattoni (1978) recruited 96 hospital patients who suffered from sleep disorders prior to elective surgery. For two consecutive nights, patients were given either a placebo or a hypnotic drug in a double-blind, randomized crossover design. Placebos and hypnotics were either blue or orange in color. Analysis of variance (ANOVA) showed a main effect of color on sleep onset time and on sleep duration, indicating enhanced hypnotic activity of the blue (active or placebo) capsules compared to the orange ones. Finally, one study provides preliminary evidence that red placebos may be most effective in treating clinical pain. In a double-blind crossover design, Huskisson (1974) administered effervescent drugs or placebos with different colors (red, blue, green, or yellow) to 22 patients with rheumatoid arthritis as on-demand analgesics. Red placebos showed an impressively higher placebo analgesic effect than the other three colors, the effect size being comparable to that of the active drugs. However, as only five patients received red placebos, the results of this study should be interpreted with caution.

From a perspective of the psychology of colors, the most widely held grouping of colors is that of "warm" and "cool" colors, with red, orange, and yellow colors seen usually as exciting and stimulating and blue, turquoise, and green colors as calming and relaxing. Several authors have investigated whether such associations are purely psychological in nature or may also have a physiological basis. Indeed, early experimental studies indicated that blue and red colors have opposite effects on physiological arousal measures (Mikellides, 1990). If brightness and saturation were controlled, however, no differential effects of the hues on physiological arousal measures could be observed (Zieliński, 2016).

The effects of colors on cognitions and emotions, however, are well documented. For example, Verhoeven, Pieterse, and Pruyn (2006) asked 90 students to imagine being hospitalized after a fall from a ladder with a broken leg. The participants were then exposed to a virtual 360° panorama of (1) an examination room and (2) a private ward room, which they could explore by moving the mouse in the desired direction. The walls of both rooms were either painted blue or white, and participants were randomly allocated to one of these conditions. Analyses of variance clearly indicated that in both types of rooms, blue walls decreased anxiety and improved pleasure compared to the white walls. This example nicely shows how colors can be used to optimize the psychosocial context of a medical intervention, and thus the well-being of patients. Colored drugs may further enhance positive associations, and such pave the way for improved outcomes of medical interventions.

Given that color has such a strong influence on our minds, one may wonder whether the effectiveness of conditioned stimuli in placebo studies is also affected by their color. Indeed, in most of these studies, red stimuli were used to indicate "pain," while green stimuli were associated with "pain relief" (Colloca & Benedetti, 2006; e.g., Colloca & Benedetti, 2009;

Colloca, Petrovic, Wager, Ingvar, & Benedetti, 2010; Colloca, Sigaudo, & Benedetti, 2008). Świder and Bąbel (2016) tested the impact of color in a placebo paradigm of social observational learning. In a between-subject design, 55 female subjects (mean age, 22 years) were randomly assigned to either a red or a green visual stimulus to signal to the participant that the observed "model" suffered less pain than in a control condition (with green and red stimuli, respectively). The results clearly showed a placebo effect induced by social observation learning. However, the placebo analgesic effect was unrelated to the color of the conditioned stimulus.

3. SIZE AND FORM OF DRUGS

In addition to color, the influence of the size and shape of the drugs on the placebo effect has been intensively studied. Buckalew and Coffield (1982b) asked 68 healthy participants to classify six capsules, each of a different size, according to their perceived effectiveness, without referring to a particular symptom. Results revealed a positive association between capsule size and perceived potency of the drug: The larger the capsule, the larger the expected effect. In a replication study in 50 black undergraduate students, Buckalew and Coffield (1982a) found an association in the opposite direction: the smaller the capsule, the larger the expected effect. This discrepancy points to possible differences in cultural learning as a possible underlying mechanisms of color–drug association. However, in none of the studies, appropriate inferential statistics were reported, and results should be regarded with caution.

In the same two experiments, Buckalew and Coffield (1982a, 1982b) asked the participants to compare the expected efficacy of a capsule and a tablet. Participants in both studies rated the capsule to be more powerful than the tablet, albeit this relationship was less pronounced in the black sample. As noted by the Buckalew and Ross (1991), capsules were fairly new during this time and were advertised as more powerful in comparison to tablets. Since the novelty of the capsules has decreased since then, one might suspect that the perceived superiority of the capsules has also decreased. Indeed, a recent study supports this view (Köteles, Fodor, Cziboly, & Bárdos, 2009). One hundred and nine female undergraduate students were asked to assign 20 pictures of five different colors (white, yellow, red, green and blue) and of four different shapes (capsule, small round/medium round/ ovale tablet) to a list of 19 possible drug effects. The pictures were presented in randomized order. Results revealed a preference for tablets over capsules with respect to six drug effects (analgesic, sedative, headache-relieving, spasmolytic, antipyretic, hypnotic), while capsules, together with ovale tablets, were preferred for antibiotic effects. Another interesting finding in this study was that in 9 out of 10 cases, the significant preferences were related to white drugs. Again, these findings suggest that cultural learning processes play a crucial role in mediating the associations between perceptual characteristics of drugs and perceived drug effects.

4. DOSING

Several studies investigated the relationship between the frequency of inert drug administration and the size of the placebo effect. For example, Blackwell et al. (1972) administered 56 medical students either one or two placebo capsules in different colors (pink or blue) in a single-blind design (see above). Students had been informed that they could receive either a stimulant or a sedative drug during the experiment. Results indicated that reported psychological changes were significantly more pronounced in participants who had been allocated to two placebo pills rather than one (Blackwell et al., 1972). De Craen et al. (1999) performed a systematic review of 79 placebo-controlled studies to test the hypothesis that the frequency of placebo administration is associated with healing rates in the treatment of duodenal ulcers. Results indicated a significantly higher 4-week healing rate in trials with a four times a day regimen (44%) as compared to trials with a twice a day regimen (36%). Sensitivity analyses confirmed the robustness of the results. However, it has to be kept in mind that such comparisons are based on nonrandomized data, and potential confounders remain uncontrolled.

The studies to date consistently suggest that a higher number of placebo pills induce a larger placebo effect than a lower number of pills. This is probably due to higher expectations in the case of higher dosing, although this has not been directly tested.

5. PRICE, DRUG LABELING, AND BRANDING

Nonverbal characteristics of medications not only refer to the perceptual properties but also to commercial features, such as price and labeling. Several studies investigated the effects of commercial factors on placebo effects.

The price of a product is usually associated with quality, and higher price of a product or a drug may therefore induce a higher expectation. As a consequence, the placebo effect would be expected to increase.

Shiv, Carmon, and Ariely (2005) tested this hypothesis and administered 193 undergraduate students an energy drink or no energy drink (control group) and asked them to perform a puzzle task. The price of the energy drink was described as either "regular" or "reduced." Performances in the puzzle task turned out to be lower in the reduced-price condition compared to the regular-price condition, the latter showing no difference to the no-treatment control condition. Results thus suggested a nocebo effect in the reduced-price condition. To explore this finding further, the authors conducted a separate study in 61 participants, in which they assessed expectancies before the puzzle-solving tasks. Mean expectations were similar in the regular-price condition and the control condition, but were lower in the reduced-price condition. These results confirm that the reduced-price condition induced a nocebo effect on task performance by eliciting negative expectations. Interestingly, drawing the participants' attention to the price–efficacy link abolished this nocebo effect, showing that the manipulation by price was unconscious. In an additional experiment, Shiv et al. (2005) compared the effects of drinks at a regular and a discounted price on actual puzzle task performance in 204 undergraduate students. In half of the subjects of each group, expectations toward the drinks were enhanced by emphasizing their positive effects on performance. Results indicated improved task performance in participants, who had received the drink at a regular-price drink together with the positive instructions, indicating that verbal suggestions were mandatory under these circumstances to induce a placebo effect.

Waber, Shiv, Carmon, and Ariely (2008) investigated whether the price of an analgesic can influence its therapeutic efficacy and informed 82 healthy volunteers (mean age, 30 years) to receive a painkiller at a regular price or at a discount price. In truth, all participants received the same placebo pill before experimental pain was induced. Results showed a remarkably smaller placebo analgesic effect in the low-priced painkiller condition compared to the regular-price condition. Due to the lack of a no-treatment control group, however, it remained unclear whether the low price reduced the analgesic effect or the high price increased it.

Plassmann, O'doherty, Shiv, and Rangel (2008) used functional magnetic resonance imaging (fMRI) study to identify brain areas that moderate the putative differential effects of "expensive" vs. "cheap" wines on experienced pleasantness. Twenty healthy volunteers participated in the experiment and had to rate the pleasantness of wine at "five different prices" while lying in the scanner. The subjects were not aware that two test pairs of wines, each at a low and a high price, were identical. As expected, the information on price modulated pleasantness ratings in the expected direction. Voxel-based morphometry (VBM) showed that these differences were moderated by the gray matter volume of three brain areas: The dorsomedial prefrontal cortex (dmPFC), which is known to be involved in cognitive appraisal of emotional experiences, the posterior insula, which is linked to somatosensory awareness, and the striatum, which is a crucial area for reward processing (Plassmann & Weber, 2015). All three areas play a role in the placebo response (Meissner et al., 2011).

Geuter et al. (2013) used fMRI to compare the neurobiological correlates of "expensive" vs. "low-priced" painkillers. They administered 52 male volunteers two times the same placebo cream together with differential information on pricing before undergoing thermal pain stimulation. In order to enhance the expectations according to the purported price, the experiments were preceded by a preconditioning phase, during which—unbeknown to the subjects—the pain intensity was lowered to a larger degree in the expensive as compared to the low-price condition. In the test phase, equally high stimulus intensities were applied. On a behavioral level, pain reduction was larger in the expensive placebo condition, as was the "willingness-to-pay" for this drug. Interestingly, blood-oxygen-level-dependent (BOLD) responses in the rostral anterior cingulate cortex (rACC) were likewise found to be graded: The "strong" placebo induced larger BOLD responses in the rACC than the "weak" placebo—an area known to be involved in endogenous pain control. The discrepancy to the findings of Plassmann et al. (2008) is most probably due to the different paradigms used (pleasantness vs. pain, expectancy vs. conditioning, VBM vs. BOLD responses).

Labeled drugs are usually more expensive than no-name drugs, and therefore a higher placebo effect of brand-labeled drugs in comparison to generic drugs may be expected. Branthwaite and Cooper (1981) investigated the analgesic effects of a popular brand name ("aspirin") in the treatment of headache. Eight hundred and thirty-five women who reported regular use of painkillers to relieve their headaches participated in the study. About half of them used the extensively advertised brand label "aspirin." Patients were randomly assigned to a package of either placebo pills or active analgesics in a double-blind design. Furthermore, half of each group received the pills in packages with the popular brand label "aspirin," the other half received the pills in black, unbranded packages labeled "analgesic tablets." All participants were instructed to take two tablets for any headache in the next 2 weeks. They were asked to rate the initial severity of headache as well as pain relief after 30 and 60 min. In addition to greater pain relief in response to the active drug, the ANOVA revealed greater pain relief in response to the brand-labeled pill as compared to the no-name pill. The effect of brand-labeling did not differ between the active drugs and the placebo pills and was the same for regular users and nonusers of aspirin. This latter finding is somewhat surprising, given that prior experience is generally regarded the strongest trigger for a placebo response (Colloca & Benedetti, 2006). Possibly, the popular brand label of "aspirin" induced such a strong placebo effect that individual experience with the drug did not surpass these improvements.

Kam-Hansen et al. (2014) further investigated whether labeling of drugs can affect clinical pain. Sixty-six patients with episodic migraine were asked to treat six consecutive migraine attacks with a pill taken from consecutively numbered envelopes and to rate pain intensities each time in a patient diary. The prepared envelopes provided information on its alleged content ("maxalt," "placebo," or "placebo or maxalt"). Furthermore, half of the envelopes in each information group contained a placebo pill, while the other half contained the active drug maxalt. Results revealed that the information to receive "maxalt," or "placebo or maxalt," was associated with larger analgesic effects in both treatment groups when compared to the "placebo" label. Remarkably, placebo labeled as "maxalt" was as effective as maxalt labeled as "placebo," and placebo labeled as "placebo" was still better than no treatment. Thus, the overt administration of a placebo pill induced a clear placebo analgesic effect, which could further be enhanced by the brand label "maxalt." The same was true for maxalt, whose action was considerably reduced when labeled as "placebo."

Also Faasse et al. (2016) compared the effects of placebos and active painkillers labeled as either generic drugs ("generic ibuprofen") or brand-labeled drugs ("nurofen"). Eighty-seven undergraduate students with frequent headaches were investigated. Placebos with the generic label were significantly less effective in reducing pain than active drugs with generic labeling. However, placebos were as effective as the active drug in the brand-label condition, suggesting a large placebo effect of the brand name. Furthermore, participants attributed more side effects to the placebo pill when it was labeled as a generic. Thus, the brand label enhanced not only the trust in the efficacy of the drug, but also in its safety.

Fehse et al. (2015) investigated 30 healthy participants reporting regular use of aspirin, who were randomly assigned to either a brand-label group ("aspirin") or a generic-label group ("1A Pharma"). In reality, all participants received the same placebo effervescent preparation together with group-specific information on history, manufacturer characteristics, and pricing of the respective "drug." On a behavioral level, a placebo effect on experimental pain was observed only in the aspirin group. On a neurofunctional level, the aspirin group showed significantly higher BOLD responses in brain areas including the dmPFC and the dorsolateral prefrontal cortex (dlPFC). Activity in the dlPFC is linked to anticipated pain relief before receiving a painkiller (Krummenacher, Candia, Folkers, Schedlowski, & Schönbächler, 2010; Wager et al., 2004; Zubieta et al., 2005). The dmPFC is involved in cognitive appraisal of emotional experience (Kalisch, Wiech, Critchley, & Dolan, 2006) and has been identified as a key area for placebo effects in chronic low back pain (Hashmi et al., 2012). The dmPFC was also shown to mediate the effects of "expensive" vs. "cheap" wines on pleasantness ratings in the study by Plassmann and Weber (2015; see above), suggesting a key role of the dmPFC for the modulation of placebo effects by the commercial features of a drug.

In conclusion, there is ample evidence that price, labeling and advertisement of a drug affect the expected and actual effects of this treatment. By manipulating such commercial characteristics, differential sizes of placebo effects can be induced. These behavioral effects are associated with graded brain responses in areas involved in placebo analgesia. A strong brand and a high price of a drug can thus act as a placebo, while a generic label and a low price can reduce the placebo response or even trigger a nocebo effect.

6. SIDE EFFECTS OF DRUGS

A further characteristic of drugs which may modulate placebo effects is the perceived bitterness of a medicine. After a famous saying, medicine that is not bitter does not work. Leslie (1954) argued that the bitterness of a medicine carries a strong placebo effect, because patients have learnt to expect this. In other words, the "side effect" of a drug such as a bitter taste may convince patients that the drug is working. The association between bitterness and efficacy is backed-up by physiological evidence, since the bitter taste of a medicine is often a sensory expression of its pharmacological activity: The more potent the drug, the more bitter it will be (Mennella, Spector, Reed, & Coldwell, 2013).

Several authors have investigated whether the efficacy of a placebo can be enhanced by using an "active" placebo which induces "side effects." This is of particular importance in the context of randomized controlled trials (RCTs), in which the side effects of active drugs carry the risk of unblinding. To avoid this, "active placebos" are recommended, such as atropine to mimic the sedative and anticholinergic effects of tricyclic antidepressants (Jensen, Bielefeldt, & Hróbjartsson, 2017). Indeed, there is evidence from placebo-controlled RCTs on tricyclic antidepressants that the placebo effect is larger (and the difference between drug and placebo smaller) when an active placebo is used (Moncrieff, Wessely, & Hardy, 2004).

Bjørkedal and Flaten (2011) were the first to test experimentally whether the perception of side effects can enhance the placebo effect. They administered 20 healthy volunteers four times a drink, two of which contained caffeine powder. Caffeine and control drinks were crossed with the information that the drink did, or did not contain a painkiller. The authors hypothesized that subjective arousal levels would increase in the caffeine condition and thus would convince participants that a painkiller has been administered, thereby enhancing the placebo analgesic effect. An unexpected finding was that caffeine reduced pain. However, information that a painkiller was administered increased the analgesic effect of caffeine, and this effect was mediated by expectancies. In contrast, information and expectancies had no effect on pain intensity when no caffeine was administered. Thus, the results partially supported the active placebo hypothesis.

Further support was provided by Rief and Glombiewski (2012), who compared the analgesic effects of an inert nasal spray with that of an active nasal placebo spray in 144 healthy volunteers. The active placebo induced a prickling nasal sensation, which was intended to mimic the side effects of an active drug. In addition, participants were allocated to different information groups with respect to the probability of receiving a real drug in relation to placebo (0%, 50%, 100%). Results revealed that in the 50% information condition, the active placebo spray was associated with a remarkably higher placebo effect on pain than was the inert placebo condition. In the 0% and 100% expectation conditions, no such differences were observed. The results show that active placebos can indeed improve the analgesic effect of placebos, at least in an uncertain situation, such as having a 50% chance of receiving the active medication.

A recent experimental study went one step further and combined diclofenac, atropine, and placebo in a double-blind double-dummy design (Berna et al., 2017). Hundred healthy participants were informed that they could receive combined analgesics that might produce dry mouth, or inert placebos. Sixty minutes after swallowing the test drugs, participants were asked to report perceived side effects and to guess whether they had received diclofenac or placebo. Results of a pain stimulation test revealed that atropine more than tripled the analgesic effect of diclofenac, while the effect of the placebo pill was not altered by atropine. A mediation path analysis showed that atropine increased dry mouth symptoms, which fostered the belief to having received the active drug. This, in turn, enhanced the analgesic response.

In sum, these findings indicate that the side effects of an intervention can indeed increase the placebo component of a treatment. However, the external validity of these results is limited by the observation that active placebos appear to work best in an uncertain situation, such as having a 50% chance to receive an active drug. Whether side effects enhance the placebo effect also in other situations remains unclear.

7. ROUTE OF DRUG ADMINISTRATION

Several studies suggest that invasive methods of drug delivery, such as injections and intravenous drug administration, induce stronger positive expectations, and thus larger placebo effects than less invasive methods, for example, oral placebos. De Craen et al. (1999) performed a systematic review of 22 RCTs in the treatment of acute migraine to compare the analgesic effects of subcutaneous and oral placebos. The average headache relief rate was significantly larger in trials using subcutaneous injection (32%) as compared to trials using oral administration (26%), suggesting a relationship between route of placebo administration and the placebo effect.

In a systematic review of 98 RCTs, Macedo, Farré, and Baños (2006) again compared the effects of different routes of placebo interventions in the acute treatment of migraine. The proportion of patients with more than 50% pain relief did not differ between subcutaneous (25%), oral (29%), and nasal (31%) routes of placebo administration, while intravenous (40%) application was associated with the largest improvement. When looking at 100% pain relief, however, subcutaneous placebos (15%) worked better than oral (9%), nasal (8%) and intravenous (11%) placebos. These inconsistencies point at the difficulty to compare placebo effects indirectly across trials, as these comparisons are based on nonrandomized data.

The question of whether such differential placebo effects exist also in fields other than pain, namely itch, was addressed in a subgroup analysis of a systematic review of 24 studies by Van Laarhoven et al. (2015).

Interestingly, oral and injected routes of placebo administration had comparable effects on itch, suggesting that for treating skin disease, invasive placebos may not be superior to oral placebos. Building upon these findings, Peerdeman et al. (2017) performed an internet-based survey with 508 participants and compared expectations about the effectiveness of oral, injected, and topical medications for relieving itch and pain. Participants were given a brief description of each mode of drug administration along with illustrative photographs and were asked to rate the expected effectiveness of these preparations. Results revealed different patterns for pain and itch: While pain-relieving medication was expected to be most effective when administered via injection and least effective when administered topically, itch-relieving medication was believed to be most effective when administered topically and least effective when administered orally. Notably, for both symptoms injected drugs were regarded more effective than oral drugs. Since itch-relieving medications are most commonly administered topically, results probably reflect the impact of knowledge and prior experience on treatment-related expectations. Such differential expectations most probably form the bases for a graded placebo response.

To conclude, evidence from systematic reviews in the field of pain suggests that invasive routes of placebo delivery are associated with stronger effects than oral and nasal administration. The interpretation of the results, however, is hampered by indirect comparison between groups, leaving possible confounders uncontrolled. Therefore, direct comparison of different placebo administration routes is needed to corroborate these findings.

A novel approach to control for possible confounders is to perform a network analysis of RCTs, in which direct and indirect comparisons within and across trials are combined (Salanti, Higgins, Ades, & Ioannidis, 2008). Bannuru et al. (2015) conducted a systematic review with network meta-analysis of 149 RCTs on chronic knee pain and compared oral, intra-articular, topical, and oral plus topical routes of placebo administration. Intra-articular as well as topical placebo interventions were both associated with larger effect sizes than oral placebos. Interestingly, when the four types of placebo administration routes were combined in a single network node, the conclusions of the network analysis regarding the best active treatment for chronic knee pain were different as compared to conclusions from a network analysis with four placebo nodes according to mode of placebo administration. These findings clearly indicate that placebo effect sizes vary systematically with the route of placebo administration and point to the importance of differential placebo effects for the interpretation of clinical trials.

8. TYPE OF TREATMENT

Placebo interventions may not only differ by route of administration but also by treatment modality. Kaptchuk et al. (2006) performed an RCT, in which 270 patients with chronic arm pain were randomized to either sham acupuncture or inert pills. Contrary to the primary study hypothesis, pain levels did not differ between the two placebo groups after 2 weeks of single-blind treatment. During the next 6 weeks, half of the patients remained on placebo, while half were re-randomized to either acupuncture or amitriptyline. Patients who remained on sham acupuncture showed significantly greater pain reduction over time than did patients, who remained on placebo pills.

Fässler, Meissner, Kleijnen, Hróbjartsson, and Linde (2015) built upon these results and performed a systematic review to retrieve RCTs, which allowed a direct comparison of different types of placebo or sham treatments. Twelve studies could be included in the review, seven of which did not find any differential placebo effects. Besides the RCT on chronic arm pain by Kaptchuk et al. (2006) as reported above, one study found a larger placebo effect on venous refill time in varicose patients treated with a topical placebo cream as compared to a placebo pill (Saradeth, Resch, & Ernst, 1994); the second study reported greater clinical improvement in Raynaud patients receiving sham biofeedback as compared to a placebo pill (Thompson, 2000); a pediatric study found stronger relieve of bronchiolitis symptoms in children receiving a placebo inhalator as compared to children receiving oral placebo pills (Gadomski et al., 1994); and Avci, Tanyildizi, and Kusku (2013) reported larger placebo effects on Wood's lamp reflection scores in children with erythrasma, who received a topical placebo cream as compared to an oral placebo. In conclusion, the review suggests that if a differential placebo effect could be found, it was always the placebo pill which induced the smaller effect.

Meissner et al. (2013) performed a meta-analysis of 79 RCTs to compare the effectiveness of different types of placebo interventions in the prophylaxis of migraine attacks. Included studies comprised a variety of placebo control groups, comprising orally administered placebos, injected placebos, sham acupuncture, sham surgery, sham electromagnetic simulation, and psychological sham treatments. In comparison to pharmacological placebo pills, sham surgery and sham acupuncture were associated with larger placebo responder rates, while there was no difference between placebo pills and sham injections. A network meta-analysis further corroborated the difference between sham acupuncture and oral pharmacological placebos (sham surgery trials were not part of the network). Remarkably, the improvement going along with sham acupuncture was as large as the improvement associated with oral pharmacological pills. The absence of a difference between oral and injected placebo treatments may best be explained by unblinding in the trials with botulinum toxin injections as the active treatment, as botox induces obvious changes in muscle tonus.

A Cochrane meta-analysis on placebo effects across conditions comprised a subgroup analysis, in which the effects of pharmacologic placebos were compared with physical and psychological placebos (Hróbjartsson & Gøtzsche, 2010). Results revealed a greater efficacy of physical placebos as compared to pharmacological placebos. Furthermore, as shown by a re-analysis by Linde, Niemann, and Meissner (2010), the most effective physical placebo was sham acupuncture.

Liu (2017) proposed a neurobiological framework for the enhanced effectiveness of invasive placebo treatments. For example in acupuncture, the imagination of needles penetrating the skin induces anxiousness, which enhances the patient's motivation to obtain the anticipated therapeutic benefit by the alleged treatment. This results in a state of "pleasant aversiveness," in which the descendent pain modulatory system is inhibited by activation of the inner brain reward system by reward expectation. The state of pleasant aversiveness, reward expectation, and the placebo analgesic effect increase with the invasiveness of an intervention, providing a putative neurobiological basis for graded placebo effects (Liu, 2017).

It should be noted that it is often unclear whether sham treatments used as "placebo" controls for nonpharmacological treatments can truly be considered inert placebos. For example, several strains of research suggest that penetrating sham acupuncture interventions have clinically relevant physiological effects (Bai & Lao, 2013). Furthermore, Chae, Lee, and Enck (2018) argued that placebo needles elicit enhanced placebo effects not only because of physiological effects but also due to the failure of blinding, as more participants in the placebo groups suspect that they have received verum acupuncture than would be expected by chance. This may further enhance positive expectations, and thus the placebo effect.

9. FUTURE DIRECTIONS

In sum, there is accumulating evidence that certain treatment characteristics, such as high price, brand label, the use of needles and high placebo doses, increase the expectancies toward the beneficial effect of a placebo treatment, which in turn boost the placebo effect. While most of the studies reported in here were performed in the field of pain, there is little doubt that other symptoms, such as anxiety, itch and nausea, can be affected differentially by placebo interventions under certain circumstances, as well. However, more studies are needed to elucidate the role of treatment characteristics for placebo effects outside the field of pain. Furthermore, most studies have been performed in experimental conditions. While the relatively few available studies performed in clinical settings suggest the relevance of treatment characteristics for modulating placebo effects under conditions of clinical practice, more research would be desirable here.

Even though the link between commercial features and placebo effects is convincing, the question arises of whether it is ethical to boost placebo effects by enhancing the price of a (placebo) treatment. A possible solution could be to activate the inner brain's reward system by ethically less doubtful reinforcers, such as social support and bonding. Finally, novel experimental approaches, from neuroimaging to -omics techniques, are warranted to better understand the neurobiological basis of enhanced placebo effects in patients and healthy controls.

REFERENCES

Avci, O., Tanyildizi, T., & Kusku, E. (2013). A comparison between the effectiveness of erythromycin, single-dose clarithromycin and topical fusidic acid in the treatment of erythrasma. *Journal of Dermatological Treatment*, *24*, 70–74.

Bai, L., & Lao, L. (2013). Neurobiological foundations of acupuncture: The relevance and future prospect based on neuroimaging evidence. *Evidence-Based Complementary and Alternative Medicine*, Article ID e812568, 9 pages.

Bannuru, R. R., Mcalindon, T. E., Sullivan, M. C., Wong, J. B., Kent, D. M., & Schmid, C. H. (2015). Effectiveness and implications of alternative placebo treatments: A systematic review and network meta-analysis of osteoarthritis trials. *Annals of Internal Medicine*, *163*, 365–372.

Berna, C., Kirsch, I., Zion, S. R., Lee, Y. C., Jensen, K. B., Sadler, P., et al. (2017). Side effects can enhance treatment response through expectancy effects: An experimental analgesic randomized controlled trial. *Pain*, *158*, 1014–1020.

Bjørkedal, E., & Flaten, M. A. (2011). Interaction between expectancies and drug effects: An experimental investigation of placebo analgesia with caffeine as an active placebo. *Psychopharmacology*, *215*, 537–548.

Blackwell, B., Bloomfield, S. S., & Buncher, C. R. (1972). Demonstration to medical students of placebo responses and non-drug factors. *The Lancet*, *1*, 1279–1282.

Branthwaite, A., & Cooper, P. (1981). Analgesic effects of branding in treatment of headaches. *British Medical Journal (Clinical Research Ed.)*, *282*, 1576–1578.

Buckalew, L. W., & Coffield, K. E. (1982a). Drug expectations associated with perceptual characteristics: Ethnic factors. *Perceptual and Motor Skills*, *55*, 915–918.

Buckalew, L. W., & Coffield, K. E. (1982b). An investigation of drug expectancy as a function of capsule color and size and preparation form. *Journal of Clinical Psychopharmacology, 2*, 245–248.

Buckalew, L. W., & Ross, S. (1991). Medication property effects on expectations of action. *Drug Development Research, 23*, 101–108.

Chae, Y., Lee, Y.-S., & Enck, P. (2018). How placebo needles differ from placebo pills? *Frontiers in Psychiatry, 9*, 243.

Colloca, L., & Benedetti, F. (2006). How prior experience shapes placebo analgesia. *Pain, 124*, 126–133.

Colloca, L., & Benedetti, F. (2009). Placebo analgesia induced by social observational learning. *Pain, 144*, 28–34.

Colloca, L., Lopiano, L., Lanotte, M., & Benedetti, F. (2004). Overt versus covert treatment for pain, anxiety, and Parkinson's disease. *The Lancet. Neurology, 3*, 679–684.

Colloca, L., & Miller, F. G. (2011). How placebo responses are formed: A learning perspective. *Philosophical Transactions of the Royal Society of London. Series B, Biological Sciences, 366*, 1859–1869.

Colloca, L., Petrovic, P., Wager, T. D., Ingvar, M., & Benedetti, F. (2010). How the number of learning trials affects placebo and nocebo responses. *Pain, 151*, 430–439.

Colloca, L., Sigaudo, M., & Benedetti, F. (2008). The role of learning in nocebo and placebo effects. *Pain, 136*, 211–218.

De Craen, A. J., Moerman, D. E., Heisterkamp, S. H., Tytgat, G. N., Tijssen, J. G., & Kleijnen, J. (1999). Placebo effect in the treatment of duodenal ulcer. *British Journal of Clinical Pharmacology, 48*, 853–860.

De la Fuente-Fernández, R., Ruth, T. J., Sossi, V., Schulzer, M., Calne, D. B., & Stoessl, A. J. (2001). Expectation and dopamine release: Mechanism of the placebo effect in Parkinson's disease. *Science, 293*, 1164–1166.

Faasse, K., Martin, L. R., Grey, A., Gamble, G., & Petrie, K. J. (2016). Impact of brand or generic labeling on medication effectiveness and side effects. *Health Psychology, 35*, 187.

Fässler, M., Meissner, K., Kleijnen, J., Hróbjartsson, A., & Linde, K. (2015). A systematic review found no consistent difference in effect between more and less intensive placebo interventions. *Journal of Clinical Epidemiology, 68*, 442–451.

Fehse, K., Maikowski, L., Simmank, F., Gutyrchik, E., & Meissner, K. (2015). Placebo responses to original vs generic ASA brands during exposure to noxious heat: A pilot fMRI study of neurofunctional correlates. *Pain Medicine, 16*, 1967–1974.

Gadomski, A. M., Aref, G. H., El Din, O. B., El Sawy, I. H., Khallaf, N., & Black, R. E. (1994). Oral versus nebulized albuterol in the management of bronchiolitis in Egypt. *The Journal of Pediatrics, 124*, 131–138.

Geuter, S., Eippert, F., Attar, C. H., & Büchel, C. (2013). Cortical and subcortical responses to high and low effective placebo treatments. *NeuroImage, 67*, 227–236.

Geuter, S., Koban, L., & Wager, T. D. (2017). The cognitive neuroscience of placebo effects: Concepts, predictions, and physiology. *Annual Review of Neuroscience, 40*, 167–188.

Hadamitzky, M., Sondermann, W., Benson, S., & Schedlowski, M. (2018). Placebo effects in the immune system. *International Review of Neurobiology, 138*, 39–59.

Hashmi, J. A., Baria, A. T., Baliki, M. N., Huang, L., Schnitzer, T. J., & Apkarian, A. V. (2012). Brain networks predicting placebo analgesia in a clinical trial for chronic back pain. *Pain, 153*, 2393–2402.

Hróbjartsson, A., & Gøtzsche, P. C. (2010). Placebo interventions for all clinical conditions. *The Cochrane Database of Systematic Reviews*, (1), CD003974.

Huskisson, E. (1974). Simple analgesics for arthritis. *British Medical Journal, 4*, 196–200.

Jacobs, K. W., & Nordan, F. M. (1979). Classification of placebo drugs: Effect of color. *Perceptual and Motor Skills, 49*, 367–372.

Jensen, J. S., Bielefeldt, A. Ø., & Hróbjartsson, A. (2017). Active placebo control groups of pharmacological interventions were rarely used but merited serious consideration: A methodological overview. *Journal of Clinical Epidemiology, 87*, 35–46.

Kalisch, R., Wiech, K., Critchley, H. D., & Dolan, R. J. (2006). Levels of appraisal: A medial prefrontal role in high-level appraisal of emotional material. *NeuroImage, 30*, 1458–1466.

Kam-Hansen, S., Jakubowski, M., Kelley, J. M., Kirsch, I., Hoaglin, D. C., Kaptchuk, T. J., et al. (2014). Altered placebo and drug labeling changes the outcome of episodic migraine attacks. *Science Translational Medicine, 6*, 218ra5.

Kaptchuk, T. J., Kelley, J. M., Conboy, L. A., Davis, R. B., Kerr, C. E., Jacobson, E. E., et al. (2008). Components of placebo effect: Randomised controlled trial in patients with irritable bowel syndrome. *BMJ (Clinical Research Ed.), 336*, 999–1003.

Kaptchuk, T. J., Stason, W. B., Davis, R. B., Legedza, A. R., Schnyer, R. N., Kerr, C. E., et al. (2006). Sham device v inert pill: Randomised controlled trial of two placebo treatments. *BMJ (Clinical Research Ed.), 332*, 391–397.

Köteles, F., Fodor, D., Cziboly, Á., & Bárdos, G. (2009). Expectations of drug effects based on colours and sizes—The importance of learning. *Clinical and Experimental Medical Journal, 3*, 99–107.

Krummenacher, P., Candia, V., Folkers, G., Schedlowski, M., & Schönbächler, G. (2010). Prefrontal cortex modulates placebo analgesia. *Pain, 148*, 368–374.

Leslie, A. (1954). Ethics and practice of placebo therapy. *The American Journal of Medicine, 16*, 854–862.

Linde, K., Niemann, K., & Meissner, K. (2010). Are sham acupuncture interventions more effective than (other) placebos? A re-analysis of data from the Cochrane review on placebo effects. *Complementary Medicine Research, 17*, 259–264.

Liu, T. (2017). Route of placebo administration: Robust placebo effects in laboratory and clinical settings. *Neuroscience and Biobehavioral Reviews, 83*, 451–457.

Lucchelli, P. E., Cattaneo, A. D., & Zattoni, J. (1978). Effect of capsule colour and order of administration of hypnotic treatments. *European Journal of Clinical Pharmacology, 13*, 153–155.

Macedo, A., Farré, M., & Baños, J.-E. (2006). A meta-analysis of the placebo response in acute migraine and how this response may be influenced by some of the characteristics of clinical trials. *European Journal of Clinical Pharmacology, 62*, 161–172.

Meissner, K. (2009). Effects of placebo interventions on gastric motility and general autonomic activity. *Journal of Psychosomatic Research, 66*, 391–398.

Meissner, K., Bingel, U., Colloca, L., Wager, T. D., Watson, A., & Flaten, M. A. (2011). The placebo effect: Advances from different methodological approaches. *The Journal of Neuroscience, 31*, 16117–16124.

Meissner, K., Fässler, M., Rücker, G., Kleijnen, J., Hróbjartsson, A., Schneider, A., et al. (2013). Differential effectiveness of placebo treatments: A systematic review of migraine prophylaxis. *JAMA Internal Medicine, 173*, 1941–1951.

Meissner, K., & Ziep, D. (2011). Organ-specificity of placebo effects on blood pressure. *Autonomic Neuroscience: Basic & Clinical, 164*, 62–66.

Mennella, J. A., Spector, A. C., Reed, D. R., & Coldwell, S. E. (2013). The bad taste of medicines: Overview of basic research on bitter taste. *Clinical Therapeutics, 35*, 1225–1246.

Meyer, B., Yuen, K. S., Ertl, M., Polomac, N., Mulert, C., Büchel, C., et al. (2015). Neural mechanisms of placebo anxiolysis. *Journal of Neuroscience, 35*, 7365–7373.

Mikellides, B. (1990). Color and physiological arousal. *Journal of Architectural and Planning Research, 7*, 13–20.

Moncrieff, J., Wessely, S., & Hardy, R. (2004). Active placebos versus antidepressants for depression. *The Cochrane Database of Systematic Reviews*, (1), CD003012.

Müller, V., Remus, K., Hoffmann, V., Tschöp, M. H., & Meissner, K. (2016). Effectiveness of a placebo intervention on visually induced nausea in women—A randomized controlled pilot study. *Journal of Psychosomatic Research, 91*, 9–11.

Pacheco-López, G., Engler, H., Niemi, M.-B., & Schedlowski, M. (2006). Expectations and associations that heal: Immunomodulatory placebo effects and its neurobiology. *Brain, Behavior, and Immunity, 20*, 430–446.

Peciña, M., Bohnert, A. S., Sikora, M., Avery, E. T., Langenecker, S. A., Mickey, B. J., et al. (2015). Association between placebo-activated neural systems and antidepressant responses: Neurochemistry of placebo effects in major depression. *JAMA Psychiatry, 72*, 1087–1094.

Peerdeman, K., Tekampe, J., Laarhoven, A., Middendorp, H., Rippe, R., Peters, M., et al. (2017). Expectations about the effectiveness of pain-and itch-relieving medication administered via different routes. *European Journal of Pain, 22*, 774–783.

Plassmann, H., O'doherty, J., Shiv, B., & Rangel, A. (2008). Marketing actions can modulate neural representations of experienced pleasantness. *Proceedings of the National Academy of Sciences of the United States of America, 105*, 1050–1054.

Plassmann, H., & Weber, B. (2015). Individual differences in marketing placebo effects: Evidence from brain imaging and behavioral experiments. *Journal of Marketing Research, 52*, 493–510.

Price, D. D., Finniss, D. G., & Benedetti, F. (2008). A comprehensive review of the placebo effect: Recent advances and current thought. *Annual Review of Psychology, 59*, 565–590.

Quinn, V. F., & Colagiuri, B. (2014). Placebo interventions for nausea: A systematic review. *Annals of Behavioral Medicine, 49*, 449–462.

Rief, W., & Glombiewski, J. A. (2012). The hidden effects of blinded, placebo-controlled randomized trials: An experimental investigation. *Pain, 153*, 2473–2477.

Rief, W., Shedden-Mora, M. C., Laferton, J. A., Auer, C., Petrie, K. J., Salzmann, S., et al. (2017). Preoperative optimization of patient expectations improves long-term outcome in heart surgery patients: Results of the randomized controlled PSY-HEART trial. *BMC Medicine, 15*, 4.

Rutherford, B. R., Wall, M. M., Brown, P. J., Choo, T.-H., Wager, T. D., Peterson, B. S., et al. (2016). Patient expectancy as a mediator of placebo effects in antidepressant clinical trials. *American Journal of Psychiatry, 174*, 135–142.

Salanti, G., Higgins, J. P., Ades, A., & Ioannidis, J. P. (2008). Evaluation of networks of randomized trials. *Statistical Methods in Medical Research, 17*, 279–301.

Sallis, R., & Buckalew, L. W. (1984). Relation of capsule color and perceived potency. *Perceptual and Motor Skills, 58*, 897–898.

Saradeth, T., Resch, K., & Ernst, E. (1994). Placebo treatment for varicosity: Don't eat it, rub it! *Phlebology, 9*, 63–66.

Schapira, K., McClelland, H., Griffiths, N., & Newell, D. (1970). Study on the effects of tablet colour in the treatment of anxiety states. *British Medical Journal, 2*, 446–449.

Schedlowski, M., Enck, P., Rief, W., & Bingel, U. (2015). Neuro-bio-behavioral mechanisms of placebo and nocebo responses: Implications for clinical trials and clinical practice. *Pharmacological Reviews, 67*, 697–730.

Shiv, B., Carmon, Z., & Ariely, D. (2005). Placebo effects of marketing actions: Consumers may get what they pay for. *Journal of Marketing Research, 42*, 383–393.

Świder, K., & Bąbel, P. (2016). The effect of the type and colour of placebo stimuli on placebo effects induced by observational learning. *PLoS One, 11*, e0158363.

Tao, D., Wang, T., & Wang, T. (2017). Effects of color on expectations of drug effects: A cross-gender cross-cultural study. *Color Research & Application, 42*, 124–130.

Thompson, B. (2000). Comparison of sustained-release nifedipine and temperature biofeedback for treatment of primary Raynaud's phenomenon. Results from a randomized clinical trial with 1-year follow-up. *Archives of Internal Medicine, 160*, 1101–1108.

Van Laarhoven, A. I., Van Der Sman-Mauriks, I. M., Donders, A. R. T., Pronk, M. C., Van De Kerkhof, P. C., & Evers, A. W. (2015). Placebo effects on itch: A meta-analysis of clinical trials of patients with dermatological conditions. *Journal of Investigative Dermatology*, *135*, 1234–1243.
Verhoeven, J. W., Pieterse, M. E., & Pruyn, A. T. (2006). Effects of interior color on health care consumers: A 360 degree photo simulation experiment. *Advances in Consumer Research*, *33*(1), 292–293. ACR North American Advances.
Waber, R. L., Shiv, B., Carmon, Z., & Ariely, D. (2008). Commercial features of placebo and therapeutic efficacy. *JAMA*, *299*, 1016–1017.
Wager, T. D., Rilling, J. K., Smith, E. E., Sokolik, A., Casey, K. L., Davidson, R. J., et al. (2004). Placebo-induced changes in FMRI in the anticipation and experience of pain. *Science*, *303*, 1162–1167.
Wan, X., Woods, A. T., Salgado-Montejo, A., Velasco, C., & Spence, C. (2015). Assessing the expectations associated with pharmaceutical pill colour and shape. *Food Quality and Preference*, *45*, 171–182.
Zieliński, P. (2016). An arousal effect of colors saturation. *Journal of Psychophysiology*, *30*, 9–16.
Zubieta, J.-K., Bueller, J. A., Jackson, L. R., Scott, D. J., Xu, Y., Koeppe, R. A., et al. (2005). Placebo effects mediated by endogenous opioid activity on μ-opioid receptors. *Journal of Neuroscience*, *25*, 7754–7762.

CHAPTER FIFTEEN

The Power of Labeling in Nocebo Effects

Kate Faasse*[,1], Leslie R. Martin[†]
*School of Psychology, University of New South Wales, Sydney, NSW, Australia
[†]Department of Psychology and Neuroscience, La Sierra University, Riverside, CA, United States
[1]Corresponding author: e-mail address: k.faasse@unsw.edu.au

Contents

Abstract

Nocebo effects comprise two broad types: primary nocebo effects, in which overall treatment efficacy is reduced; and nocebo side effects, which result in the increased experience of unpleasant secondary side effects. An important factor in generating nocebo effects of both types is the patient's expectations of how well a treatment will work, and how likely it is to cause side effects. One source of negative expectations is the presence of generic—as opposed to brand name—labeling. A medicine's labeling is likely to be one of the first aspects of a treatment that is encountered by a patient, and perhaps the most common labeling information on pharmaceuticals is the labeling that identifies the drug as being made by the originator brand manufacturer, or as a generic copy. Although generic medicines are pharmaceutically equivalent to their brand name counterparts, generics are often viewed with distrust and perceived to be inferior to branded medicines. Negative perceptions of generic pharmaceuticals may contribute to reduced treatment efficacy via enhanced primary nocebo effects, and increased nocebo side effects. This chapter reviews evidence for the role of brand and generic labeling in treatment outcomes across a range of contexts—most often laboratory research assessing pain outcomes, as well as the influence of related factors

International Review of Neurobiology, Volume 139
ISSN 0074-7742
https://doi.org/10.1016/bs.irn.2018.07.016

including price, familiarity, and treatment switches. Although increasing evidence suggests that labeling of medicines can shape nocebo effects, interventions to improve perceptions of generics do not necessarily translate into more positive treatment outcomes.

The pharmaceutical agents in medicines as well as other active components of medical treatments—for example, specific surgical techniques—are all delivered to patients within particular physical, social, and emotional environments. In addition, patients can hold a range of cognitive beliefs or expectations about aspects of medical treatments. For example, patients vary widely in their perceptions of how harmful medicines are (Horne, Weinman, & Hankins, 1999), and their beliefs about how sensitive they are to the effects of medicines (Horne et al., 2013). More negative views about medicines are associated with increased use of medical care as well as heightened perceptions of treatment side effects and the attribution of unrelated symptoms as being the result of the prescribed medicine (Faasse, Grey, Horne, & Petrie, 2015; Heller, Chapman, & Horne, 2015; Petrie, Moss-Morris, Grey, & Shaw, 2004).

Individual patient beliefs, as well as the context in which a medical treatment is delivered, can influence treatment outcomes via placebo and nocebo effects. Importantly, placebo and nocebo effects can influence responses to both inert or sham treatments, as well as active medical treatments (Bingel et al., 2011; Shedden Mora, Nestoriuc, & Rief, 2011). Thus, it is important to understand the factors that contribute to both placebo and nocebo effects, because they have the potential to influence the outcomes of almost every medical intervention in use today. The more well-known *placebo* effect refers to healing or improved health outcomes that follow treatment administration, but that are not caused by any pharmacological or active effect of the treatment itself. Instead, these improvements occur because of positive patient expectations, and learning through previous experience (Stewart-Williams & Podd, 2004).

While placebo effects involve enhanced health outcomes, nocebo effects are the experience of unpleasant or harmful treatment outcomes (Colloca & Miller, 2011; Faasse & Petrie, 2013). Nocebo effects can involve two broad classes of outcomes. First, nocebo effects can be observed as a decrease in overall treatment efficacy, reducing the intended or primary effect of a treatment, such as pain relief. For example, in a study investigating the influence of the information provided about treatment effects on the efficacy of a powerful analgesic (remifentanil), Bingel et al. (2011) manipulated the information participants received prior to receiving an infusion of remifentanil and being exposed to experimentally-induced heat pain. Participants rated their pain on a visual analogue scale (VAS) ranging from 0 (no pain) to 100 (unbearable pain), and pain intensity was assessed under four conditions. First, a no treatment administration condition, providing a baseline assessment of pain intensity with no analgesia. Second, a no information condition, in which participants were unaware that the remifentanil infusion had been started. The no information condition demonstrates the physiological effect of remifentanil with no expectations. Third, a positive information condition, in which participants were aware they were receiving a powerful analgesic drug. Finally, a negative information condition, in which participants were inaccurately informed that the drug infusion had been stopped, and that discontinuing the remifentanil could result in increased pain. The results of this study indicate that positive information can substantially enhance the overall analgesic efficacy in comparison to no information (see Fig. 1). Importantly, negative information effectively eliminated the pharmacological effects of the drug: a substantial *nocebo* effect in treatment efficacy.

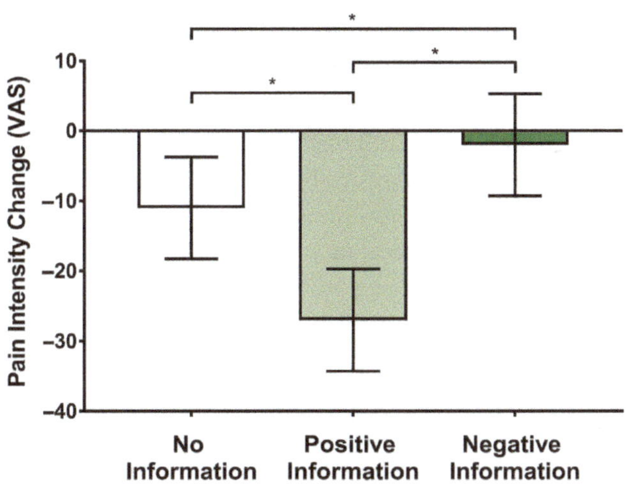

Fig. 1 Bar graph showing mean change (95% CI) in pain intensity ratings (VAS) in response to heat pain from pre- to post-analgesia under three conditions: no information, positive information, and negative information. *Adapted from Bingel, U., Wanigasekera, V., Wiech, K., Ni Mhuircheartaigh, R., Lee, M. C., Ploner, M., et al. (2011). The effect of treatment expectation on drug efficacy: Imaging the analgesic benefit of the opioid remifentanil. Science Translational Medicine, 3(70), 70ra14. https://doi.org/10.1126/scitranslmed.3001244.*

Nocebo effects can also be observed as an increase in unpleasant treatment *side effects*, or the unintended secondary effects of a treatment (Barsky, Saintfort, Rogers, & Borus, 2002; Faasse & Petrie, 2013), which can cause symptoms that are serious enough to require hospitalization and medical intervention (Reeves, Ladner, Hart, & Burke, 2007). Reeves and colleagues describe the case of a patient who took an overdose of 29 capsules that had been prescribed to him as part of a clinical drug trial of an antidepressant treatment. The patient presented to the hospital, pale and drowsy, with excessive sweating, trembling, rapid respiration rate, and hypotension requiring intravenous fluids. The clinical trial physician determined that the patient had, in fact, taken an overdose of inert placebo capsules—his symptoms were the result of a nocebo effect, and resolved rapidly after this information was revealed. Although this is a dramatic case, recent evidence indicates that nocebo effects are so common in medicine that an estimated 97% of reported pharmaceutical side effects are *not* caused by the drug itself (Mahr et al., 2017).

A larger-scale real-world example of the prevalence of nocebo side effects, and their potential implications, comes from the unpleasant side effects reported by patients taking statin medications. Statins reduce blood cholesterol, and these drugs are estimated to prevent 80,000 heart attacks and strokes each year in the United Kingdom alone (Collins et al., 2016). But, as many as one in four statin-treated patients reports experiencing unpleasant—and sometimes debilitating—muscle pain, and patients routinely discontinue these otherwise beneficial treatments because of the unpleasant nature of their side effect symptoms (Ganga, Slim, & Thompson, 2014). In contrast, in a meta-analysis of clinical trials—in which patients are unaware of whether they are taking the active statin medication or an inert placebo control treatment—the rates of reported muscle pain (and a number of other symptoms including nausea, diarrhea, and fatigue) in patients taking a statin and patients taking the inert placebo treatment are *indistinguishable* (Finegold, Manisty, Goldacre, Barron, & Francis, 2014): 7.9% of statin-treated patients ($n=22{,}058$) reported muscle aches, compared to 7.6% of placebo-treated patients ($n=21{,}624$). The experience of muscle pain is highly unlikely to be caused by the physiological effects of the drug for the majority of patients taking statins, but rather by negative expectations, fear, and other aspects of the treatment context.

Important components of the treatment context that influence nocebo outcomes include commonly held views of pharmaceutical companies (particularly those that produce generic medicines), the meaning patients ascribe to taking a brand name or generic pharmaceutical treatment, the *labeling* that is associated with these products, the cost that patients must pay in order to receive their treatments, and the physical properties associated with the treatments themselves. Pharmaceutical agents must be delivered to the patient's system through some physical pathway: this includes the physical properties of the medication itself—for example, whether the treatment comes in a capsule or a compressed tablet, the color of the medication, its taste, and how easy it is to swallow—as well as the packaging within which it is contained, and the labeling on the medication package—including pharmaceutical branding, colors, and the font used for informational text.

Whether a medicine is in the form of a capsule, topical ointment, or inhalant, the person taking it will almost always encounter its label first. Labeling information may include the brand or generic name of the medicine, a listing of active and inactive ingredients, side effect and allergy warnings, dosing information, and perhaps other things such as the product's price or colorful designs or pictures. The context surrounding medical treatments—including labeling information—can influence nocebo effects. Generic labeling, and other aspects of the treatment context related to generic labeling, can result in nocebo effects both by reducing treatment efficacy (primary nocebo effects), and by increasing the experience of unpleasant nocebo side effects. The reduced efficacy and increased side effects can occur independent of, or in addition to, those specific effects that are caused by the direct pharmacological action of a drug.

1. GENERIC MEDICINES

Perhaps most associated with the concept of pharmaceutical labeling is originator brands versus generic medicine labeling. In pharmaceutical development, the originator brand name, e.g., Prozac or Viagra, is the name that is given by the manufacturer who developed, tested, and patented the medication (Chandler & Owen, 2002). Originator brand manufacturers are typically large pharmaceutical companies with wide recognition of their company name and logo. Incorporated into brand name labeling of a medication is the manufacturer name, the originator medicine brand name, identifiable logos, direct-to-consumer advertising (in countries where it is permitted), as well as unique physical properties of the medication itself. For example, one of the most widely recognized drugs in the world is Viagra—the "little blue pill" shaped like a diamond, and imprinted with the manufacturer name: Pfizer. In contrast, generic medicines are referred to by their "non-proprietary name" which is the name of the active ingredient that they contain, e.g., fluoxetine (Prozac) or sildenafil (Viagra). Generic medicines are typically manufactured by less well-known pharmaceutical companies who do not spend as much (if any) money on advertising or public brand recognition. Generics are usually available at a reduced price: they are typically between 20% and 90% cheaper than their brand name counterparts (Dunne, Shannon, Dunne, & Cullen, 2013), and although FDA requirements state that generics must be delivered through the same route of administration (e.g., oral administration), they usually look different to the originator brand medication, for example, having a different colored capsule or a different shape or size.

Generic medicines are pharmaceutically equivalent to their brand name counterparts (Howland, 2010)—meaning that they must be shown to have the same effect in order to be brought to market. A systematic review and meta-analysis assessing the clinical equivalence of brand name and generic pharmaceuticals used to treat cardiovascular disease supports this equivalence—brand name drugs were not found to be superior to their generic counterparts (Kesselheim et al., 2008). Similarly, a retrospective analysis of 12 years of generic drugs approved by the FDA found that, on average, the difference in bioequivalence measures between generic and brand name products was less than 5%—well within safe and acceptable limits (Davit et al., 2009).

However, generic medicines are often viewed with distrust by consumers, and are perceived to be inferior to branded medicines. Branding confers the reassurance of authenticity, scientific research, and treatment efficacy—and generic drugs risk losing associations with such positive connotations (Chandler & Owen, 2002). A recent systematic review of 52 empirical studies found that patients commonly report perceiving generic medicines to be less safe, less effective, more likely to cause side effects, more risky, and of lower quality than the brand name equivalent medication (Colgan et al., 2015). These negative perceptions are held by approximately one-third of consumers. Patient concerns about effectiveness (36%) and substitution (34%) indicate that patients are most worried that generics will not work as well as their brand name counterparts, and thus that brand name and generic drugs are not interchangeable. Strikingly, negative perceptions are not limited to patients. A similar proportion of doctors and pharmacists also view generics in a negative light (Colgan et al., 2015). In particular, one-third (33%) of pharmacists' report concerns about the quality of generic medicines, while physicians appear equally concerned about the effectiveness (29%), quality

(28%), and safety (29%) of generic medicines. Medical professionals who hold these views may communicate them to patients—intentionally or otherwise—further contributing to patient perceptions that generic medicines are inappropriate substitutes for the branded equivalents. The influence of branding on treatment outcomes may be particularly robust when the drugs are used to treat more serious illnesses; generics are often viewed as less appropriate for treating illnesses that are perceived as especially grave (Figueiras, Cortes, Marcelino, & Weinman, 2010).

A case study, published in the Annals of Internal Medicine, highlights the problems that such negative perceptions of generics can have. The patient—a 69-year-old woman requiring medical treatment for a number of conditions including diabetes, arthritis, hypertension, and depression—believes that she is allergic to *all* generic medicines (Brennan & Lee, 2004). The patient's negative beliefs appear to have started 3 years earlier, after she was prescribed a generic medication (glyburide), and developed a rash that was indicative of an allergic reaction but which was unrelated to the generic formulation. However, the patient interpreted this allergic rash as evidence that she was allergic to generic formulations more broadly, and subsequently refused to take any generics that she was prescribed. On rare occasions where her physician could convince her to try a generic, she developed diffuse skin itching within a few minutes of ingestion, in an apparent experience of a nocebo side effect. The patient continued to refuse to take generic medicines, even when she was hospitalized with dyspnea, chest pressure, and supraventricular tachycardia and was prescribed a generic beta-blocker. A week later, the patient presented with a heart rate of 160 beats per minute and still refused to take the generic formulation, despite having been warned by her doctor that she was risking her life by doing so. The patient's negative beliefs about generics were so strong that she experienced unpleasant nocebo side effects and was prepared to put her life at risk in order to avoid ingesting the generic beta-blocker. The next sections of this chapter will consider the different aspects of generic labeling that may be contributing to both primary nocebo effects (i.e., reduced treatment efficacy) and secondary nocebo side effects.

2. NOCEBO EFFECTS AND ASPECTS OF GENERIC LABELING

2.1 Brand Name and Generic Labeling

As previously noted, at least one-third of the population holds one or more negative perceptions of generic medicines, including that generic drugs are less safe, less effective, of lower quality, more likely to cause side effects, and should not be substituted for their brand name equivalents (Colgan et al., 2015). Negative perceptions of generics appear to contribute to reduced treatment efficacy via primary nocebo effects—where negative beliefs reduce the primary or intended effect of a treatment. One of the central theories of how primary nocebo effects occur is via treatment expectancies, i.e., the expectation or belief that a treatment will be effective (or, in this case, less effective or even ineffective) can facilitate this outcome either directly or indirectly through a mediating physiological or psychological pathway (Kirsch, 1985; Stewart-Williams & Podd, 2004). For example, expecting increased anxiety can be directly anxiety-inducing. Anxiety may also mediate the influence of negative expectancies on the experience of pain: expecting to experience increased pain produces anticipatory anxiety, which activates the cholecystokinin system, resulting in heightened pain (Tracey, 2010). Thus, patients receiving a generic pharmaceutical may experience reduced treatment efficacy via primary nocebo effects, because their treatment expectancies are reduced.

The earliest evidence of the role of brand name labeling in shaping treatment effectiveness came from an experimental study by Branthwaite and Cooper (1981). In order to test the influence of the addition of brand name labeling to purportedly active analgesic treatments—in reality, half of all participants received inert placebo treatments, and half received active aspirin tablets (active ingredient: acetylsalicylic acid)—835 female participants were recruited to take part in a study "comparing the effectiveness of different brands of headache tablet" (p. 1576). Participants were randomized to receive one of four different tablets described as analgesics: unbranded placebo, branded placebo, unbranded aspirin, or branded aspirin. Labeling for a popular brand name was included on the packaging for the branded tablets, and the tablets themselves were imprinted with the manufacturer's logo. The unbranded tablets, similar to typical generic medicines, had plain packaging labeled simply as "analgesic tablets," and the tablets themselves were plain. Participants were asked to take the study tablets to treat the headaches that they experienced during the following 2 weeks, and complete a brief questionnaire about their initial headache pain, number of tablets taken, and their degree of pain relief experienced 1 h after taking the study tablets. The results of this study showed that the lack of branding—particularly when the tablets were inert placebos—resulted in a primary nocebo effect as evidenced by reduced treatment effectiveness. In the branded placebo condition, 50% of women reported substantial improvement in their

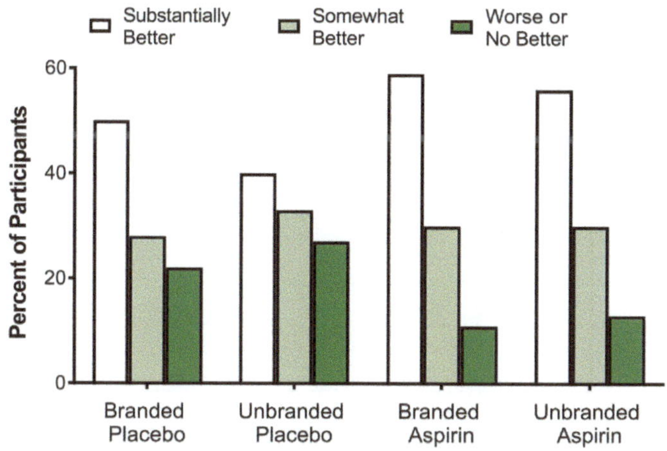

Fig. 2 Bar graph showing the percentage of patients reporting substantial improvement, moderate improvement, or no improvement in headache pain under one of four treatment conditions: branded placebo, unbranded placebo, branded aspirin, or unbranded aspirin. *Adapted from Branthwaite, A., & Cooper, P. (1981). Analgesic effects of branding in treatment of headaches. British Medical Journal, 282, 1576–1578.*

headache pain after taking the study tablets (see Fig. 2). The removal of branding reduced this rate to 40%—a 10% decrease in treatment efficacy. Similar but weaker results were seen in the active analgesia condition, with the removal of branding resulting in a 3% decrease in treatment efficacy. Overall, the study tablets were less effective in relieving headaches, i.e., they showed larger primary nocebo effects, when they were unbranded.

More recently, Faasse, Martin, Grey, Gamble, and Petrie (2016) extended this research using a similar headache treatment paradigm, explicitly labeling the treatments with either a familiar brand name, or as a generic. In addition, during the initial study session in which participants were given their study treatments, tablets were verbally referred to as being either brand name or generic versions of the pharmaceutical agent (in this case, ibuprofen). This research utilized a within-subjects design, in which all participants took each of the four treatment conditions (in random order) for their next four headaches: branded placebo, generic placebo, branded ibuprofen, and generic ibuprofen. The results of this study again support a primary nocebo effect associated with generic—as compared to brand name—labeling. Generic labeled placebo tablets were less effective at treating headache pain, compared to brand name labeled placebo tablets. To further explore this data, and compare these findings to those of Branthwaite and Cooper (1981), Fig. 3 shows the percentage of participants reporting no headache

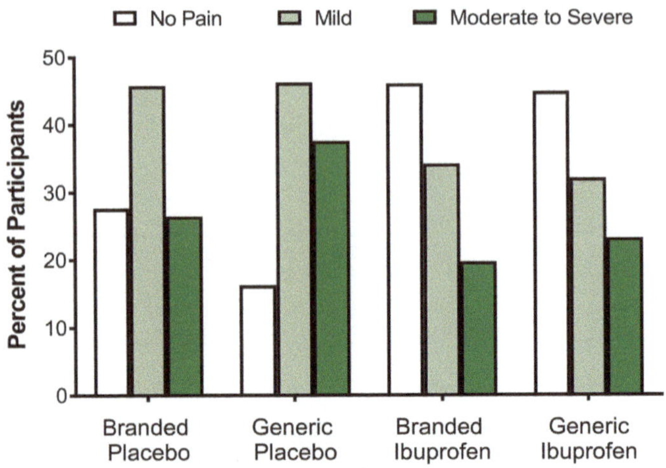

Fig. 3 Bar graph showing the percentage of participants reporting no headache pain, mild headache pain, or moderate to severe headache pain after each of the four treatment conditions: branded placebo, unbranded placebo, branded aspirin, or unbranded aspirin.

pain, mild pain, or moderate to severe pain, an hour after taking each of the study treatments. A similar pattern of results can be seen, whereby placebo-treated participants are less likely to report complete resolution of their headache when they have taken a generic (16%) compared to a brand name labeled (28%) inert tablet. Similarly, participants were more likely to report still being in moderate to severe pain after taking a generic (38%) compared to a brand name (27%) placebo. Results for the active treatments show similar, but much smaller, primary nocebo effects in treatment efficacy, with generic labeling associated with 4% more participants still being in moderate to severe pain following treatment. It seems likely that in both headache studies, there was a ceiling effect whereby both aspirin and ibuprofen (regardless of labeling) are effective in relieving day-to-day headache pain. Future research could assess the primary nocebo effect associated with generic labeling in less effective analgesia and/or in conditions that cause greater pain, for example, post-operative pain or chronic low back pain.

Faasse et al. (2016), as part of this same study, also investigated the influence of brand name and generic labeling on nocebo *side effects*, i.e., the number and intensity of unpleasant side effects that participants reported experiencing as a result of taking each of the four study treatments. In line with the results for *primary* nocebo effects (reduced efficacy of the generic labeled placebo tablets), participants also reported increased nocebo side effects after taking the generic labeled placebo tablets, in comparison to the brand name labeled placebo tablets. However, no differences were seen in response to brand name and generic labeling when participants took active ibuprofen treatments. The results of this and the previous study are limited by the lack of a no treatment control condition. The inclusion of this condition—for example, by also assessing pain and physical symptoms during one untreated headache episode (in Faasse et al., 2016), or including an additional group who did not receive any study tablets to take and who were asked to refrain from taking analgesia for an hour after the onset of a headache (in Branthwaite & Cooper, 1981)—would have allowed the authors to test whether the generic placebo treatment condition was at all effective in comparison to the natural course of a headache episode, or whether the primary nocebo effect was so powerful that participants experienced no additional pain relief from taking the generic labeled placebo tablets. Future research should consistently include no treatment control conditions, in order to understand the magnitude of nocebo effects following treatment with generic labeled compared to brand name treatments.

Recent research has also investigated the role of generic labeling and primary nocebo effects in response to experimentally-induced (rather than naturally occurring) pain, and found evidence of reduced treatment efficacy in both self-reported analgesia and in objective physiological outcomes. Fehse and Maikowski (2015) exposed 30 male participants to painful thermal heat stimuli, applied to their forearm. Participants experienced painful stimulation under two conditions: no treatment (baseline), and after placebo treatment with a tablet described as either brand name aspirin or a generic equivalent ("1A Pharma"). Participants were also given information about the manufacturer and price of the treatment they received. During painful stimulation participants rated their subjective level of pain experienced on a VAS from 0 (no pain) to 100 (worst unbearable pain), and underwent functional magnetic resonance imaging (fMRI) to assess whether they experienced placebo-induced changes in neural networks associated with the experience of pain. The results of this study indicate that participants who took brand name placebo experienced a significant reduction in pain following tablet ingestion—their pain ratings decreased by approximately 9 points on the VAS. In contrast, participants who took the generic labeled tablet experienced almost no pain relief (only a 2-point decrease), indicating that the generic labeling resulted in a primary nocebo effect in reduced treatment effectiveness. This nocebo effect was large enough to eliminate any pain relief from the inert placebo treatment. Supporting this finding, fMRI results demonstrated that brand name placebo-treated participants experienced significantly higher activation of the dorsomedial prefrontal cortex, the bilateral dorsolateral prefrontal cortex, and adjacent regions, compared to those who received the generic placebo tablet. These brain regions are thought to contribute to placebo analgesia, as well as the formation of expectations. The results of this study indicate that removing the familiar and trusted pharmaceutical branding—and replacing it with an unfamiliar generic label—resulted in a substantial primary nocebo effect that eliminated the placebo analgesia seen in responses to the branded treatment.

However, at this point it is also important to note that the effects of branding are not always evidenced; in at least one experimental study examining the efficacy of brand name versus generic labeled placebos, labeling had no impact on treatment outcomes (Faasse, Grey, Jordan, Garland, & Petrie, 2015). Interestingly, unlike all other studies discussed in this section, Faasse and colleagues (2015) did not examine the nocebo effect in pain. Instead, participants were told that they were receiving a beta-blocker treatment (actually an inert placebo) that would reduce pre-examination anxiety, as well as lowering heart rate and blood pressure. Participants were also warned that the study treatment might cause a number of unpleasant side effects. The generic labeling group did not experience either primary nocebo effects, or more nocebo side effects, when compared to the brand name labeling group. It may be that brand and generic labeling have a particular impact on pain outcomes because over-the-counter analgesics are commonly used and familiar to most participants, and thus participants may have stronger views on and perhaps personal experience of perceived differences between brand name and generic drugs in this context. Future research should seek to tease out the influence of general versus specific perceptions of and expectations about the likely effectiveness and side effects of generic drugs to treat various conditions.

The results of these studies suggest that negative perceptions of generic medicines may be contributing to reduced treatment efficacy via primary nocebo effects, and increased treatment side effects via nocebo side effects (Faasse & Petrie, 2013). However, to date no research has explicitly assessed participants' perceptions of generic pharmaceuticals, and their subsequent association with nocebo outcomes. It will be important for future research to include evaluations of self-reported general beliefs about generic medicines, for example, using questionnaires such as that by Figueiras et al. (2009), as well as assessing people's specific beliefs about a particular medicine or medicines, particularly in cases where the drug under consideration is likely to be familiar to participants. Assessing both general and specific perceptions is important in order to establish a better understanding of how people form perceptions of generics, and the influence of personal experience on such perceptions. In addition, it may be useful to include implicit measures of generic medicines in future research, for example, using the implicit association test (IAT; Greenwald, McGhee, & Schwartz, 1998). The IAT assesses the speed with which participants associate a series of positive and negative attributes with a particular construct (e.g., race, age). Ongoing work by the authors of this chapter is focused on exploring the association between attributes such as effective or ineffective, safe or unsafe, and the construct of branded or generic medicine. The difference in speed with which participants associate these positive or negative attributes with generic or brand name items is expected to represent how closely associated these things are in participants' minds—for example, taking much longer to associate generic medicines with effectiveness (compared to the time taken to associate generic medicine with ineffectiveness) would be considered indicative of more negative perceptions of generic pharmaceuticals and might partially explain persistent preferences for, and better experiences with, branded medications even among those who objectively recognize the cost-effectiveness and chemical equivalence of the generic version of a drug. Utilizing both self-report and implicit measures of general and specific beliefs about generic medicines has the potential to provide a much richer picture of the way beliefs and expectations—conscious or unconscious—can influence nocebo effects.

2.2 Treatment Price

In addition to an association with recognized pharmaceutical branding, price is an important differentiator of brand name and generic medicines: generics are up to 90% cheaper than their brand name equivalents (Dunne et al., 2013), and thus generics may be perceived as the "cheap and nasty" alternative to a higher quality and higher priced brand name pharmaceutical. In addition to the association of reduced price with generic medicines, low price more generally across a range of consumer products is associated with heightened perceived risk—i.e., cheap products are viewed as less likely to work as expected, or to produce the desired or promised benefits (Grewal, Gotlieb, & Marmorstein, 1994). In line with such beliefs, expensive placebos appear to have larger effects than their low-cost counterparts, and low price (versus high price) information provided about placebo treatments can substantially reduce their efficacy. Waber and colleagues investigated the influence of treatment price on the analgesic efficacy of two sham placebo pill conditions: one group of participants was told that their pill was priced at $2.50 (high price), and the other group was told that their pill had been discounted to $0.10 (low price; Waber, Shiv, Carmon, & Ariely, 2008). After taking the sham analgesic tablet, all participants were exposed to a series of experimentally-induced painful electric shocks. The findings indicate that low price was associated with decreased analgesic efficacy, i.e., a primary nocebo effect, in comparison to the high price condition. For those participants who believed that they had taken an analgesic pill that cost $0.10, only 61% experienced a reduction in pain (and 39% reported experiencing *increased* pain), compared to 85% of those who believed that their pill cost $2.50.

The nocebo effect associated with reduced price has also been measured in patient populations, and outside of a pain paradigm. In this study, 12 patients with Parkinson's disease were assessed under two different counterbalanced conditions: once after being treated with an "expensive" subcutaneous saline injection that was described to them as being a "novel subcutaneous dopamine agonist" that would improve their Parkinson's motor symptoms as well as underlying brain activity, and once after being treated with a "cheap" but equivalent new treatment (Espay et al., 2015). Outcomes were assessed both before and after participants received each of the sham saline injections. Although both injections resulted in improved motor function in patients, the "cheap" treatment was only half as effective as the "expensive" version, indicating a substantial primary nocebo effect of reduced price on treatment efficacy. In addition, the "cheap" treatment was significantly less effective than the administration of an active levodopa injection; in contrast, the "expensive" treatment did not differ significantly in efficacy from the active levodopa. These results indicate that highlighting the cheap or reduced price of a treatment may result in nocebo effects that can diminish the overall effectiveness of a treatment. It is currently unclear how a verbal descriptor (e.g., cheap, discounted) versus an actual numerical price value, might differ in the magnitude of resulting nocebo effects. Nor is it clear how large a numerical price difference needs to be, or where the actual price points are placed, in order to result in reduced treatment efficacy via the nocebo effect.

In a related recent study, Tinnermann, Geuter, Sprenger, Finsterbusch, and Büchel (2017) tested the influence of price on nocebo hyperalgesia in a study utilizing an inert placebo cream described as a new treatment for atopic dermatitis. Participants were told that this cream could cause increased skin sensitivity and subsequent hyperalgesia or an increase in pain in response to thermal stimulation (in reality, nocebo hyperalgesia was the primary outcome of interest). Price information was manipulated in two ways. First, through verbal information describing the treatments as either cheap or effective. Second, through visual information via the physical aspects of the labeling of the packaging of the inert cream. In a separate study, another group of participants were asked to view the two different labels. The "expensive" label included a mid and dark blue color scheme, a more visually appealing design, and a short and easy to pronounce name ("Solestan Creme") with identification as a registered trademark. In contrast, the "cheap" label used an orange and white color scheme, a plain and less appealing design, and a longer and more difficult to pronounce name ("Imotadil-LeniPharma Creme"). Importantly, no price information was provided in this study. Participants rated the "cheap" labeled treatment as costing significantly less (approximately 10 Euros) than the "expensive" labeled treatment (approximately 13 Euros).

In the main experiment, a different group of 49 participants were exposed to experimentally-induced heat pain, with and without one of the two creams. Participants provided self-report data on their subjective experience, as well as undergoing fMRI scanning at the same time. Those who received the "cheap" cream experienced significantly *reduced* nocebo hyperalgesia (i.e., a reduction in the nocebo effect on pain) as compared with those who received the "expensive" cream. In addition, the pain experienced in response to the "expensive" cream (but not the "cheap" cream) appeared to increase over time. The results of this study also indicate that prefrontal cortical activity—particularly in the rostral anterior cingulate cortex (rACC)—was important in mediating the influence of price on nocebo hyperalgesia. Previous research has supported the idea that the prefrontal cortex is implicated in encoding economic value (Chib, Rangel, Shimojo, & O'Doherty, 2009). The results of the current study also indicate that regions of the prefrontal cortex are implicated in the formation of treatment expectations based on price information, and the experience of subsequent nocebo effects.

Interestingly, in the study by Waber and colleagues discussed earlier (2008), cheaper price was associated with reduced analgesic efficacy compared to higher price, i.e., participants experienced *more* pain in response to the cheap treatment that was described as a pain relieving treatment. In contrast, the results of the study conducted by Tinnermann and colleagues (2017) found that cheaper price was associated with reduced hyperalgesic efficacy compared to higher price, i.e., participants experienced *less* pain in response to the cheap treatment that was described as a treatment for atopic dermatitis that might have the additional effect of increasing skin sensitivity and subsequent pain experience. Taken together, these results indicate that cheap price information can have opposite effects on the pain experience, depending on what participants are told about the effects of the treatment. Thus, it is not simply that high priced treatments result in reductions in unpleasant outcomes such as the experience of pain. Instead, price information appears to modulate participants' expectations of treatment outcomes, based on the information they are given about the effects of the treatment under consideration. In other words, cheap or discounted price treatments appear to be viewed as being less powerful or effective than their more expensive counterparts, with resulting nocebo effects.

Other studies looking at outcomes outside of pharmaceutical treatments also support the link between cheaper pricing and reduced efficacy, i.e., increased primary nocebo effects associated with lower price. In one study, researchers assessed the influence of either full or discounted pricing on participants' ability to successfully solve word-jumble puzzles after consuming an energy drink that was described as enhancing mental acuity (Shiv, Carmon, & Ariely, 2005). The researchers found that participants who received a "discounted" drink (costing $0.89) solved fewer puzzles than those who consumed an identical "full price" drink ($1.89). Interestingly, an additional experiment conducted as part of this series of studies found that drawing participants' attention to price-efficacy beliefs eliminated the nocebo effect associated with reduced price. This was achieved by asking participants to rate their perceived efficacy of the drink, given the price they paid for it. In addition, the finding that drawing conscious attention to price-efficacy beliefs reduced the nocebo effect of low pricing indicates that initial expectations activated by price (and perhaps brand or generic labeling) information may be formed unconsciously. This is consistent with the previously discussed hypothesis that such unconscious attributions might be effectively measured using an IAT, and future research would benefit from assessing whether similar manipulations could reduce the influence of price and brand or generic labeling on nocebo outcomes associated with pharmaceutical treatments, and whether these might be sustained over time.

Finally, in an experiment that involved blind wine tasting—drinking wine through a plastic straw while inside an fMRI machine—researchers have investigated the influence of price on participants' subjective ratings of taste pleasantness of otherwise identical wines, as well as the neurological response to these price differences (Plassmann, O'Doherty, Shiv, & Rangel, 2008). Participants consistently rated the taste of "cheaper" wine as less pleasant than identical wine presented at a higher price point. When the same wines were tasted repeatedly, but without price information, there were no differences in perceived pleasantness. In addition, lower price was associated with reduced activation in the medial orbitofrontal cortex (mOFC)—an area found to be associated with the perceived pleasantness of a range of sensory experiences. These results in which pleasantness is reduced in response to cheaper pricing information are similar to medicine-related studies, in which treatment efficacy is reduced when drugs are associated with lower or discounted price, indicating a primary nocebo effect in comparison to high price information. As yet, little is understood about how different price points—under different circumstances—influence nocebo effects. At what point does a medicine (or a wine) become "cheap and nasty"? At what price is value balanced with quality? And is there a point where price exceeds perceived benefits—potentially having a harmful effect on experiential or treatment outcomes?

2.3 Naming of Branded and Generic Medicines

Differences in how branded and generic medicines are named may also contribute to more negative expectancies for generics—or, conversely, more positive expectancies of brand name medicines—and subsequent differences in treatment outcomes. Branded or originator drugs have unique, easy to pronounce, and often memorable brand names—for example, Prozac or Viagra whereas generic medicines are labeled with their International Non-proprietary Names (INNs), which describe their active ingredient—for example, fluoxetine or sildenafil (Dunne et al., 2013). These generic drug names tend to have more syllables and are more difficult to read and pronounce. Recent evidence indicates that drugs (and food additives) with names that are more complex (e.g., Cytrigmcmium, Ribozoxtlitp, or Nxungzictrop) are viewed as being more risky, more harmful, and more likely to cause unpleasant side effects than drugs with more simple and pronounceable names (e.g., Fastinorbine, Allotoneline, or Calotropisin; Dohle & Siegrist, 2014; Song & Schwarz, 2009). Interestingly, simple and complex drug names did not produce differences in perceived efficacy, suggesting that this naming effect may be specific to nocebo outcomes related to negative treatment expectations. The influence of more complex pharmaceutical naming associated with generic medicines may have also contributed to the nocebo effects for pain seen in Tinnerman and colleagues' research (Tinnermann et al., 2017), discussed previously. However, the effect of naming conventions on either primary nocebo effects or nocebo side effects has not yet been specifically investigated in experimental research, although complex names appear to contribute to heightened expectations of negative treatment outcomes. Future research should explicitly test the influence of such naming conventions, while hold all other aspects of treatment information constant.

2.4 Familiarity and Advertising

The familiarity of a treatment—for example, through exposure to advertising or prominent placement in pharmacies—can also shape treatment outcomes. As early as the 1960s, Zajonc (1968, 2001) recognized the human tendency to like better the things that are encountered more frequently—he termed this the "mere exposure effect." In contrast, a lack of familiarity may result in less positive expectations, and, in the case of treatment efficacy, decreases due to the nocebo effect. A large body of research indicates that repeated exposure to primes, e.g., to brand labeling, brand names, or slogans via advertising, increases individuals' liking of the associated product, and this effect appears to be mediated by visual appeal—familiar products "look better" to us than unfamiliar products (Labroo, Dhar, & Schwarz, 2008). The Unites States and New Zealand are the only countries in the world that allow direct-to-consumer advertising of medicines (Ventola, 2011). As suggested by Faasse and Petrie (2016), exposure to advertising—and perhaps increased familiarity with pharmaceutical branding—may have contributed to findings that placebo effects have increased in clinical trials of treatments for neuropathic pain in the United States, but not for trials conducted in Asia or Europe where direct-to-consumer advertising is not permitted (Tuttle et al., 2015). However, familiarity can also be achieved through other pathways including prior experience, exposure to the branding of the manufacturer, visibility of products in pharmacies, and media coverage. Familiarity is also likely to be facilitated through discussions in medical consultations: across eight commonly-used medicines with both brand name and generic alternatives, physicians used the brand name of the drug in almost 80% of consultations (Steinman, Chren, & Landefeld, 2007), perhaps in part because the brand names are easier to for clinicians to pronounce and for patients to remember.

One experimental study has assessed the influence of advertising specifically on treatment outcomes. In this study, participants were exposed to television advertising for one of two different brands of allergy medication (Claratin or Zyrtec), and the results indicate that being exposed to advertising for one drug, but subsequently taking the other drug, resulted in reduced treatment efficacy. Specifically, when participants viewed ads for Zyrtec, but subsequently took Claratin (an unfamiliar treatment in this context) following a histamine challenge, they experienced reduced treatment efficacy, evidenced by a larger wheal size in response to histamine, when compared to participants who both viewed ads for Claratin and subsequently received Claratin (Kamenica, Naclerio, & Malani, 2013). In this study, the lack of familiarity of a treatment resulted in reduced physiological treatment effectiveness, i.e., an increased nocebo effect. This reduced treatment efficacy was associated with less positive expectations about the effectiveness of the Claratin medication in the group for whom this treatment was unfamiliar. The impact of advertising occurred only in naïve participants who tested negative for common allergies, and thus were unlikely to have previous direct experience with either of the antihistamine medications in the study. This finding likely indicates that participants who tested positive for common allergies were already familiar with pharmaceutical treatments for their condition, and were perhaps also more attentive to advertising for such treatments in their daily life because they have personal relevance. Given the prevalence of pharmaceutical advertising—particularly in the United States—that includes a long list of unpleasant (and sometimes life threatening) side effects, future research should also look at the influence of exposure to such advertising on the experience of nocebo side effects.

2.5 Treatment Switches

Undergoing a treatment switch—for example, from a brand name to a generic medicine—is a relatively common experience for patients on long-term medication regimens for the treatment of chronic illness. Indeed, many patients who use generic medicines first took the brand name version of the same medicine, and then switched—by choice, or because of decisions made by their insurance company or government program responsible for pharmaceutical subsidies—when the generic alternative became available. And, data suggest that this switching process itself may reduce treatment effectiveness via increased nocebo effects. Patients commonly report negative outcomes of such switches (Kjoenniksen, Lindbaek, & Granas, 2006), and the desire to remain on a familiar medication is commonly cited as a reason for refusing generic substitution (Heikkila, Mantyselka, & Ahonen, 2011). One example of a problematic treatment switch occurred in New Zealand in 2008 and 2009. Eltroxin—a thyroid-hormone replacement medication—underwent a formulation change, in which the active ingredient remained the same, but some inert binding agents were different. Despite extensive testing providing reassurance that the new formulation was bioequivalent to the old formulation, i.e., in the same dose and absorbed at approximately the same rate and in the same concentration, patients reported that the new formulation was less effective, and was causing unpleasant side effects (Faasse, Cundy, & Petrie, 2010). The problems caused by the Eltroxin formulation change were thought to be the result of the nocebo effect, related to negative expectations about the new formulation.

An experimental study was designed in order to test the influence of changing from one treatment to another on nocebo outcomes. The results of this study indicate that switching from one treatment to another may be problematic, at least in part, because it reduces the effectiveness of a treatment, as well as increasing unpleasant side effects (Faasse, Cundy, Gamble, & Petrie, 2013). In this study, participants undertook two 1-h study sessions. In the first "set-up" session, all participants took the same brand name labeled, yellow inert placebo pill named "Betaprol" which was described as an anxiolytic beta-blocker medication that would reduce anxiety, heart rate, and blood pressure, but that might also cause a range of unpleasant side effects. In the second "experimental" session, participants were randomly assigned to one of three conditions: a no change condition, in which they took the same yellow "Betaprol" tablet; a branded change condition in which they took a white "Novaprol" tablet with the same brand labeling as "Betaprol," described as a reformulated version; or a generic change condition in which they took a white "generic metoprolol" tablet with plain labeling. Although participants were informed that all the tablets were pharmacologically identical (and in reality, all tablets were inert placebos), the two change conditions experienced substantially and significantly smaller reductions in both systolic blood pressure and self-reported state anxiety (i.e., increased nocebo effects) after taking the sham beta-blocker compared to the group who remained on the familiar treatment (see Fig. 4). In addition, those who were switched to the "generic metoprolol" treatment experienced heightened unpleasant nocebo side effects in comparison to those who remained on the original and familiar "Betaprol" treatment. These results highlight, again, the likely roles of familiarity and of generic labeling in contributing to the development of primary nocebo effects and of nocebo side effects.

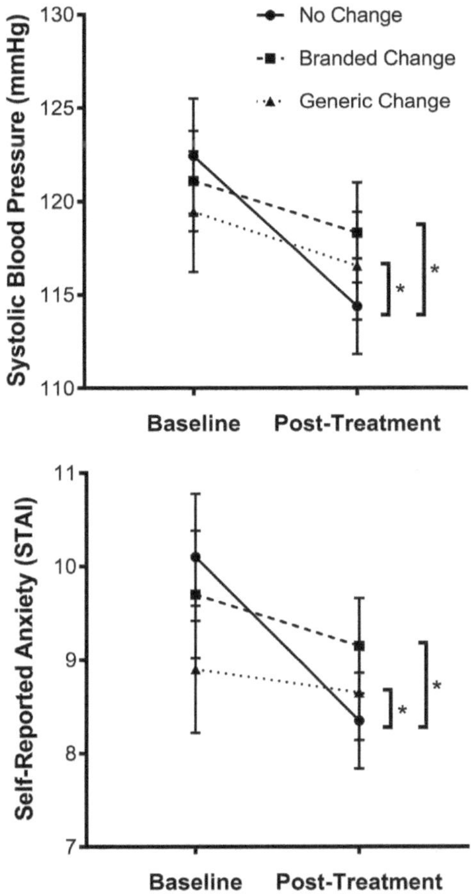

Fig. 4 Line graph showing the change in systolic blood pressure (mmHg; top panel), and self-reported anxiety (brief STAI scores; bottom panel) from baseline to post-treatment across the three groups: no change, branded change, and generic change. *Adapted from Faasse, K., Cundy, T., Gamble, G., & Petrie, K. (2013). The effect of an apparent change to a branded or generic medication on drug effectiveness and side effects. Psychosomatic Medicine, 75, 90–96. https://doi.org/10.1097/PSY.0b013e3182738826.*

3. IMPROVING PERCEPTIONS OF GENERIC MEDICINES

Although a substantial body of research has sought to investigate public and medical professionals' perceptions of generic medicines, very little research has explored the utility of interventions to improve perceptions of generics. One experimental study tested the effect of a brief intervention on both perceptions of generics, and subsequent efficacy of an analgesic (ibuprofen) labeled as either a brand name or generic, in treating headache pain

(Colgan et al., 2016). This study used a similar design to the study by Faasse et al. (2016): participants were asked to take ibuprofen tablets—with brand name and generic labeling, in counterbalanced order—to treat their next two headaches. In addition, prior to taking these analgesic treatments, half of the participants viewed an intervention video designed to improve perceptions of generic medicines, while the other half viewed a control video. The intervention video showed health specialists (physicians, pharmacists, and pharmaceutical regulatory experts) providing information that generics are as safe and effective as their brand name counterparts. The control video showed a physician talking about the epidemiology of headaches. Participants were asked about their perceptions of generic medicines both immediately following the video, and again after taking both of their study treatments.

As anticipated, the intervention group reported improved understanding of generic medicines, greater preference for generics, and they perceived generics as being more appropriate for treating serious illness. However, perceptions of safety, effectiveness, and likelihood of side effects did not change in response to the intervention. When assessing the actual treatment outcomes, i.e., the analgesic efficacy of the brand name and generic labeled treatments, despite improving some perceptions of generics, the intervention actually *reduced* the analgesic efficacy of generic compared to brand name ibuprofen in treating headaches. That is to say, that for participants who viewed the intervention video, a nocebo effect was evident in response to the generic labeled ibuprofen treatment—participants experienced *less* pain relief from the generic tablets compared to the brand name labeled tablets. In contrast—and as might be expected from the findings of both Branthwaite and Cooper (1981) and Faasse et al. (2016) which indicated that brand and generic labeling showed larger differences when participants took inert placebos rather than active analgesic treatments—those participants in the control condition experienced equivalent headache pain relief from both the brand name and generic labeled ibuprofen tablets. These results emphasize the importance of testing treatment outcomes as well as perceptions in such interventions, and highlight the complexity of designing strategies to enhance public understanding and acceptance of generic pharmaceuticals.

4. IMPLICATIONS OF GENERIC LABELING AND SUBSEQUENT NOCEBO EFFECTS

Nocebo side effects are common in medical care; in some treatments they are likely to be responsible for the majority of reported side effects

(Finegold et al., 2014; Mahr et al., 2017; Tobert & Newman, 2016). The studies reviewed in this chapter demonstrate how generic labeling and associated features of the treatment context can increase unpleasant side effects and decrease treatment efficacy. Primary nocebo effects—i.e., nocebo-induced reductions in treatment efficacy—and nocebo side effects limit the potential beneficial effects of a treatment, increase the dose needed to obtain optimal health outcomes (and thus may pose other risks associated with increased dose of the pharmacological agent), reduce patients' quality of life, add to the burden of illness, and lead to unnecessary GP visits and additional prescription medications in an attempt to achieve optimal management of the medical condition as well as to manage unpleasant symptoms (Barsky et al., 2002).

Nocebo effects also contribute substantially to treatment non-adherence or discontinuation; approximately 50% of patients do not take their medications as prescribed, with a large proportion (20%) of cases driven by the experience of unpleasant side effects (Iuga & McGuire, 2014). The direct economic impact of treatment non-adherence—from unresolved illness and additional medical care—is estimated to cost the United States health system between $100 billion and $300 billion each year (Iuga & McGuire, 2014), and these costs are avoidable. At a very conservative estimate, at least 40% of the side effects driving non-adherence are caused by the nocebo effect (Mahr et al., 2017). Considering the direct impact on non-adherence alone, the nocebo effect is costing the United States health system at least $8 billion per year—and the figure is likely much higher. And, these costs will continue to increase: people are living longer, with more chronic conditions, requiring more ongoing medical treatment than ever before.

The use of generic medicines is ubiquitous in modern medical care, and generic drugs are important in managing healthcare costs for both patients and government healthcare systems. Generic prescribing is estimated to have saved the US Government $1.67 trillion over the last decade versus identical prescription rates but with only branded medicines (Association for Accessible Medicines, 2017). However, in order for optimal cost savings associated with generic medicines to be realized, physicians must be willing to prescribe generics, and patients must be willing to fill their generic prescriptions and adhere to their prescribed regimens. Yet the experience of nocebo effects following treatment with generic medicines is likely to lead to decreased adherence, reduced trust in these medicines and in prescribers, and poorer health outcomes. Understanding the influence of generic labeling (and associated features) on negative treatment expectations and on both primary nocebo effects and nocebo side effects could help to shape best-practices recommendations for treatment labeling, patient–practitioner communication strategies, and public health information provision around generic medicines.

5. CONCLUSIONS

This chapter provides a comprehensive overview of research conducted to date examining the influence of generic labeling, and related attributes, on treatment outcomes. As more evidence becomes available, future work could usefully involve systematic reviews and meta-analyses of studies examining the influence of labeling on nocebo effects. However, at this relatively early stage, a narrative overview and discussion of this research is most appropriate. A range of aspects associated with generic pharmaceuticals—including generic labeling, reduced price, diminished familiarity and advertising, and the experience of switching treatments—can all contribute to negative perceptions of the efficacy of generic medicines, and subsequent reductions in treatment effectiveness via increased primary nocebo effects. In addition, negative expectations around medication quality, safety, and the likelihood of experiencing unpleasant side effects can contribute to increased nocebo side effects. Generic labeling, reduced price, and treatment familiarity in particular have been linked to heightened experiences of nocebo outcomes. Although the studies discussed point to the importance of designing, testing, and implementing successful interventions to improve perceptions of generic pharmaceuticals, the research to date indicates that such interventions need to be carefully considered and designed. Overt attempts to educate consumers on the equivalence of generic and brand name pharmaceuticals may backfire such that treatment efficacy of generic drugs is reduced following the intervention. Generic medicines, and their component attributes, present an interesting but complex picture of the influence of labeling on nocebo effects.

REFERENCES

Association for Accessible Medicines. (2017). *Generic drugs access and saving in the U.S.* (p. 48). Association for Accessible Medicines.

Barsky, A. J., Saintfort, R., Rogers, M. P., & Borus, J. F. (2002). Nonspecific medication side effects and the nocebo phenomenon. *The Journal of the American Medical Association*, *287*, 622–627. https://doi.org/10.1001/jama.287.5.622.

Bingel, U., Wanigasekera, V., Wiech, K., Ni Mhuircheartaigh, R., Lee, M. C., Ploner, M., et al. (2011). The effect of treatment expectation on drug efficacy: Imaging the analgesic benefit of the opioid remifentanil. *Science Translational Medicine*, *3*(70), 70ra14. https://doi.org/10.1126/scitranslmed.3001244.

Branthwaite, A., & Cooper, P. (1981). Analgesic effects of branding in treatment of headaches. *British Medical Journal*, *282*, 1576–1578.

Brennan, T. A., & Lee, T. H. (2004). Allergic to generics. *Annals of Internal Medicine*, *141*(2), 126–130. http://doi.org/141/2/126 [pii].

Chandler, J., & Owen, M. (2002). Pharmaceuticals: The new brand arena. *International Journal of Market Research*, *44*(4), 385–404.

Chib, V. S., Rangel, A., Shimojo, S., & O'Doherty, J. P. (2009). Evidence for a common representation of decision values for dissimilar goods in human ventromedial prefrontal cortex. *Journal of Neuroscience*, *29*(39), 12315–12320. https://doi.org/10.1523/JNEUROSCI.2575-09. 2009.

Colgan, S. L. E., Faasse, K., Horne, R., Pereira, J. A., Grey, A., & Petrie, K. (2016). Changing perceptions and efficacy of generic medicines: An intervention study. *Health Psychology*, *35*, 1246–1253. Advance online publication. https://doi.org/10.1037/hea0000402.

Colgan, S. L. E., Faasse, K., Martin, L. R., Stephens, M. H., Grey, A., & Petrie, K. (2015). Perceptions of generic medication in the general population, doctors and pharmacists: A systematic review. *BMJ Open 5*, e008915. https://doi.org/10.1038/clpt.2013.24.52.

Collins, R., Reith, C., Emberson, J., Armitage, J., Baigent, C., Blackwell, L., et al. (2016). Interpretation of the evidence for the efficacy and safety of statin therapy. *Lancet*, *388*, 2532–2561. https://doi.org/10.1016/S0140-6736(16)31357-5.

Colloca, L., & Miller, F. (2011). The nocebo effect and its relevance for clinical practice. *Psychosomatic Medicine*, *73*, 598–603. https://doi.org/10.1097/PSY.0b013e3182294a50.

Davit, B. M., Nwakama, P. E., Buehler, G. J., Conner, D. P., Haidar, S. H., Patel, D. T., et al. (2009). Comparing generic and innovator drugs: A review of 12 years of bioequivalence data from the United States Food and Drug Administration. *Annals of Pharmacotherapy*, *43*(10), 1583–1597. https://doi.org/10.1345/aph.1M141.

Dohle, S., & Siegrist, M. (2014). Fluency of pharmaceutical drug names predicts perceived hazardousness, assumed side effects and willingness to buy. *Journal of Health Psychology*, *19*(10), 1241–1249. https://doi.org/10.1177/1359105313488974.

Dunne, S., Shannon, B., Dunne, C., & Cullen, W. (2013). A review of the differences and similarities between generic drugs and their originator counterparts, including economic benefits associated with usage of generic medicines, using Ireland as a case study. *BMC Pharmacology and Toxicology*, *14*(1), 1. https://doi.org/10.1186/2050-6511-14-1.

Espay, A. J., Norris, M. M., Eliassen, J. C., Dwivedi, A., Smith, M. S., Banks, C., et al. (2015). Placebo effect of medication cost in Parkinson disease: A randomized double-blind study. *Neurology*, *84*(8), 794–802. https://doi.org/10.1212/WNL.0000000000001282.

Faasse, K., Cundy, T., Gamble, G., & Petrie, K. (2013). The effect of an apparent change to a branded or generic medication on drug effectiveness and side effects. *Psychosomatic Medicine*, *75*, 90–96. https://doi.org/10.1097/PSY.0b013e3182738826.

Faasse, K., Cundy, T., & Petrie, K. (2010). Thryoxine: Anatomy of a health scare. *BMJ*, *340*, 20–21.

Faasse, K., Grey, A., Horne, R., & Petrie, K. (2015). High perceived sensitivity to medicines is associated with higher medical care utilisation, increased symptom reporting and greater information-seeking about medication. *Pharmacoepidemiology and Drug Safety*, *24*(6), 592–599. https://doi.org/10.1002/pds.

Faasse, K., Grey, A., Jordan, R., Garland, S., & Petrie, K. (2015). Seeing is believing: Impact of social modeling on placebo and nocebo responding. *Health Psychology*, *34*, 880–885. https://doi.org/10.1037/hea0000199.

Faasse, K., Martin, L. R., Grey, A., Gamble, G., & Petrie, K. J. (2016). Impact of brand or generic labeling on medication effectiveness and side effects. *Health Psychology*, *35*(2), 187–190. https://doi.org/10.1037/hea0000282.

Faasse, K., & Petrie, K. (2013). The nocebo effect: Patient expectations and medication side effects. *Postgraduate Medical Journal*, *89*, 540–546. https://doi.org/10.1136/postgradmedj-2012-131730.

Faasse, K., & Petrie, K. (2016). From me to you: The effect of social modeling on treatment outcomes. *Current Directions in Psychological Science*, *25*(6), 438–443. https://doi.org/10.1177/0963721416657316.

Fehse, K., & Maikowski, L. (2015). Placebo responses to original vs generic ASA brands during exposure to noxious heat: A pilot fMRI study of neurofunctional correlates. *Pain Medicine*, *16*, 1967–1974.

Figueiras, M. J., Alves, N. C., Marcelino, D., Cortes, M. A., Weinman, J., & Horne, R. (2009). Assessing lay beliefs about generic medicines: Development of the generic medicines scale. *Psychology, Health and Medicine*, *14*(3), 311–321. https://doi.org/10.1080/13548500802613043.

Figueiras, M. J., Cortes, M. A., Marcelino, D., & Weinman, J. (2010). Lay views about medicines: The influence of the illness label for the use of generic versus brand. *Psychology and Health*, *25*(9), 1121–1128. https://doi.org/10.1080/08870440903137170.

Finegold, J. A., Manisty, C. H., Goldacre, B., Barron, A. J., & Francis, D. P. (2014). What proportion of symptomatic side effects in patients taking statins are genuinely caused by the drug? Systematic review of randomized placebo-controlled trials to aid individual patient choice. *European Journal of Preventive Cardiology*, *21*(4), 464–474. https://doi.org/10.1177/2047487314525531.

Ganga, H. V., Slim, H. B., & Thompson, P. D. (2014). A systematic review of statin-induced muscle problems in clinical trials. *American Heart Journal*, *168*(1), 6–15. https://doi.org/10.1016/j.ahj.2014.03.019.

Greenwald, A. G., McGhee, D. E., & Schwartz, J. L. (1998). Measuring individual differences in implicit cognition: The implicit association test. *Journal of Personality and Social Psychology*, *74*(6), 1464–1480. https://doi.org/10.1037/0022-3514.74.6.1464.

Grewal, D., Gotlieb, J., & Marmorstein, H. (1994). The moderating effects of message framing and source credibility on the price-perceived risk relationship. *Journal of Consumer Research*, *21*, 145–153. https://doi.org/10.1086/209388.

Heikkila, R., Mantyselka, P., & Ahonen, R. (2011). Price, familiarity, and availability determine the choice of drug—A population-based survey five years after generic substitution was introduced in Finland. *BMC Clinical Pharmacology*, *11*(1), 20. https://doi.org/10.1186/1472-6904-11-20.

Heller, M. K., Chapman, S. C. E., & Horne, R. (2015). Beliefs about medication predict the misattribution of a common symptom as a medication side effect—Evidence from an analogue online study. *Journal of Psychosomatic Research*, *79*(6), 519–529. https://doi.org/10.1016/j.jpsychores.2015.10.003.

Horne, R., Faasse, K., Cooper, V., Diefenbach, M. A., Leventhal, H., Leventhal, E., et al. (2013). The perceived sensitivity to medicines (PSM) scale: An evaluation of validity and reliability. *British Journal of Health Psychology*, *18*(1), 18–30. https://doi.org/10.1111/j.2044-8287.2012.02071.x.

Horne, R., Weinman, J., & Hankins, M. (1999). The beliefs about medicines questionnaire: The development and evaluation of a new method for assessing the cognitive representation of medication. *Psychology and Health*, *14*(1), 1–24. https://doi.org/10.1080/08870449908407311.

Howland, R. H. (2010). Are generic medications safe and effective? *Journal of Psychosocial Nursing*, *48*(3), 13–16.

Iuga, A. O., & McGuire, M. J. (2014). Adherence and healthcare costs. *Risk Management and Healthcare Policy*, 7, 35–44. https://doi.org/10.2147/RMHP.S19801.

Kamenica, E., Naclerio, R., & Malani, A. (2013). Advertisements impact the physiological efficacy of a branded drug. *Proceedings of the National Academy of Sciences of the United States of America*, *110*(32), 12931–12935. https://doi.org/10.1073/pnas.1012818110.

Kesselheim, A. S., Misono, A. S., Lee, J. L., Steadman, M. R., Brookhart, M. A., Choudhry, N. K., et al. (2008). Clinical equivalence of generic and brand-name drugs used in cardiovascular disease: A systematic review and meta-analysis. *The Journal of the American Medical Association*, *300*, 2514–2526. https://doi.org/10.1001/jama.2008.758. Clinical.

Kirsch, I. (1985). Response expectancy as a determinant of experience and behavior. *American Psychologist*, *40*(11), 1189–1202. https://doi.org/10.1037/0003-066X.40.11.1189.

Kjoenniksen, I., Lindbaek, M., & Granas, A. G. (2006). Patient's attitudes towards and experiences of generic drug substitution in Norway. *Pharmacy World & Science*, *28*(5), 284–289. https://doi.org/10.1007/s11096-006-9043-5.

Labroo, A. A., Dhar, R., & Schwarz, N. (2008). Of frog wines and frowning watches: Semantic priming, perceptual fluency, and brand evaluation. *Journal of Consumer Research*, *34*(6), 819–831. https://doi.org/10.1086/523290.

Mahr, A., Golmard, C., Pham, E., Iordache, L., Deville, L., & Faure, P. (2017). Types, frequencies, and burden of nonspecific adverse events of drugs: Analysis of randomized placebo-controlled clinical trials. *Pharmacoepidemiology and Drug Safety*, *26*, 731–741. https://doi.org/10.1002/pds.4169.

Petrie, K., Moss-Morris, R., Grey, C., & Shaw, M. (2004). The relationship of negative affect and perceived sensitivity to symptom reporting following vaccination. *British Journal of Health Psychology*, *9*, 101–111. https://doi.org/10.1348/135910704322778759.

Plassmann, H., O'Doherty, J., Shiv, B., & Rangel, A. (2008). Marketing actions can modulate neural representations of experienced pleasantness. *Proceedings of the National Academy of Sciences of the United States of America*, *105*(3), 1050–1054. https://doi.org/10.1073/pnas.0706929105.

Reeves, R. R., Ladner, M. E., Hart, R. H., & Burke, R. S. (2007). Nocebo effects with antidepressant clinical drug trial placebos. *General Hospital Psychiatry*, *29*(3), 275–277. https://doi.org/10.1016/j.genhosppsych.2007.01.010.

Shedden Mora, M., Nestoriuc, Y., & Rief, W. (2011). Lessons learned from placebo groups in antidepressant trials. *Philosophical Transactions of the Royal Society B*, *366*(1572), 1879–1888. https://doi.org/10.1098/rstb.2010.0394.

Shiv, B., Carmon, Z., & Ariely, D. (2005). Placebo effects of marketing actions: Consumers get what they pay for. *Journal of Marketing Research*, *152*, 383–393.

Song, H., & Schwarz, N. (2009). If it's difficult to pronounce, it must be risky. *Psychological Science*, *20*, 135–138.

Steinman, M. A., Chren, M.-M., & Landefeld, S. (2007). What's in a name? Use of brand versus generic drug names in United States outpatient practice. *Journal of General Internal Medicine*, *22*, 645–648.

Stewart-Williams, S., & Podd, J. (2004). The placebo effect: Dissolving the expectancy versus conditioning debate. *Psychological Bulletin*, *130*, 324–340.

Tinnermann, A., Geuter, S., Sprenger, C., Finsterbusch, J., & Büchel, C. (2017). Interactions between brain and spinal cord mediate value effects in nocebo hyperalgesia. *Science*, *358*(6359), 105–108. https://doi.org/10.1126/science.aan1221.

Tobert, J. A., & Newman, C. B. (2016). The nocebo effect in the context of statin intolerance. *Journal of Clinical Lipidology*, *10*(4), 739–747. https://doi.org/10.1016/j.jacl.2016.05.002.

Tracey, I. (2010). Getting the pain you expect: Mechanisms of placebo, nocebo and reappraisal effects in humans. *Nature Medicine, 16*(11), 1277–1283. https://doi.org/10.1038/nm.2229.

Tuttle, A. H., Tohyama, S., Ramsay, T., Kimmelman, J., Schweinhardt, P., Bennett, G. J., et al. (2015). Increasing placebo responses over time in U.S. clinical trials of neuropathic pain. *Pain, 156*, 2616–2626. https://doi.org/10.1097/j.pain.0000000000000333.

Ventola, C. L. (2011). Direct-to-consumer pharmaceutical advertising: Therapeutic or toxic? *P & T: A Peer-Reviewed Journal for Formulary Management, 36*(10), 669–684. http://doi.org/22346300.

Waber, R. L., Shiv, B., Carmon, Z., & Ariely, D. (2008). Commercial features of placebo and therapeutic efficacy. *The Journal of the American Medical Association, 299*, 1016–1017.

Zajonc, R. (1968). Attitudinal effects of mere exposure. *Journal of Personality and Social Psychology, 9*, 1–27. Monograph supplement. https://doi.org/10.1037/h0025848.

Zajonc, R. (2001). Mere exposure: A gateway to the sublimal. *Current Directions in Psychological Science, 10*(6), 224–228.

CHAPTER SIXTEEN

Placebo and Active Treatment Additivity in Placebo Analgesia: Research to Date and Future Directions

Matthew J. Coleshill[*,†,‡,1], **Louise Sharpe**[*], **Luana Colloca**[§,¶,||], **Robert Zachariae**[#], **Ben Colagiuri**[*]

[*]School of Psychology, University of Sydney, Sydney, NSW, Australia
[†]St Vincent's Clinical School, St Vincent's Hospital, University of New South Wales, Sydney, NSW, Australia
[‡]Department of Clinical Pharmacology and Toxicology, St Vincent's Hospital, Sydney, NSW, Australia
[§]Center to Advance Chronic Pain Research, University of Maryland, Baltimore, MD, United States
[¶]School of Nursing, University of Maryland, Baltimore, MD, United States
[||]School of Medicine, University of Maryland, Baltimore, MD, United States
[#]Aarhus University Hospital and Department of Psychology and Behavioural Science, Aarhus University, Aarhus, Denmark
[1]Corresponding author: e-mail address: m.coleshill@unsw.edu.au

Contents

Abstract

Placebo analgesia is a robust experimental and clinical phenomenon. While our understanding of the mechanisms of placebo analgesia has developed rapidly, some central questions remain unanswered. Among the important questions is how placebo analgesia interacts with active analgesic effects. It is an assumption underlying double-blind randomized placebo-controlled trials (RCTs) that the true effect of a treatment can be determined by examining the effect of the active treatment arm and subtracting the

International Review of Neurobiology, Volume 139
ISSN 0074-7742
https://doi.org/10.1016/bs.irn.2018.07.021

response in the placebo group ("the assumption of additivity"). However, despite the importance of this assumption for the interpretation of RCTs, it has rarely been formally examined. This article reviews the assumption of additivity in placebo analgesia by examining studies employing factorial designs manipulating both the receipt of an active analgesic and instructions about the treatment being delivered. In reviewing the literature, we identified seven studies that allowed a test of additivity. Of these, four found evidence against additivity, while the remaining three studies found results consistent with additivity. While the limited available data are somewhat mixed, the evidence suggests that at least under some conditions the assumption of additivity does not hold in placebo analgesia. The concordance between mechanisms of the active analgesic and placebo analgesia may influence whether additivity occurs or not. However, more research using factorial designs is needed to disentangle the relationship between placebo analgesia and the active effect of analgesic treatments.

1. INTRODUCTION

Double-blind randomized placebo-controlled trials (RCTs) remain the gold standard for testing the efficacy and safety of most medical treatments. The primary aim of an RCT is to isolate any improvement that is directly attributable to the specific agent being tested, above and beyond any improvement that may occur simply as a result of the treatment process itself, such as the placebo effect. The placebo effect is the portion of therapeutic improvement not directly attributable to the active effect of a treatment, instead being a product of contextual and psychobiological components following the receipt of a treatment (Benedetti, 2014). As shown in Fig. 1, RCTs aim to isolate treatment efficacy by comparing the active treatment with a matched placebo (placebo control) while keeping participants and researchers unaware of the participants' treatment allocation (blinding). If the active treatment leads to greater improvement than the placebo, then the active treatment is deemed to be efficacious. If improvement is equivalent in the active treatment and placebo arms, then the active treatment is deemed not be efficacious.

The assumption of additivity is central to RCT methodology. It entails that placebo and active treatment effects add onto each quantitatively and without interaction. In essence, this assumption means that a treatment effect and placebo effect should be entirely independent of each other. While the assumption is convenient for the purposes of RCTs, an increasing number of researchers examining the placebo effect have begun to challenge it on both theoretical and empirical grounds (Colagiuri, 2010;

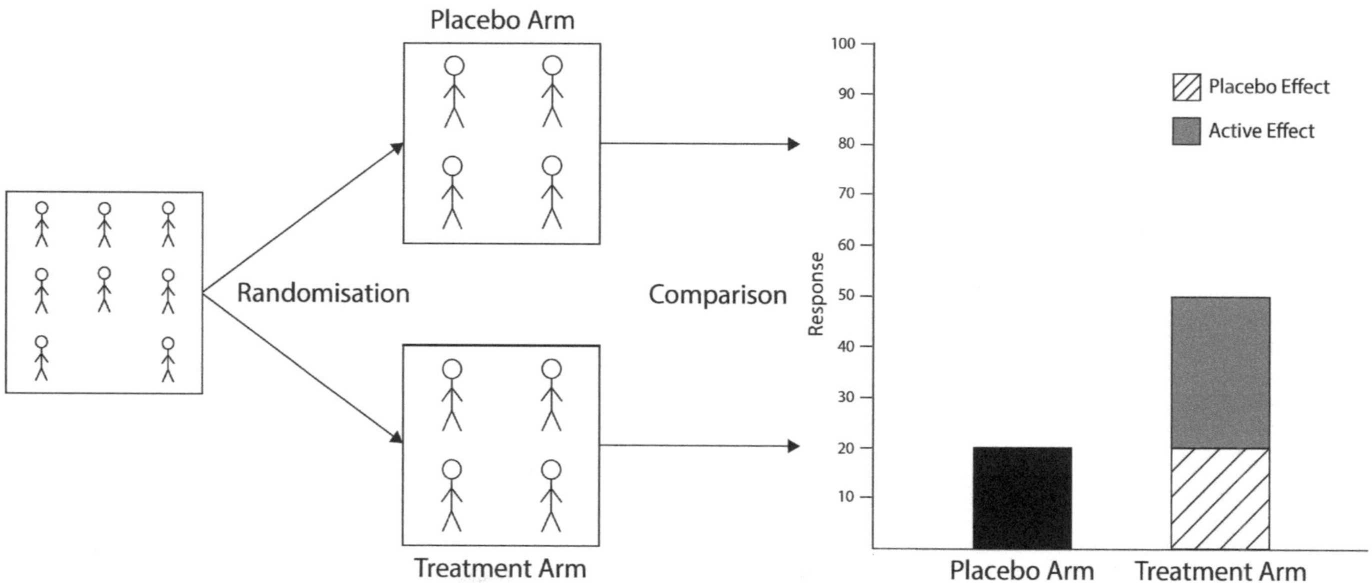

Fig. 1 A graphical depiction of a placebo-controlled RCT.

Enck, Klosterhalfen, Weimer, Horing, & Zipfel, 2011; Enck, Klosterhalfen, & Zipfel, 2011; Kirsch, 2000; Kube & Rief, 2017). If the assumption of additivity does not hold, then the difference between the active treatment and placebo arm will not represent the true efficacy of the active treatment.

1.1 Subadditivity Versus Superadditivity

There are two possible nonadditive outcomes in terms of how placebo and active treatment effects may interact. The first is *subadditivity* (Tallarida, 2001), wherein the combined treatment effect is less than the summed magnitude of the isolated active treatment effect and isolated placebo effect (Fig. 2B). In the case of subadditivity, the relative difference between active treatment and placebo in an RCT may actually underestimate the efficacy of the active treatment (Kirsch, 2000; Kube & Rief, 2017). Hence, subadditivity may result in effective treatments being dismissed as ineffective. The second possibility is *superadditivity* (Tallarida, 2001), wherein the combined treatment effect is larger than the isolated active treatment effect and isolated placebo effect (Fig. 2C). Superadditivity could suggest a synergistic interaction between the active treatment and the placebo effect and may overestimate the efficacy of the active treatment in an RCT (Kube & Rief, 2017). Hence, superadditivity may result in treatments being adopted because they appear more effective than they actually are.

1.2 Experimental Designs to Examine the Assumption of Additivity

For good ethical and practical reasons, RCTs are generally not equipped to test the assumption of additivity. Instead, additivity must be tested by a factorial design that manipulates both the treatment administered and the instructions delivered along with that treatment. Such factorial designs allow for the separation of both active treatment effects and placebo effects and importantly allow for a test of whether the two interact or not.

To date, the most common of such methods is the balanced placebo design, a 2×2 factorial design which manipulates the type of treatment (active versus placebo) and instructions (told active versus told placebo) (Enck, Klosterhalfen, & Zipfel, 2011; Fromme & Patel, 2010). As shown in Fig. 3A, this leads to four unique conditions: (a) receive active treatment and told active (combined active treatment and placebo effects), (b) receive active treatment and told placebo (isolated active treatment effect),

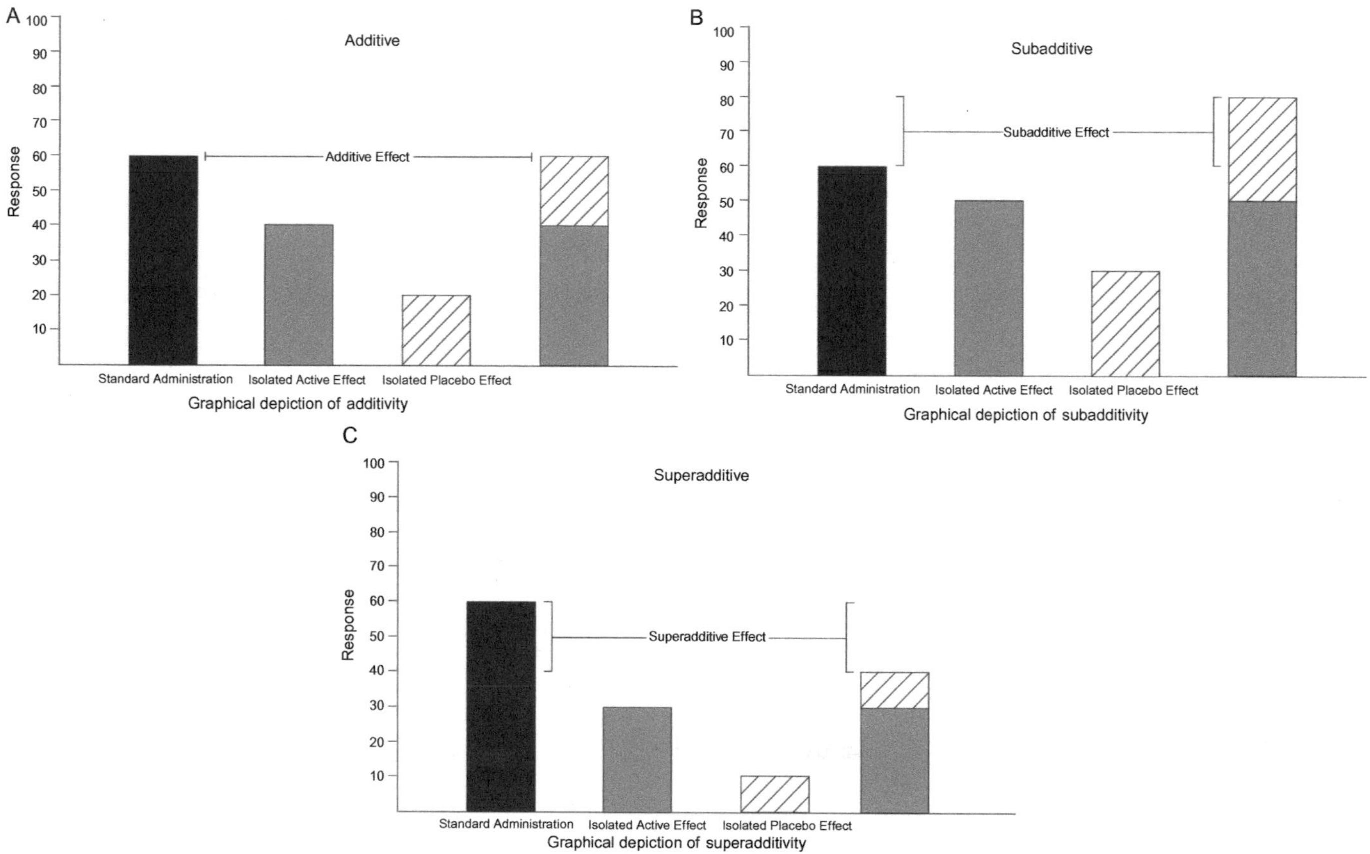

Fig. 2 Graphical representations of (A) additive, (B) subadditive, and (C) superadditive outcomes. In these outcomes, the first column indicates treatment efficacy when administered under normal circumstances (i.e., an active treatment is administered with the full awareness of the recipient). In contrast, the following two columns depict the isolated active treatment and placebo effects and the fourth column their summed total.

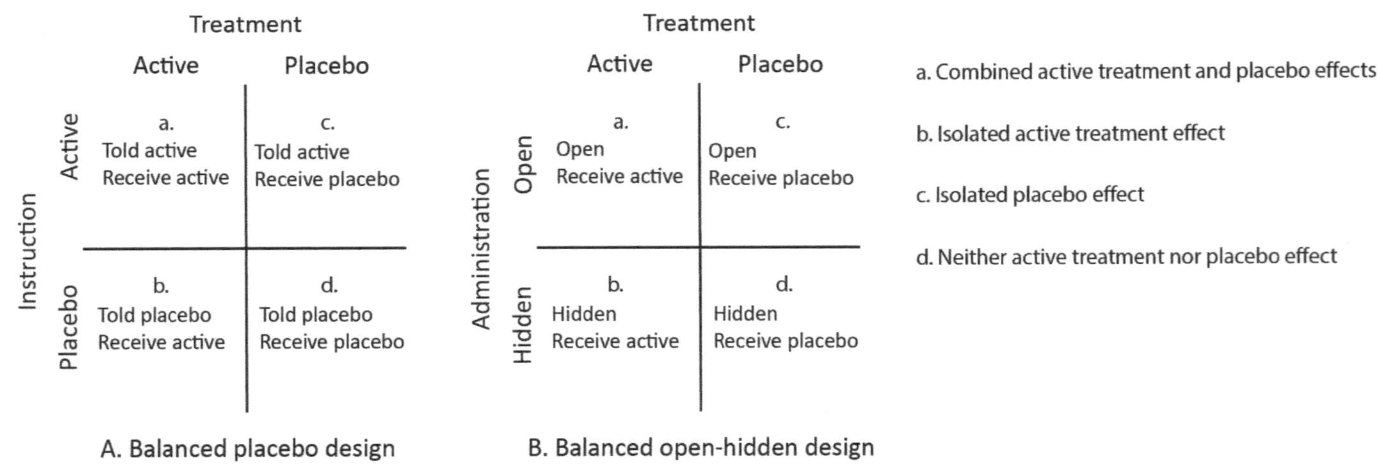

Fig. 3 The four cells of the balanced placebo (A) and open-hidden (B) designs.

(c) receive placebo treatment and told active (isolated placebo effect), and (d) receive placebo treatment and told placebo (neither active treatment nor placebo effect).

Another, more recent, design that has been used to examine placebo and drug effects is the open-hidden design, in which a drug is administered either overtly (open) or covertly (hidden, e.g., via automated intravenous infusion) in order to study its effect both with and without expectancy (Benedetti, Maggi, et al., 2003; Benedetti, Pollo, et al., 2003; Colloca, Lopiano, Lanotte, & Benedetti, 2004). A variant of this design, known as the balanced open-hidden design, compares active treatment and placebo treatment under both open and hidden administration (Atlas et al., 2012). Akin to the balanced placebo design and as shown in Fig. 3B, this paradigm has four possible conditions which allow for the interaction between active treatment and placebo effects to be examined: (a) open administration of active treatment (combined active treatment and placebo effect), (b) hidden administration of active treatment (isolated active treatment effect), (c) open administration of placebo (isolated placebo effect), and (d) hidden administration of placebo (neither active treatment nor placebo effect).

In both designs, a lack of interaction between treatment and instruction would indicate support for the assumption of additivity. Conversely, the presence of an interaction would suggest that the magnitude of the active treatment effect differs depending on the instructions accompanying it, and hence that the assumption of additivity is not supported.

1.3 Analgesia as a Model for Understanding Additivity

There are substantial differences between placebo effects in different modalities both in terms of underlying psychological and neurobiological mechanisms (Benedetti, 2008, 2009; Benedetti, Maggi, et al., 2003; Benedetti, Pollo, et al., 2003; Colagiuri, Schenk, Kessler, Dorsey, & Colloca, 2015; Colloca & Benedetti, 2005; Enck, Benedetti, & Schedlowski, 2008), as well as the degree of placebo effect itself (Hróbjartsson & Gøtzsche, 2010). Therefore, it is important to consider the question of additivity in different modalities separately because examining potential additive effects across a range of outcomes may introduce noise and conceal crucial modality-specific effects. Analgesia is a convenient and important condition to investigate additivity, not only because of the relatively well understood mechanisms of both pain and its exogenous modulation by analgesics, but also because placebo analgesia has been the most commonly studied type of placebo effect.

Numerous studies indicate that placebo treatment can ameliorate both experimental (Amanzio, Benedetti, Porro, Palermo, & Cauda, 2013) and clinical pain (Amanzio, Pollo, Maggi, & Benedetti, 2001; Benedetti, Amanzio, & Maggi, 1995; Pollo et al., 2001), with the placebo effect also comprising part of the therapeutic effect of analgesics (Benedetti, Maggi, et al., 2003; Benedetti, Pollo, et al., 2003; Bingel et al., 2011). Recent models propose that placebo analgesia is the result of expectancies induced via the treatment context, such as the verbal, social, and contextual cues surrounding the treatment (Colagiuri et al., 2015; Colloca & Benedetti, 2006, 2009; Stewart-Williams & Podd, 2004). Advances in psychopharmacology and neurobiology indicate that these expectancies trigger changes in the central nervous system that mediate placebo analgesia. Specifically, expectancies are thought to generate the placebo effect via the prefrontal cortices (PFC) (Amanzio et al., 2013; Petrovic et al., 2010). These higher order areas then engage a broad network of regions, notably the descending pain modulatory system which encompasses regions such as the rostral anterior cingulate cortex (ACC), amygdala, hypothalamus, and the periaqueductal grey (Bingel et al., 2011; Eippert et al., 2009; Tracey, 2010). Found in abundance throughout these regions are μ-opioid receptors (Baumgärtner et al., 2006; Wager, Scott, & Zubieta, 2007; Zubieta et al., 2005), which play a crucial role in inhibiting pain perception during placebo analgesia by influencing nociceptive processing at both the subcortical (Wager et al., 2004) and spinal levels in the rostral ventromedial medulla (RVM) (Basbaum & Fields, 1984; Faria, Fredrikson, & Furmark, 2008; Petrovic, Kalso, Petersson, & Ingvar, 2002). The central role of opioid transmission in placebo analgesia is further evidenced by the fact that the competitive opioid antagonist naloxone has been found to attenuate placebo analgesia (Benedetti, 1996; Grevert, Albert, & Goldstein, 1983; Levine, Gordon, & Fields, 1978).

Intriguingly, pharmacological conditioning with the nonsteroidal anti-inflammatory drug (NSAID) ketorolac produces analgesia that is only partially naloxone reversible (Amanzio & Benedetti, 1999). Instead, a placebo effect pre-conditioned in this manner has been found to be attenuated by the CB1 cannabinoid receptor antagonist rimonabant (Benedetti, Amanzio, Rosato, & Blanchard, 2011), indicating the involvement of the endocannabinoid system. This suggests that pharmacological conditioning procedures can alter the recruitment of different pain inhibitory systems during placebo analgesia.

1.4 Mechanisms of Common Analgesics

Pain is a highly heterogeneous condition, and many different types of analgesics exist. Summarizing all the various analgesics and their mechanisms of action is beyond the scope of this article; however, we briefly discuss some of the mechanistic differences here in order to highlight that additivity could be influenced by the classification of analgesic being investigated.

In the case of the centralized effect of opioid analgesics, the binding of opioid peptides to μ-opioid receptors throughout pain-processing and modulatory regions (Atlas et al., 2012; Bingel et al., 2011) acts upon pain perception both pre-synaptically and post-synaptically (Koneru, Satyanarayana, & Rizwan, 2009; Ossipov, Dussor, & Porreca, 2010; Snyder & Pasternak, 2003; Vaughan & Christie, 1997). Pre-synaptically this binding decreases the release of excitatory neurotransmitters from nociceptive neurons through hyperpolarization of the cell, leading to a reduction in evoked responses and spontaneous action potential discharge. Post-synaptically this cell hyperpolarization reduces the neurons responsiveness to excitatory input. Furthermore, administration of μ-opioid agonists activates OFF cells in the RVM (Fields, Vanegas, Hentall, & Zorman, 1983; Heinricher, Morgan, Tortorici, & Fields, 1994), while also inhibiting ON cell activity (Heinricher, Morgan, & Fields, 1992) leading to inhibition of ascending nociceptive afferents.

In contrast, other widely used analgesics have entirely distinct mechanisms of action, such as the inhibition of cyclooxygenase enzymes in NSAIDs (Seibert et al., 1994; Vane & Botting, 1996). The inhibition of cyclooxygenase enzymes suppresses prostaglandin formation, with prostaglandins contributing to inflammation through vasodilation (Green, 2001). In addition to this, prostaglandins also have pro-nociceptive effects through the sensitization of peripheral nociceptors, thus reducing the threshold to respond to excitatory input (Ito, Okuda-Ashitaka, & Minami, 2001; Omote et al., 2002). Prostaglandins have also been suggested to have some centralized effect by contributing to central sensitization and the synaptic excitability of spinal nociceptors (Minami et al., 2001; Seybold, 2009; Seybold, Jia, & Abrahams, 2003).

Another example of analgesics are sodium channel blockers such as lidocaine, which have entirely peripheral effects. These drugs modulate the voltage-gated sodium channels on the neuronal membrane, impairing the flow of sodium ions into the neuron during depolarization to a stimulus and preventing conduction or initiation of afferent nociceptive signals (Clare, Tate, Nobbs, & Romanos, 2000; Devor, 2006).

1.5 Potential Moderators of Additivity in Analgesia

The concordance between placebo analgesia and the mechanism of an analgesic is a plausible moderator of their additivity. For example, opioid treatments with similar opioid receptor affinities are generally observed to have additive outcomes (Taber, Greenhouse, Rendell, & Irwin, 1969; Tallarida, 2000), although these different opioids may still compete due to differences in receptor binding efficacy (Chen, Smith, Cahill, Cohen, & Fishman, 1993; Trescot, Datta, Lee, & Hansen, 2008). As such, endogenous opioid activity induced by placebo treatment might be either additive or nonadditive depending on the specific opioid receptor affinity of the opioid treatment. In the case of any competition, then one could expect subadditivity as endogenous and exogenous ligands vie to bind with a finite number of receptor sites.

The active effect of analgesics that act on different systems, such as NSAIDs and sodium channel blockers, may not compete with the placebo effect for opioid receptor binding and, as such, may allow for the endogenous and exogenous effects to both maximally contribute to analgesia. While in some cases this may lead to ceiling effects, and, as such, subadditivity, the use of combination analgesics, wherein a number of analgesics with distinct mechanisms of action are administered together, is common in medicine and is often found to have beneficial effects over the individual treatments (Desmeules, Rollason, Piguet, & Dayer, 2003; Raffa, 2001). Indeed, synergistic effects are frequently observed between different analgesics and may provide some evidence for superadditivity (Christie, Connor, Vaughan, Ingram, & Bagley, 2000; Christie, Vaughan, & Ingram, 1999; Cichewicz, 2004; Miranda, Silva, & Pinardi, 2004; Tallarida, 2001; Vaughan, Ingram, Connor, & Christie, 1997), in which the distinct mechanisms by which the treatments have their effect interact to produce a greater degree of analgesia when administered together than when administered individually. Such evidence adds to the notion that additivity between active and placebo effects may vary based on the congruence or incongruence of their mechanisms.

Beyond the mechanistic similarity between the active treatment and placebo effect in pain, a number of other factors may also influence whether the assumption of additivity holds in particular studies. These may include the specific dosage of the analgesic, with larger dosages increasing the likelihood that ceiling effects may contribute to a nonadditive relationship. Another factor is the time between administration and outcome measurement, as both active treatment and placebo effects contribute to analgesia at different time points (Atlas et al., 2012; Gabka & Price, 1982). Such differences in methods between studies may lead to variability in their findings regarding additivity and should be considered when interpreting the results of a study.

2. CURRENT EVIDENCE

2.1 Available Studies Testing Additivity in Placebo Analgesia

As mentioned above, despite the fact that there have been many studies exploring placebo analgesia, few studies have employed designs that allow for formal testing of additivity. To survey the literature, we conducted a systematic review (see Box 1 for search terms and selection criteria) to identify all balanced placebo designs and all balanced open-hidden designs in placebo analgesia, so that we could review the evidence for additivity. Fig. 4 shows the results of this search, in which only seven studies were found to have used one of these designs in placebo analgesia (Aslaksen, Zwarg, Eilertsen, Gorecka, & Bjørkedal, 2015; Atlas et al., 2012; Berna et al., 2017; Butcher & Carmody, 2012; Kam-Hansen et al., 2014; Lund, Vase, Petersen, Jensen, & Finnerup, 2014; Schenk, Sprenger, Geuter, & Büchel, 2014).

2.1.1 Study Characteristics

As can be seen from the study overview provided in Table 1, the seven identified studies were considerably heterogeneous with respect to the methods employed. Four of the seven studies used the balanced placebo design (Aslaksen et al., 2015; Butcher & Carmody, 2012; Kam-Hansen et al., 2014; Schenk et al., 2014), while two used the balanced open-hidden design (Atlas et al., 2012; Lund et al., 2014). The final study conducted by Berna et al. (2017) used a novel design here described as a "balanced active placebo design." In this variation of the balanced placebo design, receipt of an active treatment or placebo are factorially crossed with administration of an active placebo or placebo, with the second factor replacing the instructions of treatment allocation typical of the balanced placebo design.

The analgesics employed in these studies also varied substantially. Three studies examined lidocaine, a sodium channel blocker (Aslaksen et al., 2015; Lund et al., 2014; Schenk et al., 2014), two examined NSAIDs, namely diclofenac and ibuprofen (Berna et al., 2017; Butcher & Carmody, 2012), one examined maxalt, a triptan used to treat migraines (Kam-Hansen et al., 2014), and one examined the opioid remifentanil (Atlas et al., 2012).

BOX 1 Systematic Review Methods

Selection criteria

The principle selection criteria were the experimental design employed and the modality being investigated. Studies were required to manipulate receipt of an analgesic treatment, as indicated by the treatment itself and the outcomes, against at least two levels of expectations. These factors could be manipulated both within and between-subjects.

Search strategy

The protocol driven search strategy focused on the electronic search of four online databases, searching each from inception for articles in English:

- EMBASE (1947 to December 2016)
- PsychINFO (1887 to December 2016)
- PUBMED (1966 to December 2016)
- Web of Science (1864 to December 2016)

Registers were searched for "balanced placebo design," with a specific search strategy of 'balanced" AND "placebo" AND "design," "open-hidden design" ("open" AND "hidden" AND "design"), "open hidden administration" ("open" AND "hidden" AND "administration"), "open hidden paradigm" ("open" AND "hidden" AND "paradigm"). The latter three search terms were also performed using hyphenated "open-hidden" variations. Preliminary searches indicated that synonyms and more general criteria for the balanced placebo design failed to contribute any additional studies employing this specific experimental design. Forward and backward search strategies were also employed following the initial screening of studies retrieved by the protocol driven search. Further, this process was also completed with any review papers retrieved by the protocol driven search. Members of the Society for Interdisciplinary Placebo Studies were also contacted requesting citations for studies that met the inclusion criteria.

The time course of the treatment relative to outcome assessment was reported in most studies and varied between 13.5 min and 4 h (Aslaksen et al., 2015; Atlas et al., 2012; Berna et al., 2017; Butcher & Carmody, 2012; Kam-Hansen et al., 2014). The two exceptions were Lund et al. (2014) and Schenk et al. (2014), who did not report the time course of treatment administration and outcome assessment.

All but one of the studies examined experimentally-induced pain in healthy volunteers, with the exception being Kam-Hansen et al. (2014) who examined migraine pain in a clinical population. The selected studies also all measured pain intensity/unpleasantness through self-report, with one study also including pain threshold and tolerance (Butcher & Carmody, 2012).

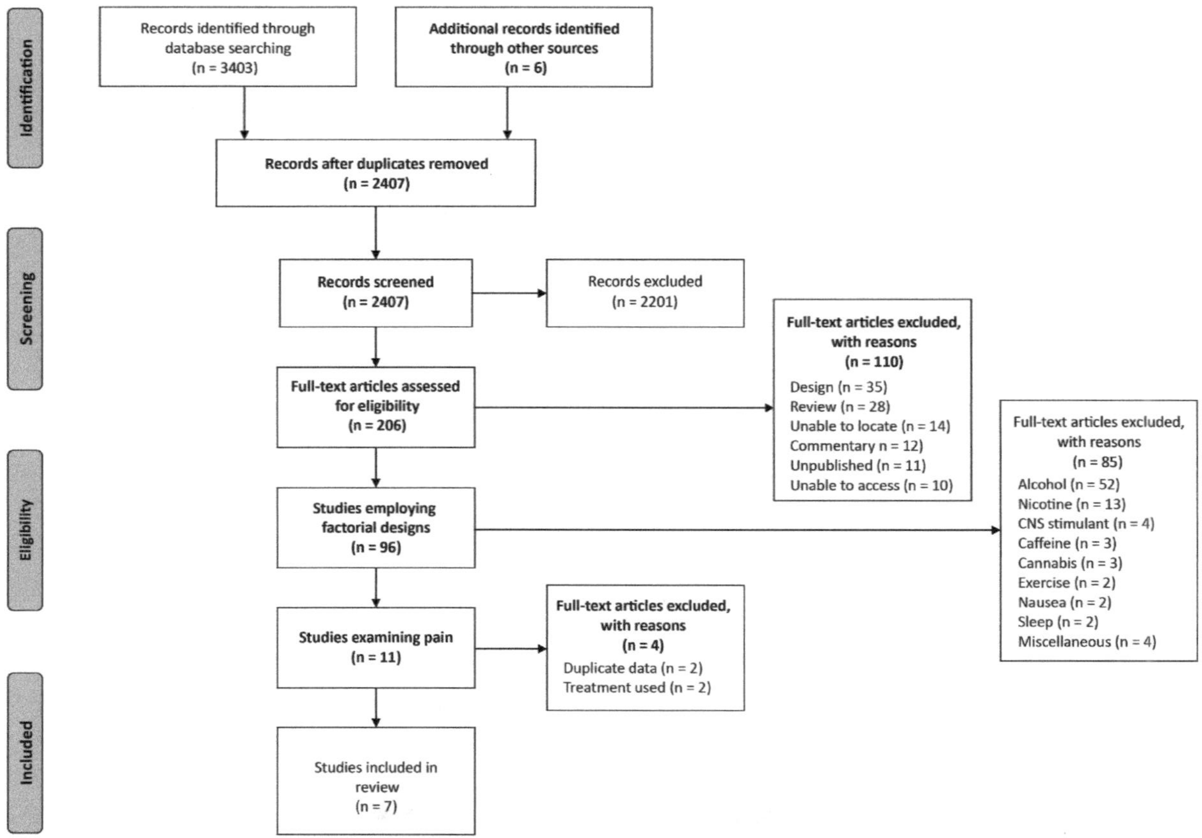

Fig. 4 Flow chart depicting the study selection process.

Table 1 Study Overview

Study	Sample	Modality	Treatment	Dosage	Administration	Classification	Design	Blinding	Measure	Stimulus
Aslaksen et al. (2015)	93	Pain	Lidocaine	3.0 g	Topical	SCB[a]	BPD[b]	–	Self-report	Thermode
Atlas et al. (2012)	14	Pain	Remifentanil	0.04 μg/kg/min	Intravenous	Opioid	BOHD[c]	–	Self-report	Thermode
Berna et al. (2017)	100	Pain	Diclofenac	100 mg	Oral (pill)	NSAID[d]	BAPD[e]	+	Self-report	Thermode
Butcher and Carmody (2012)	20	Pain	Ibuprofen	800 mg	Oral (pill)	NSAID	BPD	+	Multiple outcomes	Electrical
Kam-Hansen et al. (2014)	66	Migraine	Maxalt	10 mg	Oral (pill)	Triptan	BPD	NR[f]	Self-report	Migraine
Lund et al. (2014)	46	Pain	Lidocaine	0.1 ml	Intramuscular	SCB	BOHD	+	Self-report	Hypertonic saline
Schenk et al. (2014)	32	Pain	Lidocaine/prilocaine	NR	Topical	SCB	BPD	+	Self-report	Thermode

[a]Sodium channel blocker.
[b]Balanced placebo design.
[c]Balanced open-hidden design
[d]Nonsteroidal anti-inflammatory drug.
[e]Balanced active placebo design.
[f]Not reported.

2.1.2 Studies in Experimental Pain With Behavioral Outcomes

Butcher and Carmody (2012) employed a within-subjects balanced placebo design with the NSAID ibuprofen for experimental pain. Participants were either told they were receiving an active analgesic or placebo and then received either 800 mg ibuprofen or placebo orally. The study blinded researchers to both the participants' treatment allocation and instructions and measured pain using self-report, pain threshold, and pain tolerance across four time points (1, 2, 3, and 4 h post-administration). In the originally reported study, Butcher and Carmody (2012) also analyzed male and female participants separately. Given that there is no a priori reason for performing these analyses separately, only a reanalysis performed on the entire sample is reported here. Despite using sound methods and a design with the potential to examine additivity across the treatment's time course, this reanalysis failed to observe any main effects of treatment and instruction, or any interaction between the two across all pain outcomes.

Given the absence of an interaction, this study appears to provide evidence for additivity. However, the study also seems to find no effect of either ibuprofen or instructions on pain at all. More specifically, neither the active treatment nor instructions demonstrated any effect whatsoever on pain. Such results may be limited in terms of providing evidence for additivity. This is because the absence of any active treatment or placebo effects across conditions may have meant that the factorial manipulations were ineffective at inducing different levels of NSAID activity or expectancies, leaving no possibility for these different levels to interact. Therefore, the results of this particular study are only weakly in favor of additivity.

Lund et al. (2014) used a within-subjects balanced open-hidden design with intramuscular administration of lidocaine. The nature of the pain model, intramuscular injections of hypertonic saline, allowed for both open and hidden administration of lidocaine alongside the pain stimulus itself. Participants received either 0.1 mL lidocaine or a placebo intramuscularly either with or without information about the treatment being administered. The researcher administering the treatment was blinded to treatment allocation, but not instruction allocation, and self-reported pain was assessed. Further, a subset of the sample underwent pre-conditioning with lidocaine in order to maximize the likely placebo effect, however, no significant effect of pre-conditioning was reported.

The study observed both significant main effects of treatment and instruction, as well as an interaction between the two. In this case, post hoc simple effects analysis suggested that the interaction was indicative of a subadditive outcome. That is, the open administration of lidocaine was associated with less analgesic effect than the sum of the open administration of placebo and the hidden administration of lidocaine. These results, in the context of an RCT, would have underestimated the efficacy of lidocaine. Consistent with the strong analgesic effect of intramuscular lidocaine during both open and hidden administration, it may be that the more efficacious the medication, the more likely that ceiling effects will be observed, with these ceiling effects then contributing to a subadditive outcome.

2.1.3 Studies in Experimental Pain With Both Behavioral and Neuroimaging Outcomes

Only two of the studies found in our systematic search employed neuroimaging methods, both of which used functional magnetic resonance imaging (fMRI) to examine analgesic effects (Atlas et al., 2012; Schenk et al., 2014).

Atlas et al. (2012) used a within-subjects balanced open-hidden design to examine the opioid agonist remifentanil. Participants received an intravenous infusion of remifentanil or placebo, administered either with or without information that the treatment was being provided. The study used target-controlled infusion to maintain a stable concentration of 0.04 μg/kg/min, with 13.5-min blocks from infusion onset and a 7-min washout period between conditions. The concentration was modeled for each participant to stay below the threshold for sedation to avoid contamination during hidden administration. Researchers were not blinded to treatment allocation, and self-reported pain was assessed across a range of calibrated stimulus intensities.

In terms of the behavioral outcomes, there was a main effect of both treatment and instructions, with pain ratings being lower following remifentanil compared with placebo and instructions that an analgesic was being administered also leading to lower pain ratings. In terms of additivity, the study observed no interaction between treatment and instructions, suggesting evidence in favor of additivity.

The fMRI data were subjected to pharmacokinetic modeling to estimate the relative contribution of active treatment and placebo effects and their potential interaction. As with the behavioral data, the neuroimaging data did not reveal any results indicative of an interaction between the active effect of remifentanil and placebo analgesia, instead finding several potential dissociations in site of action. Specifically, remifentanil produced widespread reductions in activity across pain-processing regions regardless of the presence of instructions. In contrast, knowledge of treatment administration was more closely related to activity in the PFC—one of the regions that is most commonly associated with placebo analgesia (Krummenacher, Candia, Folkers, Schedlowski, & Schönbächler, 2010; Petrovic et al., 2010). Further, the time course of neural activity also varied between the active effect of remifentanil and instruction effects. With the latter being observable from the onset of infusion prior to peak remifentanil concentration. Based on these findings, Atlas et al. (2012) concluded that the active effect of remifentanil and the placebo effect produce additive and dissociable effects in the brain.

Schenk et al. (2014) employed a within-subjects balanced placebo design examining a topical formulation of lidocaine and prilocaine. A clinical model of pain was imitated by pre-treating the treatment sites with capsaicin to cause sensitization to noxious thermal stimuli. Participants received either an unreported dose of a topical lidocaine and prilocaine formulation or placebo to four areas in a double-blind manner. Participants were then instructed that some sites contained active treatment and others placebo by computer-presented instructions within the fMRI scanner before stimulation of each site. Pain outcomes were assessed using self-report. The time course between administration and outcome assessment was not reported. The study found a main effect of treatment, but no statistically significant effect of instruction. However, there was a significant interaction effect between treatment and instruction, indicative of superadditivity. Planned simple effects found that a difference was observed between the administration of the formulation with and without information that it was the active treatment, while no difference was present between the placebo administered with and without information that it was an active treatment. Thus, Schenk et al. observed that the placebo effect only influenced pain outcomes when coupled with the active treatment. As such, this study's data are inconsistent with the assumption of additivity.

In addition to interactions on the behavioral pain outcomes, Schenk et al.'s (2014) factorial design also revealed a number of patterns in the neuroimaging data. Specifically, greater neural activations were observed in the ACC and ventral striatum during the conditions that isolate the active treatment and placebo effects. Intriguingly, these effects observed in the ACC and ventral striatum appeared to be a product of the incongruence between treatment and expectations of treatment, rather than a synergistic effect. Both of these regions have also been implicated in prediction error processing (Carter et al., 1998; Hare, O'Doherty, Camerer, Schultz, & Rangel, 2008), a mechanism which has been previously suggested to play a role in the instigation of placebo analgesia (Petrovic et al., 2010; Scott et al., 2007). As such, the interactions observed by Schenk et al. (2014) appear to reflect other aspects of the balanced placebo design, that is, the incongruence between the expected and received treatment, rather than a synergistic effect between treatment and instructions on patterns of neural activity.

2.1.4 Studies With Modified Factorial Designs

Aslaksen et al. (2015) used a between-subjects variant of the balanced placebo design in which the instruction that a placebo was to be administered was replaced with information about the treatment having a hyperalgesic effect. In this manner, the study examined the effect of positive and negative instructions on the active effect of lidocaine versus placebo in a factorial design. While this study differs from other studies reviewed, its factorial design still allows the assumption of additivity to be tested. Aslaksen et al. (2015) also included two further conditions, one in which lidocaine was administered alongside no information of its specific medical properties and one where no treatment was provided. The results reported here are a reanalysis of the data including only the factorial conditions manipulating receipt of treatment (lidocaine, placebo) against instructions (analgesia, hyperalgesia).

Participants were given either 3.0 g of lidocaine or placebo topically with instructions of the treatment having either an analgesic or hyperalgesic effect. Self-reported pain outcomes were assessed 18 min after treatment administration, with researchers not being blinded to allocation. The study observed a main effect of instructions, such that those who were told to expect a hyperalgesic effect reported more pain compared to those who were told to expect an analgesic effect. There was no main effect of lidocaine on pain outcomes. However, there was an interaction between treatment and instructions, with inspection of the data suggesting that lidocaine led to reports of more (rather than less) pain in participants who were led to believe it had hyperalgesic properties.

The presence of an interaction effect indicates that the assumption of additivity in this study was not met. However, this study does not reflect a synergistic effect per se, so much as an attenuating influence of the nocebo effect on the active effect of lidocaine. The nocebo effect is the harmful counterpart to the placebo effect, where negative expectations (e.g., increased pain) lead to adverse events or a worsening of symptoms (Colloca & Miller, 2011; Enck et al., 2008; Häuser, Hansen, & Enck, 2012). The study suggests that instructions can significantly diminish the active effects of a medication such to render it ineffective. Such a finding contradicts the implicit assumptions of the additive model that expectations cannot interact with the active effect of a treatment (Kube & Rief, 2017).

The balanced active placebo design employed by Berna et al. (2017) differs substantially from the typical balanced placebo design. Their between-subjects 2 × 2 factorial design manipulated receipt of analgesic (diclofenac versus placebo) and side effect induction (atropine versus placebo). In this design, the side effects experienced as a result of the atropine were assumed to lead participants to believe that they had received the active analgesic, and as such were argued to be analogous to providing an instruction about treatment allocation in the balanced placebo design. Participants were told they would receive either active treatment or placebo and then received either 100 mg diclofenac or placebo alongside either 1.2 mg atropine or placebo in a double-blind manner. Self-reported pain was assessed 1 h after administration.

In this design, additivity may be tested by examining whether the difference in outcomes between the active treatment being administered with and without the active placebo is proportional to the difference between a placebo treatment being administered with and without the active placebo. If no difference is observed, then the increase in the placebo effect brought about by the active placebo combines with the active effect in an additive manner. Conversely, if a difference is observed then there is evidence that the increased placebo effect combines with the active effect in a nonadditive manner. Their results indicated that there were no main effects of analgesic treatment nor were there main effects of the atropine versus no atropine on pain, there was, however, an interaction between the two. By comparing the relative effect sizes of adding atropine either to diclofenac or to placebo, Berna et al. (2017) concluded that diclofenac was made substantially more effective by the addition of atropine, whereas the addition of atropine to the placebo did not increase the placebo effect. As such, this study provides evidence for superadditivity. Further mediation analysis suggested that the analgesic effect of diclofenac was mediated by subjective reports of xerostomia, and the associated beliefs about treatment assignment. That is, diclofenac only had an analgesic effect when participants experienced the side effects associated with atropine and that led them to have an expectation that they had been allocated to the active treatment condition. However, the study also failed to observe any difference in the degree of placebo effect between based on the presence of atropine. Such that, the presence of an analgesic agent was necessary for the side effect induction to induce greater placebo analgesia. As such, their results cast doubt on the assumption of additivity.

2.1.5 Studies With Behavioral Outcomes in Clinical Pain

Kam-Hansen et al. (2014) conducted the only study we could identify in a clinical model of pain. The researchers used the balanced placebo design to examine the effect of maxalt, a serotonin agonist, and the placebo effect on migraine-related pain. Using a within-subjects design, participants were given six labeled treatments which contained either 10 mg of maxalt or placebo to be taken over the course of their next six migraine episodes as a rescue medication. Four of these treatments were labeled either as maxalt or placebo. The remaining two were labeled in a double-blind manner, indicating to participants that it may be either maxalt or placebo, that is, participants were told that the medication could be either active or a placebo (i.e., RCT instructions). The order of these treatments was randomized between participants and blinding was not reported as the participant administered their own treatment and recorded their own outcome. Participants also recorded one untreated episode at the beginning of the study. The primary outcome was change in self-reported pain between 30 min after migraine onset and 2.5 h after onset. The study observed strong evidence for the assumption of additivity, finding main effects of both treatment and instruction, but no interaction between the two factors. The data suggested a robust reduction in migraine pain brought about by both the isolated effect of maxalt and the isolated the placebo effect, which additively combined when maxalt was administered alongside instructions that it was the active treatment.

2.1.6 Summary

Broadly speaking, behavioral findings regarding additivity were mixed (see Table 2 for a summary). Three studies found evidence that confirmed the assumption of additivity between active treatment and placebo effects (Atlas et al., 2012; Butcher & Carmody, 2012; Kam-Hansen et al., 2014), whereas the remaining four studies found evidence against the additivity assumption in the form of statistically significant treatment by expectancy interactions (Aslaksen et al., 2015; Berna et al., 2017; Lund et al., 2014; Schenk et al., 2014). However, the results suggested that where additivity was not supported, the direction of the effect was not consistent. That is, three of the four studies reported superadditive effects (Aslaksen et al., 2015; Berna et al., 2017; Schenk et al., 2014) and the remaining study found

Table 2 The Main Effects, Interactions, and Additive Outcomes of the Seven Selected Studies

Study	Treatment	Treatment Effect	Instruction Effect	Interaction	Additivity
Aslaksen et al. (2015)	Lidocaine	✖	✓	✓	Superadditive
Atlas et al. (2012)	Remifentanil	✓	✓	✖	Additive
Berna et al. (2017)	Diclofenac	✖	✖	✓	Superadditive
Butcher and Carmody (2012)	Ibuprofen	✖	✖	✖	Additive
Kam-Hansen et al. (2014)	Maxalt	✓	✓	✖	Additive
Lund et al. (2014)	Lidocaine	✓	✓	✓	Subadditive
Schenk et al. (2014)	Lidocaine/ prilocaine	✓	✖	✓	Superadditive

A "✓" indicates a statistically significant effect or interaction ($P < 0.05$), while a "✖" indicates an effect or interaction that failed to reach statistical significance ($P > 0.05$).

evidence of subadditivity (Lund et al., 2014). Similarly, the neuroimaging outcomes regarding additivity were also mixed. Here, one study observed additive patterns of activations between analgesic treatment and placebo effects, consistent with their behavioral findings (Atlas et al., 2012). The second study, in contrast, found nonadditive differences in neural activity between conditions representative of error processing from the incongruence between the expected and received treatment (Schenk et al., 2014). As such, while the neuroimaging outcomes reported by Schenk et al. (2014) do not appear to reflect the synergistic effect observed in their behavioral data, they still suggest that active treatment and placebo effects may nonadditively influence the neural signature of analgesia.

3. IMPLICATIONS AND LIMITATIONS

Overall, the studies considered in this review suggest that additivity between analgesic and placebo effects does not always hold in pain. To this end, one important consideration is whether additivity between placebo analgesia and an analgesic treatment varies based on the pharmacology of the treatment in question. As previously discussed, research suggests that endogenous opioids are one of the primary systems by which the placebo effect inhibits pain (Levine et al., 1978; Wager et al., 2007; Zubieta et al., 2005), the same system on which opioid analgesics have their exogenous effect. In the studies we reviewed, the one study using exogenous opioids provided strong evidence for additivity with both behavioral and neuroimaging outcomes (Atlas et al., 2012). This could suggest that exogenous and endogenous opioids may not compete with one another for μ-opioid receptor binding in such a manner as to affect their additive relationship. This finding is consistent with the pharmacology literature, where it is noted that active treatments with similar receptor agonisms tend to have additive outcomes (Taber et al., 1969; Tallarida, 2000). Despite this, given that this was the only study to have examined opioid analgesics, additional research is warranted to confirm that exogenous and endogenous opioids have additive outcomes. Further, as previously discussed, non-opioid mechanisms have been observed for placebo analgesia following conditioning with NSAIDs (Amanzio & Benedetti, 1999; Benedetti et al., 2011). While placebo analgesia still appears to be primarily mediated by endogenous opioids in the absence of pharmaceutical conditioning (Amanzio & Benedetti, 1999), the finding that non-opioid mechanisms can be engaged following pharmacological conditioning may have clinical relevance to additivity, where treatment history may influence the recruited mechanism of placebo analgesia.

In contrast, the results from studies using non-opioid analgesics largely failed to support additivity. Of the three studies examining the sodium channel blocker lidocaine, all three reported results that contradicted the assumption of additivity. One study using intravenous administration observed subadditivity (Lund et al., 2014), whereas the remaining two studies applying topical formulations found evidence for superadditivity (Aslaksen et al., 2015; Schenk et al., 2014). It is important to note that the superadditive outcome observed by Aslaksen et al. (2015) did not reflect a synergistic effect of the treatment and placebo effect, so much as an attenuation of lidocaine's analgesic effect through instructions of hyperalgesia. Regarding the remaining two studies examining lidocaine, Lund et al. (2014) found strong evidence for subadditive outcomes between the active effect of lidocaine and placebo analgesia. Conversely, Schenk et al. (2014) observed superadditive outcomes between the active effect of lidocaine and the placebo effect. One explanation for the differences between these two studies may relate to the different formulations and methods of treatment administration and their relative strength, with Lund et al. (2014) administering lidocaine intravenously and Schenk et al. (2014) a formulation of lidocaine and prilocaine topically. The potent active effect of lidocaine observed across conditions by Lund et al. may have contributed to subadditive outcomes overall. Regardless of the nature of the additive relationship, both of these studies still suggest that the additive model is not supported in the case of lidocaine. Indeed, it is potentially more problematic for the interpretation of RCTs if the assumption of additivity is violated under some conditions, and not others, and in different directions.

Of the two studies that examined NSAIDs, Berna et al. (2017) observed evidence for superadditivity, while Butcher and Carmody (2012) observed some evidence for additivity. Both studies failed to observe any effect of treatment or instruction though, which may be especially problematic when considering the data from Butcher and Carmody (2012) who used a traditional balanced placebo design. The absence of any factorial effects in a traditional balanced placebo design suggests completely ineffective active treatment and instructional manipulations because it means that all outcomes were equivalent to participants being told they were given no treatment and receiving no treatment (cell d in Fig. 3A), which is effectively a natural history group. The absence of any effects above natural history would render a lack of interaction uninformative because there are simply no effects to interact. In contrast, Berna et al.'s (2017) use of a balanced active placebo is distinct because it does not contain a natural history group. The equivalent cell in their design is placebo treatment given under double-blind instructions, which has been shown to induce placebo effects (Colagiuri & Boakes, 2010; Kam-Hansen et al., 2014; Pollo et al., 2001). Their design means that an absence of a main effect of the analgesic does not necessarily mean that the treatment had no "absolute" effect (i.e., relative to natural history). Instead, it only means that the analgesic was not more effective than the placebo effect under double-blind instructions. Similarly, the absence of a main effect of side effect manipulation simply means that the active placebo failed to increase the placebo effect above and beyond benign placebo treatment. As such, both analgesic treatment and side effect/instruction effects could still feasibly present in this design despite the lack of main effects, allowing for their interaction. In light of this, Berna et al.'s (2017) study is likely more informative in terms of evidence for testing NSAIDs' additivity and would suggest nonadditivity in the form of superadditivity, evidenced by a synergistic effect between the active effect of NSAIDs and placebo analgesia. Further research, however, is required to replicate the findings of Berna et al. (2017).

The one study examining a clinical sample in the present review—Kam-Hansen et al. (2014)—used a triptan in patients suffering from migraines. The study observed an additive effect of this treatment with the placebo effect. In contrast to the previously discussed mechanisms of placebo analgesia, less is known about the means by which the placebo effect may relieve the pain associated with migraines. The closest model currently is that of headaches, which appear to employ a similar mechanism to NSAIDs through the suppression of prostaglandins and thus inflammation (Benedetti, Durando, Giudetti, Pampallona, & Vighetti, 2015; Benedetti, Durando, & Vighetti, 2014). It should be noted, however, that these studies found that the placebo effect was either only effective in this manner after pre-conditioning (Benedetti et al., 2015) or in reducing the nocebo-provoked component of headaches (Benedetti et al., 2014), and so may not be a widely recruited mechanism for headaches generally. Despite the above evidence that a lack of correspondence between active and placebo effect mechanisms tends to produce nonadditive outcomes, the model of prostaglandin inhibition proposed here may suggest that different mechanisms produce additive outcomes in migraines. Further research is required to clarify the mechanisms of the placebo effect in migraines to better understand this outcome.

Variables such as treatment dosage and the time course in which outcomes are measured should also be considered for their potential influence on additivity. Treatment dosage, or, for that matter, route of administration, may influence additive outcomes by contributing to ceiling effects through strong isolated active treatment effects. An example of this may have been the subadditive outcome observed by Lund et al. (2014). In contrast, time course may lead studies to observe different additive outcomes depending on when outcomes are measured relative to the peak efficacy of both the active treatment and placebo effects. The small number of studies and the heterogeneity of analgesic medications employed in different studies makes extrapolating from such data difficult. Nonetheless, it remains that such information is important to the interpretation of future research and should be considered when designing, analyzing, and reporting studies.

In addition to observing a number of nonadditive outcomes in traditional factorial designs, Aslaksen et al.'s (2015) comparison of positive versus negative instructions suggests that expectations of increased pain can reduce the active effect of lidocaine. A similar finding has also been reported where the analgesic effect of remifentanil was attenuated by the instruction that no treatment had been administered (Bingel et al., 2011), though without other groups necessary to test additivity. An antithetic relationship has been observed between the opioid and cholecystokinin systems (Noble, Derrien, & Roques, 1993; Noble et al., 1999), the latter being the neurobiological system underlying nocebo hyperalgesia (Benedetti, Amanzio, Casadio, Oliaro, & Maggi, 1997; Benedetti, Amanzio, Vighetti, & Asteggiano, 2006), which may account for the influence of negative expectations on opioid analgesia. However, this does not explain the attenuation of lidocaine observed by Aslaksen et al. (2015). Such findings may instead suggest that the mechanisms of hyperalgesia may amplify pain to such an extent as to reverse an analgesic's observable effect. As such, even if further research added evidence to active and placebo analgesic effects being additive when comparing positive instructions to no treatment instructions, models of their relationship must be updated to account for the influence of negative expectations on active effects.

3.1 Limitations of Existing Research

Our review of the available literature also points to some important limitations to existing research. The first and the most obvious is the small number of studies that provide a test of additivity. The balanced placebo design was first described over 50 years ago (Ross, Krugman, Lyerly, & Clyde, 1962), and several reviews have since highlighted the importance of this design to clarify the assumptions underlying RCTs (Enck, Klosterhalfen, Weimer, et al., 2011; Enck, Klosterhalfen, & Zipfel, 2011; Kirsch, 2000; Kleijnen, de Craen, van Everdingen, & Krol, 1994). It is surprising therefore that so few studies have investigated the assumption of additivity in placebo analgesia—an assumption crucial to the interpretation of RCT outcomes. Despite placebo analgesia being the most widely examined and best understood modality of the placebo effect, to the best of our knowledge, only seven studies have been conducted that systematically manipulate both active treatment and instructions as to whether active treatment or placebo has been administered. Thus, there is currently limited data available to assess additivity.

3.1.1 Healthy Versus Clinical Samples and Ethical Considerations

Almost all (six out of seven) of the existing studies were conducted with healthy participants in acute experimental pain settings. This means that the results may not generalize to clinical patient population. Interestingly, the one study that used a clinical sample (Kam-Hansen et al., 2014) found evidence consistent with the additivity assumption. Of course, a single study cannot be taken as definitive evidence in favor of additivity in clinical outcomes. Instead, this further emphasizes the general lack of data to assess additivity. However, the lack of clinical studies assessing additivity may be due to ethical concerns. As noted by a number of reviews (Colagiuri, 2010; Enck, Klosterhalfen, Weimer, et al., 2011; Enck, Klosterhalfen, & Zipfel, 2011; Kube & Rief, 2017; Waring, 2008), the balanced placebo design poses ethical problems due to its reliance on deception. This is because the balanced placebo design entails two conditions in which participants are directly misinformed about the treatment they receive. Firstly, to establish the isolated treatment effect, participants are told that they received placebo despite receiving active treatment. Secondly, to establish the isolated placebo effect, participants who received placebo are told that they received active treatment. While the deception is necessary from a design perspective, it poses concerns, particularly in patients seeking treatment for pain-related problems, where it could be considered unethical to misinform the patients. The same issue also applies to the balanced open-hidden design, where participants are deceived in order to establish the isolated placebo effect.

Some of the ethical limitations of both the balanced placebo and open-hidden designs may be avoided in the recently developed "balanced active placebo design" (Berna et al., 2017). In this design, expectancies are manipulated by administering either an active placebo (with side effects) or a benign placebo. The principle being that the side effects brought about by the active placebo cause participants to believe they have been allocated to receive the active treatment. In this design, then, participants are all accurately and non-deceptively told that they may receive an active treatment or a placebo, which is ethically advantageous in terms of avoiding deception. However, the balanced active placebo design also introduces a new ethical problem, namely intentionally inducing side effects in participants. This might be particularly problematic for those participants who miss out on an active treatment and have side effects induced. Hence, it is questionable as to whether the balanced active placebo design provides a better ethical alternative than the traditional balanced placebo design.

3.1.2 Methodological and Statistical Concerns

Several of the available studies using the balanced placebo or other factorial designs had relatively small sample sizes. This means that these studies may have been statistically underpowered. This could be particularly problematic if the power required to detect a main effect differs from that required to detect an interaction. If so, then some of the studies may have been powered to detect an effect of either the active treatment or instructions, but not an interaction between the two. In addition to this, there are inherent problems with using null hypothesis significance testing (NHST) to test the additivity assumption in general. This is because positive evidence for additivity requires proving the null (i.e., that active treatment and placebo effects do not interact). NHST is limited in this sense because it can only provide evidence for or against an alternative hypothesis, the latter of which cannot be taken as evidence in favor of the null hypothesis (Blackwelder, 1982; Gallistel, 2009; Streiner, 2003). Hence, the reliance on NHST in the studies reviewed means that even the absence of a statistical interaction between active treatment and instruction using NHST cannot be taken as true evidence for additivity. Instead, Bayesian approaches would be needed to compare the evidence for the null versus alternative hypotheses. This is because Bayesian analysis can compare the evidence for and against two or more hypothesis (including a null hypotheses) and provides a numerical estimate of the evidence in favor one of these hypotheses relative to the other.

The balanced placebo and open-hidden designs are also subject to several issues concerning their methodology. First and foremost, these designs rely on an instructional manipulation, and, in the case of placebo analgesia at least, largely use subjective measures of self-report. In this manner, demand characteristics and response bias may contribute to the observed outcomes, such as participants reporting greater pain relief during the open conditions of the balanced open-hidden design as they believe this to be the desired outcome (Colagiuri, 2010; Colagiuri & Lovibond, 2013). Attempting to incorporate objective pain outcomes would obviously overcome this limitation, but pain, of course, is an inherently subjective experience and no good objective measures exist yet.

Necessary variations in the methods of a balanced placebo or open-hidden study based on the analgesic being examined may also introduce confounds. For example, the time course of peak analgesic effect varies between both different analgesics and the placebo effect (Atlas et al., 2012; Gabka & Price, 1982), with a study being designed to assess outcomes in relation to the analgesic being examined. This may lead to variability in additive outcomes, as studies involving different analgesics could assess pain at different time post-administration, and thus the placebo effect may be measured at different time points in relation to its peak efficacy. Ideally, outcomes should be measured across a range of time points to ensure additivity is assessed during both peak analgesic and placebo effects.

Similarly, different analgesics often require different methods of administration. A common notion in the placebo literature is that the means of administering a placebo alters the degree of placebo effect, with methods that a more complex, invasive, or intensive procedure being through to produce a greater response (Kaptchuk, Goldman, Stone, & Stason, 2000). As such, the degree of placebo effect itself may vary between different methods, introducing another source of potential variability. Indeed, the balanced open-hidden design is also often limited by the nature of the treatment administration, which must be feasibly given without the participants' awareness. The most common method being intravenously through an infusion pump, although other novel methods such as intramuscularly administering the treatment alongside the pain stimulus have been developed (Lund et al., 2014). Not only is this a limitation of the design in that it requires the use of such invasive methods, but such intravenous methods have been shown to produce a stronger placebo effect than pills (de Craen, Tijssen, de Gans, & Kleijnen, 2000). As such, these studies may produce different additive outcomes due to the presence of stronger placebo effects. Such factors should be considered not only when designing a study, but also when interpreting the additive outcomes observed.

4. CONCLUDING REMARKS AND FUTURE DIRECTIONS

The major finding of this article is that the assumption of additivity between analgesics and placebo analgesia cannot be accepted as holding true under all circumstances. There are, however, currently insufficient studies to clearly determine the conditions under which placebo and active treatment effects are additive and those where they are not. Additional studies are required to replicate and extend the current body of evidence examining placebo analgesia using the balanced placebo and other factorial designs. Such research should also be statistically powered to detect an interaction between the two factors, rather than to detect an effect of the active treatment or instruction, and be cognizant of the general problems to do with NHST and providing positive evidence for a null effect. Indeed, the use of Bayesian analyses may prove to be a more reliable method of detecting additive or nonadditive outcomes. With additional studies, meta-analysis may provide further insight into the conditions under which an additive relationship between the active effect of analgesics and placebo analgesia are observed and those when they are not.

Despite the small number of studies, the concordance between mechanisms of the analgesic and placebo analgesia did appear to moderate the additive outcome. Further research is warranted using different classifications of analgesics within the same experimental design in order to provide further evidence for the existence of synergisms between active treatment effects and placebo analgesia, as well as to understand the different instances in which subadditive outcomes may occur. These studies must be carefully designed to take into account the different pharmacodynamics of the analgesics and placebo analgesia, ideally using identical methods of administration and a range of time points to assess outcomes.

Not only does this review highlight the surprising paucity of research examining additivity between analgesics and placebo analgesia, but it also indicates that additivity has not been extensively investigated in other health-related outcomes. Indeed, we could identify only 12 studies, which examined heterogeneous samples spanning from the treatment of attention deficit hyperactivity disorder with methylphenidate (Pelham, Hoza, Kipp, Gnagy, & Trane, 1997; Pelham et al., 2002) to the sedative effects of the antihistamine hydroxyzine (Hammami et al., 2016). Given that within a single modality the placebo effect may have different additive relationships with different treatments, the application of findings from one modality to another poses substantial problems. In this manner, studies from other modalities would provide little insight into the nature of additivity between analgesics and placebo analgesia, nor can the findings from this review about pain be readily applied to the relationship between the placebo effect and treatments for other conditions. As such, the current review draws attention not only to the need for further research examining additivity in pain, but also for other health-related outcomes. In particular, examining additivity in outcomes with notably strong placebo effects, such as nausea (Quinn & Colagiuri, 2014) or sleep disturbance (Yeung, Sharpe, Glozier, Hackett, & Colagiuri, 2018) may prove to be particularly consequential for the interpretation of RCTs in these conditions.

The assumption of additivity states that the placebo effect and the active effect of a treatment are quantitatively additive and allows for the active effect to be measured by comparing the active treatment against a placebo control. Together, the studies discussed in this review suggest that there is reasonable evidence that the assumption of additivity does not hold under at least some conditions in pain. The fact that the nature of the nonadditive effect did not appear to be consistent across studies is also potentially more problematic for the interpretation of RCTs. This is because subadditive and superadditive outcomes would lead to very different interpretations of RCT data. This variation may make interpreting RCT data more difficult, as a conservative approach according to one nonadditive outcome cannot necessarily be taken. There was also some evidence that the pharmacology of the analgesic may moderate additivity, however, more rigorous research directly comparing different analgesics using appropriately designed and controlled studies are required. Finally, it was interesting to note that there was only a single study conducted in a clinical population, hence one can question whether these laboratory-based studies would generalize to patient samples. Nonetheless, our review draws attention to the overall dearth of research on pain capable of testing the assumption of additivity. Given the integral nature of this assumption to the interpretation of RCT data, there is an urgent need for more studies to be conducted using either the balanced placebo, active placebo, or open-hidden designs on a wide range of treatments in different conditions.

ACKNOWLEDGMENTS

The authors would like to extend their appreciation to the members of the Society for Interdisciplinary Placebo Studies for suggesting studies for this review to consider, as well as to Dr. Per Aslaksen and Dr. Belinda Butcher for taking the time to reanalyze their data for inclusion in this review.

REFERENCES

Amanzio, M., & Benedetti, F. (1999). Neuropharmacological dissection of placebo analgesia: Expectation-activated opioid systems versus conditioning-activated specific subsystems. *The Journal of Neuroscience*, *19*, 484–494.

Amanzio, M., Benedetti, F., Porro, C. A., Palermo, S., & Cauda, F. (2013). Activation likelihood estimation meta-analysis of brain correlates of placebo analgesia in human experimental pain. *Human Brain Mapping*, *34*(3), 738–752.

Amanzio, M., Pollo, A., Maggi, G., & Benedetti, F. (2001). Response variability to analgesics: A role for nonspecific activation of endogenous opioids. *Pain*, *90*, 205–215.

Aslaksen, P. M., Zwarg, M. L., Eilertsen, H. I. H., Gorecka, M. M., & Bjørkedal, E. (2015). Opposite effects of the same drug: Reversal of topical analgesia by nocebo information. *Pain*, *156*(1), 39–46.

Atlas, L. Y., Whittington, R. A., Lindquist, M. A., Wielgosz, J., Sonty, N., & Wager, T. D. (2012). Dissociable influences of opiates and expectations on pain. *Journal of Neuroscience*, *32*(23), 8053–8064.

Basbaum, A. I., & Fields, H. L. (1984). Endogenous pain control systems: Brainstem spinal pathways and endorphin circuitry. *Annual Review of Neuroscience*, 7, 309–338.

Baumgärtner, U., Buchholz, H. G., Bellosevich, A., Magerl, W., Siessmeier, T., Rolke, R., et al. (2006). High opiate receptor binding potential in the human lateral pain system. *NeuroImage*, *30*(3), 692–699.

Benedetti, F. (1996). The opposite effects of the opiate antagonist naloxone and the cholecystokinin antagonist proglumide on placebo analgesia. *Pain, 64*(3), 535–543.

Benedetti, F. (2008). Mechanisms of placebo and placebo-related effects across diseases and treatments. *Annual Review of Pharmacology and Toxicology, 48*, 33–60.

Benedetti, F. (2009). *Placebo effects: Understanding the mechanisms in health and disease.* New York: Oxford University Press.

Benedetti, F. (2014). Placebo effects: From the neurobiological paradigm to translational implications. *Neuron, 84*(3), 623–637.

Benedetti, F., Amanzio, M., Casadio, C., Oliaro, A., & Maggi, G. (1997). Blockade of nocebo hyperalgesia by the cholecystokinin antagonist proglumide. *Pain, 71*, 135–140.

Benedetti, F., Amanzio, M., & Maggi, G. (1995). Potentiation of placebo analgesia by proglumide. *The Lancet, 346*, 1231.

Benedetti, F., Amanzio, M., Rosato, R., & Blanchard, C. (2011). Nonopioid placebo analgesia is mediated by CB1 cannabinoid receptors. *Nature Medicine, 17*(10), 1228.

Benedetti, F., Amanzio, M., Vighetti, S., & Asteggiano, G. (2006). The biochemical and neuroendocrine bases of the hyperalgesic nocebo effect. *The Journal of Neuroscience, 26*(46), 12014–12022.

Benedetti, F., Durando, J., Giudetti, L., Pampallona, A., & Vighetti, S. (2015). High-altitude headache: The effects of real vs sham oxygen administration. *Pain, 156*(11), 2326–2336.

Benedetti, F., Durando, J., & Vighetti, S. (2014). Nocebo and placebo modulation of hypobaric hypoxia headache involves the cyclooxygenase-prostaglandins pathway. *Pain, 155*(5), 921–928.

Benedetti, F., Maggi, G., Lopiano, L., Lanotte, M., Rainero, I., Vighetti, S., et al. (2003). Open versus hidden medical treatments: The patient's knowledge about a therapy affects the therapy outcome. *Prevention & Treatment, 6*(1), 1a.

Benedetti, F., Pollo, A., Lopiano, L., Lanotte, M., Vighetti, S., & Rainero, I. (2003). Conscious expectation and unconscious conditioning in analgesic, motor, and hormonal placebo/nocebo responses. *Journal of Neuroscience, 23*(10), 4315–4323.

Berna, C., Kirsch, I., Zion, S. R., Lee, Y. C., Jensen, K. B., Sadler, P., et al. (2017). Side effects can enhance treatment response through expectancy effects: An experimental analgesic randomized controlled trial. *Pain, 158*(6), 1014–1020.

Bingel, U., Wanigasekera, V., Wiech, K., Mhuircheartaigh, R., Ploner, M., & Tracey, I. (2011). The effect of treatment expectation on drug efficacy: Imaging the analgesic benefit of the opioid remifentanil. *Science Translational Medicine, 3*(70), 70ra14.

Blackwelder, W. C. (1982). "Proving the null hypothesis" in clinical trials. *Controlled Clinical Trials, 3*(4), 345–353.

Butcher, B. E., & Carmody, J. J. (2012). Sex differences in analgesic response to ibuprofen are influenced by expectancy: A randomized, crossover, balanced placebo-designed study. *European Journal of Pain, 16*(7), 1005–1013.

Carter, C. S., Braver, T. S., Barch, D. M., Botvinick, M. M., Noll, D., & Cohen, J. D. (1998). Anterior cingulate cortex, error detection, and the online monitoring of performance. *Science, 280*(5364), 747–749.

Chen, J. C., Smith, E. R., Cahill, M., Cohen, R., & Fishman, J. B. (1993). The opioid receptor binding of dezocine, morphine, fentanyl, butorphanol and nalbuphine. *Life Sciences, 52*(4), 389–396.

Christie, M. J., Connor, M., Vaughan, C. W., Ingram, S. L., & Bagley, E. E. (2000). Cellular actions of opioids and other analgesics: Implications for synergism in pain relief. *Clinical and Experimental Pharmacology and Physiology, 27*(7), 520–523.

Christie, M. J., Vaughan, C. W., & Ingram, S. L. (1999). Opioids, NSAIDs and 5-lipoxygenase inhibitors act synergistically in brain via arachidonic acid metabolism. *Inflammation Research, 48*(1), 1–4.

Cichewicz, D. L. (2004). Synergistic interactions between cannabinoid and opioid analgesics. *Life Sciences*, *74*(11), 1317–1324.

Clare, J. J., Tate, S. N., Nobbs, M., & Romanos, M. A. (2000). Voltage-gated sodium channels as therapeutic targets. *Drug Discovery Today*, *5*(11), 506–520.

Colagiuri, B. (2010). Participant expectancies in double-blind randomized placebo-controlled trials: Potential limitations to trial validity. *Clinical Trials*, 7(3), 246–255.

Colagiuri, B., & Boakes, R. A. (2010). Perceived treatment, feedback, and placebo effects in double-blind RCTs: An experimental analysis. *Psychopharmacology*, *208*(3), 433–441.

Colagiuri, B., & Lovibond, P. F. (2013). Psychological processes that can bias responses to placebo treatment for pain. In L. Colloca, M. A. Flaten, & K. Meissner (Eds.), *Placebo and pain: From bench to bedside*: Elsevier Inc. Chapters.

Colagiuri, B., Schenk, L. A., Kessler, M. D., Dorsey, S. G., & Colloca, L. (2015). The placebo effect: From concepts to genes. *Neuroscience*, *307*, 171–190.

Colloca, L., & Benedetti, F. (2005). Placebos and painkillers: Is mind as real as matter? *Nature Reviews. Neuroscience*, *6*(7), 545.

Colloca, L., & Benedetti, F. (2006). How prior experience shapes placebo analgesia. *Pain*, *124*(1), 126–133.

Colloca, L., & Benedetti, F. (2009). Placebo analgesia induced by social observational learning. *Pain*, *144*(1–2), 28–34.

Colloca, L., Lopiano, L., Lanotte, M., & Benedetti, F. (2004). Overt versus covert treatment for pain, anxiety, and Parkinson's disease. *The Lancet. Neurology*, *3*(11), 679–684.

Colloca, L., & Miller, F. G. (2011). The nocebo effect and its relevance for clinical practice. *Psychosomatic Medicine*, *73*(7), 598.

de Craen, A. J., Tijssen, J. G. P., de Gans, J., & Kleijnen, J. (2000). Placebo effect in the acute treatment of migraine: Subcutaneous placebos are better than oral placebos. *Journal of Neurology*, *247*(3), 183–188.

Desmeules, J., Rollason, V., Piguet, V., & Dayer, P. (2003). Clinical pharmacology and rationale of analgesic combinations. *European Journal of Anaesthesiology*, *20*(S28), 7–12.

Devor, M. (2006). Sodium channels and mechanisms of neuropathic pain. *The Journal of Pain*, 7(1), S3–S12.

Eippert, F., Bingel, U., Schoell, E. D., Yacubian, J., Klinger, R., Lorenz, J., et al. (2009). Activation of the opioidergic descending pain control system underlies placebo analgesia. *Neuron*, *63*(4), 533–543.

Enck, P., Benedetti, F., & Schedlowski, M. (2008). New insights into the placebo and nocebo responses. *Neuron*, *59*(2), 195–206.

Enck, P., Klosterhalfen, S., Weimer, K., Horing, B., & Zipfel, S. (2011). The placebo response in clinical trials: More questions than answers. *Philosophical Transactions of the Royal Society of London. Series B, Biological Sciences*, *366*(1572), 1889–1895.

Enck, P., Klosterhalfen, S., & Zipfel, S. (2011). Novel study designs to investigate the placebo response. *BMC Medical Research Methodology*, *11*(1), 90.

Faria, V., Fredrikson, M., & Furmark, T. (2008). Imaging the placebo response: A neurofunctional review. *European Neuropsychopharmacology*, *18*(7), 473–485.

Fields, H. L., Vanegas, H., Hentall, I. D., & Zorman, G. (1983). Evidence that disinhibition of brain stem neurones contributes to morphine analgesia. *Nature*, *306*, 684–686.

Fromme, K., & Patel, A. B. (2010). Balanced placebo design. In I. P. Stolerman (Ed.), *Encyclopedia of psychopharmacology*. Berlin, Heidelberg: Springer.

Gabka, J., & Price, R. K. (1982). Tooth pulp stimulation: A method of determining the analgesic efficacy of meptazinol in man. *British Journal of Clinical Pharmacology*, *14*(1), 104–106.

Gallistel, C. R. (2009). The importance of proving the null. *Psychological Review*, *116*(2), 439.

Green, G. A. (2001). Understanding NSAIDs: From aspirin to COX-2. *Clinical Cornerstone*, *3*(5), 50–59.

Grevert, P., Albert, L. H., & Goldstein, A. (1983). Partial antagonism of placebo analgesia by naloxone. *Pain, 16*(2), 129–143.

Hammami, M. M., Hammami, S., Al-Swayeh, R., Al-Gaai, E., Farah, F. A., & De Padua, S. J. (2016). Drug*placebo interaction effect may bias clinical trials interpretation: Hybrid balanced placebo and randomized placebo-controlled design. *BMC Medical Research Methodology, 16*(1), 166.

Hare, T. A., O'Doherty, J., Camerer, C. F., Schultz, W., & Rangel, A. (2008). Dissociating the role of the orbitofrontal cortex and the striatum in the computation of goal values and prediction errors. *Journal of Neuroscience, 28*(22), 5623–5630.

Häuser, W., Hansen, E., & Enck, P. (2012). Nocebo phenomena in medicine: Their relevance in everyday clinical practice. *Deutsches Ärzteblatt International, 109*(26), 459.

Heinricher, M. M., Morgan, M. M., & Fields, H. L. (1992). Direct and indirect actions of morphine on medullary neurons that modulate nociception. *Neuroscience, 48*(3), 533–543.

Heinricher, M. M., Morgan, M. M., Tortorici, V., & Fields, H. L. (1994). Disinhibition of off-cells and antinociception produced by an opioid action within the rostral ventromedial medulla. *Neuroscience, 63*(1), 279–288.

Hróbjartsson, A., & Gøtzsche, P. C. (2010). Placebo interventions for all clinical conditions. *The Cochrane Database of Systematic Reviews,* (1), CD003974.

Ito, S., Okuda-Ashitaka, E., & Minami, T. (2001). Central and peripheral roles of prostaglandins in pain and their interactions with novel neuropeptides nociceptin and nocistatin. *Neuroscience Research, 41*(4), 299–332.

Kam-Hansen, S., Jakubowski, M., Kelley, J. M., Kirsch, I., Hoaglin, D. C., Kaptchuk, T. J., et al. (2014). Altered placebo and drug labeling changes the outcome of episodic migraine attacks. *Science Translational Medicine, 6*(218), 218ra5.

Kaptchuk, T. J., Goldman, P., Stone, D. A., & Stason, W. B. (2000). Do medical devices have enhanced placebo effects? *Journal of Clinical Epidemiology, 53*(8), 786–792.

Kirsch, I. (2000). Are drug and placebo effects in depression additive? *Biological Psychiatry, 47*(8), 733–735.

Kleijnen, J., de Craen, A. J., van Everdingen, J., & Krol, L. (1994). Placebo effect in double-blind clinical trials: A review of interactions with medications. *The Lancet, 344*(8933), 1347–1349.

Koneru, A., Satyanarayana, S., & Rizwan, S. (2009). Endogenous opioids: Their physiological role and receptors. *Global Journal of Pharmacology, 3*(3), 149–153.

Krummenacher, P., Candia, V., Folkers, G., Schedlowski, M., & Schönbächler, G. (2010). Prefrontal cortex modulates placebo analgesia. *Pain, 148*(3), 368–374.

Kube, T., & Rief, W. (2017). Are placebo and drug-specific effects additive? Questioning basic assumptions of double-blinded randomized clinical trials and presenting novel study designs. *Drug Discovery Today, 22*(4), 729–735.

Levine, J. D., Gordon, N. C., & Fields, H. L. (1978). The mechanisms of placebo analgesia. *Lancet, 2,* 654–657.

Lund, K., Vase, L., Petersen, G. L., Jensen, T. S., & Finnerup, N. B. (2014). Randomised controlled trials may underestimate drug effects: Balanced placebo trial design. *PLoS One, 9*(1), e84104.

Minami, T., Nakano, H., Kobayashi, T., Sugimoto, Y., Ushikubi, F., Ichikawa, A., et al. (2001). Characterization of EP receptor subtypes responsible for prostaglandin E2-induced pain responses by use of EP1 and EP3 receptor knockout mice. *British Journal of Pharmacology, 133*(3), 438–444.

Miranda, H. F., Silva, E., & Pinardi, G. (2004). Synergy between the antinociceptive effects of morphine and NSAIDs. *Canadian Journal of Physiology and Pharmacology, 82*(5), 331–338.

Noble, F., Derrien, M., & Roques, B. P. (1993). Modulation of opioid antinociception by CCK at the supraspinal level: Evidence of regulatory mechanisms between CCK and enkephalin systems in the control of pain. *British Journal of Pharmacology, 109*(4), 1064–1070.

Noble, F., Wank, S. A., Crawley, J. N., Bradwejn, J., Seroogy, K. B., Hamon, M., et al. (1999). International Union of Pharmacology. XXI. Structure, distribution, and functions of cholecystokinin receptors. *Pharmacological Reviews, 51*(4), 745–781.

Omote, K., Kawamata, T., Nakayama, Y., Yamamoto, H., Kawamata, M., & Namiki, A. (2002). Effects of a novel selective agonist for prostaglandin receptor subtype EP4 on hyperalgesia and inflammation in monoarthritic model. *Anesthesiology: The Journal of the American Society of Anesthesiologists, 97*(1), 170–176.

Ossipov, M. H., Dussor, G. O., & Porreca, F. (2010). Central modulation of pain. *The Journal of Clinical Investigation, 120*(11), 3779–3787.

Pelham, W. E., Hoza, B., Kipp, H. L., Gnagy, E. M., & Trane, S. T. (1997). Effects of methylphenidate and expectancy on ADHD children's performance, self-evaluations, persistence, and attributions on a cognitive task. *Experimental and Clinical Psychopharmacology, 5*(1), 3.

Pelham, W. E., Hoza, B., Pillow, D. R., Gnagy, E. M., Kipp, H. L., Greiner, A. R., et al. (2002). Effects of methylphenidate and expectancy on children with ADHD: Behavior, academic performance, and attributions in a summer treatment program and regular classroom settings. *Journal of Consulting and Clinical Psychology, 70*(2), 320.

Petrovic, P., Kalso, E., Petersson, K. M., Andersson, J., Fransson, P., & Ingvar, M. (2010). A prefrontal non-opioid mechanism in placebo analgesia. *Pain, 150*(1), 59–65.

Petrovic, P., Kalso, E., Petersson, K. M., & Ingvar, M. (2002). Placebo and opioid analgesia—Imaging a shared neuronal network. *Science, 295*(5560), 1737–1740.

Pollo, A., Amanzio, M., Arslanian, A., Casadio, C., Maggi, G., & Benedetti, F. (2001). Response expectancies in placebo analgesia and their clinical relevance. *Pain, 93*, 77–84.

Quinn, V. F., & Colagiuri, B. (2014). Placebo interventions for nausea: A systematic review. *Annals of Behavioral Medicine, 49*(3), 449–462.

Raffa, R. B. (2001). Pharmacology of oral combination analgesics: Rational therapy for pain. *Journal of Clinical Pharmacy and Therapeutics, 26*(4), 257–264.

Ross, S., Krugman, A. D., Lyerly, S. B., & Clyde, A. D. J. (1962). Drugs and placebos: A model design. *Psychological Reports, 10*(2), 383–392.

Schenk, L. A., Sprenger, C., Geuter, S., & Büchel, C. (2014). Expectation requires treatment to boost pain relief: An fMRI study. *Pain, 155*(1), 150–157.

Scott, D. J., Stohler, C. S., Egnatuk, C. M., Wang, H., Koeppe, R. A., & Zubieta, J. K. (2007). Individual differences in reward responding explain placebo-induced expectations and effects. *Neuron, 55*(2), 325–336.

Seibert, K., Zhang, Y., Leahy, K., Hauser, S., Masferrer, J., Perkins, W., et al. (1994). Pharmacological and biochemical demonstration of the role of cyclooxygenase 2 in inflammation and pain. *Proceedings of the National Academy of Sciences of the United States of America, 91*(25), 12013–12017.

Seybold, V. S. (2009). The role of peptides in central sensitization. In B. Canning & D. Spina (Eds.), *Sensory nerves. Handbook of experimental pharmacology.* Berlin, Heidelberg: Springer.

Seybold, V. S., Jia, Y. P., & Abrahams, L. G. (2003). Cyclo-oxygenase-2 contributes to central sensitization in rats with peripheral inflammation. *Pain, 105*(1–2), 47–55.

Snyder, S. H., & Pasternak, G. W. (2003). Historical review: Opioid receptors. *Trends in Pharmacological Sciences, 24*(4), 198–205.

Stewart-Williams, S., & Podd, J. (2004). The placebo effect: Dissolving the expectancy versus conditioning debate. *Psychological Bulletin, 130*(2), 324.

Streiner, D. L. (2003). Unicorns do exist: A tutorial on "proving" the null hypothesis. *The Canadian Journal of Psychiatry*, *48*(11), 756–761.

Taber, R. I., Greenhouse, D. D., Rendell, J. K., & Irwin, S. (1969). Agonist and antagonist interactions of opioids on acetic acid-induced abdominal stretching in mice. *Journal of Pharmacology and Experimental Therapeutics*, *169*(1), 29–38.

Tallarida, R. J. (2000). *Drug synergism and dose-effect data analysis.* CRC Press.

Tallarida, R. J. (2001). Drug synergism: Its detection and applications. *Journal of Pharmacology and Experimental Therapeutics*, *298*(3), 865–872.

Tracey, I. (2010). Getting the pain you expect: Mechanisms of placebo, nocebo and reappraisal effects in humans. *Nature Medicine*, *16*, 1277–1283.

Trescot, A. M., Datta, S., Lee, M., & Hansen, H. (2008). Opioid pharmacology. *Pain Physician*, *11*, S133–S153.

Vane, J. R., & Botting, R. M. (1996). Mechanism of action of anti-inflammatory drugs. *Scandinavian Journal of Rheumatology*, *25*, 9–21.

Vaughan, C. W., & Christie, M. J. (1997). Presynaptic inhibitory action of opioids on synaptic transmission in the rat periaqueductal grey in vitro. *The Journal of Physiology*, *498*, 463–472.

Vaughan, C. W., Ingram, S. L., Connor, M. A., & Christie, M. J. (1997). How opioids inhibit GABA-mediated neurotransmission. *Nature*, *390*(6660), 611.

Wager, T. D., Rilling, J. K., Smith, E. E., Sokolik, A., Casey, K. L., Davidson, R. J., et al. (2004). Placebo-induced changes in FMRI in the anticipation and experience of pain. *Science*, *303*(5661), 1162–1167.

Wager, T. D., Scott, D. J., & Zubieta, J. K. (2007). Placebo effects on human μ-opioid activity during pain. *Proceedings of the National Academy of Sciences of the United States of America*, *104*(26), 11056–11061.

Waring, D. R. (2008). The antidepressant debate and the balanced placebo trial design: An ethical analysis. *International Journal of Law and Psychiatry*, *31*(6), 453–462.

Yeung, V., Sharpe, L., Glozier, N., Hackett, M. L., & Colagiuri, B. (2018). A systematic review and meta-analysis of placebo versus no treatment for insomnia symptoms. *Sleep Medicine Reviews*, *38*, 17–27.

Zubieta, J. K., Bueller, J. A., Jackson, L. R., Scott, D. J., Xu, Y., Koeppe, R. A., et al. (2005). Placebo effects mediated by endogenous opioid activity on μ-opioid receptors. *The Journal of Neuroscience*, *25*(34), 7754–7762.

CHAPTER SEVENTEEN

Nocebo Responses in Brain Diseases: A Systematic Review of the Current Literature

Panagiotis Zis*, Dimos-Dimitrios Mitsikostas†,1
*Academic Department of Neurosciences, Sheffield Teaching Hospitals NHS Foundation Trust, Sheffield, United Kingdom
†1st Neurology Department, Aeginition Hospital, National and Kapodistrian University of Athens, Athens, Greece
[1]Corresponding author: e-mail address: dmitsikostas@med.uoa.gr

Contents

Abstract

Placebo is an intervention with no therapeutic effect that is used as a control in randomized controlled trials (RCTs). Placebo effects and responses can produce a beneficial effect that cannot be attributed to the properties of the intervention itself, since it is usually inactive, and should, therefore, be due to the patient's expectations about treatment (placebo effects), or confounding factors such as natural history, co-interventions,

International Review of Neurobiology, Volume 139
ISSN 0074-7742
https://doi.org/10.1016/bs.irn.2018.07.025

biases, among other co-factors (placebo responses). However, adverse events (AEs) may occur when using a placebo intervention, a phenomenon that is called nocebo. Like placebo, the nocebo effect is a cognitive and idiosyncratic phenomenon with specific biological bases, controlled by distinctive neurotransmitters within mapped brain areas most likely located within the network of the limbic system.

Nocebo responses has been found to be very prevalent in various neurological conditions, in particular, in many brain disorders including headache, Parkinson's disease, Alzheimer's disease, depression, epilepsy, multiple sclerosis and motor neuron disease. Pooled AE rates in the placebo groups (nocebo AE rates) vary from 25% in the symptomatic treatment for multiple sclerosis RCTs to almost 80% in motor neuron disease RCTs. Pooled dropout rates because of AEs in the placebo groups (i.e., nocebo dropout rates) vary from 2% in multiple sclerosis RCTs to almost 10% in PD RCTs. Across all brain disorders, the nature of AEs reported in the placebo-treated subjects mirrors those reported by active drug-treated subjects, suggesting that awareness of drug side-effect profiles might have influenced patient expectations and, thus, nocebo responses.

Unexplored brain diseases where nocebo should be studied further include mental disorders (i.e., schizophrenia and bipolar disorder), vascular disorders (i.e., acute ischemic stroke, vascular dementia), degenerative disorders (i.e., frontotemporal dementia, Lewy body dementia) and other systemic atrophies of the brain (i.e., hereditary ataxias).

1. INTRODUCTION

The term placebo in this article refers to any intervention with no therapeutic effect that is used as a control in randomized controlled trials (RCTs). Placebo can produce a beneficial effect that cannot be attributed to the properties of the placebo itself entirely (since it is inactive), but potentially are due to the patient's belief and expectations about treatment (placebo effects) or confounding factors such as regression to the mean, natural history, biases among other co-factors (placebo responses) (Colloca, 2017a, 2017b).

The exact pathophysiological mechanism of the placebo effect remains unknown in its details, though in several brain disorders, such as Parkinson's disease (PD), neuroimaging studies have demonstrated that placebos stimulate the release of dopamine in the striatum of patients, and can alter the activity of dopamine neurons (Lidstone, 2014). In depression and anxiety, placebo effects cause changes in electrical and metabolic activity in different brain regions (i.e., the ventral striatum and the amygdala, respectively) (Benedetti, Carlino, & Piedimonte, 2016). In Alzheimer's disease, prefrontal executive control and functional connectivity of prefrontal areas seem to play a significant role in the placebo phenomena (Benedetti et al., 2016).

In pain, placebo effects produce activity changes and enhanced functional coupling in areas of the brain such as the dorsolateral prefrontal cortex, the anterior cingulate cortex, and subcortical regions including the hypothalamus, amygdala, the periaqueductal gray and the spinal cord (Colloca & Grillon, 2014).

In contrast, negative outcomes have been observed when patients receive placebos. Respectively, the nocebo response refers to the phenomenon of experiencing adverse events (AEs) when treated with a substance with no active therapeutic effect (placebo). In contrast, the nocebo effect is partially the result of negative expectations by patients that medical treatment will probably harm rather than heal (Mitsikostas, Chalarakis, Mantonakis, Delicha, & Sfikakis, 2012; Mitsikostas, Mantonakis, & Chalarakis, 2014). Nocebo is associated with lower adherence to the therapeutic intervention and higher rates of treatment withdrawal, as well as significant difficulty in assessing the efficacy and the safety profile of a drug (Barsky, Saintfort, Rogers, & Borus, 2002; Enck, Benedetti, & Schedlowski, 2008). Previous negative treatment experiences (Benedetti, Lanotte, Lopiano, & Colloca, 2007), along with several psychological factors such as stress and anxiety (Elsenbruch et al., 2012; Häuser, Bartram, Bartram-Wunn, & Tölle, 2012; Manchikanti, Giordano, Fellows, & Hirsch, 2011), often control nocebo.

Nocebo responses have been found to be very prevalent in various neurological conditions, including epilepsy (Zis, Shafiq, & Mitsikostas, 2017), motor neuron disease (MND) (Shafiq, Mitsikostas, & Zis, 2017) multiple sclerosis (MS) (Papadopoulos & Mitsikostas, 2010), headache (Mitsikostas, 2012; Mitsikostas, Mantonakis, & Chalarakis, 2011), neuropathic pain (Papadopoulos & Mitsikostas, 2012), fibromyalgia (Mitsikostas et al., 2012), diabetic peripheral neuropathy (Häuser et al., 2012), chronic inflammatory demyelinating polyneuropathy (Zis, Hadjivassiliou, Sarrigiannis, Jenkins, & Mitsikostas, 2018), Ménièere's disease (Dimitriadis & Zis, 2017), restless leg syndrome (Silva et al., 2017), PD (Stathis, Smpiliris, Konitsiotis, & Mitsikostas, 2013), depression (Mitsikostas et al., 2014; Rutherford, Wall, Glass, & Stewart, 2014), and Alzheimer's disease (AD) (Zis & Mitsikostas, 2015). This indicates significant implications for clinical practice related to treatment adherence and outcomes.

Herein, we conducted a systematic review in which nocebo responses in non-traumatic brain diseases are discussed. Only previous systematic reviews per disease and meta-analyses have been considered for the purposes of this paper, as such methodologies can provide the highest quality of evidence (Burns, Rohrich, & Chung, 2011).

2. LITERATURE SEARCH STRATEGY

A computer-based literature search was conducted on May 12th, 2018 on PubMed. For the search we used two Medical Subject Headings (MeSH) terms in the title and/or the abstract. Term A was "nocebo" and Term B was "brain" or "neurological" or "neurology." To be included in this review, papers should have met the following inclusion criteria: (1) be written in English, (2) refer to a brain disease and (3) study nocebo phenomena.

This search strategy resulted in the identification of 65 articles. Of these 54 were excluded as summarized in the PRISMA chart (Moher et al., 2009) (Fig. 1). After perusing the reference lists of the papers we identified another nine eligible articles to be included. In total we used 20 papers for this review.

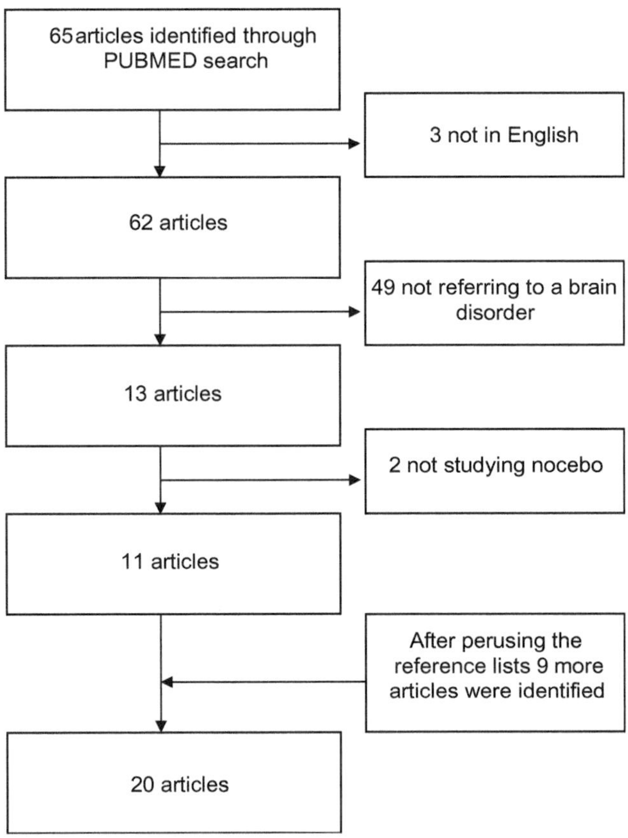

Fig. 1 PRISMA chart.

3. CLASSIFICATION OF BRAIN DISORDERS

Based on the International Classification of Diseases version 10 (ICD-10) (WHO, 1992), the non-traumatic brain disorders are broadly classified into demyelinating, degenerative, episodic and paroxysmal, infectious, neoplastic, vascular, systematic atrophies affecting the brain, and psychological factors. Table 1 summarizes this classification. To date, systematic reviews and meta-analyses of the nocebo responses have been published in the following brain diseases: headaches, depression, AD, PD, MS, epilepsy and MND.

4. SYNOPSIS OF NOCEBO PHENOMENA IN NON-TRAUMATIC BRAIN DISORDERS

4.1 Primary Headache Disorders

Primary headache disorders are chronic brain disorders, which include migraine, tension-type headaches (TTHs), the trigeminal autonomic cephalgias (TACs), and other primary headache disorders. The clinical phenotype is often

Table 1 Classification of the Non-Traumatic Brain Disorders Based on the ICD-10

Classification	ICD-10 Chapter	ICD-10 Block(s)	Example(s)
Neoplastic	II	C71	Malignant neoplasm of brain (C71)
Psychiatric	V	F00–F99	Schizophrenia (F20)
			Depressive episode (F32)
Infectious	VI	G00–G09	Bacterial meningitis (G00)
Degenerative		G20–G32	Parkinson's disease (G20)
			Alzheimer's disease (G30)
Demyelinating		G35–G37	Multiple sclerosis (G35)
Episodic and paroxysmal		G40–G47	Epilepsy (G40)
			Migraine (G43)
Systemic atrophies affecting the brain		G10–G13	Hereditary ataxia (G11)
			Motor neuron disease (G12.2)
Vascular	I	I61–I63	Intracerebral hemorrhage (I61)
			Cerebral infarction (I63)

paroxysmal. Several environmental and genetic factors play a significant role in their pathogenesis. The term "primary" refers to a lack of an underlying structural pathology, trauma, or systemic disease. Several dynamic-like changes have been reported in patients suffering from primary headaches, for example, the cortical spreading depression during the aura phase of migraine (Hadjikhani et al., 2001). Although not life-threating, primary headaches, and migraine in particular, are leading causes of disability worldwide, affecting people in their activities of daily living (GBD 2015 Neurological Disorders Collaborator Group, 2017). Migraine alone is the first cause of disability in people younger than 50 years (Steiner, Stovner, Vos, Jensen, & Katsarava, 2018).

In a meta-analysis of 56 RCTs studying acute symptomatic treatment (ST) for migraine, conducted by Mitsikostas et al. (2011), the nocebo AE rate (defined as the percentage of patients who experienced at least one AE) and the nocebo dropout rate (defined as the percentage of patients who discontinued the treatment because of AE) were 18.5% and 0.3%, respectively (Mitsikostas et al., 2011). After meta-analyzing 45 RCTs focusing on migraine prevention, the nocebo AE rate and the nocebo dropout rate were estimated to be 42.8% and 4.8%, respectively (Mitsikostas et al., 2011).

Pooling the data from four RCTS studying TTH prevention, Mitsikostas et al. (2011) showed that the nocebo dropout rate is similar (5.4%) but the nocebo AE rate is reduced when compared to migraines (24.0%). A possible explanation for the increased nocebo AE rate in patients with migraines is that a common symptom of the migraine itself is nausea, which in RCTs is being reported as AEs. Thus, a proportion of the AEs can be attributed to the disease itself rather than being the nocebo phenomenon.

4.2 Depression

Depression is a leading cause of disability worldwide, and a major contributor to the overall global burden of any disease. The prevalence of depression increases with age (Zis, Daskalaki, et al., 2017). A diagnosis of a major depressive disorder requires the presence of symptoms such as depressed mood, sleep cycle disturbances, fatigue and poor concentration for at least 2 weeks, causing clinically significant distress or impairment in social functioning (Baquero & Martín, 2015).

In a meta-analysis of 21 RCTs of various antidepressants conducted by Mitsikostas et al. (2014), the nocebo AE rate in depression was estimated to be 44.7%. The nocebo dropout rate was estimated to be 4.5%. In that study, the difference between the nocebo rates and the respective rates of the active drug arms (active drug AE rate 40.9% and active drug dropout rate 6.9%) was very narrow.

Univariate analysis identified three factors significantly correlating with the dropout rate because of AEs: age, placebo response, and duration of depression. However, multivariate analysis failed to identify any particular factor contributing to nocebo responses.

In a smaller meta-analysis (Dodd et al., 2015) only of duloxetine RCTs, the nocebo AE rate was estimated to be higher (63.9%), whereas the nocebo dropout rate was very similar (4.7%). In that study there was no evidence to support the claim that reported AEs are influenced by the AEs mentioned in the clinical trials participant information and consent forms, or that reported AEs would be influenced by AE profiles of previous antidepressant medications used by these study participants. However, patients who had previously used complementary medications were more likely to report AEs (Dodd et al., 2015).

4.3 Alzheimer's Disease

AD is a progressive neurodegenerative disease that is the cause of 60–70% of cases of dementia. Advancing age is a primary risk factor for the disease, with the majority of cases being diagnosed after the sixth decade of life. However, in patients with familial causes or subjects with trisomy 21 the age of onset of AD can be earlier (Zis, Dickinson, Shende, Walker, & Strydom, 2012; Zis, McHugh, et al., 2014; Zis, Strydom, Buckley, Adekitan, & McHugh, 2017; Zis & Strydom, 2018). The most common early symptom is short-term memory loss, however, as the disease advances, symptoms can include problems with language, disorientation, mood swings, loss of motivation, inability to attend to activities of daily living, and behavioral issues (Burns & Iliffe, 2009). AD is characterized by loss of neurons and synapses in the cerebral cortex and certain subcortical regions, resulting in atrophy of the affected regions, predominantly the temporal lobe and parietal lobe. Both amyloid plaques and neurofibrillary tangles are clearly visible by microscopy in brains of those afflicted by AD (Kotzbauer, Trojanowsk, & Lee, 2001).

In a meta-analysis, which included 20 RCTs and was conducted by Zis and Mitsikostas (2015), the nocebo AE rate in AD was estimated to be 57.8%. The nocebo dropout rate was estimated to be 6.6%. In that study, the difference between the nocebo rates and the respective rates of the active drug arms (active drug AE rate 61.6% and active drug dropout rate 8.8%) was very narrow.

Univariate analysis showed that sample size of the study, body mass index (BMI) and mini-mental state examination (MMSE) score were potential factors negatively correlating with AEs and dropouts among the placebo-treated populations, whereas age was a potential factor positively correlating with AEs and dropouts among the placebo-treated populations (Zis & Mitsikostas, 2015). In addition, year of publication was negatively correlated with the nocebo AE rate, disease duration was positively correlated with nocebo AE rate, and male sex was negatively correlated with nocebo dropout rate. However, meta-regression analysis showed that only sample size was negatively correlated with both the nocebo AEs and dropout rates (Zis & Mitsikostas, 2015).

4.4 Parkinson's Disease

PD is also a chronic degenerative disorder, predominantly affecting the dopamine-secreting neurons in the basal ganglia. The classical clinical manifestation of PD includes resting tremor, rigidity, bradykinesia and difficulty with walking (Kalia & Lang, 2015). However, apart from the motor symptoms, patients with PD present with a wide range of non-motor symptoms, including cognitive dysfunction, dysautonomia and pain (Adewusi et al., 2018a, 2018b; Zis, Erro, Walton, Sauerbier, & Chaudhuri, 2015; Zis, Grünewald, Chaudhuri, & Hadjivassiliou, 2017; Zis, Rizos, et al., 2014) at all stages of the disease (Zis, Martinez-Martin, et al., 2015).

In a meta-analysis, which included 41 RCTs and was conducted by Stathis et al. (2013) the nocebo AE rate in PD was estimated to be 64.7%. The nocebo dropout rate was estimated to be 8.8%. In that study, the difference between the nocebo rates and the respective rates of the active drug arms (active drug AE rate 73.3% and active drug dropout rate 9.8%) was very narrow, showing that nocebo is a very important confounding factor of the reported AEs in RCTs as well as in clinical practice when treating patients with AD. A very strong correlation was also found between the nocebo AEs and nocebo dropout rates, and the active drug AEs and active drug dropout rates, respectively.

Mestre, Lang, and Okun (2016) suggested that AE reporting in deep brain stimulation (DBS) studies could be potentially influenced by investigators and also by staff members who may unintentionally communicate a lack of understanding of a therapy, or alternatively a lack of acceptance or confidence for a particular intervention. Such verbal and nonverbal messages can result in negative effects of expectation in the same way that they can encourage a greater expectation of benefit (Mestre et al., 2016).

Keitel, Ferrea, Südmeyer, Schnitzler, and Wojtecki (2013) showed that tremor is a Parkinsonian symptom responsive to nocebo interventions in their DBS study. Negative expectations might counteract the therapeutic effect of DBS, undermining the potency of patients' expectation and its relevance for therapeutic outcomes (Keitel, Ferrea, et al., 2013). Of interest is the fact that placebo DBS responders with an improvement of $\geq$25% show impaired verbal fluency (nocebo). These results indicate that positive motor expectations exert both motor placebo and cognitive nocebo responses. Such nocebos are likely due to implicit learning mechanisms associated with an expectation-induced placebo response on motor performance (Keitel, Wojtecki, et al., 2013).

4.5 Multiple Sclerosis

MS is a demyelinating immune-mediated disease of the central nervous system (CNS). While the cause is unclear, the underlying pathogenetic mechanism is considered to involve myelin destruction by the immune system or failure of the myelin-producing cells (Nakahara, Maeda, Aiso, & Suzuki, 2012). MS lesions may affect the brain, the cerebellum and the spinal cord, resulting in a range of signs and symptoms, including physical, and sometimes mental health problems. Common symptoms include double vision, unilateral loss of vision, weakness, altered sensation, or poor coordination (Compston & Coles, 2008). MS can take several forms. In the relapsing remitting forms, the new symptoms occur as an isolated attack, whereas in the progressive forms, there is a progressive decline in the patient's functionality as a result of accumulating lesions (Lublin & Reingold, 1996).

Using a meta-analytic approach, Papadopoulos and Mitsikostas (2010) estimated that in MS the pooled incidence of nocebo responses is 74.4% in disease-modifying treatment (DMT) trials, and 25.3% in ST trials. The pooled nocebo dropout rate is 2.1% in DMT and 2.3% in ST trials.

Meta-regression analysis revealed a higher nocebo incidence in parallel design ST studies compared to crossover ones, and a higher nocebo dropout rate in phase II ST studies compared to phase III ones. Nocebo dropout rates in DMT trials exhibited an association with the year of study publication and the frequency of drug administration.

4.6 Epilepsy

Epilepsy is one of the most common chronic neurological diseases. An epileptic seizure results from transient abnormal synchronization of neurons in the brain that disrupts normal patterns of neuronal communication (Moshé, Perucca, Ryvlin, & Tomson, 2015). Cases of epilepsy may be organized into epilepsy syndromes based on the specific features that are present, such as age of onset of seizures and seizure type. The majority of patients with epilepsy suffer from idiopathic epilepsy, however secondary epilepsy is not uncommon and is often a result of brain injury, stroke or brain tumors.

Zaccara, Giovannelli, Cincotta, Loiacono, and Verrotti (2015) have investigated the nocebo effect by analyzing placebo-treated, drug-resistant, focal epileptic patients in RCTs. They estimated that the proportion of placebo-treated patients withdrawing because of AEs is 3.9%, and the proportion of patients with AEs is 60.3%. The drawback of this study was that the majority of the included RCTs were studies where placebo was compared to an active drug as an add-on treatment. Therefore, patients that were treated with placebo were already on one or more other antiepileptic drugs.

In an attempt to estimate the nocebo response in patients not receiving any active antiepileptic treatment, Zis, Shafiq, and Mitsikostas (2017) conducted a meta-analysis of four RCTs in refractory partial epilepsy during pre-surgical monitoring. The authors reported that the pooled estimate of the percentage of placebo-treated patients who withdrew treatment was 76.8%, and the pooled estimate of the percentage of placebo-treated patients who withdrew from treatment because of AEs related to treatment (other than seizures, which were considered to be a disease related event) was 3.2%. The pooled estimate of the percentage of active drug-treated patients who withdrew treatment was 52.1% and the pooled estimate of the percentage of active drug-treated patients who withdrew from treatment because of AEs related to treatment was 8.3%.

Differences in nocebo responses among the different types of epilepsy have not been studied, as there are no RCTs with pure placebo arms in types of epilepsy other than refractory partial epilepsy.

4.7 Motor Neuron Disease

MND is a term referring to progressive degenerative diseases, including amyotrophic lateral sclerosis (ALS), primary lateral sclerosis (PLS), progressive muscular atrophy (PMA), and progressive bulbar palsy (PBP). In MND both the upper and the lower motor neurons are usually affected, thus although it is not exclusively a brain disorder, MND can affect the motor cortex.

A meta-analysis of 12 RCTs conducted by Shafiq et al. (2017), showed that approximately 8 in 10 placebo-treated patients (78.3%) reported at least one AE and approximately 1 in 12 placebo-treated patients discontinued placebo treatment because of AEs other than death (8.4%). The respective rates for the active treatment groups were 84.7% and 12.6%. The fact that these differences are very narrow demonstrates that nocebo is an important confounding factor of reported AEs in MND trials.

4.8 Restless Leg Syndrome

Restless legs syndrome (RLS) is a disorder characterized by a strong urge to move the lower limbs, preceded by an unpleasant feeling that improves to a degree with moving them. Occasionally the upper limbs may also be affected. The symptoms are presented usually when at rest and therefore can impact the patient's sleep (Trenkwalder, Paulus, & Walters, 2005). The causes of RLS are many, though >60% of cases of RLS are familial and are inherited in an autosomal dominant fashion with variable penetrance (Trenkwalder et al., 2005). Both the dopaminergic system and iron insufficiency in the substantia nigra play a significant role in the pathogenesis of RLS (Connor et al., 2003).

A meta-analysis of 72 RCTs conducted by Silva et al. (2017), showed that overall, 45.4% of placebo-treated patients experience AEs, and approximately 2.1% of the placebo-treated patients discontinued placebo treatment because of AEs. The proportion of patients experiencing AE is greater in idiopathic RLS.

5. NATURE OF AEs ACROSS THE BRAIN DISORDERS

It is clear that across all disorders, the nature of AEs reported in the placebo-treated subjects, mirrors those reported by active drug-treated subjects, suggesting that awareness of drug side-effect profiles might have influenced patient expectations and, thus, nocebo responses (Shafiq et al., 2017). The AEs across brain disorders other than headache, where meta-analyses have been performed to date are summarized in Table 2. However, the inherent difficulty in attributing non-specific symptoms (i.e., headache and dizziness) has to be recognized as a potential source of bias in these meta-analyses.

An important observation is that it can be difficult to distinguish whether AEs arise subsequent to drug administration or as a consequence of disease worsening. In particular, dyskinesias reported in both the placebo and the active-treated patients with PD, urinary tract infections reported in the placebo-treated patients with Alzheimer's disease, and all the commonest

Table 2 Adverse Events (AEs) Reported in Placebo-Treated Groups (Nocebo AEs) and Active-Treated Groups Across Brain Disorders, Where Pooled Data Were Available

	Placebo-Treated Patients	Active Drug-Treated Patients
Depression	Headache	Dizziness
	Nausea	Somnolence/sedation
	Somnolence/sedation	Headache
Alzheimer's disease	Diarrhea	Diarrhea
	Headache	Headache
	Urinary tract infection	Dizziness
Parkinson's disease	Nausea	Nausea
	Dizziness	Dizziness
	Dyskinesia	Dyskinesia
Epilepsy	Headache	Headache
	Dizziness	Nausea
		Dizziness
Motor neuron disease	Fatigue/asthenia	Respiratory disorders
	Respiratory disorders	Fatigue/asthenia
	Dysphagia	Dysphagia

The three commonest AEs are mentioned in each disorder (most common on top).

AEs reported across both the placebo and the active-treated patients with MND, may equally be part of the disease or the disease progression. Theoretically, the only way to estimate genuine nocebo rates would be by comparing a group of patients not treated at all with a group of patients treated with an inactive substance (placebo). The difference of the rates of new symptoms occurring during such a study, between the two arms, would be the true nocebo effect. However, since there are available treatments for these diseases, such studies are unethical and therefore unrealistic (Shafiq et al., 2017).

6. COMPARISON AMONG BRAIN DISORDERS

Although direct comparison with other neurological diseases is difficult because of the reasons related to trial populations, primary end-points, severity, progression, and pathophysiology, it can nonetheless give an overall picture of the magnitude of nocebo responses across the various brain disorders.

The occurrence of nocebo AE and nocebo dropout rates is relevant as suggested by this article that summarizes all neurological disorders where the nocebo response has been studied, using an identical methodology (see Table 3).

The highest nocebo dropout rate has been observed in PD (Stathis et al., 2013). Human experimental evidence suggests that negative expectations could result in motor deterioration in patients with PD (Scott et al., 2008). PET studies showed that high placebo effects were associated with greater dopamine (DA) and opioid activity in the nucleus accumbens, whereas nocebo effects were associated with a deactivation of DA and opioid release (Scott et al., 2008). Both systems modulate a number of processes, including the regulation of reward and affective states. Thus, increased nocebo should be expected in PD, although DA replacement therapy results in changes in many aspects of neural activity within the entire basal ganglia cortical networks that are not yet fully understood (Amanzio, Corazzini, Vase, & Benedetti, 2009; The Parkinson Study Group, 2004).

Table 3 Nocebo AE Rates and Nocebo Dropout Rates in All Brain Disorders Where Nocebo Effect Has Been Studied to Date

Disorder	AE Rate (%)	Dropout Rate (%)
Motor neuron disease	78.3	8.4
Multiple sclerosis		
Disease-modifying trials	74.4	2.1
Symptomatic therapy trials	25.3	2.3
Parkinson's disease	64.7	8.8
Refractory partial epilepsy	60.8	4.0
Alzheimer's disease	57.8	6.6
Rest less leg syndrome	45.5	2.1
Depression	44.7	4.5
Headache		
Preventative treatment	42.8	4.8
Symptomatic treatment	18.5	0.3

AE, adverse event.

Not considering the nocebo dropout rate during ST for headache and MS, which are short-lived interventions, the lowest nocebo dropout rate has been observed in RLS (Silva et al., 2017) and during DMTs in MS (Papadopoulos & Mitsikostas, 2010). A possible explanation for the latter is that trial participants are not necessarily representative of the general population of MS patients, as the wide range of available MS treatments today has reduced the pool of untreated, potentially eligible patients for trials. Those MS patients who decide to participate in trials are likely to be more motivated and committed to adhere to the treatment regime (recruitment bias) (Papadopoulos & Mitsikostas, 2010).

The highest nocebo AE rate has been observed in MND (Shafiq et al., 2017) Although AEs that are reported in RCTs are classified as drug-related, a potential source of bias is that it can be difficult to distinguish whether symptoms arise subsequent to placebo administration or as a consequence of disease worsening. Especially, AEs of respiratory origin, which have been documented as drug-related in the respective RCTs, might as well be a disease complication per se (Shafiq et al., 2017).

The lowest nocebo AE rate has been observed during ST for headaches (Mitsikostas et al., 2011), closely followed by the nocebo AE rate during ST for MS. An interesting finding was that the nocebo AE rate during DMT is almost three times higher compared to ST for MS. This may reflect the fact that the meta-analysis included RCTs of already licensed drugs for ST (Papadopoulos & Mitsikostas, 2010).

7. NOCEBO: DISEASE OR TREATMENT SPECIFIC?

One of the major determinants of nocebo, if not the most important, is the uncertainty of drug action, which pervades all clinical trials, and particularly trials of brain diseases (Benedetti et al., 2016).

Although the nocebo rates are similar, the nature of nocebo AEs experienced might differ among placebo groups, and depends on the active drug (Amanzio et al., 2009; Mitsikostas et al., 2011). For example, in anti-migraine RCTs, anorexia and memory difficulties, which are typical AEs of anticonvulsants, were present only in the placebo arm of these trials. This suggests that the AEs in the placebo arms of clinical trials of anti-migraine medications might depend on the AEs of the active medication against which the placebo is compared, which is in accordance with the fact that alerting patients about side effects of the medication can turn into a fulfilling-prophecy (Amanzio et al., 2009).

On the other hand, Zaccara, Giovannelli, Giorgi, Franco, and Gasparini (2016) showed that nocebo rates in RCTs are affected by the clinical condition for which the experimental drug is given. For example, placebo-treated patients with obesity and binge eating disorders, and patients with pain, have significantly higher proportions of intolerable AEs requiring drug withdrawal than placebo-treated epilepsy patients in anticonvulsant RCTs.

8. THE BRAIN'S KEY ROLE IN NOCEBO

The neurobiological mechanisms of nocebo effects are still poorly understood (Blasini, Corsi, Klinger, & Colloca, 2017; Schienle, Höfler, Übel, & Wabnegger, 2018), however, there is evidence indicating that the brain plays a significant role in nocebo effects and responses (Benedetti et al., 2016). Experimental and clinical evidence suggests that an important component of nocebo effects is the uncertainty in clinical trials. Nocebo and placebo effects are related to expectations, such that the effect of an active ingredient (the drug) on a biological system (the patient) can be affected by the act of administering the treatment. Thus, the patient's expectations of clinical improvement can interfere with the specific therapeutic effects of both pharmacological and non-pharmacological treatments (Benedetti et al., 2016; Blasini et al., 2017).

Interestingly, CNS diseases show significantly higher nocebo AE rates compared to peripheral nervous system diseases (Dimitriadis & Zis, 2017; Häuser et al., 2012; Zis et al., 2018), which indirectly suggests that nocebo phenomena are centrally mediated, and might depend on the underlying pathophysiology and the disability caused by the disease itself.

The dynamic between nocebo and the brain is bidirectional. In an interesting electrophysiological study, Albu and Meagher (2016) showed that negative expectations can induce specific electroencephalography (EEG) changes. In brief, healthy subjects were randomly assigned to a nocebo or a control group. A five-minute EEG was recorded under resting state, tonic innocuous heat, and tonic noxious heat before and after the application of a sham inert cream to the non-dominant volar forearm, combined with cognitive manipulation. The intensity and unpleasantness of heat-induced pain increased after cognitive manipulation in the nocebo group compared to control, and was associated with enhanced low alpha (8–10 Hz) activity. However, changes in alpha activity were predicted by catastrophizing but not by pain intensity or unpleasantness, which suggests that low alpha power might reflect brain activity related to negative cognitive-affective responses to pain.

9. DIRECTIONS FOR THE FUTURE

Although the main brain disorders have been studied to date (headache, PD Alzheimer's disease, depression, epilepsy, MS and MND), there are still unexplored areas which should be investigated further. These include psychiatric (i.e., schizophrenia and bipolar disorder), vascular (i.e., acute ischemic stroke, vascular dementia), degenerative (i.e., frontotemporal dementia, Lewy body dementia) and other systemic atrophies of the brain (i.e., hereditary ataxias). Studying nocebo in neoplastic and infectious brain disorders is an unrealistic aim, since there are already available treatments for such disorders and, therefore, conducting new RCTs with a pure-placebo arm raises ethical issues.

Studying nocebo in clinical practice might also prove to be useful for the appropriate choice of treatment. To date, Q-No is the only validated tool that can be used in daily practice to predict nocebo responses (Mitsikostas & Deligianni, 2015).

10. CONCLUSIONS

Nocebo has been found to be very prevalent in various brain disorders, including headache, PD, Alzheimer's disease, depression, epilepsy, MS and MND. There are still unexplored areas which should be investigated further, given the clinical significance of nocebo effects and responses.

The consequences of nocebo effects and responses in clinical practice are important, as current treatments are based upon evidence confounded by the nocebo phenomenon. The available meta-analyses do not provide any direct evidence for nocebo in clinical practice, but as long as RCT findings predict the effectiveness and the safety of a treatment in clinical practice, post hoc safety analyses may provide important considerations relevant to daily clinical practice as well. In any case, clinicians should be aware that drug intolerance and treatment failure might be caused by nocebo, and should recruit individualized strategies to limit it.

The nocebo phenomenon has, however, implications for RCT design. Modifying informed consents (Colloca, 2017b; Colloca & Miller, 2011) for the tested pharmaceutical agents to include the chance of nocebo, blinding the investigators to the analysis of recorded AEs, paying more attention to the AEs reported by the participants, and requesting more detailed safety reports by the investigators, are potential interventions in future RCTs.

REFERENCES

Adewusi, J. K., Hadjivassiliou, M., Vinagre-Aragón, A., O'Connor, K. R., Khan, A., Grünewald, R. A., et al. (2018a). Sensory neuropathic symptoms in idiopathic Parkinson's disease: Prevalence and impact on quality of life. *Acta Neurologica Belgica*. https://doi.org/10.1007/s13760-018-0947-3 (May 23).

Adewusi, J. K., Hadjivassiliou, M., Vinagre-Aragón, A., O'Connor, K. R., Khan, A., Grünewald, R. A., et al. (2018b). Peripheral neuropathic pain in idiopathic Parkinson's disease: Prevalence and impact on quality of life; a case controlled study. *Journal of the Neurological Sciences*, *392*, 3–7. https://doi.org/10.1016/j.jns.2018.06.022.

Albu, S., & Meagher, M. W. (2016). Expectation of nocebo hyperalgesia affects EEG alpha-activity. *International Journal of Psychophysiology*, *109*, 147–152.

Amanzio, M., Corazzini, L. L., Vase, L., & Benedetti, F. (2009). A systematic review of adverse events in placebo groups of anti-migraine clinical trials. *Pain*, *146*(3), 261–269.

Baquero, M., & Martín, N. (2015). Depressive symptoms in neurodegenerative diseases. *World Journal of Clinical Cases*, *3*(8), 682–693.

Barsky, A. J., Saintfort, R., Rogers, M. P., & Borus, J. F. (2002). Nonspecific medication side effects and the nocebo phenomenon. *Journal of the American Medical Association*, *287*(5), 622–627.

Benedetti, F., Carlino, E., & Piedimonte, A. (2016). Increasing uncertainty in CNS clinical trials: The role of placebo, nocebo, and Hawthorne effects. *The Lancet. Neurology*, *15*(7), 736–747.

Benedetti, F., Lanotte, M., Lopiano, L., & Colloca, L. (2007). When words are painful: Unraveling the mechanisms of the nocebo effect. *Neuroscience*, *147*, 260–271.

Blasini, M., Corsi, N., Klinger, R., & Colloca, L. (2017). Nocebo and pain: An overview of the psychoneurobiological mechanisms. *Pain Reports*, *2*(2). https://doi.org/10.1097/PR9.0000000000000585 (pii: e585).

Burns, A., & Iliffe, S. (2009). Alzheimer's disease. *British Medical Journal*, *338*, b158.

Burns, P. B., Rohrich, R. J., & Chung, K. C. (2011). The levels of evidence and their role in evidence-based medicine. *Plastic and Reconstructive Surgery*, *128*(1), 305–310.

Colloca, L. (2017a). Nocebo effects can make you feel pain. *Science*, *358*(6359), 44. https://doi.org/10.1126/science.aap8488.

Colloca, L. (2017b). Tell me the truth and I will not be harmed: Informed consents and nocebo effects. *The American Journal of Bioethics*, *17*(6), 46–48. https://doi.org/10.1080/15265161.2017.1314057.

Colloca, L., & Grillon, C. (2014). Understanding placebo and nocebo responses for pain management. *Current Pain and Headache Reports*, *18*(6), 419.

Colloca, L., & Miller, F. G. (2011). The nocebo effect and its relevance for clinical practice. *Psychosomatic Medicine*, *73*(7), 598–603. https://doi.org/10.1097/PSY.0b013e3182294a50.

Compston, A., & Coles, A. (2008). Multiple sclerosis. *The Lancet*, *72*(9648), 1502–1517.

Connor, J. R., Boyer, P. J., Menzies, S. L., Dellinger, B., Allen, R. P., Ondo, W. G., et al. (2003). Neuropathological examination suggests impaired brain iron acquisition in restless legs syndrome. *Neurology*, *61*(3), 304–309.

Dimitriadis, P. A., & Zis, P. (2017). Nocebo effect in Meniere's disease: A meta-analysis of placebo-controlled randomized controlled trials. *Otology and Neurotology*, *38*(9), 1370–1375.

Dodd, S., Schacht, A., Kelin, K., Dueñas, H., Reed, V. A., Williams, L. J., et al. (2015). Nocebo effects in the treatment of major depression: Results from an individual study participant-level meta-analysis of the placebo arm of duloxetine clinical trials. *Journal of Clinical Psychiatry*, *76*(6), 702–711.

Elsenbruch, S., Schmid, J., Bäsler, M., Cesko, E., Schedlowski, M., & Benson, S. (2012). How positive and negative expectations shape the experience of visceral pain: An experimental pilot study in healthy women. *Neurogastroenterology and Motility*, *24*(10), 914-e460.

Enck, P., Benedetti, F., & Schedlowski, M. (2008). New insights into the placebo and nocebo responses. *Neuron, 59*, 195–206.

GBD 2015 Neurological Disorders Collaborator Group. (2017). Global, regional, and national burden of neurological disorders during 1990-2015: A systematic analysis for the Global Burden of Disease Study 2015. *The Lancet. Neurology, 16*(11), 877–897.

Hadjikhani, N., Sanchez Del Rio, M., Wu, O., Schwartz, D., Bakker, D., Fischl, B., et al. (2001). Mechanisms of migraine aura revealed by functional MRI in human visual cortex. *Proceedings of the National Academy of Sciences of the United States of America, 98*(8), 4687–4692.

Häuser, W., Bartram, C., Bartram-Wunn, E., & Tölle, T. (2012). Adverse events attributable to nocebo in randomized controlled drug trials in fibromyalgia syndrome and painful diabetic peripheral neuropathy: Systematic review. *Clinical Journal of Pain, 28*(5), 437–451.

Kalia, L. V., & Lang, A. E. (2015). Parkinson's disease. *The Lancet, 386*(9996), 896–912.

Keitel, A., Ferrea, S., Südmeyer, M., Schnitzler, A., & Wojtecki, L. (2013). Expectation modulates the effect of deep brain stimulation on motor and cognitive function in tremor-dominant Parkinson's disease. *PLoS One, 8*(12), e81878.

Keitel, A., Wojtecki, L., Hirschmann, J., Hartmann, C. J., Ferrea, S., Südmeyer, M., et al. (2013). Motor and cognitive placebo-/nocebo-responses in Parkinson's disease patients with deep brain stimulation. *Behavioural Brain Research, 250*, 199–205.

Kotzbauer, P. T., Trojanowsk, J. Q., & Lee, V. M. (2001). Lewy body pathology in Alzheimer's disease. *Journal of Molecular Neuroscience, 17*(2), 225–232.

Lidstone, S. C. (2014). Great expectations: The placebo effect in Parkinson's disease. In F. Benedetti, P. Enck, E. Frisaldi, & M. Schedlowski (Eds.), *Placebo. Handbook of experimental pharmacology:* Vol. 225. Berlin, Heidelberg: Springer.

Lublin, F. D., & Reingold, S. C. (1996). Defining the clinical course of multiple sclerosis: Results of an international survey. National Multiple Sclerosis Society (USA) Advisory Committee on Clinical Trials of New Agents in Multiple Sclerosis. *Neurology, 46*(4), 907–911.

Manchikanti, L., Giordano, J., Fellows, B., & Hirsch, J. A. (2011). Placebo and nocebo in interventional pain management: A friend or a foe—Or simply foes? *Pain Physician, 14*(2), E157–E175.

Mestre, T. A., Lang, A. E., & Okun, M. S. (2016). Factors influencing the outcome of deep brain stimulation: Placebo, nocebo, lessebo, and lesion effects. *Movement Disorders, 31*(3), 290–296.

Mitsikostas, D. D. (2012). Nocebo in headaches: Implications for clinical practice and trial design. *Current Neurology and Neuroscience Reports, 12*(2), 132–137.

Mitsikostas, D. D., Chalarakis, N. G., Mantonakis, L. I., Delicha, E., & Sfikakis, P. P. (2012). Nocebo in fibromyalgia: Meta-analysis of placebo-controlled clinical trials and implications for practice. *European Journal of Neurology, 19*, 672–680.

Mitsikostas, D. D., & Deligianni, C. I. (2015). Q-No: A questionnaire to predict nocebo in outpatients seeking neurological consultation. *Neurological Sciences, 36*(3), 379–381.

Mitsikostas, D. D., Mantonakis, L. I., & Chalarakis, N. G. (2011). Nocebo is the enemy, not placebo. A meta-analysis of reported side effects after placebo treatment in headaches. *Cephalalgia, 31*(5), 550–561.

Mitsikostas, D. D., Mantonakis, L., & Chalarakis, N. (2014). Nocebo in clinical trials for depression: A meta-analysis. *Psychiatry Research, 215*(1), 82–86.

Moher, D., Liberati, A., Tetzlaff, J., Altman, D. G., & The PRISMA Group. (2009). Preferred reporting items for systematic reviews and meta-analyses: The PRISMA statement. *BMJ, 339*, b2535.

Moshé, S. L., Perucca, E., Ryvlin, P., & Tomson, T. (2015). Epilepsy: New advances. *The Lancet, 385*(9971), 884–898.

Nakahara, J., Maeda, M., Aiso, S., & Suzuki, N. (2012). Current concepts in multiple sclerosis: Autoimmunity versus oligodendrogliopathy. *Clinical Reviews in Allergy & Immunology*, *42*(1), 26–34.

Papadopoulos, D., & Mitsikostas, D. D. (2010). Nocebo effects in multiple sclerosis trials: A meta-analysis. *Multiple Sclerosis*, *16*(7), 816–828.

Papadopoulos, D., & Mitsikostas, D. D. (2012). A meta-analytic approach to estimating nocebo effects in neuropathic pain trials. *Journal of Neurology*, *259*(3), 436–447.

Rutherford, B. R., Wall, M. M., Glass, A., & Stewart, J. W. (2014). The role of patient expectancy in placebo and nocebo effects in antidepressant trials. *Journal of Clinical Psychiatry*, *75*(10), 1040–1046.

Schienle, A., Höfler, C., Übel, S., & Wabnegger, A. (2018). Emotion-specific nocebo effects: An fMRI study. *Brain Imaging and Behavior*, *12*(1), 180–187.

Scott, D. J., Stohler, C. S., Egnatuk, C. M., Wang, H., Koeppe, R. A., & Zubieta, J. K. (2008). Placebo and nocebo effects are defined by opposite opioid and dopaminergic responses. *Archives of General Psychiatry*, *65*(2), 220–231.

Shafiq, F., Mitsikostas, D. D., & Zis, P. (2017). Nocebo in motor neuron disease: Systematic review and meta-analysis of placebo-controlled clinical trials. *Amyotrophic Lateral Sclerosis and Frontotemporal Degeneration*, *18*(7–8), 576–582.

Silva, M. A., Duarte, G. S., Camara, R., Rodrigues, F. B., Fernandes, R. M., Abreu, D., et al. (2017). Placebo and nocebo responses in restless legs syndrome: A systematic review and meta-analysis. *Neurology*, *88*(23), 2216–2224.

Stathis, P., Smpiliris, M., Konitsiotis, S., & Mitsikostas, D. D. (2013). Nocebo as a potential confounding factor in clinical trials for Parkinson's disease treatment: A meta-analysis. *European Journal of Neurology*, *20*(3), 527–533.

Steiner, T. J., Stovner, L. J., Vos, T., Jensen, R., & Katsarava, Z. (2018). Migraine is first cause of disability in under 50s: Will health politicians now take notice? *The Journal of Headache and Pain*, *19*(1), 17.

The Parkinson Study Group. (2004). Levodopa and the progression of Parkinson's disease. *New England Journal of Medicine*, *351*, 2498–2508.

Trenkwalder, C., Paulus, W., & Walters, A. S. (2005). The restless legs syndrome. *The Lancet. Neurology*, *4*(8), 465–475.

World Health Organization. (1992). *International classification of diseases, 10th revision (ICD-10).* Geneva, Switzerland: World Health Organization.

Zaccara, G., Giovannelli, F., Cincotta, M., Loiacono, G., & Verrotti, A. (2015). Adverse events of placebo-treated, drug-resistant, focal epileptic patients in randomized controlled trials: A systematic review. *Journal of Neurology*, *262*(3), 501–515.

Zaccara, G., Giovannelli, F., Giorgi, F. S., Franco, V., & Gasparini, S. (2016). Analysis of nocebo effects of antiepileptic drugs across different conditions. *Journal of Neurology*, *263*(7), 1274–1279.

Zis, P., Daskalaki, A., Bountouni, I., Sykioti, P., Varrassi, G., & Paladini, A. (2017). Depression and chronic pain in the elderly: Links and management challenges. *Clinical Interventions in Aging*, *12*, 709–720.

Zis, P., Dickinson, M., Shende, S., Walker, Z., & Strydom, A. (2012). Oxidative stress and memory decline in adults with Down syndrome: Longitudinal study. *Journal of Alzheimer's Disease*, *31*(2), 277–283.

Zis, P., Erro, R., Walton, C. C., Sauerbier, A., & Chaudhuri, K. R. (2015). The range and nature of non-motor symptoms in drug-naive Parkinson's disease patients: A state-of-the-art systematic review. *NPJ Parkinson's Disease*, *1*, 15013.

Zis, P., Grünewald, R. A., Chaudhuri, R. K., & Hadjivassiliou, M. (2017). Peripheral neuropathy in idiopathic Parkinson's disease: A systematic review. *Journal of the Neurological Sciences*, *378*, 204–209.

Zis, P., Hadjivassiliou, M., Sarrigiannis, P. G., Jenkins, T. M., & Mitsikostas, D. D. (2018). Nocebo in chronic inflammatory demyelinating polyneuropathy; a systematic review and meta-analysis of placebo-controlled clinical trials. *Journal of the Neurological Sciences, 15*(388), 79–83.

Zis, P., Martinez-Martin, P., Sauerbier, A., Rizos, A., Sharma, J. C., Worth, P. F., et al. (2015). Non-motor symptoms burden in treated and untreated early Parkinson's disease patients: Argument for non-motor subtypes. *European Journal of Neurology, 22*(8), 1145–1150.

Zis, P., McHugh, P., McQuillin, A., Praticò, D., Dickinson, M., Shende, S., et al. (2014). Memory decline in Down syndrome and its relationship to iPF2alpha, a urinary marker of oxidative stress. *PLoS One, 9*(6), e97709.

Zis, P., & Mitsikostas, D. D. (2015). Nocebo in Alzheimer's disease; meta-analysis of placebo-controlled clinical trials. *Journal of the Neurological Sciences, 355*(1–2), 94–100.

Zis, P., Rizos, A., Martinez-Martin, P., Pal, S., Silverdale, M., Sharma, J. C., et al. (2014). Non-motor symptoms profile and burden in drug naïve versus long-term Parkinson's disease patients. *Journal of Parkinson's Disease, 4*(3), 541–547.

Zis, P., Shafiq, F., & Mitsikostas, D. D. (2017). Nocebo effect in refractory partial epilepsy during pre-surgical monitoring: Systematic review and meta-analysis of placebo-controlled clinical trials. *Seizure, 45*, 95–99.

Zis, P., & Strydom, A. (2018). Clinical aspects and biomarkers of Alzheimer's disease in Down syndrome. *Free Radical Biology and Medicine, 114*, 3–9.

Zis, P., Strydom, A., Buckley, D., Adekitan, D., & McHugh, P. C. (2017). Cognitive ability in Down syndrome and its relationship to urinary neopterin, a marker of activated cellular immunity. *Neuroscience Letters, 636*, 254–257.

FURTHER READING

Heimer, G., Rivlin-Etzion, M., Bar-Gad, I., Goldberg, J. A., Haber, S. N., & Bergman, H. (2006). Dopamine replacement therapy does not restore the full spectrum of normal pallidal activity in the 1-methyl-4-phenyl-1,2,3,6-tetra-hydropyridine primate model of Parkinsonism. *The Journal of Neuroscience, 26*, 8101–8114.

CHAPTER EIGHTEEN

The Ethics of Placebo Effects in Clinical Practice and Research

Marco Annoni[1]
Consiglio Nazionale delle Ricerche (CNR)—Istituto di Tecnologie Biomediche (ITB), Rome, Italy
Fondazione Umberto Veronesi, Milan, Italy
[1]Corresponding author: e-mail address: marco.annoni@cnr.itb.it

Contents

Abstract

This chapter provides a synthetic overview of the contemporary debate over the ethics of placebos and placebo effects in both clinical and research contexts. Section 1 briefly reconstructs how ethical attitudes toward the use of placebos have changed during the last century following the emergence of autonomy in medical ethics. Next, Sections 2–4 chart the main ethical issues concerning the use of placebos in clinical settings, examining: the ethics of deceptive placebos; the ethics of placebos without deception; and the ethics of modulating placebo and nocebo effects without placebos. Finally, Section 5 outlines the role of placebos in research, discussing the ethics of placebos as controls in clinical experiments.

1. THE ETHICS OF PLACEBOS IN HISTORY: FROM THE "PIOUS FRAUD" TO INFORMED CONSENT

For centuries, the administration of deceptive placebos has been frequently practiced and yet rarely condemned. In 1807, Thomas Jefferson observed that "one of the most successful physicians I have ever known has assured me that he used more bread pills, drops of colored water, and

International Review of Neurobiology, Volume 139
ISSN 0074-7742
https://doi.org/10.1016/bs.irn.2018.07.031

powder of hickory ashes, than of all other medication put together"; he notoriously defined this practice as the "pious fraud" (Jefferson, 1898, p. 199). Similarly, almost a century later, Harvard physician Cabot (1903, p. 348) noted that "I was brought up, as I suppose every physician is, to use placebo, bread pills, water subcutaneously, and other devices." Still, in 1954, a *Lancet* article described the placebo as a common "means of reinforcing a patient's confidence in his recovery, when the diagnosis is undoubted and no more effective treatment is possible" (Anon., 1954, p. 321).

This favorable attitude toward benevolent deception was rooted in the classical paternalistic paradigm that dominated traditional medicine. On this view, the ethics of the doctor–patient relationship was grounded primarily on considerations of beneficence and non-maleficence. Accordingly, the ethics of doctor–patient communication was inspired by the same moral principle that has guided medicine since the times of Hippocrates: "Help, or at least do not harm." As the use of "bread pills" and "drops of colored water" was deemed sometimes helpful and rarely harmful, the administration of deceptive placebos has often been excused, sometimes expected, and generally considered one of the hallmarks of the doctor–patient relationship (Annoni & Miller, 2014). The same was true also with respect to the use of placebos in medical research. Since most medical experimentations was conducted by doctors on their patients, the use of placebos for scientific purposes was regulated according to the same moral principles adopted in clinical medicine: if a covert placebo was deemed useful for advancing the doctor's knowledge, and sufficiently innocuous or beneficial for the patient, its administration was generally considered morally permissible, even without the patient's awareness and explicit consent.

After the Second World War, two factors conspired in revolutionizing ethical attitudes toward the use of placebos in medicine. The first was the "rediscovery" of autonomy as one of the core principles of medical ethics (Faden & Beauchamp, 1986). The 1950s—and in partial response to events such as the Nuremberg Trial (Weindling, 2011) and the Tuskegee study (Jones, 2011)—saw the establishing of the fundamental ethical principle that every person has the right to refuse participation in research and medical treatments. In the following decades, the consolidation of this principle led to overcoming the old model based on medical paternalism, replacing it with a new model grounded in respect for autonomy and informed consent. As a consequence, the practice of administering placebos without patients' consent began to attract ethical scrutiny and criticism.

A second factor, then, has been the series of empirical studies that, in the last decades, have unraveled the neurobiological mechanisms of placebo and nocebo effects (Benedetti, 2015; Kaptchuk & Miller, 2015). These studies have demonstrated not only that placebo and nocebo effects are ubiquitous in research and clinical settings, but also that they can modulate symptoms across highly prevalent conditions such as pain, depression, migraine, Parkinson's disease, or irritable bowel syndrome (Benedetti, 2015; Colloca, Flatten, & Meissner, 2013; Wager & Atlas, 2015). Together, these findings have revolutionized the prior understanding of the placebo phenomenon, posing the issue of how doctors and researchers could ethically harness placebo and nocebo effects in medical settings.

Against this background, this chapter provides a synthetic overview of the contemporary debate over the ethics of placebos and placebo effects in clinical and research contexts. What follows is divided in two conceptual parts. The first three sections chart the main ethical issues concerning the use of placebos in clinical settings, examining: the ethics of deceptive placebos, the ethics of placebos without deception, and the ethics of modulating placebo and nocebo effects without placebos. Finally, the last section focuses on the role of placebos in research, outlining the main ethical issues pertaining to the use of placebos as controls in randomized clinical trials.

2. THE ETHICS OF DECEPTIVE PLACEBOS

In the last decade a series of empirical surveys have investigated clinicians' attitudes toward the use of placebos in clinical settings. A 2008 study among US internists and rheumatologists concluded that between 46% and 58% had prescribed placebo treatments at least once (Tilburt, Emanuel, Kaptchuk, Curlin, & Miller, 2008). Similarly, a 2013 survey among UK doctors found that 97% of the participants reported having used placebos, and 77% use them at least once a week (Howick et al., 2013) The only systematic review that has been published (22 studies published between 1973 and 2009) concluded that between 41% and 99% of health professionals use deceptive placebos at least once a year (Fässler, Meissner, Schneider, & Linde, 2010). Perhaps unsurprisingly, these studies suggest that covert and deceptive placebos are still widely used across clinical settings, especially in general practice.

The use of deceptive placebos in clinical medicine, however, is ethically problematic (Annoni, 2015; Colloca, 2014; Miller & Colloca, 2009).

Consider a clinician administering a pill by saying "This is a powerful analgesic containing morphine: it will ease your pain." After taking the pill, the patient reports a decrease in pain; the pill, however, was just a placebo. This case exemplifies the ethical dilemma at the center of the classic placebo debate. On the one hand, placebos may benefit patients by positively modulating their symptoms through placebo effects and by providing psychological relief; on the other hand, their prescription involves deception. Is it ever ethical for a clinician to administer deceptive placebos for the patients' own good? To chart the main answers to this question, at the outset it is useful to appreciate the main rationales that are usually invoked to criticize the use of deceptive placebos.

First, there are considerations related to autonomy and informed consent. Autonomy is one of the core values in contemporary medical ethics, law and political theory (Faden & Beauchamp, 1986). In medical ethics, the concept of autonomy is usually defined in terms of "self-rule that is free from both controlling interferences by others and from certain limitations such as an inadequate understanding that prevents meaningful choice" (Beauchamp & Childress, 2009, p. 99). Accordingly, the principle of respect for autonomy is understood as the conjunction of two main ethical obligations (Coggon & Miola, 2011). First, clinicians have an obligation to respect their patients' *liberty*, i.e., to refrain from interfering with the patients' decisions and actions. Second, clinicians have an obligation to respect patients' *agency*, i.e., to provide all the necessary information in order to foster patients' understanding. Respecting both obligations is a precondition for securing the patients' informed consent: obviously, one cannot autonomously consent to (or veto) a treatment if one is coerced, manipulated, or has not been adequately informed about the clinical situation and the available therapeutic options. For instance, in the example above, the doctor's deception prevented the patient from autonomously consenting or refusing the placebo. Thus, prescribing deceptive placebos infringes on patients' autonomy and their right to informed consent.

Second, there are considerations of trust. Trust is important in all cooperative settings, but it is especially important in medicine (Pellegrino, 1991). Without trust, it would be impossible to have meaningful therapeutic relationships as well as to secure the recruitment of volunteers in clinical research (Eyal, 2014). Moreover, trust is essential in all therapeutic encounters with non-autonomous agents, such as infants or persons suffering from severe cognitive impairments. Therefore, cultivating trusting relationships is a key to achieve the goals of medicine. However, trust-based relationships are also frail. Once shattered, trust may be difficult if not impossible to repair. Patients who discover that they have received a deceptive placebo might legitimately feel duped and manipulated. This might lead them to question clinicians' moral and professional integrity, with negative consequences for the quality of the therapeutic relationship. For instance, Cabot (1903) recounted:

> *[...] one day a patient caught me in the attempt to put her to sleep by means of a subcutaneous injection of water. 'I saw you get that ready', said she, 'and there is no morphin in it; you were just trying to deceive me'. I was fairly caught and there was no use trying to bluff it out, so I merely protested that my deception was well meant. [...] 'Of course I see that', she said, 'but how am I to know in future what other tricks you will think it best to play me for my good? How am I to believe anything you say from now on?' (p. 348).*

As this example shows, just one uncovered placebo—even if "well-meant"—may irremediably jeopardize patients' trust.

Third, there are considerations of non-maleficence. It has been argued that "since placebos are inert, it is natural to expect that they are completely harmless, unlike active medications that frequently elicit unwanted side effects" (Foddy, 2009, p. 5). The idea that placebos are "inert" and therefore "harmless," however, is misleading: deceptive placebos may harm individual patients and society in different and important ways. First, placebos are always relatively "inert," but never absolutely so: sugar pills are not "inert" for diabetics, and lactose tablets are not "inert" for lactose-intolerant patients. Also, placebos can induce psychological addiction. Another risk is for patients to overlook their symptoms: by "walking away" with a prescription, one may not look for a second opinion or further diagnostic examinations. Also, many unnecessary diagnostic procedures "are undertaken to give patients a sense that efforts are being made on their behalf" (Bok, 1974, p. 21). Many involve the expenditure of time and money, and some carry non-negligible risks. Prescribing deceptive placebos has also implications for public health: for instance, antibiotics prescribed as placebos may increase bacterial resistance. Finally, deceptive placebos may contribute to the medicalization of society by promoting the false belief that there is "a pill for every ill" (Miller & Colloca, 2009). In sum, the administration of deceptive placebos may harm both patients and society in multiple ways.

Against this backdrop, one can argue either that deceptive placebos are always unethical; or that sometimes deceptive placebos are morally justifiable provided certain conditions are met. Today ethical guidelines tend to endorse the first alternative, considering deceptive placebos always unethical without prior consent. This is the position, for instance, of the American Medical Association (AMA), in which it is stated that "In the clinical setting, the use of a placebo without the patient's knowledge may undermine trust, compromise the patient-physician relationship, and result in medical harm to the patient. Physicians may use placebos for diagnosis or treatment only if they… obtain the patient's general consent" (American Medical Association, 2018, Opinion 2.1.4). According to this view, the use of deceptive placebos is *never* permissible without patients' consent. Aside from the AMA, other scholars and institutions have defended similar categorical views (Asai & Kadooka, 2013).

The proposal to categorically ban deceptive placebos has the advantage of simplicity: in any case, doctors will always know what to do—i.e., avoid deceptive placebos without consent. However, this position has also important limitations. In particular, advocates of a categorical ban may justify it either by arguing that all instances of paternalistic deception are unethical, and thus also the paternalistic administration of deceptive placebos; or by arguing that a categorical ban is justified *only* for deceptive placebos. Both strategies face additional problems and counter-objections.

The first possibility is to argue that deception—or, at least, lying—is always unethical. Usually, this position is defended by scholars endorsing a Kantian approach for which the respect of autonomy is paramount (Koorsgaard, 1996). These categorical positions, however, lead to counter-intuitive consequences in medical contexts. Consider the case of a patient who has a heart-attack upon learning that his son has had an accident. Before the doctor intervenes, the patient asks whether his son has survived; unfortunately, he has not. What should the doctor say? If the doctor tells the truth, the patient's condition will likely worsen, increasing his chances of dying. If, instead, the doctor lies, he might increase the chances of saving the patient's life. With respect to these scenarios, the prevailing position in medical ethics is that doctors' duties of beneficence and maleficence may legitimately override the duty of veracity (Annoni, 2018; Beauchamp & Childress, 2009; Sokol, 2007). On this view, telling the truth is an important ethical obligation in clinical settings; yet it is not as important as saving patients' lives. In sum, the view that deception is never morally permissible in medicine is seldom endorsed in medical ethics, and never without important qualifications.

The second alternative is to defend a categorical ban that applies only to deceptive placebos. Advocates of this position may justify the categorical prohibition on precautionary grounds: given their potentially harmful consequences for trust, patients' health, and society—it is argued—it is better to prohibit deceptive placebos altogether in clinical contexts. Whether this position sounds appealing depends on the perceived harm and benefits of deceptive placebos. The greater the harm with respect to the benefits, the more the categorical ban appears reasonable; vice versa, the greater the benefits with respect to the harm, the less it appears justifiable.

In difference to categorical views, in recent years several scholars have argued that sometimes deceptive placebos are justifiable in clinical contexts. Generally, the crux of these arguments consists in showing that the clinical benefits of deceptive placebos clearly outweigh their potential harm and negative implications for autonomy and trust. To do so, proponents of these views usually stress the therapeutic and diagnostic utility of deceptive placebos.

As for therapeutic utility, robust empirical evidence supports the claim that the effectiveness of deceptive placebos may sometimes match or surpass that of conventional medications for conditions like pain, depression, recurring migraine, and irritable bowel syndrome (Benedetti, 2015). Deceptive placebos may be a viable therapeutic option, especially when other conventional therapies have already proven ineffective (Cohen & Shapiro, 2013; Foddy, 2009). A possible reply to this view is that the evidence supporting the therapeutic utility of placebo is not conclusive and should not be overestimated (Castelnuovo et al., 2018). Placebo and nocebo effects are still phenomena not entirely understood, and they are far too unpredictable and variable between individuals to be considered a reliable therapeutic option (Benedetti, 2015; Miller & Colloca, 2011).

A second possibility is to argue that deceptive placebos may have significant diagnostic utility. For example, it has been argued that deceptive placebos can be used to distinguish patients with epilepsy from patients with pseudo-seizures, which are attacks indistinguishable from epileptic seizures but have a psychological origin (Kolber, 2007; Slater, Brown, Jacobs, & Ramsay, 1995). Empirical studies have demonstrated that doctors can use deceptive placebos to induce seizures in patients with pseudo-seizures, but not in patients with epilepsy (Slater et al., 1995). Moreover, standard diagnostic tests may be prohibitively expensive, as they require observing the encephalography of patients during actual seizures, while the use of deceptive placebos has been found superior to standard clinical observation, which yields high risks of misdiagnosis. Thus, in this case, it could be argued that recommending a deceptive placebo could lead to significant clinical benefits for the patient who receives an otherwise difficult diagnosis. Yet performing this diagnostic test would be unethical under a categorical placebo policy like that of the AMA.

In general, arguments based on the clinical utility of placebos may provide a compelling case to reject categorical bans: if one concedes that paternalistic deception may be ethical where its benefits outweigh its potential harm; and if deceptive placebos may have significant clinical benefits and negligible harm, then one could argue that deceptive placebos may be morally justified. The strength of these arguments, however, depends both on the empirical evidence supporting the case for the therapeutic utility of deceptive placebos, and on the meta-ethical view which one endorses: for instance, supporters of categorical bans for reasons of autonomy will likely not be persuaded by arguments regarding the clinical benefits of deceptive placebos.

A final argument in favor of deceptive placebos is that they might be consistent with patients' preferences (Kolber, 2007). Some patients may actually "want to be locally manipulated [through a placebo] for the sake of achieving [a] larger goal," i.e., symptomatic relief (Cohen & Shapiro, 2013, p. 703). At present, empirical evidence on patients' attitudes is scarce. A 1993 Swedish study found that 25% of interviewed patients agreed completely or for the most part that physicians ought to prescribe more placebos, while 63% agreed that it is acceptable to administer placebos to dying cancer patients who will not discover the truth (Lynöe, Mattsson, & Sandlund, 1993). A survey of US patients found that "most respondents (50%–84%) judged it acceptable for doctors to recommend placebo treatments"; only 21.9% judged that it was never acceptable for doctors to recommend placebos (Hull et al., 2013). More recently, a large survey with 853 US patients found that most respondents felt it was acceptable for a physician to use placebos (50%–85%), citing the lack of harm and the potential benefits as major justifications. Interestingly, of the minority of respondents who judged it unacceptable to administer placebo treatments, the most cited rationale was not the doctors' duty of veracity, but their obligation to do more—e.g., more diagnostic examinations or more use of active rather than placebo therapies (Ortiz, Hull, & Colloca, 2016).

These findings may reinforce the case for deceptive placebos in two ways. First, if doctors' decisions are aligned with patients' preferences, it may be argued that recommending deceptive placebos constitutes only a minor infringement of patients' autonomy (Kolber, 2007). Second, such placebos could mitigate the concern for trust. Patients who favorably consider the prescription of deceptive placebos may not consider the discovery of a placebo as a betrayal of their trust; perhaps some may even consider it a sign of doctor's commitment to their well-being. However, two cautionary remarks are in order. First, meeting patients' unexpressed preferences is not the same as respecting patients' autonomy. As noted by Barnhill (2011, p. 228), the fact "that a patient's unexpressed preferences are met doesn't mean her autonomy is respected because respecting her autonomy requires letting her make an actual informed decision about her treatment." Thus, deceptive placebos without prior consent may be in line with patients' unexpressed preferences and still violate their autonomy and right to informed consent. Second, patients' unexpressed preferences may not be predictive of their actual reactions. Patients may agree that, in general, physicians should use more placebos; nonetheless, they may still feel betrayed when they learn that *they* were the ones deceived. Importantly, one of the above surveys found that respondents valued honesty by physicians regarding the use of placebos and believed that non-transparent use could undermine the relationship between patients and physicians (Hull et al., 2013). Therefore, even if some patients might not consider a deceptive placebo as a betrayal of their trust, some will, with potentially negative consequences for their therapeutic relationships.

3. THE ETHICS OF PLACEBOS WITHOUT DECEPTION

In recent years researchers and scholars have proposed different ways to avoid the ethical hurdles traditionally associated with deceptive placebos. In particular, three strategies have been proposed: (i) to use neither "deceptive" nor "transparent" verbal disclosures; (ii) to prescribe "open-label" placebos—i.e., placebos without deception; and (iii) to implement forms of "negative informed consent" or "authorized concealment." Not all these strategies, however, are equally successful.

The first possibility is that of questioning the standard definition of "deception." Usually, deception is defined as "intentionally causing someone to have a false belief that the deceiver believes to be false" (Carson, 2010, p. 46). However, consider the following disclosure as a way of introducing a placebo:

> *(a) "I am prescribing you a pill which research suggests can be of benefit to you. In your circumstances I have reason to believe that it will work, with a minimum of side effects"*
>
> ***(Gold & Lichtenberg, 2014, p. 221).***

This disclosure is not transparent, as the patient is not informed that the pill is a placebo. Yet—it has been argued—it is also not deceptive, as it is factually true that placebos may positively modulate patients' symptoms (Cohen & Shapiro, 2013; Gold & Lichtenberg, 2014). If this argument holds, then it would be possible to prescribe placebos, avoiding the classic ethical dilemmas associated with deception.

This line of reasoning, however, is problematic for three reasons (Annoni, 2018; Annoni & Miller, 2014). First, it assumes a too narrow conception of doctors' duty of veracity. To respect the patients' autonomy and their right to informed consent, doctors must tell patients the truth. Yet the commandment "tell the truth!" is ambiguous, as it can be understood in two senses: as a negative obligation not lie and deceive (a duty of truthfulness); and as a positive obligation to disclose all the information that could be relevant for respecting and fostering patients' autonomy (a duty to inform). Arguably, disclosing the nature of prescribed medications does make a conceivable difference for patients' autonomy. For instance, in an emergency situation, not knowing what medicines a patient is taking may lead a clinician to withhold a treatment in order to avoid harmful drug interactions (Kolber, 2007). Thus, although (a) may not contain false information, it still violates the duty to inform patients in order to respect their autonomy and right to informed consent.

Furthermore, it is also possible to deceive by omission, i.e., by failing to correct a false belief that we know is false (or, at least, not true) (Carson, 2010; Chisholm & Feehan, 1977). In contemporary clinical settings patients may reasonably expect that all prescribed medicines have been tested and approved for their specific efficacy. To deliberately contravene this widely held belief qualifies as deception. As noted by Bok (1974):

> *The statement that a placebo may help a patient is not a lie or even, in itself, deceitful. Yet the circumstances in which a placebo is prescribed introduce an element of deception [...] they convey the impression that the treatment prescribed will have the ingredients necessary to improve the patient's condition. The actions of the physician are therefore deceptive even if the words are so general as not to be lies. (p. 20)*

Thus, in today's clinical contexts, it can be argued that (a) violates clinicians' duty of truthfulness.

A third reason is that omissions, deceptions, and lies are often part of one and the same continuum of dishonesty, and they tend to overlap (Annoni, 2018; Bok, 1978). For instance, what if a patient starts asking questions about the prescribed pill? How will the doctor explain that the pill is "for free," it comes without a labeled bottle or a package leaflet, and it is impossible to find more information on it online? Presumably, at some point, the doctor will be forced either to disclose the attempted deception or to lie. Thus, although (a) may not contain false information per se, it could easily lead to more dishonesty on doctor's part if patients ask for more information. In sum, the administration of a placebo can be either honest or dishonest: there is no moral "middle ground" that can be clinically exploited without also violating the clinicians' duty of veracity.

A more promising strategy, then, is to recommend "open-label placebos," or placebos without deception (Colloca & Howick, 2018; Kaptchuk et al., 2010). Traditionally, it has been held that placebos must be deceptive in order to be effective. However, in the last decade, a growing number of empirical studies have demonstrated that placebos may be effective even without deception (Charlesworth et al., 2017; Kaptchuk et al., 2010). Yet, while these studies provide an important "proof of principle," two cautionary considerations are in order. First, all these studies had a small sample size and often lacked adequate masking procedures (Charlesworth et al., 2017). Before considering open-label placebos a viable therapeutic option, further studies are needed to elucidate their mechanisms of action. For example, at present it is still unclear whether open-label placebos are as effective as deceptive placebos, or whether they imply a trade-off between veracity and effectiveness. Second, it is still debated whether, and to what extent, open-label placebos can be introduced to patients without equivocation or unethical manipulation of their expectations to promote placebo responses (Blease, Colloca, & Kaptchuk, 2016; Charlesworth et al., 2017; Justman, 2013). Therefore, at present, it is still premature to consider open-label placebos as viable therapeutic options.

A third strategy is to introduce concealed placebos in a way compatible with the respect for autonomy. For instance, the AMA (2018) suggests that a placebo

> *may still be effective if the patient knows it will be used but cannot identify it and does not know the precise timing of its use...The physician does not need to identify precisely when the placebo will be administered. In this way, the physician respects the patient autonomy and fosters a trusting relationship, while the patient may still benefit from the placebo effect. (Opinion 2.1.4)*

Patients may consent in advance to the use of placebos for diagnostic and therapeutic purposes, while at the same time waiving their right to be specifically informed about the timing of the administration. In this way, it could be argued, the patients' autonomy is respected because patients have actually consented to the use of placebos and not to receive specific information about their use.

Shaw (2009) has developed a similar view according to which placebos can be ethically administered if patients have previously and explicitly consented not to receive certain information about the chemical composition or efficacy of their prescriptions. Shaw's view is grounded in Kihlbom's proposal (2008) of complementing the framework of informed consent with the possibility for patients to express a preliminary "negatively informed consent," to give their voluntary and explicit consent to undergo the treatment, and to expresses their voluntary and explicit wish not to have more information. By implementing this approach, patients would entrust their clinicians with the decision about what needs to be revealed in each case. This approach may thus open new options for physicians considering the clinical use of placebos. However, it has two main limitations. First, warning patients in advance that they could receive a placebo may be self-defeating, as it might lower the patient's expectations, thus reducing the diagnostic and therapeutic utility of covert placebos. Moreover, increasing the suspicion that they are being treated with placebos may reduce the patients' compliance with effective medications.

4. THE ETHICS OF MODULATING PLACEBO AND NOCEBO EFFECTS WITHOUT PLACEBOS

Modulating patients' health outcomes through placebo and nocebo effects represents a relatively new, and partly uncharted, territory for empirical and ethical inquiry. Although empirical studies have shown that both contextual and behavioral variables may be associated with placebo and nocebo effects, to date scholars have focused mainly on the ethics of manipulating verbal communications for the sake of maximizing placebo and minimizing nocebo effects (Kam-Hansen et al., 2014). In addition, the AMA, at the end of its placebo *Opinion*, suggests that: "Physicians can produce a placebo-like effect through the skillful use of reassurance and encouragement, thereby building respect and trust, promoting the patient-physician relationship, and improving health outcomes" (American Medical Association, 2018, Opinion 2.1.4). Accordingly, the main question at the center of the present debate is: under which conditions, and to what extent, is it ethical for doctors to intentionally manipulate verbal communications to modulate placebo and nocebo effects?

A first important issue regards informed consent and nocebo effects. The clinicians' duty to inform comprises also the side effects of prescribed therapies and procedures, some of which may be prone to nocebo modulation, such as pain (Colloca, 2017; Kaptchuk et al., 2006; Miller & Colloca, 2011).

This raises an ethical dilemma: on the one hand, disclosing these side effects is required to secure informed consent; on the other hand, such disclosure will increase the likelihood of nocebo-related side effects. A similar dilemma emerges also in the case of "procedural disclosure," i.e., the practice of warning patients about the sensations that they are likely to experience while undergoing therapeutic or diagnostic procedures (Krauss, 2015). Like disclosures for informed consent, procedural disclosures may involve different declarative and descriptive statements (e.g., "I will give you x, and you will feel y") in definitive and qualified forms (e.g., "you will feel x" or "you may feel y"); and like disclosures for informed consent, procedural disclosures may elicit nocebo-related effects where the anticipated events are prone to nocebo modulation.

With respect to these cases, scholars have proposed three complementary strategies. First, doctors may omit information out of considerations of non-maleficence (Fortunato, Wasserman, & Menkes, 2017). This strategy, however, violates clinicians' duty to inform; as such, it may be ethically justifiable only to prevent severe or irreversible nocebo effects (Colloca, 2017). A second strategy is to minimize nocebo effects without forestalling informed consent. To this end, doctors may implement an "authorized concealment" approach, asking patients in advance whether they do not wish to be informed about possible side effects (Colloca & Miller, 2011; Miller & Colloca, 2011). Additionally, clinicians may resort to framing effects to lower the chances of eliciting nocebo-related effects (Colloca & Finniss, 2012), as well as considering more neutral formulations without precise descriptions (Krauss, 2015). This second strategy is viable where paternalistic omission is unethical, due to the mild and transient nature of nocebo-related side effects. Finally, a third strategy consists in educating patients more about placebo and nocebo effects, so as to aid them to "understand and contextualize the potential for anxiety and other psychophysiological mechanisms that could lead to the occurrence of unwanted nocebo effects" (Colloca, 2017, p. 48).

Another interesting issue regards the skillful use of verbal communication to maximize placebo effects (Annoni & Miller, 2016; Castelnuovo et al., 2018). Consider the following statements:

(b) This pill is a painkiller containing morphine;

(c) We want to minimize your pain by giving you this pill called X. X contains morphine. Morphine is a very powerful painkiller. Thus, you can expect your pain to be relieved soon. Please let me know if you continue to feel pain.

Both (b) and (c) are truthful statements; yet (c) is likely to elicit stronger placebo effects than (b), as it provides additional information meant to strengthen the patient's positive expectations. Other things being equal, doctors may resort to the skillful use of "therapeutic communication" to enhance placebo effects while administering therapies, provided that they do not resort to misleading communication; that the disclosure targets outcomes prone to placebo modulation; that the patient's unique clinical profile and agency are taken into account; and that the information disclosure has negligible risks for patients and negligible costs for doctors (Annoni & Miller, 2016).

A final issue, then, concerns the prescription of "placebo-based interventions"; that is to say, of therapies that have been found to be slightly more effective than placebos, and yet significantly better than no treatment or the standard of care (Annoni & Miller, in press). Consider the case of acupuncture to reduce migraine frequency. A 2016 Cochrane review sampled 22 trials with 4985 participants in total, and found that acupuncture was significantly more effective than standard drug prophylaxis after treatment, and yet only marginally more effective than sham acupuncture (Linde et al., 2016). This suggests that acupuncture is an effective treatment to reduce migraine frequency, even though most of its efficacy may be due to placebo effects. This leads to an ethical conundrum: if, during information disclosure, doctors mention the prominence of the placebo component, they may lower the patient's expectations and the associated placebo effects; if, instead, doctors omit this information, they could violate their duty to inform. In such cases, it has been argued, doctors should carefully consider the patient's situation in the context of the doctor–patient relationship, and adopt a pragmatic criterion, explicitly mentioning such information only if it makes a conceivable difference for the patient's autonomy; otherwise, they can omit it out of considerations of beneficence without violating their duty to inform (Annoni & Miller, in press).

5. THE ETHICS OF PLACEBO CONTROLS IN RANDOMIZED CLINICAL TRIALS

In the last 50 years, the double-blind, placebo-controlled randomized controlled trial (RCT) has been widely hailed as the "gold-standard" of clinical research (Howick, 2011; Schultz & Grimes, 2006). In a standard RCT, a group of participants receives the experimental intervention, while another receives an indistinguishable placebo control. Placebo controls serve at least two important epistemic functions in RCTs. First, they allow testing for the absolute efficacy of treatments where no standard of care is available (Miller, 2008). Second, placebo controls protect the masked allocation of research subjects in a trial, preserving its internal validity (Howick, 2011).

The use of placebo controls in research, however, raises a plethora of complex and intertwined epistemic and ethical issues (Howick, 2011; Miller, 2008). In particular, one of the major ethical controversies over the use of placebos as controls in research concerns the so-called RCT dilemma (Gifford, 1986; Miller, 2008). This dilemma is related to the inherent tension between the ends of clinical research and the ends of clinical practice. On the one hand, clinical research aims at acquiring new generalizable knowledge; on the other hand, clinical medicine aims at restoring or improving the health and well-being of individual patients. Sometimes these two ends may conflict, for the best procedure to test for the efficacy of a treatment may not be the best procedure to treat the patients' individual condition. In research ethics, this tension emerges especially with respect to the choice between placebo-controlled trials (PCTs) and active-controlled trials (ACTs)—i.e., trials in which the experimental treatment is tested versus a proven effective treatment.

Current codes of ethics and regulatory standards differ in their recommendations, thus leaving room for debate (Miller, 2008; Skierka & Michels, 2018). For instance, The Declaration of Helsinki, in its 2013 revision, states:

> *The benefits, risks, burdens and effectiveness of a new intervention must be tested against those of the best proven intervention(s), except in the following circumstances:*
> *Where no proven intervention exists, the use of placebo, or no intervention, is acceptable; or Where for compelling and scientifically sound methodological reasons the use of any intervention less effective than the best proven one, the use of placebo, or no intervention is necessary to determine the efficacy or safety of an intervention and the patients who receive any intervention less effective than the best proven one, placebo, or no intervention will not be subject to additional risks of serious or irreversible harm as a result of not receiving the best proven intervention. (Article 33)*

This formulation does not clarify situations in which placebo controls are acceptable, and thus is open to interpretation and disagreement (Skierka & Michels, 2018).

Over the years, two opposing opinions have emerged: some have argued that PCTs may be ethical even when proven remedies exist; others consider PCTs unethical in such cases, and advocate the use of ACTs instead. Those defending the primacy of ACTs usually justify their position by endorsing one version of the principle or doctrine of *clinical equipoise*. The term "clinical equipoise" was proposed in a landmark article by Freedman (1987) to indicate to the principle according to which RCTs are ethical only insofar as there is a genuine medical uncertainty about whether one treatment is inferior to another (or to a placebo). Thus, the principle of clinical equipoise forecloses "the use of placebos in the face of established treatment, because enrolling in a trial would imply that a proportion of the enrollees will receive medical attention currently considered inferior by the expert community" (Freedman, Glass, & Weijer, 1996, p. 253). The doctrine of clinical equipoise is premised on the recognition that clinical research and clinical practice ought to be grounded on the same moral principles. Since it is unethical in clinical practice to administer an inferior treatment (or a placebo) to a patient instead of the best proven therapy—it is argued—likewise it is unethical to withhold a proven effective treatment in clinical research in the absence of clinical equipoise.

In contrast, other scholars have convincingly argued that, notwithstanding its popularity, the doctrine of clinical equipoise is fundamentally flawed. As noted by Miller (2008, p. 266), "the presumption that randomized clinical trials must be compatible with the ethics of the physician-patient relationship assumes erroneously that the randomized clinical trial is a form of therapy"; however, "it is impossible to maintain strict fidelity to doing what is best medically for individual patients in the context of the ethical randomized clinical trial, because they are not designed for, and may conflict with, personalized care" (Miller, 2008). For instance, unlike routine medical care, in clinical trials the dosages and access to concomitant treatments are restricted according to the study protocol; a "drug washout" period often precedes the randomization in the trial; and participants must undergo a series of clinically unnecessary monitoring procedures to measure the trial outcomes, such as blood draw, biopsies, lumbar punctures, or imaging techniques, all of which carry some degree of risk and harm. These restrictions and procedures are essential for conducting PCTs, and yet they would be unethical in clinical settings, where the clinicians' therapeutic obligation mandates that each patient must receive individualized care in line with professional standards. For the critics of clinical equipoise these obvious practical differences demonstrate that clinical research and clinical care are fundamentally different practices governed by different moral principles. In clinical practice, the principles of beneficence and non-maleficence are always patient-centered; in clinical research, instead, beneficence concerns the promotion of the social good of improving health by means of generalizable knowledge, while non-maleficence limits the risks to which participants can be exposed for the sake of this goal (Miller, 2008).

Therefore, critics and advocates of clinical equipoise disagree on whether placebo controls are ethical in case a proven therapy exists. Notwithstanding this crucial difference, both sides agree that in some cases placebo controls may be justified—for instance, when no proven therapy exists. Furthermore, both sides agree that in order to be justifiable, placebo controls must satisfy certain ethical conditions. While scholars have articulated different threshold conditions for PCTs to be ethical, three aspects are common across all accounts.

First, PCTs must have a valid scientific rationale (Miller, 2008). Thus, when placebo controls are superfluous, they are also unethical.

Second, PCTs must not expose trial participants to excessive risks. These include the risks of receiving a placebo control as well as of foregoing a proven effective therapy. At present, there are no objective criteria for deciding what constitutes an "excessive risk" in PCTs. Accordingly, a careful risk assessment of receiving the placebo control should be carried out for each individual trial (Miller, 2008). Nevertheless, a few general aspects are worth mentioning. As for the risks of placebo controls, they obviously depend on the nature of the tested treatment. Placebo controls may differ in kind and nature, with some carrying potential for considerable harm—especially in the case of "sham" or "placebo surgery" that requires invasive procedures and anesthesia (Miller & Colloca, 2011). As for the risks of foregoing proven medications, at the outset researchers may consider alternative PCT design, such as "add-on" placebo-controlled studies, in which all participants receive a treatment (Schultz & Grimes, 2006). Furthermore, adequate safeguards to minimize risk ought to be implemented, especially for severe conditions for which constant monitoring and rescue medications should be foreseen.

Finally, PCTs pose important concerns for informed consent (Miller, 2008). In particular, empirical surveys suggest that roughly half of the volunteers in clinical trials do not adequately understand the purpose and nature of the study in which they participate, as well as other salient aspects of the experimentation, including the rationale and scope of randomization, double-blinding and placebo controls (Lidz, Appelbaum, Grisso, & Renaud, 2004). Importantly, patients often fall prey to the "therapeutic misconception," as they mistakenly believe that clinical trials have a primary therapeutic purpose. A recent study of cancer patients has found that only a quarter of the respondents understood correctly that the benefits were likely to be negligible or non-existent for them, despite 90% having declared that they were satisfied with having understood the aims of the trial (Joffe, Cook, Cleary, Clark, & Weeks, 2001). It is thus imperative for researchers to improve information disclosure to prevent therapeutic misconceptions and to respect the patients' right to informed consent. To this end, it is also important to educate participants more about the role of placebo controls as well as about the nature of placebo and nocebo effects (Blease, Bishop, & Kaptchuk, 2017).

6. CONCLUSIONS

Over the last decades, a wave of empirical studies has begun to unravel the neurological mechanisms of placebo and nocebo effects. These findings have shown that placebo and nocebo effects may significantly modulate clinical outcomes across clinical and research settings. Nevertheless, the deliberate modulation of placebo and nocebo effects in medicine raises a host of ethical issues.

In clinical settings, doctors have traditionally resorted to deceptive placebos to induce symptomatic relief in patients as well as to achieve other clinical aims. Recommending deceptive placebos, however, is ethically problematic—even if it is meant to be for the patient's good—as deception violates patients' autonomy and their right to informed consent, may jeopardize the bond of trust between doctor and patient, and could harm individuals and society in different ways. Against this background, current ethical guidelines tend to endorse a policy of categorical prohibition of deceptive placebos without prior consent. This position, however, is difficult to justify on both categorical and precautionary grounds.

Thus, three other strategies have been proposed. One popular strategy is to argue that deceptive placebos may sometimes be justifiable where their prospective benefits outweigh the concerns for autonomy, trust, and non-maleficence. In contrast, the other two strategies consist in exploring alternatives such as placebos without deception, or the harnessing of placebos and nocebo effects through the skillful use of verbal communication. Each of these alternative poses its distinct set of ethical questions and indicates a new direction for the future of the debate over the ethics of placebo and placebo effects.

In research settings, however, placebo controls are one of the hallmarks of the contemporary evidence-based biomedical enterprise. While it is generally agreed that placebo-controlled trials offer a robust methodology to assess the efficacy of experimental interventions, it is still debated whether it is ethical to use placebos instead of active controls. Critics of placebo controls often justify their position by endorsing the doctrine of clinical equipoise, according to which a placebo could be used as a control only if there is a genuine uncertainty over the efficacy of the experimental treatment in the scientific community. Yet this view is problematic and difficult to justify, as it seems to conflate the ethics of clinical care with the ethics of research. Notwithstanding such disagreements, a wide consensus exists that where no proven therapy exists, the use of placebo controls may be morally justifiable, provided that it is supported by a valid methodological rationale, does not expose trial participants to excessive risks, and is accompanied by adequate information disclosure to avoid therapeutic misconceptions.

REFERENCES

American Medical Association. (2018). *Code of medical ethics Opinion 2.1.4: Use of placebo in clinical practice.* Retrieved June 15, 2018, from American Medical Association. https://www.ama-assn.org/delivering-care/use-placebo-clinical-practice.

Annoni, M. (2015). Exceptional lies: The ethics of deceptive placebos in clinical settings. *Biblioteca della Libertà, 213*, 1–17.

Annoni, M. (2018). Lying, deception, equivocation and the ethics of prescribing placebos in clinical practice. *Rivista Italiana di Filosofia del Linguaggio, 12*, 30–41.

Annoni, M., & Miller, F. G. (2014). Placebo in clinical practice: An ethical overview. *Douleur et Analgésie, 27*, 215–220.

Annoni, M., & Miller, F. G. (2016). Placebo effects and the ethics of therapeutic communication: A pragmatic perspective. *Kennedy Institute of Ethics Journal, 26*, 79–103.

Annoni, M., & Miller, F. G. (in press). Informed consent and the ethics of placebo-based interventions for migraine in clinical practice. Placebo and nocebo in headache. London and Berlin: Springer.

Anon. (1954). The humble humbug. *The Lancet, ii.* 321.

Asai, A., & Kadooka, Y. (2013). Reexamination of the ethics of placebo use in clinical practice. *Bioethics, 27*(4), 186–193.

Barnhill, A. (2011). What it takes to defend deceptive placebo use. *Kennedy Institute of Ethics Journal, 21*(3), 219–250.

Beauchamp, T. L., & Childress, J. F. (2009). *Principles of biomedical ethics.* New York and Oxford: Oxford University Press.

Benedetti, F. (2015). *Placebo effects: Understanding the mechanisms in health and disease* (2nd ed.). New York: Oxford University Press.

Blease, C. R., Bishop, P. L., & Kaptchuk, T. J. (2017). Informed consent and clinical trials: Where is the placebo effect? *British Medical Journal, 356*, j463.

Blease, C. R., Colloca, L., & Kaptchuk, T. J. (2016). Are open-label placebos ethical? Informed consent and ethical considerations. *Bioethics, 30*, 407–414.

Bok, S. (1974). The ethics of giving placebos. *Scientific American, 231*(5), 17–23.

Bok, S. (1978). *Lying: Moral choices in public and private life.* New York: Vintage Books.

Cabot, R. C. (1903). The use of truth and falsehood in medicine: An experimental study. *American Medicine, 5*, 3444–3449.

Carson, T. L. (2010). *Lying and deception. Theory and practice*. New York: Oxford University Press.
Castelnuovo, G., Giusti, E. M., Manzoni, G. M., Saviola, D., Gabriell, S., Lacerenza, M., et al. (2018). What is the role of the placebo effect for pain relief in neurorehabilitation? Clinical implications from the Italian Consensus Conference on Pain in Neurorehabilitation. *Frontiers in Neurology*, *9*, 310. https://doi.org/10.3389/fneur.2018.00310.
Charlesworth, J. E. G., Petkovic, G., Kelley, J. M., Hunter, M., Onakpoya, I., Roberts, N., et al. (2017). Effects of placebos without deception compared with no treatment: A systematic review and meta-analysis. *Journal of Evidence Based Medicine*, *10*, 97–107.
Chisholm, R. M., & Feehan, T. D. (1977). The intent to deceive. *Journal of Philosophy*, *74*, 143–159.
Coggon, J., & Miola, J. (2011). Autonomy, liberty, and medical decision making. *Cambridge Law Journal*, *3*, 523–547.
Cohen, S., & Shapiro, H. (2013). "Comparable placebo treatment" and the ethics of deception. *Journal of Medicine and Philosophy*, *38*, 696–709.
Colloca, L. (2014). Ethics of placebos. In M. Simon & J. Abelson (Eds.), *Encyclopaedia of human biology* (3rd ed.). Oxford: Elsevier.
Colloca, L. (2017). Tell me the truth and I will not be harmed: Informed consents and nocebo effects. *American Journal of Bioethics*, *17*, 46–48.
Colloca, L., & Finniss, D. (2012). Nocebo effects, patient-clinician communication, and therapeutic outcomes. *Journal of the American Medical Association*, *307*, 567–568.
Colloca, L., Flatten, M. A., & Meissner, K. (Eds.), (2013). *Placebo and pain: From bench to bedside*. Amsterdam: Academic Press.
Colloca, L., & Howick, J. (2018). Placebos without deception: Outcomes, mechanisms, and ethics. *International Review of Neurobiology*, *138*, 219–240.
Colloca, L., & Miller, F. G. (2011). The nocebo effect and its relevance for clinical practice. *Psychosomatic Medicine*, *73*(7), 598–603.
Eyal, N. (2014). Using informed consent to save trust. *Journal of Medical Ethics*, *40*(7), 437–444.
Faden, R. R., & Beauchamp, T. L. (1986). *A history and theory of informed consent*. New York: Oxford University Press.
Fässler, M., Meissner, K., Schneider, A., & Linde, K. (2010). Frequency and circumstances of placebo use in clinical practice—A systematic review of empirical studies. *BMC Medicine*, *8*, 15.
Foddy, A. (2009). A duty to deceive: Placebos in clinical practice. *American Journal of Bioethics*, *9*(12), 4–12.
Fortunato, J., Wasserman, J., & Menkes, D. L. (2017). When respecting autonomy is harmful: A clinically useful approach to the nocebo effect. *American Journal of Bioethics*, *17*, 36–42.
Freedman, B. (1987). Equipoise and the ethics of clinical research. *New England Journal of Medicine*, *317*, 141–145.
Freedman, B., Glass, K. C., & Weijer, C. (1996). Placebo orthodoxy in clinical research. II: Ethical, legal and regulatory myths. *Journal of Law, Medicine, and Ethics*, *24*(3), 252–259.
Gifford, F. (1986). The conflict between randomized clinical trials and the therapeutic obligation. *Journal of Medicine and Philosophy*, *11*, 347–366.
Gold, A., & Lichtenberg, P. (2014). The moral case for the clinical placebo. *Journal of Medical Ethics*, *40*, 219–224.
Howick, J. (2011). *The philosophy of evidence-based medicine*. Oxford: Wiley-Blackwell.
Howick, J., Bishop, F. L., Heneghan, C., Wolstenholme, J., Stevens, S., Hobbs, F. D. R., et al. (2013). Placebo use in the United Kingdom: Results from a national survey of primary care practitioners. *PLoS One*, *8*(3), e58247.

Hull, S., Colloca, L., Avins, A., Gordon, N. P., Somkin, C. P., Kaptcuk, T., et al. (2013). Patients' attitudes about the use of placebo treatments: Telephone survey. *British Medical Journal, 346*, f3757.

Jefferson, T. (1898). *The writings of Thomas Jefferson*. New York: Putnam.

Joffe, S., Cook, E. F., Cleary, P. D., Clark, J. W., & Weeks, J. C. (2001). Quality of informed consent in cancer clinical trials: A cross-sectional survey. *The Lancet, 358*(9295), 1772–1777.

Jones, J. H., & Crouch, R. A. (2011). The Tuskegee syphilis Experiment. In E. J. Emanuel, C. C. Grady, R. K. Lie, F. G. Miller, & D. W. Wendler (Eds.), *The Oxford handbook of clinical research ethics* (pp. 86–96). New York: Oxford University Press.

Justman, S. (2013). Deceit and transparency in placebo research. *Yale Journal of Biology and Medicine, 86*, 323–331.

Kam-Hansen, S., Jakubowski, M., Kelley, J. M., Kirsch, I., Hoaglin, D. C., Kaptchuk, T. J., et al. (2014). Altered placebo and drug labeling changes the outcome of episodic migraine attacks. *Science Translational Medicine, 6*. 218ra215.

Kaptchuk, T. J., & Miller, F. G. (2015). Placebo effects in medicine. *New England Journal of Medicine, 373*, 8–9.

Kaptchuk, T. J., et al. (2006). Sham device v inert pill: Randomised controlled trial of two placebo treatments. *British Medical Journal, 332*(7538), 391–397.

Kaptchuk, T. J., et al. (2010). Placebos without deception: A randomized controlled trial in irritable bowel syndrome. *PLoS One, 5*(12), e15591.

Kihlbom, U. (2008). Autonomy and negatively informed consent. *Journal of Medical Ethics, 34*(3), 146–149.

Kolber, A. J. (2007). A limited defense of clinical placebo deception. *Yale Policy Review, 26*, 75–134.

Koorsgaard, C. (1996). *Creating the kingdom of ends*. Cambridge: Cambridge University Press.

Krauss, B. S. (2015). "This may hurt": Predictions in procedural disclosure may do harm. *British Medical Journal, 350*, h649.

Lidz, C. W., Appelbaum, P. S., Grisso, T., & Renaud, M. (2004). Therapeutic misconception in clinical research: Frequency and risk factors. *IRB: Ethics and Human Research, 26*, 1–8.

Linde, K., et al. (2016). Acupuncture for the prevention of episodic migraine. *Cochrane Database of Systematic Reviews, 6*, CD001218.

Lynöe, N., Mattsson, B., & Sandlund, M. (1993). The attitudes of patients and physicians towards placebo treatment—A comparative study. *Social Science and Medicine, 36*, 767–774.

Miller, F. G., & Crouch, R. A. (2008). The ethics of placebo-controlled trials. In E. J. Emanuel, C. C. Grady, R. K. Lie, F. G. Miller, & D. W. Wendler (Eds.), *The Oxford handbook of clinical research ethics* (pp. 261–272). New York: Oxford University Press.

Miller, F. G., & Colloca, L. (2009). The legitimacy of placebo treatments in clinical practice: Evidence and ethics. *American Journal of Bioethics, 9*, 39–47.

Miller, F. G., & Colloca, L. (2011). The placebo phenomenon and medical ethics: Rethinking the relationship between informed consent and risk-benefit assessment. *Theoretical Medicine and Bioethics, 32*, 229–243.

Ortiz, R., Hull, C. S., & Colloca, L. (2016). Patient attitudes about the clinical use of placebo: Qualitative perspectives from a telephone survey. *British Medical Journal Open, 6*, e011012.

Pellegrino, E. (1991). Trust and distrust in professional ethics. In E. D. Pellegrino, R. M. Veatch, & J. P. Langan (Eds.), *Ethics, trust, and the professions: Philosophical and cultural aspects* (pp. 69–89). Washington DC: Georgetown University Press.

Schultz, K., & Grimes, D. A. (Eds.), (2006). *The Lancet handbook of essential concepts in clinical research*. Spain: Elsevier.

Shaw, D. M. (2009). Prescribing placebos ethically: The appeal of negatively informed consent. *Journal of Medical Ethics, 35*(2), 97–99.
Skierka, A. S., & Michels, K. B. (2018). Ethical principles and placebo-controlled trials—Interpretation and implementation of the Declaration of Helsinki's placebo paragraph in medical research. *BMC Medical Ethics, 19*, 24.
Slater, J. D., Brown, M. C., Jacobs, W., & Ramsay, R. E. (1995). Induction of pseudoseizures with intravenous saline placebo. *Epilepsia, 36*(6), 580–585.
Sokol, D. K. (2007). Can deceiving patients be morally acceptable? *British Medical Journal, 334*, 984–986.
Tilburt, J. C., Emanuel, E. J., Kaptchuk, T. J., Curlin, F. A., & Miller, F. G. (2008). Prescribing 'placebo treatments': Results of national survey of US internists and rheumatologists. *British Medical Journal, 337*, a1938.
Wager, T. D., & Atlas, L. (2015). The neuroscience of placebo effects: Connecting context, learning and health. *Nature Reviews Neuroscience, 16*, 403–417.
Weindling, P. J. (2011). The Nazi medical experiments. In E. J. Emanuel, C. C. Grady, R. A. Crouch, R. K. Lie, F. G. Miller, & D. W. Wendler (Eds.), *The Oxford handbook of clinical research ethics* (pp. 18–30). New York: Oxford University Press.

FURTHER READING

Miller, F. G., Colloca, L., & Kaptchuk, T. J. (2009). The placebo effect: Illness and interpersonal healing. *Perspectives in Biology of Medicine, 52*, 518–539.
World Medical Association. (2013). *Declaration of Helsinki. 64th World Medical Association Assembly*. Retrieved June 15, 2018, from http://www.wma.net/en/30publications/10policies/b3/.

FSC
www.fsc.org
MIX
Papier | Fördert
gute Waldnutzung
FSC® C083411

Kolibri 360 GmbH
Ferdinand-Jühlke-Straße 7
99095 Erfurt, Deutschland
produktsicherheit@kolibri360.de